T0250765

Lecture Notes in Computer Science

855

Edited by G. Goos, J. Hartmanis and J. van Leeuwen

Advisory Board: W. Brauer   D. Gries   J. Stoer

Jan van Leeuwen (Ed.)

Algorithms – ESA '94

Second Annual European Symposium
Utrecht, The Netherlands, September 26–28, 1994
Proceedings

Springer-Verlag
Berlin Heidelberg New York
London Paris Tokyo
Hong Kong Barcelona
Budapest

Jan van Leeuwen (Ed.)

Algorithms – ESA '94

Second Annual European Symposium
Utrecht, The Netherlands, September 26-28, 1994
Proceedings

Springer-Verlag

Berlin Heidelberg New York
London Paris Tokyo
Hong Kong Barcelona
Budapest

Series Editors

Gerhard Goos
Universität Karlsruhe
Postfach 69 80, Vincenz-Priessnitz-Straße 1, D-76131 Karlsruhe, Germany

Juris Hartmanis
Department of Computer Science, Cornell University
4130 Upson Hall, Ithaka, NY 14853, USA

Jan van Leeuwen
Department of Computer Science, Utrecht University
Padualaan 14, 3584 CH Utrecht, The Netherlands

Volume Editor

Jan van Leeuwen
Department of Computer Science, Utrecht University
Padualaan 14, 3584 CH Utrecht, The Netherlands

CR Subject Classification (1991): F.2, G.2-4, I.3.5, E.1, E.5, I.7.3, B.4.3, C.1.2, C.2.1

ISBN 3-540-58434-X Springer-Verlag Berlin Heidelberg New York

CIP data applied for

This work is subject to copyright. All rights are reserved, whether the whole or part of the material is concerned, specifically the rights of translation, reprinting, re-use of illustrations, recitation, broadcasting, reproduction on microfilms or in any other way, and storage in data banks. Duplication of this publication or parts thereof is permitted only under the provisions of the German Copyright Law of September 9, 1965, in its current version, and permission for use must always be obtained from Springer-Verlag. Violations are liable for prosecution under the German Copyright Law.

© Springer-Verlag Berlin Heidelberg 1994
Printed in Germany

Typesetting: Camera-ready by author
SPIN: 10478962   45/3140-543210 - Printed on acid-free paper

Preface

The 2nd Annual European Symposium on Algorithms (ESA'94) was held September 26-28, 1994, near Utrecht (the Netherlands), in the conference facilities of the 'National Sports Centre Papendal' of the Dutch National Sports Federation. This volume contains all contributed papers presented at the symposium, together with extended abstracts of the invited lectures by Andrew V. Goldberg (Stanford University) and Gaston H. Gonnet (ETH Zürich).

ESA was established in 1993 as a main annual event for all researchers interested in algorithms, theoretical as well as applied. The international symposium covers all research on algorithms and their analysis as it is carried out in the fields of computer science, discrete applied mathematics and all other areas of algorithm-oriented research and its application. The proceedings of ESA'93 appeared as Volume 726 in the series Lecture Notes in Computer Science.

In response to the Call for Papers for ESA'94, the program committee received 159 submissions, indicating a strong and growing interest in the symposium. The program committee for ESA'94 consisted of:

Helmut Alt (Berlin)   Thomas Lengauer (Sankt Augustin)

Giuseppe di Battista (Rome)   Jan Karel Lenstra (Eindhoven)

Philippe Flajolet (Paris)   Andrzej Lingas (Lund)

Alon Itai (Haifa)   Bill McColl (Oxford)

Lefteris Kirousis (Patras)   Friedhelm Meyer a. d. Heide (Paderborn)

Jan van Leeuwen (Utrecht, chairman)   Wojciech Rytter (Warsaw)

The program committee met on May 16, 1994, in Utrecht and selected 42 contributed papers for inclusion in the scientific program of ESA'94. The selection was based on originality, quality, and relevance to the study of algorithms, and reflects many of the current directions in algorithms research. We wish to thank all those who submitted extended abstracts for consideration, and all subreferees and colleagues who helped in the extensive evaluation process.

The scientific program of ESA'94 evidently shows the gradual expansion of algorithms research into new areas of computational endeavor in science and industry, and the program committee hopes that this component will continue to manifest itself as part of the ESA program in future years.

Many thanks are due to the organizing committee of ESA'94 for making the symposium happen. The organizing committee consisted of:

Hans Bodlaender, Marko de Groot, Jan van Leeuwen, Margje Punt, and
Marinus Veldhorst,

all from the Department of Computer Science of Utrecht University. ESA'94 was sponsored by the Association for Computing Machinery (ACM), the ACM Special Interest Group on Algorithms and Computation Theory (SIGACT), the European Association for Theoretical Computer Science (EATCS) and the European Research Consortium for Informatics and Mathematics (ERCIM). We are grateful to the Royal Dutch Academy of Sciences (KNAW) and Utrecht University for supporting ESA'94.

We thank the 'National Sports Centre Papendal' of the Dutch National Sports Federation for allowing the use of its excellent conference facilities for the symposium, the Department of Computer Science of Utrecht University for supporting the organization of ESA'94, and Margje Punt for invaluable assistance in all matters related to the symposium.

Utrecht, September 1994   Jan van Leeuwen

Contents

Invited Lectures

Optimization Algorithms for Large Networks 1
  A.V. Goldberg

Wanna Buy an Algorithm? Cheap! or: Algorithms for Text Searching 10
Which Could Have Commercial Value
  G.H. Gonnet

Automatic Graph Drawing and Rendering

Planar Drawings and Angular Resolution: Algorithms and Bounds 12
  A. Garg, R. Tamassia

A Better Heuristic for Orthogonal Graph Drawings 24
  T. Biedl, G. Kant

Hamiltonian Triangulations for Fast Rendering 36
  E.M. Arkin, M. Held, J.S.B. Mitchell, S.S. Skiena

Spanners and Steiner Trees

Efficient Construction of a Bounded Degree Spanner with Low Weight 48
  S. Arya, M. Smid

Approaching the 5/4-Approximation for Rectilinear Steiner Trees 60
  P. Berman, U. Fössmeier, M. Karpinski, M. Kaufmann, A. Zelikovsky

Efficient Datastructures and Complexity Analysis

Membership in Constant Time and Minimum Space 72
  A. Brodnik, J.I. Munro

Faster Searching in Tries and Quadtrees – An Analysis of Level Compression 82
  A. Andersson, S. Nilsson

The Analysis of a Hashing Scheme by the Diagonal Poisson Transform 94
  P.V. Poblete, A. Viola, J.I. Munro

Some Lower Bounds for Comparison-Based Algorithms 106
  S. Carlsson, J. Chen

Graph Algorithms I

An Efficient Algorithm for Edge-Ranking Trees 118
  X. Zhou, T. Nishizeki

Edge-Disjoint (s,t)-Paths in Undirected Planar Graphs in Linear Time 130
  K. Weihe

A Simple Min Cut Algorithm 141
  M. Stoer, F. Wagner

Approximation Algorithm on Multi-Way Maxcut Partitioning 148
  J.D. Cho, S. Raje, M. Sarrafzadeh

A Linear-Time Algorithm for Finding a Central Vertex of a Chordal Graph 159
  V. Chepoi, F. Dragan

Distributed Algorithms

The Time Complexity of Updating Snapshot Memories 171
  A. Israeli, A. Shirazi

Non-Exploratory Self-Stabilization for Constant-Space Symmetry-Breaking 183
  G. Parlati, M. Yung

On-Line Distributed Data Management 202
  C. Lund, N. Reingold, J. Westbrook, D. Yan

Computational Geometry

A Unified Scheme for Detecting Fundamental Curves in Binary Edge Images 215
  T. Asano, N. Katoh, T. Tokuyama

How to Compute the Voronoi Diagram of Line Segments: Theoretical 227
and Practical Results
  C. Burnikel, K. Mehlhorn, S. Schirra

Range Searching and Point Location among Fat Objects 240
  M.H. Overmars, A.F. van der Stappen

Computational Geometry and Robotics

Convex Tours of Bounded Curvature 254
  J.-D. Boissonnat, J. Czyzowicz, O. Devillers, J.-M. Robert, M. Yvinec

Optimal Shortest Path and Minimum-Link Path Queries in the Presence 266
of Obstacles
  Y.-J. Chiang, R. Tamassia

Fast Algorithms for Collision and Proximity Problems Involving 278
Moving Geometric Objects
  P. Gupta, R. Janardan, M. Smid

Operations Research and Combinatorial Optimization

Reverse-Fit: A 2-Optimal Algorithm for Packing Rectangles 290
  I. Schiermeyer

An Optimal Algorithm for Preemptive On-Line Scheduling 300
  B. Chen, A. van Vliet, G.J. Woeginger

Tight Approximations for Resource Constrained Scheduling Problems 307
  A. Srivastav, P. Stangier

An Algorithm for 0-1 Programming with Application to Airline Crew Scheduling .. 319
  D. Wedelin

Graph Algorithms II

An $o(n)$ Work EREW Parallel Algorithm for Updating MST 331
  S.K. Das, P. Ferragina

On the Structure of DFS-Forests on Directed Graphs and the 343
Dynamic Maintenance of DFS on DAG's
  P.G. Franciosa, G. Gambosi, U. Nanni

Finding and Counting Given Length Cycles 354
  N. Alon, R. Yuster, U. Zwick

Parallel Computation and Interconnection Networks

Greedy Hot-Potato Routing on the Mesh 365
  I. Ben-Aroya, A. Schuster

Desnakification of Mesh Sorting Algorithms 377
  J.F. Sibeyn

Tight Bounds on Deterministic PRAM Emulations with Constant Redundancy 391
  A. Pietracaprina, G. Pucci

PRAM Computations Resilient to Memory Faults 401
  B.S. Chlebus, A. Gambin, P. Indyk

An Area Lower Bound for a Class of Fat-Trees 413
  G. Bilardi, P. Bay

Complexity Theory

A Unified Approach to Approximation Schemes for NP- and PSPACE-Hard 424
Problems for Geometric Graphs
  H.B. Hunt III, M.V. Marathe, V. Radhakrishnan, S.S. Ravi,
  D.J. Rosenkrantz, R.E. Stearns

The Parallel Complexity of Eden Growth, Solid-on-Solid Growth and 436
Ballistic Deposition
  R. Greenlaw, J. Machta

A New Approach to Resultant Computations and Other Algorithms 448
with Exact Division
  A. Schönhage, E. Vetter

Text Processing Algorithms

Testing Equivalence of Morphisms on Context-Free Languages 460
  W. Plandowski

Work-Time Optimal Parallel Prefix Matching 471
  L. Gąsieniec, K. Park

On the Exact Complexity of the String Prefix-Matching Problem 483
  D. Breslauer, L. Colussi, L. Toniolo

Incremental Text Editing: A New Data Structure 495
  P. Ferragina

Erratum to the ESA '93 Proceedings 508

Authors Index 509

Optimization Algorithms For Large Networks

Invited Lecture

Andrew V. Goldberg

Computer Science Department, Stanford University, Stanford, CA 94305, USA

1 Introduction

Every year, computers get fast r. Given this trend, one could assume that as computers get faster, program efficiency becomes less important. In many applications, however, program efficiency is now more important than ever. This is because computer memory is getting bigger and applications are getting more sophisticated. As a result, the need to solve bigger and harder problems arises. Hard problems require superlinear time, and thus big hard problems require significantly more time even though computers get faster.

In this talk we discuss recent developments in design of efficient algorithms for several important network optimization problems, including maximum and minimum-cost flow, assignment, and shortest paths problems. Polynomial-time algorithms for some of these classical problems were developed in the fifties and early sixties [5, 28, 54, 57], even before the importance of polynomial time was formalized by Edmonds. Since then, numerous algorithms achieving better and better theoretical time bounds were developed.

Theoretical analysis of algorithms is extremely useful in developing and refining algorithmic ideas and in understanding the strong and weak points of an algorithm. Theoretical analysis, however, has its limitations. The theoretical time bounds are usually asymptotic worst-case bounds. For many algorithms, their typical running time in practice is much better than the worst case running time. Also, some algorithms with good asymptotic worst-case bounds are impractical because of large constant factors and low order terms.

An alternative is to analyze the expected running time. For this probabilistic analysis, however, one has to know the input distribution, and for nontrivial algorithms such analysis is very hard even for the simplest distributions. Furthermore, the distribution of problems coming form a given application is rarely known. As in the worst-case analysis, large constants and low order terms are also a problem. Current state of the art results on probabilistic analysis of the network algorithms discussed in this paper cannot be used to predict behavior of the algorithms in practice, but they may give evidence that practical behavior of an algorithm is much better than the worst-case behavior. See *e.g.* [50, 58].

* Supported in part by ONR Young Investigator Award N00014-91-J-1855 and NSF Grant CCR-9307045.

The best test of the practical algorithm efficiency is experimental evaluation. Such an evaluation is especially convincing when conducted for a known problem class or distribution. Since relative behavior of algorithms may depend on a problem class, careful experimentation of a wide class of problems is required before the best algorithm for a problem can be determined.

The results of an experimental study of an algorithm are implementation-dependent. A right data structure or a heuristic that improves typical running time of an algorithm but not the worst-case running time can make the difference between a successful and an unsuccessful implementations. Unlike low level implementation details, which are also important but can be taken care of using standard techniques [6], selection and development of data structures and heuristics requires good understanding of the underlying algorithm. Development of a good code is an interactive process: an implementation is evaluated to determine the bottlenecks and data structures are changed or heuristics added to improve performance.

The First DIMACS Implementation Challenge, organized by Johnson and McGeoch [49], was devoted to experimental evaluation of flow and matching algorithms. The Challenge stimulated work in the area and set new standards for experimental evaluation of algorithms. Some promising algorithms were implemented for the first time. A number of generators and diverse test families were developed for the maximum flow, minimum-cost flow, assignment, and weighted matching problems.

For the maximum flow problem, the current consensus is that the push-relabel method of Goldberg and Tarjan [36, 43], in combination with certain heuristics, works best in practice. We discuss these results in Section 3.

Although significant progress has been made on the minimum-cost flow and assignment problems, no single algorithm evaluated during the Challenge outperformed the competition on all problem classes. Our subsequent research [37, 38, 40] significantly improved performance of the scaling push-relabel method for these problems. We discuss these results and argue the practical importance of the this method in Sections 4 and 5.

The shortest paths problem is one of the most common and widely studied network optimization problem. Many theoretically good algorithms were developed for this problem since late fifties. A number of algorithms were developed and claimed to perform well in practice in spite of inferior theoretical bounds. In our resent study [15, 16] we implemented these algorithms as well as some new ones and compared these implementations on a wide class of problems families. These results are discussed in Section 6.

Recent experimental results on the algorithms for the maximum flow, minimum-cost flow, assignment, and shortest paths problems show that some resent methods with theoretically good behavior are practical. These methods are robust and outperform the previously used methods on most problem classes, in some cases by a wide margin, while never loosing by a wide margin. The relative performance of the new codes is especially good on large hard problems, where the need for speed is most critical.

2 Problem Definition

In this section we define the problems under consideration. All problems deal with a graph $G = (V, E)$; we define n and m by $|V| = n$ and $|E| = m$.

The input to the *maximum flow problem* is a directed graph G, two distinguished nodes $s, t \in V$ (the source and the sink), and a *capacity function* $u : E \to \mathbf{R}_+$. When the input capacities are integral, we denote the biggest capacity by U. A *pseudoflow* is a function $f : E \to \mathbf{R}_+$ satisfying *capacity constraints* $f(a) \le u(a)$ for all $a \in E$. A *flow* is a pseudoflow satisfying *conservation constraints*

$$\sum_{(u,v)\in E} f(u, v) = \sum_{(v,w)\in E} f(v, w)$$

for all $v \in V - \{s, t\}$. The *value* of a flow f is given by $|f| = \sum_{(s,w)\in E} f(s, w)$. The goal is to find a flow of maximum value.

The input to the *minimum-cost flow problem* is the same as for the maximum flow problem with an additional *cost function* $c : E \to \mathbf{R}$. The cost of a flow f is defined by $c(f) = \sum_{a\in E} f(a)c(a)$. The goal is to find a maximum flow of minimum cost.

Given an undirected graph G, a *matching* is a set of edges, $M \subseteq V$, no two elements of which are adjacent. Given a cost function c on edges, the *cost* of a matching is the sum of costs of edges in the matching. The *assignment problem* is to find a matching of minimum cost in a bipartite graph.

The single-source shortest paths problem is as follows: given a graph G with a source $s \in V$ and a length function $c : E \to \mathbf{R}$, find either shortest paths from s to all other nodes or a negative length cycle. Dijkstra's shortest paths problem is the special case of this problem with nonnegative c.

In the last three problems, when the cost (length) function is integral, we denote the largest absolute value of this function by C.

3 Maximum Flow Problem

The basic methods for the maximum flow problem include the network simplex method of Dantzig [18, 19], the augmenting path method of Ford and Fulkerson [28], the blocking flow method of Dinitz [25], and the push-relabel method of Goldberg and Tarjan [36, 43]. The best theoretical time bounds for the maximum flow problem, based on the latter method, are as follows. An algorithm of Goldberg and Tarjan [43] runs in $O(nm \log(n^2/m))$ time, an algorithm of King et. al. [53] runs in $O(nm + n^{2+\epsilon})$ time for any constant $\epsilon > 0$, an algorithm of Cheriyan et. al. [12] runs in $O(nm + (n \log n)^2)$ time with high probability, and an algorithm of Ahuja et. al. [3] runs in $O\left(nm \log\left(\frac{n}{m\sqrt{U}} + 2\right)\right)$ time.

Prior to the push-relabel method, several studies have shown that Dinitz' algorithm [25] is in practice superior to other methods, including the network simplex method [18, 19], Ford-Fulkerson algorithm [28, 27], and Karzanov's algorithm [51]. See *e.g.* [13, 45]. Several recent studies (*e.g.* [4, 21, 22, 59]) show that the push-relabel method is superior to Dinitz' method in practice.

The push-relabel method is based on two operations, *push* and *relabel*, used to update flow and distance functions on the basis of local information. (See [43] for details.) These operations are very efficient and allow a great deal of flexibility in the operation ordering and the use of heuristics. For a good implementation, it is necessary to use heuristics that update the distance function of the basis of global information. Two heuristics are useful: *global relabeling* (see *e.g.* [4, 36, 59]) and *gap relabeling* (due to Derigs and Meier [21]).

The results of the DIMACS Challenge seem to indicate that the maximum distance node selection and global relabeling wins most of the time [4, 59]. Preliminary results of our current study [14] suggest that the maximum distance selection in combination with both global relabeling and gap relabeling works best even better.

4 Minimum-Cost Flow Problem

The minimum-cost flow problem has a rich history. The classical out-of-kilter method for the problem, due to Fulkerson [31] and Minty [56], was developed in the early sixties. The first polynomial-time algorithm for the problem is due to Edmonds and Karp [27] and to Dinitz [26]; they also introduce the important technique of *capacity scaling*. The first *cost scaling* algorithm for the minimum-cost flow problem is due to Röck [65] and Bland and Jensen [10]. A generalization of the cost scaling approach was developed by Goldberg and Tarjan [36, 44]. The fastest known algorithms for the minimum-cost flow problem achieve the following time bounds: $O(m(m + n \log n) \log U)$ [26, 27], $O(nm \log(n^2/m) \log(nC))$ [44], $O(nm \log(nC) \log \log U)$ [1], and $O(m(m + n \log n) \log n)$ [60].

For a long time the network simplex method was the most robust code to solve the problem in practice. Two widely used network simplex codes are RNET [46] and NETFLO [52]. However, it was observed that on some classes of problems other algorithms outperform network simplex. See *e.g.* [8, 10]. Similar results were obtained during the DIMACS Challenge [9, 41, 64]. Each of the the minimum-cost flow codes evaluated during the Challenge was losing to another code by a wide margin on some problem families.

Continuing the work of [41], we produced an efficient implementation of the scaling push-relabel method called CS [37, 38]. The key to improved performance was a heuristic for global price updates which is related to the global relabeling heuristic for maximum flows. Compared to RNET, NETFLO, and RELAX [8], CS performs very robustly. In the tests on DIMACS problem families, CS was hundreds of times faster than the other codes on several problem families while never loosing by more than a factor of six. The scaling push-relabel method is relatively new and is likely to lead to even more efficient implementations.

5 Assignment Problem

For the assignment problem, the classical Hungarian method Kuhn [54] achieves the best known strongly polynomial bound of $O(nS(n, m, C))$, where $S(n, m, C)$

is the time required to solve Dijkstra's shortest paths problem (see Section 6). The fastest weakly polynomial algorithm for the problem, due to Gabow and Tarjan [32], runs in $O(\sqrt{nm}\log(nC))$ time. This algorithm uses cost scaling.

The Hungarian method is closely related to the successive shortest augmenting path algorithm. These methods were for a long time considered the best for solving assignment problems in practice (see e.g. [20]). For some problem classes, the auction algorithm [7] has been shown to outperform the classical methods. Recent studies show that the scaling push-relabel method and the very similar scaling auction method [11] are more efficient than the shortest augmenting path method. In particular, our scaling push-relabel code CSA (developed jointly with Kennedy) [40] usually outperforms good shortest augmenting path codes [47, 48], often by a wide margin. CSA is usually faster on small problems and on many problem families CSA exhibits smaller asymptotic growth rate. It is always competitive with the shortest augmenting path implementations on large problems and often wins by a wide margin. CSA is also significantly faster than the interior-point implementation of [63] on all problem instances we considered.

6   Shortest Paths Problem

Theoretically efficient algorithms for the shortest paths problem were developed in the late fifties, by Bellman [5], Ford [28], and Moore [57] for the general problem and by Dijkstra [24] for the problem with nonnegative length function. The best current theoretical time bounds for the general problem are $O(nm)$ (achieved by the Bellman-Ford-Moore algorithm), and $O(\sqrt{nm}\log(N))$ [39]. The latter bound assumes that the arc lengths are integers greater than $-N$. If the arc lengths are nonnegative, implementations of Dijkstra's algorithm achieve better bounds. An implementation of [29] runs in $O(m + n\log n)$ time. An improved time bound of $O(m + n\log n/\log\log n)$ [30] can be obtained in a random access machine computation model that allows certain word operations. Under the assumption that arc lengths are integers in the interval $[0, \ldots, C]$, the implementation of [2] runs in $O(m + n\sqrt{\log C})$ time. On acyclic graphs, the problem can be solved in linear time.

A number of algorithms with worse (often exponential) running time bounds have been proposed. These algorithms were claimed to be efficient in practice in spite of the time bounds. See e.g. [23, 55, 61, 62]. These claims were often made on the basis of very limited computational experiments. A good description of the classical algorithms and their implementations appears in [33].

Jointly with Cherkassky and Radzik [15, 16], we undertook a careful study of shortest paths algorithms. Our study included algorithms by Bellman-Ford-Moore [5, 28, 57], Pape-Levit [55, 62], Pallottino [61], Glover et. al. [34] (see also [33, 35], and Goldberg and Radzik [42]. We also studied a number of variations of Dijkstra's algorithm, including implementations using k-ary heaps (see e.g. [17]), Fibonacci heaps [29], R-heaps [2], buckets [23], a new double-bucket implementation, and several other variations of the bucket implementation.

As a part of the experimental study, we developed problem generators and test families. Relative performance of many algorithms varies from one family to another. In particular, algorithms with poor time bounds, such as the Pape-Levit algorithm, exhibit the most severe variations in performance.

Although the study did not produce a clear winner, two implementations behaved robustly in their domain: the double-bucket implementation of Dijkstra's algorithm for networks with nonnegative arc lengths and a variant of the Goldberg-Radzik algorithm (suggested by Cherkassky) for networks with negative-length arcs.

7 Concluding Remarks

Some of the recently developed network optimization algorithms lead to the best code for solving practical problem instances. In particular, good implementations of the push-relabel method work well for the maximum flow, minimum-cost flow, and assignment problems. The new implementations can often solve large problems that previously available codes cannot solve in reasonable time. The ideas leading to theoretically fast algorithms often lead to practically fast algorithms, even though additional ideas may be needed to make an algorithm efficient in the worst-case sense efficient in practice.

Sometimes, experimental work leads to theoretical results as well. Our scaling shortest paths algorithm [39] was motivated by dual variable update procedures developed for the CS code. In [15], we introduce a notion of *potential invariant* shortest paths algorithm and show that many algorithms, including the classical Bellman-Ford-Moore algorithm, are potential invariant. This result was suggested by the data from a well-designed experiment.

Finally, we would like to mention that some of the programs mentioned above are in public domain. The list of programs and generators available from DIMACS and the instructions for obtaining them appear in [49]. Our programs can be obtained by sending mail to `ftp-request@theory.stanford.edu`. Put `send csmin.tar`, `send csas.tar`, and `splib.tar` to get the minimum-cost flow code, the assignment code, and the shortest paths codes and generators.

References

1. R. K. Ahuja, A. V. Goldberg, J. B. Orlin, and R. E. Tarjan. Finding Minimum-Cost Flows by Double Scaling. *Math. Prog.*, 53:243–266, 1992.
2. R. K. Ahuja, K. Mehlhorn, J. B. Orlin, and R. E. Tarjan. Faster Algorithms for the Shortest Path Problem. Technical Report CS-TR-154-88, Department of Computer Science, Princeton University, 1988.
3. R. K. Ahuja, J. B. Orlin, and R. E. Tarjan. Improved Time Bounds for the Maximum Flow Problem. *SIAM J. Comput.*, 18:939–954, 1989.
4. R. J. Anderson and J. C. Setubal. Goldberg's Algorithm for the Maximum Flow in Perspective: a Computational Study. In D. S. Johnson and C. C. McGeoch, editors, *Network Flows and Matching: First DIMACS Implementation Challenge*, pages 1–18. AMS, 1993.

5. R. E. Bellman. On a Routing Problem. *Quart. Appl. Math.*, 16:87–90, 1958.

6. J. L. Bentley. *Writing Eficient Programs*. Prentice-Hall, 1982.

7. D. P. Bertsekas. A Distributed Algorithm for the Assignment Problem. Unpublished Manuscript, Lab. for Decision Systems, M.I.T., 1979.

8. D. P. Bertsekas and P. Tseng. Relaxation Methods for Minimum Cost Ordinary and Generalized Network Flow Problems. *Oper. Res.*, 36:93–114, 1988.

9. R. G. Bland, J. Cheriyan, D. L. Jensen, and L. Ladányi. An Empirical Study of Min Cost Flow Algorithms. In D. S. Johnson and C. C. McGeoch, editors, *Network Flows and Matching: First DIMACS Implementation Challenge*, pages 119–156. AMS, 1993.

10. R. G. Bland and D. L. Jensen. On the Computational Behavior of a Polynomial-Time Network Flow Algorithm. *Math. Prog.*, 54:1–41, 1992.

11. D. A. Castañon. Reverse Auction Algorithms for the Assignment Problems. In D. S. Johnson and C. C. McGeoch, editors, *Network Flows and Matching: First DIMACS Implementation Challenge*, pages 407–430. AMS, 1993.

12. J. Cheriyan, T. Hagerup, and K. Mehlhorn. Can a Maximum Flow be Computed in $o(nm)$ Time? In *Proc. ICALP*, 1990.

13. B. V. Cherkassky. Personal communication. 1991.

14. B. V. Cherkassky and A. V. Goldberg. Efficient Implementation of Push-Relabel Method for the Maximum Flow Problem. In preparation.

15. B. V. Cherkassky, A. V. Goldberg, and T. Radzik. Shortest Paths Algorithms: Theory and Experimental Evaluation. Technical Report STAN-CS-93-1480, Department of Computer Science, Stanford University, 1993.

16. B. V. Cherkassky, A. V. Goldberg, and T. Radzik. Shortest Paths Algorithms: Theory and Experimental Evaluation. In *Proc. 5th ACM-SIAM Symposium on Discrete Algorithms*, pages 516–525, 1994.

17. T. H. Cormen, C. E. Leiserson, and R. L. Rivest. *Introduction to Algorithms*. MIT Press, Cambridge, MA, 1990.

18. G. B. Dantzig. Application of the Simplex Method to a Transportation Problem. In T. C. Koopmans, editor, *Activity Analysis and Production and Allocation*, pages 359–373. Wiley, New York, 1951.

19. G. B. Dantzig. *Linear Programming and Extensions*. Princeton Univ. Press, Princeton, NJ, 1962.

20. U. Derigs. The Shortest Augmenting Path Method for Solving Assignment Problems – Motivation and Computational Experience. *Annals of Oper. Res.*, 4:57–102, 1985/6.

21. U. Derigs and W. Meier. Implementing Goldberg's Max-Flow Algorithm — A Computational Investigation. *ZOR — Methods and Models of Operations Research*, 33:383–403, 1989.

22. U. Derigs and W. Meier. An Evaluation of Algorithmic Refinements and Proper Data-Structures for the Preflow-Push Approach for Maximum Flow. In *ASI Series on Computer and System Sciences*, volume 8, pages 209–223. NATO, 1992.

23. R. B. Dial. Algorithm 360: Shortest Path Forest with Topological Ordering. *Comm. ACM*, 12:632–633, 1969.

24. E. W. Dijkstra. A Note on Two Problems in Connection with Graphs. *Numer. Math.*, 1:269–271, 1959.

25. E. A. Dinic. Algorithm for Solution of a Problem of Maximum Flow in Networks with Power Estimation. *Soviet Math. Dokl.*, 11:1277–1280, 1970.

26. E. A. Dinic. Metod porazryadnogo sokrashcheniya nevyazok i transportnye zadachi. In *Issledovaniya po Diskretnoĭ Matematike*. Nauka, Moskva, 1973. In Russian. Title translation: Excess Scaling and Transportation Problems.

27. J. Edmonds and R. M. Karp. Theoretical Improvements in Algorithmic Efficiency for Network Flow Problems. *J. Assoc. Comput. Mach.*, 19:248–264, 1972.

28. L. R. Ford, Jr. and D. R. Fulkerson. *Flows in Networks*. Princeton Univ. Press, Princeton, NJ, 1962.

29. M. L. Fredman and R. E. Tarjan. Fibonacci Heaps and Their Uses in Improved Network Optimization Algorithms. *J. Assoc. Comput. Mach.*, 34:596–615, 1987.

30. M. L. Fredman and D. E. Willard. Trans-dichotomous Algorithms for Minimum Spanning Trees and Shortest Paths. In *Proc. 31st IEEE Annual Symposium on Foundations of Computer Science*, pages 719–725, 1990.

31. D. R. Fulkerson. An Out-of-Kilter Method for Minimal Cost Flow Problems. *SIAM J. Appl. Math*, 9:18–27, 1961.

32. H. N. Gabow and R. E. Tarjan. Faster Scaling Algorithms for Network Problems. *SIAM J. Comput.*, pages 1013–1036, 1989.

33. G. Gallo and S. Pallottino. Shortest Paths Algorithms. *Annals of Oper. Res.*, 13:3–79, 1988.

34. F. Glover, R. Glover, and D. Klingman. Computational Study of an Improved Shortest Path Algorithm. *Networks*, 14:25–37, 1984.

35. F. Glover, D. Klingman, and N. Phillips. A New Polynomially Bounded Shortest Paths Algorithm. *Oper. Res.*, 33:65–73, 1985.

36. A. V. Goldberg. *Efficient Graph Algorithms for Sequential and Parallel Computers*. PhD thesis, M.I.T., January 1987. (Also available as Technical Report TR-374, Lab. for Computer Science, M.I.T., 1987).

37. A. V. Goldberg. An Efficient Implementation of a Scaling Minimum-Cost Flow Algorithm. Technical Report STAN-CS-92-1439, Department of Computer Science, Stanford University, 1992.

38. A. V. Goldberg. An Efficient Implementation of a Scaling Minimum-Cost Flow Algorithm. In *Proc. 3rd Integer Prog. and Combinatorial Opt. Conf.*, pages 251–266, 1993.

39. A. V. Goldberg. Scaling Algorithms for the Shortest Paths Problem. In *Proc. 4th ACM-SIAM Symposium on Discrete Algorithms*, pages 222–231, 1993.

40. A. V. Goldberg and R. Kennedy. An Efficient Cost Scaling Algorithm for the Assignment Problem. Technical Report STAN-CS-93-1481, Department of Computer Science, Stanford University, 1993.

41. A. V. Goldberg and M. Kharitonov. On Implementing Scaling Push-Relabel Algorithms for the Minimum-Cost Flow Problem. In D. S. Johnson and C. C. McGeoch, editors, *Network Flows and Matching: First DIMACS Implementation Challenge*, pages 157–198. AMS, 1993.

42. A. V. Goldberg and T. Radzik. A Heuristic Improvement of the Bellman-Ford Algorithm. *Applied Math. Let.*, 6:3–6, 1993.

43. A. V. Goldberg and R. E. Tarjan. A New Approach to the Maximum Flow Problem. *J. Assoc. Comput. Mach.*, 35:921–940, 1988.

44. A. V. Goldberg and R. E. Tarjan. Finding Minimum-Cost Circulations by Successive Approximation. *Math. of Oper. Res.*, 15:430–466, 1990.

45. D. Goldfarb and M. D. Grigoriadis. A Computational Comparison of the Dinic and Network Simplex Methods for Maximum Flow. *Annals of Oper. Res.*, 13:83–123, 1988.

46. M. D. Grigoriadis. An Efficient Implementation of the Network Simplex Method. *Math. Prog. Study*, 26:83–111, 1986.

47. J. Hao and G. Kocur. An Implementation of a Shortest Augmenting Path Algorithm for the Assignment Problem. In D. S. Johnson and C. C. McGeoch, editors, *Network Flows and Matching: First DIMACS Implementation Challenge*, pages 453–468. AMS, 1993.

48. D. B. Johnson and S. Venkatesan. Using Divide and Conquer to Find Flows in Directed Planar Netwroks in $O(n^{1.5} \log n)$ Time. In *Proc. 20th Annual Allerton Conf. on Communication, Control, and Computing*, pages 898–905, 1982.

49. D. S. Johnson and C. C. McGeoch, editors. *Network Flows and Matching: First DIMACS Implementation Challenge*. AMS, 1993.

50. R. M. Karp, R. Motwani, and N. Nisan. Probabilistic Analysis of Network Flow Algorithms. *Math. of Oper. Res.*, 18:71–97, 1993.

51. A. V. Karzanov. Determining the Maximal Flow in a Network by the Method of Preflows. *Soviet Math. Dok.*, 15:434–437, 1974.

52. J. L. Kennington and R. V. Helgason. *Algorithms for Network Programming*. John Wiley and Sons, 1980.

53. V. King, S. Rao, and R. Tarjan. A Faster Deterministic Maximum Flow Algorithm. In *Proc. 3rd ACM-SIAM Symposium on Discrete Algorithms*, pages 157–164, 1992.

54. H. W. Kuhn. The Hungarian Method for the Assignment Problem. *Naval Res. Logist. Quart.*, 2:83–97, 1955.

55. B. Ju. Levit and B. N. Livshits. *Neleneinye Setevye Transportnye Zadachi*. Transport, Moscow, 1972. In Russian.

56. G. J. Minty. Monotone Networks. *Proc. Roy. Soc. London*, A(257):194–212, 1960.

57. E. F. Moore. The Shortest Path Through a Maze. In *Proc. of the Int. Symp. on the Theory of Switching*, pages 285–292. Harvard University Press, 1959.

58. R. Motwani. Expanding Graphs and the Average-case Analysis of Algorithms for Matchings and Related Problems. In *Proc. 21st Annual ACM Symposium on Theory of Computing*, pages 550–561, 1989.

59. Q. C. Nguyen and V. Venkateswaran. Implementations of Goldberg-Tarjan Maximum Flow Algorithm. In D. S. Johnson and C. C. McGeoch, editors, *Network Flows and Matching: First DIMACS Implementation Challenge*, pages 19–42. AMS, 1993.

60. J. B. Orlin. A Faster Strongly Polynomial Minimum Cost Flow Algorithm. In *Proc. 20th Annual ACM Symposium on Theory of Computing*, pages 377–387, 1988.

61. S. Pallottino. Shortest-Path Methods: Complexity, Interrelations and New Propositions. *Networks*, 14:257–267, 1984.

62. U. Pape. Implementation and Efficiency of Moore Algorithms for the Shortest Root Problem. *Math. Prog.*, 7:212–222, 1974.

63. K. G. Ramakrishnan, N. K. Karmarkar, and A. P. Kamath. An Approximate Dual Projective Algorithm for Solving Assignment Problems. In D. S. Johnson and C. C. McGeoch, editors, *Network Flows and Matching: First DIMACS Implementation Challenge*, pages 431–452. AMS, 1993.

64. M. G. C. Resende and G. Veiga. An Efficient Implementation of a Network Interior Point Method. In D. S. Johnson and C. C. McGeoch, editors, *Network Flows and Matching: First DIMACS Implementation Challenge*, pages 299–348. AMS, 1993.

65. H. Röck. Scaling Techniques for Minimal Cost Network Flows. In U. Pape, editor, *Discrete Structures and Algorithms*, pages 181–191. Carl Hansen, München, 1980.

Wanna Buy an Algorithm? Cheap!
or: Algorithms for Text Searching Which Could Have Commercial Value

Invited Lecture

Gaston H. Gonnet

Informatik, E.T.H. Zürich
ETH Zentrum, CH-8092 Zürich, Switzerland

In recent years, fueled by the availability of large text databases and by the problems posed by molecular biology, significant research has been done in the area of text searching. In this talk we would like to explore which algorithms solve real problems and which do not (in the authors humble opinion).

While at the University of Waterloo, the author was involved with the computerization of the Oxford English Dictionary, which not only proved to be successful (in the sense that a dictionary was eventually produced) in the academic world, but was also the seed for a spin-off company which sells large text database software. More recently, while the author was in Zürich, he formed a research group in Computational Biochemistry and had the chance of attacking problems in this area. This experience had also a large practical component, as our group has the policy that the algorithms there developed are placed under an automatic e-mail server which is available worldwide.

This talk will describe a few algorithms which are intuitively a good choice in the important area of prefix searching. The problem of index building, when done for real databases, becomes a major obstacle. Here we find counter-intuitive examples. For example, a quadratic algorithm which will always be preferable to a linear algorithm. The notions of complexity would have to be adapted to a new reality if they were to be used by the commercial world.

It is interesting to give an idea of the sizes of databases that we may be searching. For example, a protein database will contain about 40M amino acids, i.e. 40M index points. A DNA/RNA database will require about 200M index points. The Oxford English Dictionary, with 570Mb of text requires 120M index points. A patent database (patents relating to computers and electronics) will have about 50Gb and about 5000M index points. The complete e-mail historical file of a large insurance company had about 600M index points. These represent very typical figures and applications of text searching.

Index building is a serious problem. We can illustrate this with a very simple example. If we were to use a linear time algorithm (like McCreight suffix trees), and further assume that each index point requires a single disk access (normal implementations will require two, but with clever rearrangements one is achievable), building the patent database mentioned above will require 3.2 years of disk accessing time. Obviously even linear time algorithms using random access are not feasible. An algorithm, which places as much as possible in internal memory and makes incremental passes, will require quadratic time. This quadratic time though, is divided by the amount of main memory available. In this case, the indexing of the patent database, with a large main memory (1Gb) can be done in about 23 days of computing.

The challenge for algorithm design that will have application value, is to properly identify which are the most significant complexity measures. For example, it is not acceptable to compare algorithms which do sequential access to disks with algorithms which do random access. Once that all these measures are identified, the resulting better algorithms are probably quite different from the ones touted as best nowadays. We will explore this topic with further detail. In summary, we hope to give a peek of reality and compare these two worlds (commercial and theoretical) which are probably following their own diverging paths.

Further Reading. A comprehensive description of basic text searching algorithms can be found in [*Handbook of Algorithms and Data Structures In Pascal and C, Second edition*, G. Gonnet and R. Baeza-Yates, Addison-Wesley, 1991, Chapter 7]. The book [Information Retrieval: Algorithms and Data Structures, edited by Frakes, W. and Baeza-Yates, R., Prentice-Hall, 1992] contains a collection of contributed chapters, some of which are particularly relevant to this topic. The yearly conference "Combinatorial Pattern Matching" (last proceedings published by Springer Verlag, CPM-93, Padova, Italy, and CPM-94, Asilomar, California) contain a variety of papers covering the more theoretical fringe of this topic. Information on the algorithms used by commercial text searching systems is very scarce. The field considers itself very competitive and internal information is plainly not available in general. The vast majority of the systems use inverted indices. These inverted indices are often tagged with positional information (position within the structure, phrase, etc.) and frequency information to allow more sophisticated searches. A few systems use (or complement their main search algorithms with) hashed-signatures files.

Planar Drawings and Angular Resolution: Algorithms and Bounds[*]

(Extended Abstract)

Ashim Garg and Roberto Tamassia

Department of Computer Science
Brown University
Providence, RI 02912–1910, USA

Abstract. We investigate the problem of constructing planar straight-line drawings of graphs with large angles between the edges. Namely, we study the angular resolution of planar straight-line drawings, defined as the smallest angle formed by two incident edges. We prove the first nontrivial upper bound on the angular resolution of planar straight-line drawings, and show a continuous trade-off between the area and the angular resolution. We also give linear-time algorithms for constructing planar straight-line drawings with high angular resolution for various classes of graphs, such as series-parallel graphs, outerplanar graphs, and triangulations generated by nested triangles. Our results are obtained by new techniques that make extensive use of geometric constructions.

1 Introduction

Coping with the finite resolution of display devices and of the human eye is a fundamental problem in graph drawing. Namely, in visualization applications, it is important to construct drawings of graphs that avoid placing vertices and edges too close. In this paper we investigate the problem of constructing planar straight-line drawings with with large angles between the edges. Namely, we study the angular resolution of straight-line drawings, defined as the smallest angle formed by two incident edges. Besides visualization applications (see, e.g., [5]), constructing drawing with high angular resolution is important in the design of optical communications networks (see, e.g., [8]).

The study of the angular resolution of drawings has attracted considerable interest in the last years. Formann, Hagerup, Haralambides, Kaufmann, Leighton, Simvonis, Welzl, and Woeginger [8] were the first to study the angular resolution of (generally *nonplanar*) straight-line drawings of various classes of graphs. They show that a degree-d graph has a straight-line drawing with angular resolution $\Omega(1/d^2)$, and every degree-d planar graph has a straight-line (*nonplanar*)

---

[*] Research supported in part by the National Science Foundation under grant CCR-9007851, by the U.S. Army Research Office under grants DAAL03-91-G-0035 and DAAH04–93–0134, and by the Office of Naval Research and the Defense Advanced Research Projects Agency under contract N00014-91-J-4052, ARPA order 8225. The Email addresses of the authors are ag@cs.brown.edu and rt@cs.brown.edu.

drawing with angular resolution $\Omega(1/d)$. (Note that the above bounds are independent from the number of vertices of the graph.) They also pose the open problem of characterizing the angular resolution of *planar* straight-line drawings.

Malitz and Papakostas [15] make further progress on the above open problem by proving a lower bound on the angular resolution dependent only on the degree of the graph. Namely, they show that a degree-d planar graph admits a planar straight line drawing with angular resolution $\Omega(1/7^d)$, irrespectively of the number of vertices. They also analyze a linear program associated with two necessary conditions on the angles of a planar triangulation (i.e., the sum of the angles around a vertex is 2π, and the sum of the angles inside each triangle is π), and find that such a linear program admits a solution with $\Omega(1/d)$ angles. This leads them to conjecture that the trivial $\Omega(1/d)$ lower bound on the angular resolution may be achievable for planar straight-line drawings.

Kant [10, 11] shows that testing whether a biconnected planar graph admits a planar straight-line drawing with angular resolution greater than or equal to a given constant is NP-hard, thus providing further motivation to the study of asymptotic bounds. He also considers polyline drawings (where edges are drawn as polygonal chains), and shows that a degree-d triconnected planar graph admits a planar polyline grid drawing with angular resolution $\Omega(1/d)$.

Di Battista and Vismara [7] provide a characterization of the angles in planar straight-line drawings of maximal planar graphs through a nonlinear system of inequalities. As a consequence, they show that testing whether a prescribed assignment of angles to a maximal planar graph is achievable in a planar straight-line drawing can be done in linear time, and that constructing a planar straight-line drawing with maximum angular resolution can be reduced to the solution of a nonlinear optimization problem.

Our results can be summarized as follows:

In Section 3, we prove the first nontrivial upper bound on the angular resolution of planar straight-line drawings by exhibiting a family of degree-d planar graphs that require angular resolution $O(\sqrt{\log d/d^3})$ in any planar straight-line drawing. Previously, only the trivial $O(1/d)$ bound was known, and determining whether it is achievable was posed as an open problem in [15, 11].

In Section 4, we show a continuous trade-off between the area and the angular resolution of planar straight-line drawings. Namely, there exists a constant $c > 1$ such that, for any $n > d > 6$, with $n - 3$ a multiple of $d - 4$, there exists a planar graph G_n^d with n vertices and degree d such that any planar straight-line drawing of G_n^d with angular resolution ρ has area $\Omega(c^{\rho n})$. In particular, this implies that there exist bounded-degree planar graphs for which asymptotically optimal resolution can be achieved only at the expense of an exponential increase in the area. Previously, area/trade-off results were proved for planar drawings (see, e.g., [1, 2, 3, 6, 4, 9, 10, 11, 12, 13, 14, 16, 17, 18, 19]), depending on various restrictions on the representation (e.g., upward, straight-line, orthogonal, polyline). Our result is the first "continuous" area/trade-off result for planar drawings.

In Section 5, we give linear-time algorithms for constructing planar straight-line drawings with high resolution of various classes of graphs, such as series-parallel graphs, outerplanar graphs, and nested-star graphs (i.e., maximal planar graphs generated from a triangle by repeated insertions of stars). The angular resolution is shown to be $\Omega(1/d^2)$ for series-parallel and nested-star graphs, and $\pi/(d-1)$ for outerplanar graphs. Previously, Malitz and Papakostas [15] showed that angular resolution $\pi/2(d-1)$ can be achieved for outerplanar graphs. No polynomial bounds in $1/d$ were known for series-parallel graphs and nested-star graphs. We leave it as an open problem either proving or disproving that $\Omega(1/d^2)$ is a tight lower bound on the angular resolution of the series-parallel and nested-star graphs. Note that the family of graphs requiring angular resolution $O(\sqrt{\log d/d^3})$ in any planar straight-line drawing are actually nested-star graphs.

The significance of our result on nested-star graphs is motivated by the following observation: Malitz and Papakostas [15] show that a disk packing of a planar degree-d graph yields a planar straight-line drawing with angular resolution $\Omega(1/7^d)$, and they exhibit a family of nested-star graphs to prove that this exponential bound is the best that can be obtained with the disk packing method. Our results imply that the disk packing method yields suboptimal results and leave as an open problem whether the angular resolution of planar straight-line drawings can be bounded from below by a polynomial in $1/d$.

Our algorithms for drawing nested-star and series-parallel graphs exploit various geometric properties of the angles in certain triangulations. Since most existing algorithms for drawing planar graphs rely on graph-theoretic properties (e.g., orientation, coloring, flow), the techniques in this paper appear to be interesting in the way they blend geometric and graph-theoretic methods. We believe that our results will stimulate further research in this direction.

Due to the space limitations, in this extended abstract we have either sketched the proofs or omitted them.

2 Definitions

First, we review basic graph drawing terminology [5]. A *straight-line* drawing maps each edge into a straight-line segment. The *angular resolution* of a straight-line drawing is the smallest angle formed by two edges incident on the same vertex. A drawing is *planar* if no two edges intersect. The *area* of a drawing is the area of the smallest convex polygon covering the drawing. Whenever we give lower bounds on the area, we assume that the drawing is constrained by some rule that prevents it from being arbitrarily scaled down (e.g., integer coordinates for the vertices, or a minimum unit distance between any two vertices). The results of Section 4 are fully general, since they hold under any rule that implies a finite minimum area for the drawing of a graph.

The *degree* of a graph is the maximum degree of any vertex in the graph. The notation $d_G(u)$ indicates the degree of vertex u in graph (or subgraph) G. Given a line segment with endpoints A and B, AB denotes either the line segment itself or its length. Given two line segments AB and BC, $\angle ABC$ denotes either the angle between AB and BC or its measure. Finally, $\triangle ABC$ denotes the triangle with vertices A, B and C.

3 Upper Bound on the Angular Resolution

In this section we show that there exists an infinite family of degree-d graphs that require angular resolution $O(\sqrt{\log d}/d^3)$ in any planar straight-line drawing. Thus, for these graphs the linear program of [15] gives an inconsistent drawing.

First, we give the following geometric lemma (see Fig. 1(a)).

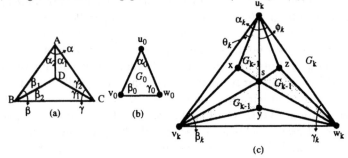

Figure 1: (a) Dividing a Triangle; (b) Graph G_0; (c) Graph G_k;

Lemma 1. *Let D be a point inside a triangle $\triangle ABC$. Let α_1, α_2, β_1, β_2, γ_1 and γ_2 be the angles defined in Fig. 1(a). If $\alpha \leq \pi/2$ and $\alpha_2/\alpha_1 \geq 1$, then*

$$\min(\frac{\beta_2}{\beta_1}, \frac{\gamma_2}{\gamma_1}) \leq \frac{\pi^2}{4}\sqrt{\frac{\alpha_1}{\alpha_2}} \tag{1}$$

We are now ready to present the main theorem of this section.

Theorem 2. *There exists an infinite family of planar graphs with degree d and $\Theta(3^{d/9})$ vertices that require angular resolution $O(\sqrt{(\log d)}/d^3)$ in any planar straight-line drawing.*

Proof. Consider graph G_k, which is recursively defined as follows: G_0 is a triangle with vertices u_0, v_0 and w_0 (see Fig. 1(b)). Graph G_k is assembled from three copies of G_{k-1}, denoted H_1, H_2 and H_3, as shown in Fig. 1(c)). Simple recurrence relations can be used to show that G_k has degree $9k$ and $\Theta(3^k)$ vertices.

Given a straight line drawing of G_k with $u_k v_k w_k$ on the external face, we define the following angles (see Fig. 1(c)): $\alpha_k = \angle w_k u_k v_k$, $\beta_k = \angle u_k v_k w_k$, $\gamma_k = \angle v_k w_k u_k$, $\phi_k = \angle w_k u_k s$ and $\theta_k = \angle v_k u_k x$.

We now claim that the angular resolution of any planar straight-line drawing of G_m with $\triangle u_m v_m w_m$ as the external face is $O(\sqrt{(\log m)/m^3})$. Assume without any loss of generality that $\alpha_m \leq \pi/2$ because in any triangle at least one internal angle has value not exceeding $\pi/2$. If α_0 is $O(\sqrt{(\log m)/m^3})$ the claim holds. So suppose $\alpha_0 = \Omega(1/m^{3/2})$. We shall find another angle in the drawing that is $O(\sqrt{(\log m)/m^3})$. Namely, for any $k \leq m$, let $r_k = \alpha_{k-1}/(\phi_k + \theta_k)$. Thus $\alpha_{k-1} = \alpha_k r_k/(r_k + 1)$. We have $\alpha_0 = \alpha_m \prod_1^k r_k/(r_k + 1)$. Since $\alpha_m < \pi$ and $\alpha_0 = \Omega(1/m^{3/2})$, we obtain

$$\prod_{k=1}^{m} r_k/(r_k + 1) = \Omega(1/m^{3/2}) \tag{2}$$

We now prove that there exists an $i \geq m/2$ such that $r_i = \Omega(m/\log m)$. For any k, from Eq. 2 we get that $\prod_{j=k}^{m} r_j/(r_j + 1) = \Omega(1/m^{3/2})$ because

$r_q/(r_q+1) < 1$ for any q. If $r(k) = \max_{j=k}^m(r_j)$, we get $\{r(k)/(r(k)+1)\}^{m-k} = \Omega(1/m^{3/2})$. Consequently

$$\left(1 + \frac{1}{r(k)}\right)^{m-k} = O(m^{3/2}) \tag{3}$$

From Eq. 3, with simple manipulations we obtain $r(m/2) = \Omega(m/\log m)$. Thus there is an index $i \geq m/2$ such that $r_i = \Omega(m/\log m)$. By Lemma 1, either $\beta_{i-1} \leq \beta_i \pi^2/4(\sqrt{r_i} + \pi^2/4)$ or $\gamma_{i-1} \leq \gamma_i \pi^2/4(\sqrt{r_i} + \pi^2/4)$. Since $d_{i-1}(v_{i-1}) = d_{i-1}(w_{i-1}) = 3i-1$, both β_{i-1} and γ_{i-1} include $3i-2 \geq 3m/2-2$ angles in their interior. Hence for sufficiently large m, there is an angle inside either β_{i-1} or γ_{i-1} that is $O(\sqrt{(\log m)/m^3})$. This concludes the proof of the claim.

It can be readily seen that a planar straight-line drawing of G_m where $\triangle u_m v_m w_m$ is not the external face contains as a subdrawing a drawing of G_{m-1} with external face $\triangle u_{m-1} v_{m-1} w_{m-1}$. Hence, any planar straight-line drawing of G_m has angular resolution $O(\sqrt{(\log(m-1))/(m-1)^3}) = O(\sqrt{(\log m)/m^3})$. We conclude by recalling that graph G_m has degree $\Theta(m)$ and $\Theta(3^m)$ vertices.

4 Tradeoff between Area and Angular Resolution

In this section we show that there exists a continuous tradeoff between the area requirement and the angular resolution of a planar straight-line drawing. Namely, for every $n > d > 0$, we show that there exists a planar graph G_n^d with n vertices and degree d such that any planar straight-line drawing of G_n^d with resolution ρ requires area $\Omega(c^{\rho n})$, where c is a constant greater than 1.

For a fixed d, we recursively define graph G_k as follows. Graph G_0 is a triangle consisting of vertices A_0, B_0, and C_0. Graph G_k is obtained from G_{k-1} as shown in Fig. 2(a). For $k \geq 2$, G_k has $k(d-4)+3$ vertices and degree d. Graph G_n^d is defined as $G_{\frac{n-3}{d-4}}$, whenever $n-3$ is a multiple of $d-4$.

First, we give a technical lemma, which uses the notation of Fig. 2.

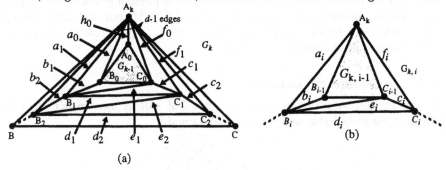

(a)

(b)

Figure 2: (a) Graph $G(d,n)$; (b) Introducing vertices and edges

Lemma 3. *Given a planar straight line drawing of G_k, let $\triangle_i^k$ be the area of $\triangle A_k B_i C_i$, with $1 \leq i \leq (d-4)/2$, then in at least one of the following pairs of edges, both edges have length greater than $\sqrt{2\triangle_i^k}$: (a_i, a_{i+1}), (f_i, f_{i+1}), (d_i, e_{i+1}), and (d_i, b_{i+1}).*

We are now ready to state the main result of this section:

Theorem 4. *There exists a constant $c > 1$ such that, for any $n > d > 6$, with $n - 3$ a multiple of $d - 4$, there exists a planar graph G_n^d with n vertices and degree d such that any planar straight-line drawing of G_n^d with angular resolution ρ has area $\Omega(c^{\rho n})$.*

Proof. Consider a planar straight line drawing of G_k with $A_k B_k C_k$ as the external face for some $k = (n - 3)/(d - 4)$. We denote with Δ_k the area of G_k, and with Δ_i^k the area of triangle $A_k B_i C_i$ (see Fig. 2). By Lemma 3, we have

$$\Delta_{i+1}^k > \Delta_i^k + \frac{1}{2}\sqrt{2\Delta_i^k}\sqrt{2\Delta_i^k}\sin\rho > \Delta_i^k(1 + \sin\rho). \tag{4}$$

Since $\Delta_k = \Delta_{\frac{d-4}{2}+1}^k$ and $\Delta_1^k = \Delta_{k-1}$, we obtain the recurrence

$$\Delta_k > \Delta_{k-1}(1 + \sin\rho)^{\frac{d-4}{2}}. \tag{5}$$

Recalling the recursive construction of G_k, we get

$$\Delta_k > \Delta_0(1 + \sin\rho)^{\frac{d-4}{2}k} = \Delta_0(1 + \sin\rho)^{\frac{d-4}{2}\frac{n-3}{d-4}} = \Delta_0(1 + \sin\rho)^{\frac{n-3}{2}} \tag{6}$$

Since $\sin\rho \geq \frac{1}{2}\rho$ and $\rho \leq \pi/3$, we have that $\Delta_k > r^{\rho n}$ for some constant $r > 1$. Finally, we observe that in any planar drawing of G_k with $A_k B_k C_k$ not being the external face, there is a subgraph of G_k isomorphic to $G_{k/2}$, that is drawn with $A_{k/2} B_{k/2} C_{k/2}$ as the external face. We conclude that any planar straight line drawing of G_k has area $\Omega(c^{\rho n})$ for some constant $c > 1$.

Note that Theorem 4 is fully general, since it holds under any rule that implies a finite minimum area for the drawing of a graph.

Corollary 5. *There exists an infinite family of bounded-degree graphs that require exponential area in any planar straight-line drawing with bounded (from below) angular resolution.*

5 Constructing Drawings with High Angular Resolution

In this section we give linear-time algorithms for constructing planar straight-line drawings with good angular resolution for various classes of planar graphs. We show that every nested-star graph and series-parallel graph admits a planar straight-line drawing with angular resolution $\Omega(1/d^2)$, and that every outerplanar graph has a planar straight-line drawing with angular resolution $\pi/(d - 1)$, where d denotes the degree of a graph.

5.1 A Geometric Lemma

In this section we give a geometric lemma that will be used in our algorithms for drawing nested-star and series-parallel graphs.

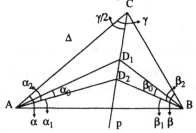

Figure 3: Triangle Δ in Lemma 6

Lemma 6. *Let $\Delta = \Delta ABC$ be a triangle with internal angles α, β and γ where $\alpha \leq \beta \leq \pi/2$, as shown in Fig. 3. Let p be the bisector of angle γ, and D_1 and D_2 be any two points on p inside Δ. Let $\alpha_0 = \angle D_1 A D_2$ and $\beta_0 = \angle D_1 B D_2$. Then*

$$\frac{\beta_0}{\beta} \geq \frac{1}{4\pi^2} \frac{\alpha_0}{\alpha}$$

5.2 Nested-Star Graphs

A nested-star graph G is defined recursively as follows: G has three distinguished vertices called its *apex* vertices (denoted by u, v and w in Fig. 4). G either consists of a single triangle (see Fig. 4(a)) or it is assembled from three nested-star graphs G_1, G_2 and G_3, as shown in Fig. 4(b). Lemma 7, which can be easily proved by induction on the number of vertices in G, states a property of nested-star graphs that is used by our drawing algorithm.

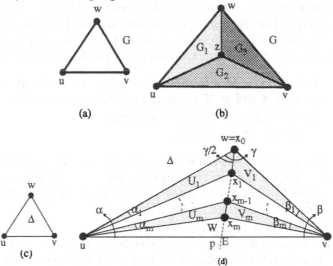

Figure 4: A nested-star graph G: (a) as a single triangle; (b) formed by three nested-star graphs; Drawing G when (c) it is a single triangle; (d) it consists of nested-star subgraphs

Lemma 7. *Let G be a nested-star graph with apex vertices u, v and z. The vertices of G that are common neighbors of u and v form a path.*

We now describe a recursive algorithm that constructs a drawing of a nested-star graph G with angular resolution $1/(24\pi d^2)$, where d is the degree of G.

Let u, v and w be the apex vertices of G. Given a triangle Δ, which is said to be *allocated* to G, we show how to draw G inside Δ (see Fig. 4(c)–(d)). The vertices of Δ are u, v and w. If G has three edges, then its drawing is Δ (see Fig. 4(c)). Otherwise, we draw G as follows (see Fig. 4(d)): Let α, β and γ be the internal angles of Δ at u, v and w respectively. We assume without loss of generality that angle $\gamma \geq \beta \geq \alpha$. By Lemma 7, let $x_0 = w, x_1, x_2, \ldots, x_m$ be the path formed by the common neighbors of u and v. Let U_i be the nested-star subgraph of G with x_{i-1}, u and x_i as its apex vertices. Let V_i be the nested-star subgraph of G with x_{i-1}, v and x_i as its apex vertices. Notice that ux_mv is a face of G. Let W be the nested-star subgraph of G which consists only of the triangle ux_mv. Let p be a line through w that bisects angle γ. For each $i \geq 1$, place x_i on line p in the interior of Δ such that $\angle x_{i-1}ux_i$ is $(d_{U_i}(u) - 1)\alpha/(d_G(u) - 1)$. (Recall that $d_G(u)$ denotes the degree of vertex u in graph G.) We allocate triangle $\Delta x_{i-1}ux_i$ to U_i and triangle $\Delta x_{i-1}vx_i$ to V_i, and then draw U_i and V_i recursively in their allocated triangles.

Theorem 8. *A nested-star graph with degree d admits a planar straight-line drawing with angular resolution $1/(24\pi d^2)$ that can be constructed in linear time.*

Proof. Allocate an equilateral triangle Δ to the graph and then apply the above algorithm. We prove that the resulting drawing has angular resolution $1/(24\pi d^2)$. Intuitively, when allocating a triangle to a nested-star subgraph H of G, the algorithm allows the internal angle at an apex vertex of H to become "small", but no smaller than $1/(24\pi d)$. The algorithm then divides the angle evenly between the edges incident on the vertex in H, so that the minimum angle is at least $1/(24\pi d^2)$. The proof is based over maintaining the following invariant:

Invariant 1: Let Δuvw be a triangle allocated to a nested-star graph G with apex vertices u, v and w. If u is designated the *critical vertex* of G then

- $(d_G(u) - 1)/(24\pi d^2) \leq \angle vuw \leq \pi/6$, and
- $\angle uvw \geq \pi/6$ and $\angle uwv \geq \pi/6$.

Intuitively, the algorithm divides the angle at a critical vertex evenly between the incident edges by allocating an angle to each U_i proportional to the degree of the vertex in U_i. The key to proving Invariant 1 is showing that the angle allocated to each V_i is at least $1/(24\pi d)$.

We now prove that Invariant 1 holds for every nested-star subgraph of the input graph. Initially designate any apex vertex of the input nested-star graph to be the critical vertex. Invariant 1 holds trivially for the input graph. we now prove that the Invariant 1 is maintained at each recursion. Suppose Invariant 1 holds for a nested-star subgraph G with apex vertices u, v and w. Let Δuvw be the allocated triangle of G. Let u be the critical vertex of G. Suppose we denote the internal angles of Δuvw at u, v and w by α, β and γ respectively (see Fig. 4(d)). Let $\gamma \geq \beta$. Therefore $\gamma \geq \pi/3$ and $\beta \leq \pi/2$. The algorithm allocates triangles $\Delta x_{i-1}ux_i$ and $\Delta x_{i-1}vx_i$ to the nested-star subgraphs U_i and V_i respectively. Let E be the intersection of the bisector p of γ with the line segment uv.

We first show that Invariant 1 holds for U_i. Let $\alpha_i = \angle x_{i-1}ux_i$. $\angle ux_{i-1}x_i \geq \angle uwx_1 = \gamma/2 \geq \pi/6$. $\angle ux_ix_{i-1} \geq \angle uEw = \pi - \gamma/2 - \alpha \geq \pi/6$. $\alpha_i = (d_{U_i}(u) - 1)\alpha/(d_G(u) - 1)$. We designate u as the critical vertex of U_i. Because Invariant 1 holds for G, $\alpha \geq (d_G(u) - 1)/24\pi d^2$. Hence $\alpha_i \geq d_{U_i}(u)/24\pi d^2$. Therefore Invariant 1 holds for U_i.

Now we show that Invariant 1 holds for V_i. We designate v as the critical vertex of V_i. Let $\beta_i = \angle x_{i-1}vx_i$. $\angle vx_{i-1}x_i \geq \angle vwx_1 = \gamma/2 \geq \pi/6$. $\angle vx_ix_{i-1} \geq$

$\angle vEw = \pi - \gamma/2 - \beta \geq \pi/6$. From Lemma 6, $\beta_i \geq (\alpha_i/4\pi^2\alpha)\beta$. Since $\beta \geq \pi/6$ and $\alpha_i = (d_{U_i}(u) - 1)\alpha/(d_G(u) - 1)$,

$$\beta_i \geq \frac{1}{4\pi^2} \frac{d_{U_i}(u) - 1}{d_G(u) - 1} \frac{\pi}{6}$$

since $d_G(u) \leq d$, $d_{U_i}(u) \geq 2$ and $d_{V_i}(v) \leq d$, We have $\beta_i \geq (1/(24\pi d) \geq d_{V_i}(v)/(24\pi d^2)$ Therefore Invariant 1 holds for V_i. We can similarly show using Lemma 6 that Invariant 1 holds for W.

Finally we show using induction over the number of vertices in G that the minimum angle in the drawing of G constructed by the algorithm is at least $1/24\pi d^2$. The base case is when G consists of a single triangle Δ. Each internal angle of Δ is at least $1/24\pi d^2$ because of Invariant 1. In a general inductive step let G consists of nested-star graphs $W, U_1, U_2, \ldots, U_m$ and $V_1, V_2, \ldots, V_m$, where U_i V_i, and W are as described in the algorithm. Since the drawing of each U_i and V_i, and W has minimum angle at least $1/24\pi d^2$ by our inductive hypothesis, we conclude that the minimum angle in the drawing of G is at least $1/24\pi d^2$.

5.3   Series-Parallel Graphs

A series-parallel graph G is defined recursively as follows [1]: It contains a pair of distinguished vertices called its *poles*. It consists of either a single edge, or of a series composition, or of a parallel composition of two series-parallel graphs. For us the following equivalent definition is more useful: It consists of either a single edge, or a series composition of two series-parallel graphs, or a parallel composition of series-parallel graphs $G_1, G_2, \ldots, G_m$ where each G_i is either a single edge or a series composition of two series-parallel graphs. A *series-parallel subgraph* of G is a subgraph of G that is also a series parallel graph.

We now describe a recursive algorithm that constructs a drawing of a series-parallel graph G with angular resolution $1/(48\pi d^2)$, where d is the degree of G. Let s and t be the poles of G. We allocate a triangle Δ to G and draw G recursively inside Δ. Two vertices of Δ are s and t (see Fig 5(a)). Let the third vertex of Δ be r. One internal angle of Δ is $\pi/3$ ($\angle rts$ in Fig 5(a)).

We consider three cases:

- G is a single edge (s, t): We draw it as a straight line between s and t (see Fig 5(b)).

- G is a series composition of two series-parallel graphs G_1 and G_2, where s is a pole of G_1 and t is a pole of G_2: We draw G as shown in Fig. 5(c). We assume without any loss of generality that $\angle rst \leq \angle rts$. Let r_2 be the midpoint of the line segment rt. Let λ_2 be the line passing through r_2 making an angle $\pi/3$ with the line st. Let u be the intersection of λ_2 and st. Let λ_1 be the line parallel to the line rt passing through u. r_1 is the intersection of λ_1 and the line sr. We allocate triangle Δsur_1 to G_1 and triangle Δtur_2 to G_2 and draw G_1 and G_2 recursively in their allocated triangles.

- G is a parallel composition of the series-parallel graphs $G_1, G_2, \ldots, G_m$: Since G does not have multiple edges, there can be only two subcases: (i) No G_i consists of a single edge, or (ii) one and only one G_i consists of a single edge

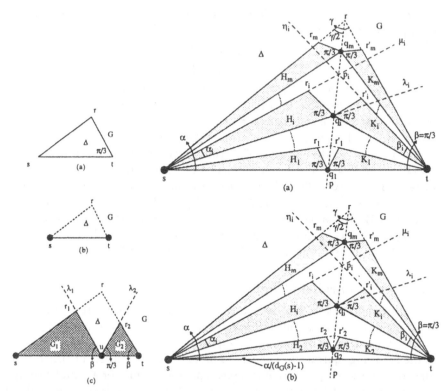

Figure 5: Drawing a series-parallel graph G: (a) Triangle Δ allocated to G; (b) Drawing G when G is a single edge; (c) Drawing G when G is a series composition;

Figure 6: Drawing a series-parallel graph G when it is a parallel composition of sp-graphs $G_1, G_2, \ldots, G_m$: (a) Case i; (b) Case ii.

and without any loss of generality let G_1 be that subgraph. In both subcases **i** and **ii**, if G_i does not consist of a single edge, it is a series composition of two series parallel subgraphs. Let us denote the subgraph whose pole is s by H_i and the other subgraph whose pole is t by K_i. Let q_i be the pole common to H_i and K_i. Fig. 6(a) shows the drawing of G in case **i** and Fig. 6(b) shows the drawing of G in case **ii**. In both the cases, if u_i does not consist of a single edge (i.e., it is a series composition), we allocate a triangle $\Delta s r_i q_i$ to H_i and a triangle $\Delta s r_i' q_i$ to K_i, and recursively draw H_i and K_i in their allocated triangles. The position of each r_i, q_i and r_i' is determined as follows: Assume without any loss of generality that $\angle rst \leq \angle rts$. Let $\alpha = \angle rst$. Let p be the bisector of angle $\angle srt$. Draw lines λ_i and μ_i from s such that (a) the angle between λ_i and μ_i is $(2d_{H_i}(s) - 1)\alpha/(2d_G(s) - 1)$ and (b) angle between μ_i and λ_{i+1} is $\alpha/(2d_G(s) - 1)$. (Recall that we denote by $d_G(u)$, the degree of vertex u in G.) If u_1 consists of a single edge then G_1 is drawn as a straight line between s and t and $\angle q_2 st$ is $\alpha/(2d_G(s) - 1)$ (Fig. 6(b)). q_i is the intersection of λ_i with p. r_i is a point on μ_i such that $\angle r_i q_i s$ is $\pi/3$. Let p_i be the intersection of p and μ_i. Let η_i be the line passing through t and

p_i. r'_i is a point on η_i such that $\angle r'_i q_i t$ is $\pi/3$.

Theorem 9. *A series-parallel graph with degree d admits a planar straight-line drawing with angular resolution $1/(48\pi d^2)$ that can be constructed in linear time.*

Proof. The proof is similar to the proof of Theorem 8 and uses Lemma 6.

5.4 Outerplanar Graphs

In this section we give a linear time algorithm that constructs a drawing of a maximal outerplanar graph G with angular resolution $\pi/(d-1)$, where d is the degree of G. This improves the lower bound given in [15] by a factor of two.

Our algorithm constructs a breadth-first search tree $T(G)$ of G and places the vertices on concentric circles in the drawing so that the vertices equidistant from the root of $T(G)$ are placed on the same circle. A sketch of the algorithm is as follows:

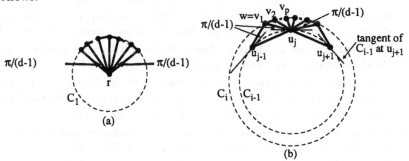

Figure 7: Drawing an outerplanar graph G: (a) Level 1; (b) Level i

1. Construct a breadth-first search tree and assign vertices to levels such that vertices equidistant from the root of the tree get placed at the same level.

2. Construct a drawing as follows (see Fig. 7): Let C_1 be a circle. Place the root of the tree at the center of C_1. Place the vertices of level 1 at equidistant positions on C_1 such that the angle subtended on the center of the circle by consecutive vertices is $\pi/(d-1)$ (see Fig. 7(a)).

   Now, suppose we have already placed the vertices of levels 1 to $i-1$ on concentric circles $C_1, C_1, \ldots, C_{i-1}$. Let $u_1, u_2, \ldots, u_m$ be the left-to-right order of the vertices on C_{i-1} (see Fig. 7(b)). Draw lines from each u_j making angles $\pi/(d-1)$ and $-\pi/(d-1)$ respectively with the tangents at u_j. If u_j and u_{j-1} have a common neighbor (e.g., vertex w in Fig. 7(b)) on level i, then place it at the point of intersection of the lines drawn from u_j and u_{j-1}. Let C_i be the circle concentric to C_{i-1} passing through these points. Let $v_1, v_2, \ldots, v_p$ be the neighbors on level i of u_j. Place each v_k on C_i such that if there is an edge (v_k, v_{k+1}) in G, then $\angle v_k u_j v_{k+1}$ is $\pi/(d-1)$.

Theorem 10. *A maximal outerplanar graph with degree d admits a planar straight-line drawing with angular resolution $\pi/(d-1)$ that can be constructed in linear time.*

References

1. P. Bertolazzi, R. F. Cohen, G. Di Battista, R. Tamassia, and I. G. Tollis. How to draw a series-parallel digraph. *Proc. 3rd Scand. Workshop Algorithm Theory*, vol. 621 of *Lecture Notes in Computer Science*, pp. 272–283. Springer-Verlag, 1992.

2. R. P. Brent and H. T. Kung. On the area of binary tree layouts. *Inform. Process. Lett.*, 11:521–534, 1980.

3. P. Crescenzi, G. Di Battista, and A. Piperno. A note on optimal area algorithms for upward drawings of binary trees. *Comp. Geom. Theory Appl.*, 2:187–200, 1992.

4. H. de Fraysseix, J. Pach, and R. Pollack. Small sets supporting Fary embeddings of planar graphs. *Proc. 20th ACM Sympos. Theory Comput.*, pp. 426–433, 1988.

5. G. Di Battista, P. Eades, R. Tamassia, and I. G. Tollis. Algorithms for drawing graphs: an annotated bibliography. Preprint, Dept. Comput. Sci., Brown Univ., Providence, RI, November 1993. To appear in *Comput. Geom. Theory Appl.* Preliminary version available via anonymous ftp from wilma.cs.brown.edu, gdbiblio.tex.Z and gdbiblio.ps.Z in /pub/papers/compgeo.

6. G. Di Battista, R. Tamassia, and I. G. Tollis. Area requirement and symmetry display of planar upward drawings. *Discrete Comput. Geom.*, 7:381–401, 1992.

7. G. Di Battista and L. Vismara. Angles of planar triangular graphs. *Proc. 25th ACM Sympos. Theory Comput. (STOC 93)*, pp. 431–437, 1993.

8. M. Formann, T. Hagerup, J. Haralambides, M. Kaufmann, F. T. Leighton, A. Simvonis, E. Welzl, and G. Woeginger. Drawing graphs in the plane with high resolution. *Proc. 31th IEEE Sympos. Found. Comput. Sci.*, pp. 86–95, 1990.

9. A. Garg, M. T. Goodrich, and R. Tamassia. Area-efficient upward tree drawings. *Proc. 9th ACM Sympos. Comput. Geom.*, pp. 359–368, 1993.

10. G. Kant. Drawing planar graphs using the *lmc*-ordering. *Proc. 33th IEEE Sympos. Found. Comput. Sci.*, pp. 101–110, 1992.

11. G. Kant. *Algorithms for Drawing Planar Graphs*. PhD thesis, Dept. Comput. Sci., Univ. Utrecht, Utrecht, Netherlands, 1993.

12. G. Kant. A more compact visibility representation. *Proc. 19th Internat. Workshop Graph-Theoret. Concepts Comput. Sci. (WG'93)*, 1993.

13. G. Kant, G. Liotta, R. Tamassia, and I. Tollis. Area requirement of visibility representations of trees. *Proc. 5th Canad. Conf. Comp. Geom.*, pp. 192–197, 1993.

14. C. E. Leiserson. Area-efficient graph layouts (for VLSI). *Proc. 21st IEEE Sympos. Found. Comput. Sci.*, pp. 270–281, 1980.

15. S. Malitz and A. Papakostas. On the angular resolution of planar graphs. *Proc. 24th ACM Sympos. Theory Comput.*, pp. 527–538, 1992.

16. P. Rosenstiehl and R. E. Tarjan. Rectilinear planar layouts and bipolar orientations of planar graphs. *Discrete Comput. Geom.*, 1(4):343–353, 1986.

17. W. Schnyder. Embedding planar graphs on the grid. *Proc. 1st ACM-SIAM Sympos. Discrete Algorithms*, pp. 138–148, 1990.

18. R. Tamassia and I. G. Tollis. A unified approach to visibility representations of planar graphs. *Discrete Comput. Geom.*, 1(4):321–341, 1986.

19. L. Valiant. Universality considerations in VLSI circuits. *IEEE Trans. Comput.*, C-30(2):135–140, 1981.

A Better Heuristic for Orthogonal Graph Drawings*

Therese Biedl[1] and Goos Kant[2]

[1] RUTCOR – Rutgers University, P.O. Box 5062, New Brunswick, NJ 08903-5062, USA, therese@rutcor.rutgers.edu
[2] Department of Computer Science, Utrecht University, Padualaan 14, 3584 CH Utrecht, the Netherlands, goos@cs.ruu.nl

Abstract. An orthogonal drawing of a graph is an embedding in the plane such that all edges are drawn as sequences of horizontal and vertical segments. We present a linear time and space algorithm to draw any connected graph orthogonally on a grid of size $n \times n$ with at most $2n + 2$ bends. Each edge is bent at most twice.

In particular for non-planar and non-biconnected planar graphs, this is a big improvement. The algorithm is very simple, easy to implement, and it handles both planar and non-planar graphs at the same time.

1 Introduction

The research area of *graph drawing* has become an extensively studied field of interest which presents an exciting connection between computational geometry and graph theory. The wide spectrum of applications includes VLSI-layout, software engineering, and project management (see [1] for an up-to-date overview with more than 300 references). The aesthetical quality of a drawing cannot be precisely defined, and depending on the application different criteria have been used. Important characteristics of a "readable" representation are the number of bends, number of crossings, the sizes of the angles, and the required area.

In this paper we study the problem of *orthogonal drawings*, i.e. the drawing of a graph $G = (V, E)$ where each edge is represented by a sequence of alternatingly horizontal and vertical segments. Such a representation is possible only when every vertex has at most 4 incident edges (so-called *4-graphs*). In [11, 7] it was shown that deciding whether G can be embedded in a grid of prescribed area is NP-complete. Lengauer [12] posed it as an exercise to show that G can be embedded in a $2n \times 2n$ grid such that there are at most five bends per edge. In [15] this was improved to an embedding of same size with at most two bends per edge. If G is planar it can be embedded in an $n \times n$ grid with $2n + 4$ bends if it is biconnected, and $2.4n + 2$ bends otherwise [16, 18]. The number of bends along each edge is at most 4. For 3-planar and triconnected 4-planar graphs better bounds are known [10].

---

* Research of the second author was supported by the ESPRIT Basic Research Actions program of the EC under contract No. 7141 (project ALCOM II).

In [19] a lower bound of $2n - 2$ bends is presented for biconnected planar graphs. If a combinatorial embedding of a planar graph is given, an orthogonal representation of it with minimal number of bends can be computed in $\mathcal{O}(n^2 \log n)$ time [17]. However, the number of bends per edge can be large which makes the drawing unattractive. If the planar embedding is not given, the problem is polynomial time solvable for 3-planar graphs [2] and NP-hard for 4-planar graphs [8]. In particular, Garg & Tamassia showed that it is even NP-hard to approximate the minimum number of bends in a planar orthogonal drawing with an $O(n^{1-\epsilon})$ error for any $\epsilon > 0$ [8].

The latter motivates the research for a simple general heuristic to construct orthogonal representations of planar and non-planar graphs. In this paper we present a new algorithm that runs in $\mathcal{O}(n)$ time and produces orthogonal drawings of connected planar and non-planar graphs with the following properties: (i) the total number of bends is at most $2n + 2$; (ii) the number of bends along each edge is at most 2 (unless the graph is the octahedron) (iii) the area of the embedding is $n \times n$. In particular for non-planar and non-biconnected planar graphs, this is a big improvement. The result is obtained by constructing orthogonal drawings of the biconnected components using the so-called st-ordering and merging all these drawings into one orthogonal drawing of the entire graph.

The paper is organized as follows: Section 2 gives some definitions and introduces the st-ordering. In Section 3 the algorithm for biconnected graphs is explained. It is extended to non-biconnected graphs in Section 4. In Section 5 we explain how to come to a linear time implementation. Section 6 contains remarks and open problems.

2   Definitions

Let $G = (V, E)$ be a graph, $n = |V|, m = |E|$. We consider only 4-graphs, i.e. graphs of maximum degree 4. Here $m = \frac{1}{2} \sum_{v \in V} deg(v) \leq 2n$. Call such a graph 4-regular if all vertices have exactly 4 neighbors. An orthogonal drawing of G is an embedding of G in the plane such that all edges are drawn as sequences of horizontal and vertical lines. A point where the drawing of an edge changes its direction is called a bend of this edge. We call such a drawing an embedding in the (rectangular) grid if all vertices and bendpoints are drawn on integer points. If the drawing can be enclosed by a quadrangle of width n_1 and height n_2 we call it an embedding with gridsize $n_1 \times n_2$ and area $n_1 \cdot n_2$.

In this paper we consider only simple graphs, i.e. graphs without loops and multiple edges. Call a simple graph biconnected if removing any vertex and its incident edges leaves a connected graph. The biconnected components (or blocks) of a connected graph are (a) its maximal biconnected subgraphs, and (b) its bridges together with their endpoints. If removing $\{v_1, v_2\}$ disconnects the graph we call $\{v_1, v_2\}$ a cutting pair of G. A (combinatorial) embedding of a planar graph is a representation in which at every vertex all edges are sorted in clockwise order when visiting them around the vertex with respect to the planar embedding. Testing biconnectivity, finding all cutting pairs, and constructing an embedding of a planar graph can be done in linear time (see e.g. [3, 9]). The octahedron is the unique planar 4-regular graph with 6 vertices. An st-ordering is an ordering $\{v_1, v_2, \ldots, v_n\}$ of the vertices such that every v_j $(2 \leq j \leq n-1)$ has at least one *predecessor* and at least one *successor*, i.e. neighbors v_i, v_k with $i < j < k$.

Theorem 1. (Lempel, Even & Cederbaum [13], Even & Tarjan [6]) *Let G be biconnected and $s, t \in V$. Then there exists an st-ordering such that s is the first and t is the last vertex. It can be computed in $\mathcal{O}(m)$ time.*

If G is planar in any *st*-ordering $\{v_1, \ldots, v_n\}$ the predecessors of v_i appear as an interval in any embedding, as do the successors [13]. The edges from v_i to its predecessors (successors) are called *incoming* (*outgoing*) edges of v_i. Their number is the backdegree $bdeg(v_i)$ (forwarddegree $fdeg(v_i)$). A straightforward calculation shows $\sum_{i=1}^{n} bdeg(v_i) = \sum_{i=1}^{n} fdeg(v_i) = m$.

3 Drawing biconnected graphs

3.1 Embedding in an $(m - n + 1) \times n$-grid

The basic idea for drawing biconnected graphs is the same for planar and non-planar graphs. Given a biconnected 4-graph $G = (V, E)$, obtain an *st*-ordering for it and embed the vertices consecutively. At every stage every uncompleted edge (i.e. an edge where exactly one endpoint is drawn) is associated to a column. The accomplished drawing of the edge ends in that column which is empty above that point.

Fig. 1. Embedding of the first two vertices and appearance at some stage

We maintain this invariant by adding as many new columns for each vertex as necessary. If G is planar then we add the new columns directly neighbored to the current vertex. Otherwise they are added at the left resp. right border. The following pictures show this in more detail for vertices of degree 4. Vertices of lower degree are handled according to their backdegree where only as many columns as necessary are added.

Lemma 2. *The gridsize is at most $(m - n + 1) \times n$*

Proof. We use a height of one for v_1 and v_2. Every following vertex increases the height by one, the last vertex by at most two. Therefore the height is at most n. When embedding v_1 and v_2 we use a width of $fdeg(v_1) + fdeg(v_2) - 2$ (observe that the width is #{columns}−1). Every following vertex $v_i \neq v_n$ increases the width by $fdeg(v) - 1$ and v_n increases it by $0 = fdeg(v_n) - 1 + 1$. So the width is $\sum_{v \in V}(fdeg(v) - 1) + 1 = m - n + 1$. □

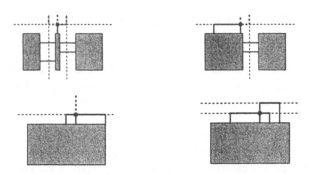

Fig. 2. Embedding v in the planar case ($bdeg(v) = 1, 2, 3, 4$)

Lemma 3. *There are at most $2m - 2n + 4$ bends; at most one edge has three bends, all others have at most two bends.*

Proof. With the embedding of $v \neq v_1, v_2, v_n$ there are $bdeg(v) - 1$ and $fdeg(v) - 1$, hence $deg(v) - 2$ new bends. Embedding v_1 and v_2 gives $fdeg(v_1) + fdeg(v_2) - 1$ bends, and v_n requires $bdeg(v_n)$ bends if $bdeg(v_n) = 4$ and $bdeg(v_n) - 1$ bends otherwise. As $bdeg(v_1) = 0, bdeg(v_2) = 1$, and $fdeg(v_n) = 0$ we have $\sum_{v \in V}(deg(v) - 2) + 4 = 2m - 2n + 4$ bends if $deg(v_n) = 4$ and $2m - 2n + 3$ bends otherwise.

Every edge $(v_i, v_j), i < j$ is bent at most once when v_i is embedded. Completing the edge needs at most one additional bend if $v_j \neq v_n$. Embedding v_n bends one edge twice, all others at most once, thus only this edge can have three bends. □

Lemma 4. *The thrice-bent edge can be avoided unless the graph is the octahedron.*

Proof. Consider first graphs with less than $2n$ edges. These have at least one vertex of degree less than 4. If we choose this vertex to be v_n no edge is bent thrice.

For non-planar graphs we can (by drawing differently) choose which edge we want to be bent twice with the embedding of v_n. If we choose the edge that is outgoing from v_{n-1} (this exists by the property of an st-ordering) it was not bent at v_{n-1} (remark that $fdeg(v_{n-1}) = 1$ by simplicity of G). Therefore this edge has only two bends in total. For planar graphs with at least 7 vertices we show in Section 5 that we can assume that the edge (v_{n-1}, v_n) is on the outerface. This edge then is leftmost or rightmost incoming edge of v_n. So again we can bent this edge twice with the embedding of v_n.

The only graphs which are not covered by these arguments are those which are planar, 4-regular, and have at most 6 vertices. But there is only one such graph: the octahedron. □

Lemma 5. *For planar graphs the resulting drawing is planar.*

Proof. Remark that every edge is drawn with at most two horizontal lines. We show that we produce no crossing when embedding either horizontal line. This is clear for the first: As we add new columns directly neighbored to the vertex there is no vertical line that could be crossed.

For the second horizontal line remark the following property of an st-ordering in a planar graph (see [14, 18]): The predecessors of vertex v_i always lie on the outerface of the graph G_i induced by $v_1, \ldots, v_{i-1}$. This implies that for each vertex v_i the predecessors must form an interval on the outerface of G_i. If v_n and the edge (v_1, v_2) lie on the outerface the columns of incoming edges of v_i build an interval in the set of columns of unfinished edges. In Section 5 we show that we can assume this. Therefore we do not produce a crossing when we draw the second horizontal part of an edge and consequently produce no crossing at all. □

3.2 Embedding in an $(m - n) \times n$-grid

In order to embed non-biconnected graphs we need to reduce the size for biconnected 4-graphs with many edges. Hereto we embed them in an $(m - n) \times n$-grid. This is not possible for all graphs (take e.g. a simple cycle, in this case $m = n$). We show it for graphs with at most one vertex of degree 2. Compute an st-ordering such that the last vertex has minimal degree (this implies $deg(v_i) \geq 3$ for $i < n$). Let l be the smallest index such that $bdeg(v_l) \geq 2$. The vertices $v_1, \ldots, v_{l-1}$ induce a tree. If now $l = n$ then by biconnectivity v_l must be adjacent to all leaves of the tree. But these leaves then have degree 2, a contradiction. So $l < n$.

As we show in Section 5 we can change the ordering in such a way that $bdeg(v_l) = 2$. Let v_i, v_j, $i < j$ be the predecessors of v_l. Draw v_l at the crossing of the row of v_j and the column of the unfinished edge (v_i, v_l) (note that in case $j = 2$ we have to change the embedding of v_2, so we do not necessarily save a row). We then need a width of $fdeg(v_j) + fdeg(v_l) - 3$ and $deg(v_l) + deg(v_j) - 6$ bends to embed v_j and v_l. Embedding all other vertices as before and recalculating gives an $(m - n) \times n$ grid with $2m - 2n + 2$ bends.

Fig. 3. Drawing v_j and v_l, especially if $j = 2$

Applying this for 4-regular graphs we get an $n \times n$-grid. For $m = 2n - 1$ we have at most one vertex of degree 2 and hence a width of $n - 1$. For $m \leq 2n - 2$ the width is $m - n + 1 \leq n - 1$. Also we can then choose v_n to have $deg(v_n) \leq 3$. In this case we need a height of one and $deg(v_n) - 1$ bends to embed v_n, hence the height and the total number of bends is reduced by one each. This proves the following theorem.

Theorem 6. *Let $G = (V, E)$ be a biconnected simple 4-graph. Then G can be embedded in an $n \times n$ grid with at most $2n + 2$ bends. If G is not 4-regular an $(n-1) \times (n-1)$ grid and $2n - 1$ bends suffice. Each edge is bent at most twice unless G is the octahedron. If G is planar then so is the drawing.*

4 Non-biconnected graphs

In this section we describe how to embed graphs which are not biconnected with similar bounds as in Section 3. To do so we split the graph into its blocks and embed them separately. The main result is the following theorem.

Theorem 7. *Let $G = (V, E)$ be a non-biconnected simple 4-graph. Then G can be embedded in an $n \times n$ grid with at most $2n$ bends. If G is not 4-regular an $(n-1) \times (n-1)$ grid and $2n - 1$ bends suffice. Each edge is bent at most twice. If G is planar then so is the drawing.*

Proof. We prove this theorem by giving an embedding of width n for not 4-regular graphs. This bound is tight only for graphs with many edges. We then apply the idea of Section 3.2 to reduce the width by one. The embedding also fulfills the conditions on height and bends, but the proofs are similar and we therefore omit them here. Define [...] to be 1 if the expression inside the brackets is true and 0 otherwise.

Invariant *Let v with $deg(v) \leq 3$ be given. G can be drawn in a grid of width $n - 1 - [deg(v) = 1]$ with v as "final vertex", i.e. only v and its incident edges are drawn in the last row.*

We show the invariant by induction on the number of vertices. In the base case G is biconnected and we apply the algorithm of Section 3 with v as last vertex. By $deg(v) \leq 3$ the drawing fulfills the conditions. For the induction step assume first that $deg(v) = 1$ and let w be the neighbor of v. Apply induction on $G - v$ with w as final vertex and draw it in a grid of width $(n-1) - 1$. Add one row on top of the drawing and place v above w. This drawing fulfills the conditions.

If $deg(v) \geq 2$ let G_0 be a block containing v. Let $v_1, \ldots, v_s$ be the cutvertices of G in G_0. Let G_i be the subgraph of G consisting of v_i and the connected components of $G - v_i$ not containing G_0. Notice that the intersection of G_0 and G_i ($1 \leq i \leq s$) is v_i, and that the intersection of G_i and G_j is empty if $i \neq j$. Hence if we define $n_i = |V(G_i)|$ then $\sum_{i=0}^{s} n_i = n + s$. G_i has less vertices than G, so by induction it can be embedded in a grid of width $n_i - 1 - [deg_{G_i}(v_i) = 1]$ with v_i as final vertex.

Compute an st-ordering of G_0 with v as last vertex. Every vertex $w \in G_0$ which is no cutvertex is embedded as in Section 3. For embedding a cutvertex v_i we add $n_i - 1$ rows and $n_i - [deg_{G_i}(v_i) = 1]$ columns to the side of the incoming edges of v_i. Again we add the columns next to the incoming edges of v in the planar case and at the boundary of the drawing in the non-planar case. Place the drawing of G_i in them, leaving out the last row and hence the drawing of v_i. Now v_i has incoming edges from both G_0 and G_i and we embed it as in Section 3.

Lemma 8. *The width is at most n.*

Proof. To embed a vertex of G_0 we need an increase in width of $fdeg(w) - 1$ if $w \neq v$ and $0 = fdeg(v)$ otherwise. So for G_0 we need a width of $m_0 - n_0 + 1$. As G is not 4-regular we know $m_0 \leq 2n_0 - 1$. Even more, every vertex of degree 2 decreases this bound by 1.

If v is no cutvertex we can estimate $m_0 \leq 2n_0 - 1 - \sum_{i=1}^{s}[deg_{G_0}(v_i) \leq 2] = 2n_0 - 1 - s + \sum_{i=1}^{s}[deg_{G_0}(v_i) = 3]$. If v is a cutvertex it is (by $deg(v) \leq 3$) contained in a bridge which we choose to be G_0. Then $m_0 = 1, n_0 = 2$ and again $m_0 \leq 2n_0 - 1 - s + \sum_{i=1}^{s}[deg_{G_0}(v_i) = 3]$.

In order to merge $G_1, \ldots, G_s$ we increase the grid by $\sum_{i=1}^{s}(n_i - [deg_{G_i}(v_i) = 1])$. As $deg_G(v_i) \leq 4$ we know that this is not larger than $\sum_{i=1}^{s} n_i - \sum_{i=1}^{s}[deg_{G_0}(v_i) = 3]$. Altogether we bound the width by $m_0 - n_0 + 1 + \sum_{i=1}^{s} n_i - \sum_{i=1}^{s}[deg_{G_0}(v_i) = 3] \leq n_0 - s + \sum_{i=1}^{s} n_i = n$. □

We need to decrease the width by one when the bound of Lemma 8 is tight, i.e. when the estimation of m_0 is tight. In this case all vertices of G_0 have either $deg_{G_0}(w) \geq 3$ or they are a cutvertex. With the algorithm from Section 5 we can find an ordering $\{w_1, \ldots, w_{n_0}\}$ of the vertices of G_0 such that the first vertex with backdegree exceeding 1 has backdegree 2. Also every w_j, $1 < j < n_0$ has a predecessor, and it either has a successor or it is a cutvertex. We want to embed the first vertex with backdegree 2 together with its last predecessor to save one column.

To be able to get the proper estimations we then need another way of merging G_i: We connect v_i with its predecessors as in Section 3, add sufficiently many columns for the outgoing edges and G_i, flip the drawing of G_i, and embed it in the added columns. We again need $n_i - [deg_{G_i}(v_i) = 1] + fdeg(v_i) - 1$ columns to embed v_i, and this is true even if $fdeg(v_i) = 0$.

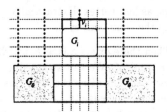 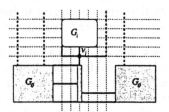

Fig. 4. Merging G_i when embedding v_i – two possibilities

In our ordering every vertex w with $fdeg(w) = 0$ was either the last vertex or a cutvertex. So with the trick of Section 3.2 we need a width of $\sum_{w \in G_0}(fdeg(w) - 1) + 1 - 1 + \sum_{i=1}^{s} n_i - [deg_{G_i}(v_i) = 1] \leq n - 1$ to embed our graph.

 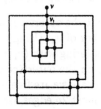

Fig. 5. A non-biconnected graph and its orthogonal drawing

The proof of the width is finished with considering 4-regular non-biconnected graphs. We split G at a cutvertex v into two components and embed them as given by the invariant. Place the drawings next to each other, leaving out both last lines, and add v as in Section 3. This gives a width of at most $n_1 + n_2 - 1 = n$. □

5 Linear time complexity

In this section we describe how to implement the algorithm such that it works in linear time. There are three core items to consider here: (i) how to embed G and to find an st-ordering for a planar G such that (v_1, v_2) and (v_{n-1}, v_n) are on the outerface, (ii) how to find an st-ordering such that the first vertex with $bdeg \geq 2$ has $bdeg = 2$, and (iii) how to add the vertices such that all coordinates can be computed efficiently.

5.1 Computing the st-ordering for planar graphs

The first problem is how to compute an st-ordering such that the edges (v_1, v_2) and (v_{n-1}, v_n) are on the outerface. We assume that v_n is given in advance (important for Section 4, the non-biconnected case) and belongs to the outerface. We first find an ordering with edge (v_1, v_2) lying on the outerface. Let v_1 be a neighbor of v_n. Let $v^* \neq v_n$ be the other neighbor of v_1 on the outerface. If $\{v_1, v^*\}$ is not a cutting pair we can contract edge (v_1, v^*) without losing biconnectivity. Compute an st-ordering for G with the contracted vertex as first and v_n as last vertex. Extract edge (v_1, v^*), giving vertex v^* number 2 and increasing the other numbers by 1.

Assume now that $\{v_1, v^*\}$ is a cutting pair. Then there exists a face F which is not the outerface and which is incident to both v_1 and v^*. We can swap edge (v_1, v^*) into F and repeat the argument with the new neighbor of v_1 that appears on the outerface. Remark that even after repeated iteration the edge (v_1, v^*) never appears at the outerface again. Also not all neighbors of v_1 can form a cutting pair with v_1. Hence after at most $deg(v_1)$ steps we have a neighbor v' of v_1 on the outerface such that $\{v_1, v'\}$ is no cutting pair. Use the argument given above to find an st-ordering with edge $(v_1, v_2 = v')$ on the outerface.

This algorithm obviously works in linear time: In the beginning we mark all vertices which form a cutting pair with v_1. Also for each such vertex v we store two faces containing v_1 and v. Hence we only have to traverse the neighbors of v_1 and, if we find a cutting pair, swap the edge into one face which is not the outerface.

Assume now that the outerface contained at least four vertices and let $\hat{v} \neq v_1, v^*$ be the other neighbor of v_n. Using the same argument with v_n and $\hat{v}$ we can show that there is an st-ordering with (v_1, v_2) and (v_{n-1}, v_n) on the outerface. We needed the edge (v_{n-1}, v_n) on the outerface only to avoid the thrice-bent edge for biconnected graphs in which case we are free to choose v_n. Every planar graph with at least seven vertices and at most $2n$ edges has a face with at least four vertices. So we can choose v_n to belong to this face and avoid for all such graphs the thrice-bent edge.

5.2 Changing the ordering such that $bdeg(w_l) = 2$

In this subsection we concentrate on computing an ordering such that the first vertex of backdegree exceeding 1 has backdegree 2. We assume that the last vertex v is prescribed. Also some vertices $v_1, \ldots, v_s$, $s \geq 0$ are given. We want our ordering to be an st-ordering with the only exception that the v_i's are allowed not to have predecessors. So for Section 3.2 we set $s = 0$ and our ordering is an st-ordering. For Section 4 the v_i's correspond to the cutvertices in G_0.

This is possible for biconnected graphs where all vertices but $\{v, v_1, \ldots, v_s\}$ have $deg \geq 3$. Compute an st-ordering $\{w_1, \ldots, w_n = v\}$. Define l to be the minimal index with $bdeg(w_l) \geq 2$ and set $u^* = w_l$. As seen in Section 3 we know $l < n$. Orient all edges (w_i, w_j), $i < j$ from w_i to w_j. We now reverse the orientation of some edges and get an acyclic orientation such that all sinks are in $\{v, v_1, \ldots, v_n\}$. By traversing it in topological order with v as last vertex we get a suitable ordering.

The vertices $\{w_1, \ldots, w_{l-1}\}$ obviously induce a tree T. Adding w_l produces cycles. Let v^* be the vertex of smallest index contained in a cycle. Choose an incident edge e^* which belongs to a cycle. Especially, if $v^* = w_1$ and the edge (w_1, w_2) belongs to a cycle then choose e^* to be this edge. Let C be a cycle containing e^* such that u^* has exactly two neighbors in C (this can e.g. be done by taking a shortest such cycle). Reverse the orientation of all incoming edges of u^* that do not belong to C.

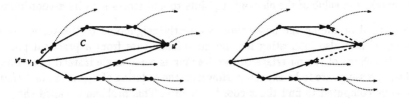

Fig. 6. How the orientation is changed

We show that the orientation obtained is the desired one. u^* has at least one incoming edge from C and one outgoing edge to $G - T$. A neighbor w of u^* not in C has exactly two incoming edges (one from T, one from u^*). But either $w = v_i$ and w is allowed to be a sink, or $deg(w) \geq 3$ and w has at least on outgoing edge. All other vertices have incoming and outgoing edges as before. Assume we have a directed cycle D (this must contain u^*). The only incoming edges of u^* now belong to C, so D must contain vertices of C. There can be at most one edge directed from $G - C$ to C, namely the incoming edge of v^*. But by definition of v^* its predecessor does not belong to a cycle, so we have no such D.

Traverse now first the graph induced by $w_1, \ldots, w_l$ and then the rest of G in topological order. By definition u^* has exactly two neighbors in C, hence $bdeg(u^*) = 2$ and all vertices before it have at most one predecessor. By choice of C the unique source is w_1 and no vertex is topologically between w_1 and w_2. Therefore when traversing G in topological order we can choose w_2 directly after the source w_1, and then traverse all other vertices. In this way w_1 and w_2 are again first and second vertex and in the planar case the edge (w_1, w_2) is still on the outerface.

5.3 Computing the coordinates

In the biconnected case remark first that the y-coordinate of a vertex is never changed later. The same holds for the x-coordinates in the non-planar case, so we have to worry only about the x-coordinates in the planar case. Let v be the vertex we are dealing with. Let $e_1, \ldots, e_s$ be the neighbors of v in clockwise order in the adjacency list of v. Let $\{x_1, \ldots, x_s\}$ be the x-coordinates associated with $e_1, \ldots, e_s$. Notice that if the graph is planar then $x_1, \ldots, x_s$ are descending (unfortunately this is not necessarily true in the non-planar case). The x-coordinate of v is x_k where $k = \lceil \frac{s}{2} \rceil$.

The crucial observation now is that we need not know the values of x_i to find the x-coordinate of v. Hence we can do the following trick: Throughout the algorithm maintain a list *Columns*. Every embedded vertex v contains a pointer $x(v)$ to one element of *Columns*. When embedding v we let $x(v)$ point to the same element in *Columns* as x_k points to. This means that later all vertices with a pointer to the same element of *Columns* receive the same x-coordinate. Whenever we want to add a column we add a new element in *Columns*. By storing a list as a sequence of pointers we can do so without changing any of the x-values of prior vertices. Moreover the columns are added directly left and right of v we are adding. Since we know this place in *Columns* this requires $\mathcal{O}(1)$ time. The final x-coordinates are computed by traversing *Columns* and assigning ascending values to each element. Every vertex then checks the value of the element it points to and stores it as its x-coordinate.

The method of adding the new columns directly neighbored to the vertex we are dealing with is also interesting for the non-planar case from a practical point of view: the shape of the outerface of the drawing is a rectangle instead of a triangle and less crossings are to be expected. However, the ordering of $x_1, \ldots, x_s$ in *Columns* has to be computed to find the x-coordinate of v. This problem is called the *order maintenance problem*: determining which of two elements comes first in a list under a sequence of Insert and Delete operations. Dietz & Sleator [4] presented a linear space data structure for this problem, answering the order queries in $\mathcal{O}(1)$ time. This yields a linear time algorithm for the non-planar drawing algorithm (though less simple), where new columns are added directly neighbored to the vertex.

For non-biconnected graphs we follow the notations of Section 4. We store the coordinates implicitly by variables $dist_x(w), dist_y(w)$ for every vertex $w \in G_0$. They denote the distance between w and v. With every cutvertex v_i we also store in $swap(v_i)$ whether G_i has to be swapped or not. When re-embedding v_i with G_0 we may have to update $dist_x(u)$ for all vertices in the topmost block of G_i. The distance values for each vertex are hence touched at most twice and can be calculated in linear time.

Evaluating the final coordinates happens in top-down fashion. Assume that v is already in its final position (x_v, y_v). For every vertex w in G_0 we add $dist_x(w)$ to x_v to get the x-coordinate and add resp. subtract (depending on $swap(v)$) $dist_y(w)$ to y_v for the y-coordinate. If $swap(v)$ is true we also set $swap(v_i)$ to its reverse for all cutvertices v_i. Then we continue in the subgraphs G_i. Doing so we can compute all final coordinates in linear time.

6 Conclusion

In this paper we have considered the problem of orthogonal drawings of graphs. A general linear time algorithm has been presented to construct an orthogonal representation of a connected graph, with the gridsize and number of bends shown below and at most two bends per edge. In particular for non-biconnected planar graphs this is a considerable improvement in the number of bends.

		$m \leq 2n - 2$	$m = 2n - 1$	4-regular
Biconnected	Gridsize	$(m - n + 1) \times (n - 1)$	$(n - 1) \times (n - 1)$	$n \times n$
	Bends	$2m - 2n + 3$	$2n - 1$	$2n + 2$
Connected	Gridsize	$(n - 1) \times (n - 1)$	$(n - 1) \times (n - 1)$	$n \times n$
	Bends	$2n - 1$	$2n - 1$	$2n$

Fig. 7. Overview of the results

In our algorithm the 3-bent edge only occurs for the octahedron. Even & Granot [5] proved that this graph indeed requires a 3-bent edge. Hence the octahedron is the only 4-graph which cannot be drawn with at most two bends per edge.

We end this paper with the mention of some open problems and directions for further research:

- Very little is known about lower bounds on the grid-area for non-planar graphs. Most results for planar graphs (see e.g. [19]) can not be transferred to non-planar graphs as they demand planar embeddings. Our algorithm yields an embedding of area n^2. It is known that the random degree three graph has minimum crossing number $\Omega(n^2)$ and therefore needs a grid of area $\Omega(n^2)$. Are there better lower bounds? Which methods exist to prove them?
- Lower bounds on the number of bends, as presented in [19] for instance, are not transferable to non-planar drawings. We managed to find a class of graphs which need at least n bends in any orthogonal drawing. Is there something better?
- Storer [16] gave embedded 4-planar graphs requiring an $(n - 2) \times (n - 2)$ grid. Tamassia, Tollis & Vitter [19] proved that there exist embedded 4-planar graphs requiring $2n - 2$ bends. However, when changing the embedding, a lower number of bends can be obtained. It is unclear what the precise lower bounds are for orthogonal drawings of not-embedded 4-planar graphs. Notice that finding the minimum number for this class of graphs is also NP-hard [8].
- Whenever a planar biconnected graph contains a separating triangle at least one edge of the separating triangle requires two bends in any planar drawing. Also each 4-graph with at least $2n - 2$ edges requires a twice-bent edge for (w.l.o.g.) the outgoing edge in left direction of the leftmost vertex. It is unknown which planar graphs not containing a separating triangle and having at most $2n - 3$ edges can be drawn with each edge bent at most once. For 3-planar graphs the answer is positive as shown by Kant [10].

References

1. G. Di Battista, P. Eades, R. Tamassia, and I.G. Tollis Algorithms for Automatic Graph Drawing: An Annotated Bibliography, to appear in *Comp. Geom.: Theory and Applications*, Preliminary version also available via anonymous ftp from wilma.cs.brown.edu (128.148.33.66), files /pub/gdbiblio.tex.Z and /pub/gdbiblio.ps.Z.

2. G. Di Battista, G. Liotta, and F. Vargiu, Spirality of orthogonal representations and optimal drawings of series-parallel graphs and 3-planar graphs, In: *Proc. Workshop on Algorithms and Data Structures*, Lecture Notes in Computer Science 709, Springer-Verlag, 1993, pp. 151–162.

3. N. Chiba, T. Nishizeki, S. Abe, and T. Ozawa, A linear algorithm for embedding planar graphs using PQ-trees, *J. of Computer and System Sciences* 30 (1985), pp. 54–76.

4. P.F. Dietz, and D.D. Sleator, Two algorithms for maintaining order in a list, in: *Proc. 19th Annual ACM Symp. Theory of Computing*, 1987, pp. 365–372.

5. S. Even, and G. Granot, *Rectilinear Planar Drawings with Few Bends in Each Edge*, Manuscript, Faculty of Comp. Science, the Technion, Haifa (Israel), 1993.

6. S. Even, and R.E. Tarjan, Computing an *st*-numbering, *Th. Comp. Science* 2 (1976), pp. 436–441.

7. M. Formann, and F. Wagner, The VLSI layout problem in various embedding models, *Graph-Theoretic Concepts in Computer Science (16th Workshop WG'90)*, Springer-Verlag, Berlin/Heidelberg, 1992, pp. 130–139.

8. A. Garg, and R. Tamassia, *On the Computational Complexity of Upward and Rectilinear Planarity Testing*, Tech. Report CS-94-10, Dept. of Comp. Science, Brown University, Providence, 1994.

9. J. Hopcroft, and R.E. Tarjan, Dividing a graph into triconnected components, *SIAM J. Comput.* 2 (1973), pp. 135–158.

10. G. Kant, Drawing planar graphs using the *lmc*-ordering, Extended Abstract in: *Proc. 33th Ann. IEEE Symp. on Found. of Comp. Science*, Pittsburgh, 1992, pp. 101-110. Extended and revised version in:
    G. Kant, *Algorithms for Drawing Planar Graphs*, PhD thesis, Dept. of Computer Science, Utrecht University, 1993.

11. M.R. Kramer, and J. van Leeuwen, The complexity of wire routing and finding minimum area layouts for arbitrary VLSI circuits. In: F.P. Preparata (Ed.), *Advances in Computer Research, Vol. 2: VLSI Theory*, JAI Press, Reading, MA, 1992, pp. 129–146.

12. Th. Lengauer, *Combinatorial Algorithms for Integrated Circuit Layout*, Teubner/Wiley & Sons, Stuttgart/Chichester, 1990.

13. A. Lempel, S. Even, and I. Cederbaum, An algorithm for planarity testing of graphs, *Theory of Graphs, Int. Symp. Rome* (1966), pp. 215–232.

14. P. Rosenstiehl, and R.E. Tarjan, Rectilinear planar layouts and bipolar orientations of planar graphs, *Discr. and Comp. Geometry* 1 (1986), pp. 343–353.

15. M. Schäffter, Drawing graphs on rectangular grids, *Discr. Appl. Math.* (to appear).

16. J. Storer, On minimal node-cost planar embeddings, *Networks* 14 (1984), pp. 181–212.

17. R. Tamassia, On embedding a graph in the grid with the minimum number of bends, *SIAM J. Comput.* 16 (1987), pp. 421-444.

18. R. Tamassia, and I.G. Tollis, Efficient embedding of planar graphs in linear time, *Proc. IEEE Int. Symp. on Circuits and Systems*, Philadelphia, pp. 495–498, 1987.

19. R. Tamassia, I.G. Tollis, and J.S. Vitter, Lower bounds for planar orthogonal drawings of graphs, *Inf. Proc. Letters* 39 (1991), pp. 35–40.

Hamiltonian Triangulations for Fast Rendering

Esther M. Arkin*[1] and Martin Held[1,2]** and Joseph S. B. Mitchell[1]*** and Steven S. Skiena[3]†

[1] Applied Math, SUNY, Stony Brook, NY 11794-3600, USA
[2] Computerwissenschaften, Univ. Salzburg, A-5020 Salzburg, Austria
[3] Computer Science, SUNY, Stony Brook, NY 11794-4400, USA

Abstract. High-performance rendering engines in computer graphics are often pipelined, and their speed is bounded by the rate at which triangulation data can be sent into the machine. To reduce the data rate, it is desirable to order the triangles so that consecutive triangles share a face, meaning that only one additional vertex need be transmitted to describe each triangle. Such an ordering exists if and only if the dual graph of the triangulation contains a Hamiltonian path. In this paper, we consider several problems concerning triangulations with Hamiltonian duals and a related class of "sequential triangulations".

1 Introduction

The speed of high-performance rendering engines on triangular meshes in computer graphics can be bounded by the rate at which triangulation data is sent into the machine. Obviously, each triangle can be specified by three data points, but to reduce the data rate, it is desirable to order the triangles so that consecutive triangles share a face. Using such an ordering, only the incremental change of one vertex per triangle need be specified, potentially reducing the rendering time by a factor of three by avoiding redundant clipping and transformation computations.

A perfect ordering exists if and only if the dual graph of the triangulation contains a Hamiltonian path. We say that a triangulation is *Hamiltonian* if its dual graph contains a Hamiltonian path.

Even if a given triangulation of n vertices and m faces is Hamiltonian, the topology of the triangulation is not necessarily specified by the encoding sequence of vertices $v_0, \ldots, v_{m+1}$ defining the incremental changes, since, in general, v_i can form a triangle with either v_{i-1}, v_{i-2} or v_{i-1}, v_{i-3}. Figure 1a shows a case in which the next triangle is determined by the geometry, whereas both extensions are possible in Figure 1b.

Therefore, to completely specify the topology of the triangulation, we must specify the insertion order of the new vertices. At least two different models are

 * Partially supported by NSF (CCR-9204585).
 ** Partially supported by Boeing, NSF (DMS-9312098).
 *** Partially supported by Boeing, NSF (ECSE-8857642, CCR-9204585).
 † Partially supported by NSF (CCR-9109289) and ONR (award 400x116yip01).

currently in use, both assuming that all turns alternate from left to right. The Silicon Graphics triangular-mesh renderer *OPenGL* [17] demands that the user issue a *swaptmesh* call whenever the insertion order deviates from alternating left-right turns. In general, this topology can be transmitted at the cost of one extra bit per triangle. The *IGL* graphics library [3] expects triangulations as "vertex-strips", lists of vertices without turn specifications. To get two consecutive left or right turns the vertex must be sent twice, creating an empty triangle, which is discarded. For efficiency, it would be highly desirable to eliminate these extra bits or vertices by finding a path using an implied turn-order. We call a triangulation *sequential* if its dual graph contains a Hamiltonian path such that no three triangulation edges consecutively crossed by this path are incident upon the same vertex of the triangulation. Note that a sequential triangulation has an implied left-right turn-order.

In this paper, we consider the problem of constructing Hamiltonian and sequential triangulations, towards the goal of increasing the efficiency of rendering in computer graphics. Two different classes of problems are of interest:

1. Given a point set or polygon, construct a Hamiltonian or sequential triangulation for it.
2. Given a triangulation T, find a Hamiltonian or sequential path for it, or failing that, either find a decomposition of T into a minimal number of Hamiltonian or sequential paths or add Steiner points such that a Hamiltonian or sequential path can be generated.

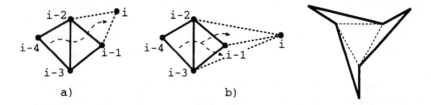

Fig. 1. The new triangle need not be uniquely determined.

Fig. 2. Straight walkability does not imply a Hamiltonian triangulation.

We are unaware of any previous results concerning Hamiltonian paths in the *dual* of triangulations. However, the graph theoretic properties of triangulations have been extensively studied. Dillencourt exhibited a point set whose Delaunay triangulation does not contain a Hamiltonian cycle [7], and showed that testing for the existence of such a cycle is NP-complete [9]. Hamiltonian cycles in Delaunay triangulations form the heart of the Connect-the-Dots heuristic [13] for extracting the shape of points in the plane. Further results on Hamiltonian cycles in Delaunay triangulations include [5, 8]; the length of the longest cycle in simple *d*-polytopes with *n* vertices is considered by [14].

Gray codes [11] are sequences of combinatorial objects which minimize the maximum incremental change between them. Hamiltonian triangulations can be considered as triangulations satisfying a certain "Gray code" property. Gray codes are known for various combinatorial objects, including subsets [19] and partitions [15].

In this paper, we consider several problems concerning triangulations with Hamiltonian duals. Specifically, we

- Show that *any* set of n points in general position in the plane has a Hamiltonian triangulation, and give an optimal $\Theta(n \log n)$ algorithm for constructing such a triangulation. We have implemented and tested our algorithm.
- Prove that certain non-degenerate point sets in the plane do not admit a sequential triangulation, and give efficient algorithms for testing whether a given triangulation or polygon is sequential.
- Show how to test whether a given n-gon P has a Hamiltonian triangulation in time linear in the size of its visibility graph, and show that the problem is NP-complete for polygons with holes.
- Show how to add Steiner points to a given triangulation in order to create Hamiltonian triangulations which avoid narrow angles, thereby yielding guaranteed-quality Hamiltonian mesh generation.
- Give an encoding sequence for any triangulation whose length is at most 9/4 that of optimal.

2   Hamiltonian Triangulations of Polygons

In this section, we consider the problem of constructing Hamiltonian triangulations of simple polygons. The dual of any triangulation of a polygon is a tree, where the leaf nodes of the dual represent ears of the polygon, triangles with two of its three faces defined by the boundary of the polygon. A triangulation of a polygon is Hamiltonian if and only if it contains exactly two ears – i.e., if and only if the dual graph is a path. Toussaint has dubbed such polygons "serpentine polygons".

Convex polygons trivially have Hamiltonian triangulations – simply triangulate with all chords incident upon a single vertex. However, the nonconvex simple polygon in Figure 2 has no Hamiltonian triangulation. We present an efficient algorithm for finding a Hamiltonian triangulation of a polygon, if it exists. We also consider the problem of adding few Steiner points to polygons so as to permit Hamiltonian triangulations.

Deciding whether a polygon has a Hamiltonian triangulation is related to the concepts of LR-visibility and straight walkability. A polygon P is *LR-visible* if P can be partitioned into two subchains, called L and R, such that every point of L can be seen by at least one point on R and vice versa. Das, Heffernan, and Narasimhan [6] show how to test for LR-visibility in linear time. Any polygon with a Hamiltonian triangulation is clearly LR-visible, but Figure 2 shows the converse is not true. A polygon P is *straight walkable* if it has vertices s and g

which partition P into two chains, such that two "guards" walking from s to g on different chains without backtracking are mutually visible at all times. Tseng and Lee [18] give an $O(n \log n)$ time algorithm to find all pairs of points s and g for which an n sided polygon is straight walkable. Any polygon that is straight walkable is clearly LR-visible, and any polygon with a Hamiltonian triangulation is clearly straight walkable. However, Figure 2 shows a straight walkable polygon with a unique triangulation, which is not Hamiltonian. (Note that this example polygon is also monotone and star-shaped.)

We say that a polygon is *discretely straight walkable* if it has vertices s and g which partition P into two chains, such that two "guards" can walk from s to g on different chains without backtracking, where only one of the guards is allowed to walk at a time, while the other guard rests at a vertex, and the two guards are mutually visible whenever both guards are at vertices of the polygon. Clearly, a polygon is discretely straight walkable if and only if it has a Hamiltonian triangulation.

Theorem 1. *The problem of testing whether a given simple n-gon P has a Hamiltonian triangulation can be solved in $O(|E|)$ time, where $|E|$ is the number of visibility graph edges in the polygon.*

Proof. (Belleville and Shermer independently arrived at a similar result [16].) We use dynamic programming. Assume that the vertices of P are numbered in clockwise order. Any visible chord (i, j) of P represents a potential triangulation edge, and partitions P into two subpolygons, one to the left of (i, j) whose vertices are $(i, i+1, \ldots, j)$, and one to the right of (i, j) whose vertices are $(j, j+1, \ldots, i)$, where all additions and subtractions are modulo n. Let $D[i, j]$ be a Boolean function on a pair of vertices i and j which is true if and only if the subpolygon to the left of (i, j) has a Hamiltonian triangulation ending with (i, j). Note that it suffices to define $D[i, j]$ for all pairs of vertices i, j which are mutually visible.

Observe that for all $1 \leq i \leq n$, $D[i, i+2]$ is true iff $(i, i+2)$ is a chord of P. The dynamic programming recursion for $D[i, j]$ is

$$D[i, j] = \{D[i, j-1] \text{ and } (i, j-1) \text{ visible}\} \text{ or } \{D[i+1, j] \text{ and } (i+1, j) \text{ visible}\}.$$

P has a Hamiltonian triangulation if and only if there exists a pair of vertices i, j such that both $D[i, j]$ and $D[j, i]$ are true. Once the visibility graph of P has been constructed in $O(|E|)$ time using Hershberger's algorithm [12], we maintain the visibility graph edges in buckets numbered $k = 2, 3, \ldots, n-2$ for $i - j = k$, and process them in order of increasing k. (Edges of the polygon are *not* considered to be edges of the visibility graph.) It is easy to see that each of the $O(|E|)$ computations of the recurrence takes constant time, since Hershberger's algorithm yields a data structure in which the visible edges from each vertex are sorted in order around the polygon. To answer whether $(i, j-1)$ is visible, we need only ask whether the previous (clockwise) visible vertex from i before j is $j - 1$. Similarly we can decide in constant time whether $(i+1, j)$ is visible. $\square$

Theorem 2. *Given a polygon with (polygonal) holes, it is NP-complete to de-termine whether there exists a triangulation of the interior whose dual graph is Hamiltonian.*

Proof. We use a reduction from the known NP-complete problem of determining whether a planar cubic graph is Hamiltonian [10]. To avoid confusion, we refer to graph vertices as *nodes*, polygon vertices as *vertices*, graph edges as *arcs*, and polygon or hole edges as *edges*.

Given a straight line plane drawing of an arbitrary planar cubic graph G, we construct a polygon with holes corresponding to the interior faces of the planar graph. Each arc of G is mapped to a narrow "V"-shaped tunnel. Thus, for every arc we have 4 edges that are part of the boundary of the polygon or its holes. Each node in G will have three corresponding vertices, which we refer to as *node-vertices*. For every tunnel there are thus two pairs of node-vertices at the "mouths" of the tunnel. Every arc of G has two additional vertices at the bend of the "V", which we call *arc-vertices*. See Figure 3 for an illustration, where vertices A–F are node-vertices, B–E are at the mouths of the tunnel, and vertices G and H are arc-vertices.

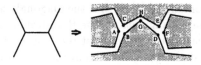

Fig. 3. Transforming a graph to a polygon with holes.

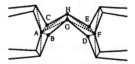

Fig. 4. Hamiltonian triangulations imply Hamiltonian paths.

It is not difficult to see that we can construct such a polygon with holes so that each of the two arc-vertices are visible to both sets of 3 node-vertices corresponding to nodes incident to that arc. Furthermore, no node-vertex is visible to another node-vertex unless they correspond to the same node in the graph. This completes the construction of a polygon with holes P, which can clearly be done in polynomial time.

Note that any triangulation of P must use all diagonals connecting pairs of arc-vertices, for otherwise certain triangles would have interior angles larger that 180 degrees. We refer to these diagonals as *forced* diagonals; in Figure 4, forced diagonals are shown by solid lines whereas the other diagonals are depicted using dotted lines. Thus every triangle in the triangulation is formed by 3 node-vertices, or 2 node-vertices and 1 arc-vertex, or 1 node-vertex and 2 arc-vertices, where the node-vertices correspond to the same node in the graph, and the arc-vertices correspond to a one of the arcs incident to that node.

We now show that if G has a Hamiltonian path (circuit) then there must be a triangulation whose dual contains a Hamiltonian path (circuit). After inserting all forced diagonals, we complete the triangulation as follows. For arcs in the Hamiltonian path, we triangulate the corresponding tunnel into 4 triangles, by connecting each pair of node-vertices at the mouth of the tunnel, obtaining 2 rectangles that are triangulated by one additional (arbitrary) diagonal for each. For arcs not in the Hamiltonian path, we have 6 triangles obtained by connecting the two arc-vertices to both node-vertices (whose corresponding nodes are incident to the corresponding edge) that are not at the mouth of the tunnel. In Figure 4 the path does not pass through the tunnel corresponding to the horizontal edge. The Hamiltonian path in G has a corresponding Hamiltonian path in the dual of the triangulation described, which goes through the four triangles of the arcs in the graph-path, and before proceeding to the next arc in the graph-path goes through 3 of the 6 triangles of the arc not in the graph-path.

To show the converse, namely that if there exists a Hamiltonian triangulation in P then G is Hamiltonian, we recall that the diagonals connecting pairs of arc-vertices are forced in any triangulation. The Hamiltonian path in the dual to the triangulation crosses each such diagonal at most once. The arcs corresponding to forced diagonals that the path crosses once form a Hamiltonian path in the original graph. □

While not all polygons admit a Hamiltonian triangulation, it turns out that any polygon can be made Hamiltonian by adding Steiner points. We have two results – the first demonstrating that few Steiner points always suffice, the second showing that a modest number of Steiner points can result in a Hamiltonian triangulation that avoids slivery triangles.

Theorem 3. *Let P be a simple n-gon which is partitioned into $k > 1$ convex regions in its minimum convex decomposition. Then a Hamiltonian cycle (resp. path) triangulation can be constructed for P using $k - 1$ (resp. $k - 2$) Steiner points in $O(n)$ time. Furthermore, there exist simple n-gons that require $k - 1$ (resp. $k - 2$) Steiner points for a Hamiltonian triangulation.*

Theorem 4. *Every triangulation T containing k triangles can be converted into a Hamiltonian cycle triangulation by adding k Steiner points such that the minimum angle of the Hamiltonian triangulation is not less than half of the minimum angle of the original triangulation.*

Note that this result enables us to postprocess any of the known "guaranteed-quality" triangulations and to convert them into a Hamiltonian triangulation, thereby roughly preserving the features of these triangulations. For instance, we may want to minimize the maximum angle or maximize the minimum angle. We refer to Bern and Eppstein [2] for an extensive review of techniques for generating guaranteed-quality triangulations. Of course, such a postprocessing enlarges the triangulation, which works against our goal of minimizing the data to transmit.

In "conformal" mesh-generation, some of the edges of the original triangulation may not be flipped or removed. In this case, a Hamiltonian triangulation can be generated by placing additional Steiner points on the midpoints of the conformal edges. By observing that the center of the largest inscribed circle of a triangle is constrained to the interior of a similar subtriangle formed by the midpoints of the triangle's edges it can be shown that the introduction of these additional Steiner points does not yield any new angle smaller than one of the already existing angles. Alternatively, care can be taken during the generation of the spanning tree of the dual graph of the original triangulation that no edge which may not be flipped needs to be crossed. We omit details in this abstract.

3   Hamiltonian Triangulations of Point Sets

We now consider the question of whether or not a set of points has a Hamiltonian triangulation. We begin by observing that the Delaunay triangulation is not necessarily Hamiltonian; see Figure 5. We will show that, surprisingly, *any* non-degenerate point set in the plane has a Hamiltonian triangulation, and give an elegant and efficient algorithm for constructing it. In fact, we have computational experience with algorithms that we have implemented to compute Hamiltonian triangulations; we show an example output in Figure 6. Our triangulation algorithm proves to be similar to one of Avis and ElGindy [1] (also Yvinec [20]), although they do not consider the question of finding Hamiltonian cycles in the dual graph.

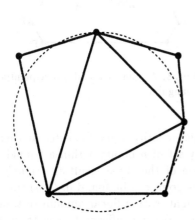

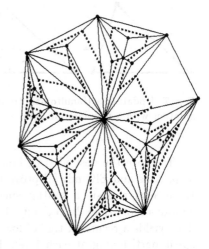

Fig. 5. The Delaunay triangulation is not necessarily Hamiltonian.

Fig. 6. A Hamiltonian triangulation of a point set.

Theorem 5. *Let S be a set of n points in the plane. If S is convex, it has a Hamiltonian path triangulation. Otherwise, S has a Hamiltonian cycle triangulation, and such a triangulation can be constructed in $O(n \log n)$ time.*

Proof. For simplicity, we will assume that S is in general position (no 3 collinear points); J. Mittleman has observed that this assumption is not necessary. If S is convex, then it clearly has a Hamiltonian path triangulation – simply add all chords incident upon any single point to the convex hull of S, $conv(S)$. Therefore, we may assume that there is at least one point v in the interior of $conv(S)$.

Adding chords from v to each vertex of $conv(S)$ yields a triangulation, which has a Hamiltonian cycle in its dual. Each of the remaining points of S lies within a face of this triangulation.

We will now add any additional interior points in an incremental fashion. Because of the non-degeneracy assumption, each point lies in the interior of one face of the current triangulation. Adding three triangulation edges from the new point to the three vertices of its enclosing triangle T divides the triangle into three new triangles. One of the new triangles will contain the edge crossed entering T by the Hamiltonian cycle, while another will contain the exit edge of T. As shown in Figure 7, three triangles can be connected to respect this ordering and preserve the Hamiltonian cycle.

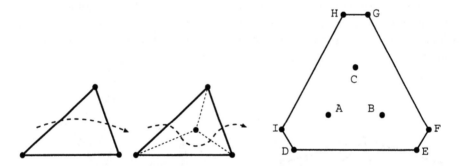

Fig. 7. Updating a Hamiltonian triangulation with a new interior point.

Fig. 8. A point set with no sequential triangulation

To prove the desired time bound, we must identify a good splitter at each step. Avis and ElGindy [1] prove that in any set of n points within a triangle, there always exists a splitter such that none of the three resulting triangles contains more than $2n/3$ points. Further, this point can be found in $O(n)$ time. This yields a recursion tree of logarithmic height, with a linear amount of time spent partitioning at each level. Thus $\Theta(n \log n)$ time suffices to construct a Hamiltonian cycle triangulation of n points. □

The original deterministic algorithm by Avis and ElGindy does not lead to an easy implementation, as it relies on linear-time median finding in one dimension.

Therefore, it is interesting to note that a simple randomized algorithm achieves the same expected time bound:

Lemma 6. *Using a random splitter in the algorithm of Theorem 5 produces a Hamiltonian triangulation in expected $O(n \log n)$ time.*

4 Sequential Triangulations

We say that a triangulation is *sequential* if its dual graph contains a Hamiltonian cycle (resp., path) such that no three triangulation edges consecutively crossed by the Hamiltonian cycle (resp., path) are incident upon the same vertex of the triangulation. This implies that all turns alternate left-right, so sequential triangulations can be described without the extra bit-per-face needed to specify a general Hamiltonian triangulation.

Sequential triangulations represent a significant restriction. For example, convex polygons have an exponential number of Hamiltonian triangulations but have only a linear number of sequential triangulations. Also, point sets that admit Hamiltonian cycle triangulations may not have sequential cycle triangulations, as illustrated by a triangle with an interior point. The most tantilizing question is whether sequential path triangulations always exist for any non-degenerate point set, as do Hamiltonian path triangulations.

After an extensive computer search (of over 300,000 randomly generated configurations of point sets), followed by analysis, we have resolved this question in the negative (see Figure 8):

Theorem 7. *For any $n \geq 9$, there exists a set of n points in general position that does not admit a sequential triangulation (of the convex hull).*

Lemma 8. *Let S be a set of points (in general position). If at most two points of S lie strictly in the interior of the convex hull of S then S has a sequential triangulation.*

We may ask the same questions for sequential triangulations that we have asked for Hamiltonian triangulations: For a given point set or polygon, does a sequential triangulation exist and, if so, how can we compute one? While testing whether a triangulation is Hamiltonian is hard (see Section 5), a fast algorithm exists to test whether it is sequential:

Theorem 9. *Testing whether a given triangulation on n points is sequential can be done in $O(n)$ time.*

Proof. Observe that from a specific ordering of three consecutive vertices of a sequential triangulation, the potential predecessor and successor vertices are completely defined. Therefore, any given triangle has only six possible orientations.

Select an arbitrary triangle as a starting point, and for each orientation compute the sequence of predecessors until it self-intersects, ie. a triangle occurs twice among the predecessors. After constructing a similar sequence among the successors, by concatenating the two sequences together we have a chain of length at most $2m$, where m is the number of triangles. This orientation of the triangle is involved in a sequential triangulation iff there is a substring of the chain consisting of m triangles without repetition.

Clearly, each of the six complete chains can be constructed in linear time. To test for an appropriate substring, we sweep from left to right, maintaining an array of how many times each triangle has occurred in the last m elements, and a counter of how many distinct triangles have been covered in the last m elements. For each new element, we delete one from the count of the triangle leaving the length-m window, decrement the distinct triangle counter if this has gone to zero, increment the count for the new triangle, and increment the distinct triangle counter if this has changed to one. We report a sequential triangulation iff the distinct count has gone to m. Each of the $O(n)$ updates can be performed in constant time. □

Testing whether a polygon has a sequential triangulation can be done within the same time bound as testing whether it has a Hamiltonian triangulation:

Theorem 10. *The problem of testing whether a given polygon P has a sequential Hamiltonian triangulation can be solved in $O(|E|)$ time, where $|E|$ is the number of visibility graph edges for P.*

Proof. Assume that the vertices of the polygon are numbered in clockwise order. Note that once we pick a vertex k of the polygon to be numbered 1, there are only two choices for the vertices to be numbered 2 and 3, namely $k-1$ and $k+1$, after which the remaining numbering is unique. We thus restrict our attention to finding a suitable choice for the first vertex, and which of its neighbors will be numbered 2.

Observe that each visibility graph edge (i, j) can be an edge in a sequential Hamiltonian triangulation for at most two vertices of the subchain $(i, i+1, \ldots, j)$ being assigned the number 1 (all additions and subtractions are modulo n). When $i + j$ is odd, the two vertices are $\lfloor (i + j)/2 \rfloor$ and $\lceil (i + j)/2 \rceil$, denoted type L and R, respectively. When $i + j$ is even, this vertex is given by $(i + j)/2$, and we refer to it as both a type L and R vertex. We initialize our algorithm by setting the number of 'votes of confidence' of type L and R for each vertex to 0. For each visibility graph edge, we increase the number of type L and R votes for the appropriate one or two vertices, since each such edge increases two counts. Finally, a vertex can be numbered 1 in a sequential Hamiltonian triangulation if and only if it received exactly $n - 3$ such type L (or R) votes, where $n - 3$ is the number of diagonals in a triangulation of an n-gon. □

5 Path Coverings and Partitionings of Triangulations

We have seen that any set of points admits a Hamiltonian triangulation (Theorem 5) and that, by adding Steiner points to an existing triangulation, we are always able to make a triangulation Hamiltonian (Theorem 4). However, in some applications, we are required to work with a given triangulation and are not free to add Steiner points. Hence, we are faced with the problem of ordering the faces of an *existing* triangulation to minimize the transmission rate.

The problem of testing whether a given triangulation is Hamiltonian is easily seen to be NP-complete, by reduction from Hamiltonian circuit/path in planar cubic graphs [10]: Given an instance of a planar cubic graph, embed it in the plane, take its dual, and then redraw the dual using straight line segments. This can be done so that the exterior face is convex; hence, we obtain a triangulation. Therefore, the problem of finding the minimum path cover or minimum path partition is also hard. The problem is easy for polygons, however:

Lemma 11. *A minimum path partition of a simple n-gon triangulation can be constructed in $O(n)$ time.*

Proof. The dual graph of a polygon triangulation is always a tree; hence, this problem reduces to finding a minimum path cover or partition of a tree. Clearly, half the number of odd-degree vertices represents a lower bound on the number of paths in a partition or cover. This bound can be realized by a greedy strategy – any path that originates and ends on odd-degree vertices can be deleted to leave a graph with two fewer odd-degree vertices. □

For our motivating application, we seek a description of a triangulation that minimizes the encoding length. The Silicon Graphics triangular-mesh renderer *OPenGL* mentioned in the introduction demands a sequential representation; i.e., all turns alternate from left to right. To get two consecutive left or right turns the vertex must be sent twice, creating an empty triangle, which is discarded, since there is no mechanism to specify turn bits.

By specifying empty triangles and redrawing previously rendered triangles, every triangulation has a legal encoding sequence of vertices. To show this, consider a spanning tree of the dual of the triangulation, and a depth-first traversal of this tree, giving a walk which visits each of the m triangles twice. Empty triangles can be used to ensure that turns alternate left and right. Our goal is to build as short an encoding sequence as possible for a given triangulation.

An ear of a point-set triangulation is a triangle defined by two convex hull edges. Ears lead to long encoding sequences, for once an ear has been entered, two vertices must be repeated in order to leave the ear. We say a triangulation is *deaf* if it has no ears.

Lemma 12. *For any encoding sequence of a deaf triangulation which visits m' non-empty triangles, there is an encoding sequence that has at most $m'/2$ empty triangles.*

Theorem 13. *Let S_{opt} be an encoding sequence of minimum length for a given triangulation T on m triangles. An encoding sequence for T of length at most $(9/4) \cdot |S_{opt}| + O(1)$ can be constructed in $O(n^3)$ time.*

47

Acknowledgements: We thank Jarek Rossignac for inspiring this work with his talk at the Third ARO-MSI Stony Brook Workshop on Computational Geometry, and both Claudio Silva and Chris Smith for helpful discussions. We also thank Josh Mittleman, Joe O'Rourke, Tom Shermer, and Godfried Toussaint for comments and suggestions that improved the paper.

References

1. D. Avis and H. ElGindy. Triangulating point sets in space. *Discrete Comput. Geom.*, 2:99–111, 1987.
2. M. Bern and D. Eppstein. Mesh generation and optimal triangulation. In D.-Z. Du and F. K. Hwang, editors, *Computing in Euclidean Geometry*, LNCS Vol. 1, pp. 23–90. World Scientific, Singapore, 1992.
3. R. Cassidy, E. Gregg, R. Reeves, and J. Turmelle. IGL: The graphics library for the i860, March 22, 1991.
4. N. Christofides. Worst-case analysis of a new heuristic for the traveling salesman problem. In J. F. Traub, editor, *Sympos. on New Directions and Recent Results in Algorithms and Complexity*, page 441, New York, NY, 1976. Academic Press.
5. H. Crapo and J-P. Laumond. Hamiltonian cycles in Delaunay complexes. *Geometry and Robotics Workshop Proceedings*, Springer-Verlag LNCS Vol. 391, pp. 292–305, 1989.
6. G. Das, P. Heffernan, and G. Narasimhan. LR-Visibility in polygons. In *Proc. 5th Canad. Conf. Comput. Geom.*, pages 303–308, Waterloo, Canada, 1993.
7. M. B. Dillencourt. A non-Hamiltonian, nondegenerate Delaunay triangulation. *Inform. Process. Lett.*, 25:149–151, 1987.
8. M. B. Dillencourt. Hamiltonian cycles in planar triangulations with no separating triangles. *J. Graph Theory*, 14:31–49, 1990.
9. M. B. Dillencourt. Finding Hamiltonian cycles in Delaunay triangulations is NP-complete. In *Proc. 4th Canad. Conf. Comput. Geom.*, pages 223–228, 1992.
10. M. Garey, D. Johnson, and R. Tarjan. The planar Hamiltonian circuit problem is NP-complete. *SIAM J. Computing*, 5:704–714, 1976.
11. F. Gray. Pulse code communication. United States Patent Number 2,632,058, March 17, 1953.
12. J. Hershberger. An optimal visibility graph algorithm for triangulated simple polygons. *Algorithmica*, 4:141–155, 1989.
13. J. O'Rourke, H. Booth, and R. Washington. Connect-the-dots: a new heuristic. *Comput. Vision Graph. Image Process.*, 39:258–266, 1987.
14. N. Prabhu. Hamiltonian simple polytopes. Technical Report 90-17, DIMACS, Rutgers Univ., 1990.
15. C. Savage. Gray code sequences of partitions. *J. Algorithms*, 10:577–595, 1989.
16. T. Shermer, private communication, April 1994.
17. Silicon Graphics, Inc. Graphics library programming guide, 1991.
18. L. H. Tseng and D. T. Lee, "Two-Guard Walkability of Simple Polygons", manuscript, 1992.
19. H. Wilf. *Combinatorial Algorithms: An Update.* Society for Industrial and Applied Mathematics, Philadelphia, 1989.
20. M. Yvinec. Triangulation in 2D and 3D space. *Geometry and Robotics Workshop Proceedings*, Springer-Verlag LNCS Vol. 391, pp. 275–291, 1989.

Efficient Construction of a Bounded Degree Spanner with Low Weight

Sunil Arya*   Michiel Smid*

Abstract

Let S be a set of n points in $\mathbb{R}^d$ and let $t > 1$ be a real number. A t-spanner for S is a graph having the points of S as its vertices such that for any pair p, q of points there is a path between them of length at most t times the Euclidean distance between p and q. An efficient implementation of a greedy algorithm is given that constructs a t-spanner having bounded degree such that the total length of all its edges is bounded by $O(\log n)$ times the length of a minimum spanning tree for S. The algorithm has running time $O(n \log^d n)$. Applying recent results of Das, Narasimhan and Salowe to this t-spanner gives an $O(n \log^d n)$ time algorithm for constructing a t-spanner having bounded degree and whose total edge length is proportional to the length of a minimum spanning tree for S. Previously, no $o(n^2)$ time algorithms were known for constructing a t-spanner of bounded degree. In the final part of the paper, an application to the problem of distance enumeration is given.

1 Introduction

Given a set S of n points in $\mathbb{R}^d$ and a real number $t > 1$, a t-spanner for S is a graph having the points of S as its vertices such that for any pair p, q of points there is a path between them having total length at most t times the Euclidean distance between p and q. Much research has been recently done on the problem of efficiently constructing spanners that satisfy additional constraints. Quantities that are of interest are the number of edges in the spanner, the maximum degree, and the weight, which is defined as the total length of all edges. It is clear that each t-spanner must have at least $n - 1$ edges. Also, the weight must be at least equal to the weight of a minimum spanning tree for S. We denote the latter by $wt(MST)$.

Feder and Nisan gave a simple $O(n^2 \log n)$ time algorithm for constructing spanners with bounded degree. (See [13, 16].) However, these spanners can have a very large weight. Chandra et al.[1] present a path greedy algorithm for constructing a spanner with bounded degree. Recent results of Das et al. [3, 5] prove that this spanner has weight $O(wt(MST))$. The algorithm of [1] has running time $O(n^3 \log n)$.

Das and Narasimhan [4] present a fast implementation of a variant of the path greedy algorithm using graph clustering techniques that runs in $O(n \log^2 n)$ time. Their algorithm works as follows. Given the set S, they compute any $\sqrt{t}$-spanner G having $O(n)$ edges. Then, in the second step, they compute, in $O(n \log^2 n)$ time, a $\sqrt{t}$-spanner G' of G, which, clearly, is a t-spanner for S. Das and Narasimhan partition the edges of G' into two sets E_0 and E_1, such that the total weight of E_0 is at most equal to $wt(MST)$, and the edges of E_1 satisfy the so-called leap-frog property. It was shown recently by Das, Heffernan,

*Max-Planck-Institut für Informatik, Im Stadtwald, D-66123 Saarbrücken, Germany. E-mail: {arya,michiel}@mpi-sb.mpg.de. This work was supported by the ESPRIT Basic Research Actions Program, under contract No. 7141 (project ALCOM II).

Narasimhan and Salowe [3, 5] that then the weight of E_1 is bounded by $O(wt(MST))$. Hence, the weight of the t-spanner G' is proportional to $wt(MST)$. Das and Narasimhan take for G the $\sqrt{t}$-spanner of [11, 16], which can be constructed in $O(n \log n)$ time. As a result, in $O(n \log^2 n)$ time, they construct a t-spanner G' of weight $O(wt(MST))$. Since the spanners of [11, 16] are not of bounded degree, however, the t-spanner G' will, in general, also not have bounded degree.

In [13], it is shown that there exists a t such that a t-spanner of degree four can be constructed. In [2], the analogous result is proved for degree-3 spanners. Hence, there has been much interest in spanners of small degree.

In this paper, we present an $O(n \log^d n)$ time algorithm for constructing a bounded degree spanner having weight $O(wt(MST))$. The importance of this result lies in the fact that this is the first algorithm that constructs such a spanner in $o(n^2)$ time. In fact, it is even the first $o(n^2)$ time algorithm for constructing a spanner of bounded degree.

A set of directed edges is said to possess the *gap property* if the sources and sinks of any two edges in the set are separated by a distance at least proportional to the length of the shorter of the two edges. Chandra et al.[1] have shown that if the edges of a graph can be partitioned into a constant number of subsets such that within each subset the gap property holds, then the weight of the graph is bounded by $O(wt(MST) \log n)$ and it has bounded degree.

The idea of the path greedy algorithm is to consider pairs of points in order of increasing distance, adding an edge (p, q) if and only if the partial spanner built until then does not already contain a path between p and q of length at most t times the distance between p and q. It is obvious that the resulting graph is a t-spanner. Additionally, Chandra et al. prove that the edges in this spanner can be partitioned into a constant number of subsets such that each subset satisfies the gap property. Hence, it has bounded degree and weight $O(wt(MST) \log n)$.

In this paper we show that we can in some sense reverse the emphasis of this greedy strategy. We consider pairs of points in order of increasing distance, adding an edge (p, q) if and only if it does not violate the gap property. More precisely, the edges of the partial spanner built until then can be partitioned into a constant number of subsets such that within each subset the gap property holds. (We call this the *gap greedy* strategy). It is obvious that the resulting graph has weight $O(wt(MST) \log n)$ and bounded degree. We are able to show that this graph is also a t-spanner.

The major advantage of the gap greedy approach is that we can give an efficient implementation for a minor variant of it that runs in $O(n \log^d n)$ time. One of the main ideas is that we do not have to consider the pairs in increasing order of their exact distance. It suffices to consider them in increasing order of their approximate distance. If an edge (p, q) is added to the spanner, then several points become "forbidden" as source or destination end points for later edges. Using range trees, we can implicitly maintain the non-forbidden points and their approximate distances. In each iteration, we then take a pair p, q of non-forbidden points having "minimal approximate" distance, add this pair as an edge to the graph, determine the points that become forbidden and remove the approximate distances they induce from the data structure.

Hence, in $O(n \log^d n)$ time, we construct a spanner of bounded degree having weight $O(wt(MST) \log n)$. By applying the above mentioned results of [4] to this spanner, we get an $O(n \log^d n)$ time algorithm for constructing a spanner of bounded degree with weight $O(wt(MST))$.

In the final part of this paper we show how spanners can be used to enumerate distances efficiently. More precisely, given the spanner that results from our algorithm, we can enumerate the k smallest distances in the set S in sorted order, in time $O(n + k \log k)$. The value of k need not be known at the start of the enumeration. We show similar results for enumerating approximate distances.

For the problem of enumerating the k smallest distances, the following was known. Salowe [12] and Lenhof and Smid [8] achieve $O(n \log n + k)$ time for any dimension, but in both algorithms, the value of k must be known in advance and the distances are not enumerated in sorted order. In the plane, Dickerson et al.[7] show that given the Delaunay triangulation, the k smallest distances can be enumerated in $O(n + k \log k)$ time. In this algorithm, the value of k need not be known in advance and the distances are enumerated in sorted order. Hence our spanner can be regarded as an efficient data structure that can be used for distance enumeration.

2 Preliminaries

Let S be a set of n points in $\mathbb{R}^d$. We will consider graphs having the points of S as their vertices. For convenience, we only consider directed graphs. The *weight* of an edge (p, q) is defined as the Euclidean distance between p and q. The *weight of a path* in a graph is defined as the sum of the weights of all edges on the path. If (p, q) is an edge, then p is called its *source* and q is called its *sink*.

The Euclidean distance between the points p and q in $\mathbb{R}^d$ is denoted by $|pq|$. We denote by $|pq|_\infty$ the L_∞-distance between p and q, i.e., $|pq|_\infty = \max_{1 \leq i \leq d} |p_i - q_i|$.

Let $t > 1$. A graph $G = (S, E)$ is called a *t-spanner* for S if for any pair p, q of points of S there is a path in G from p to q having weight at most t times the Euclidean distance between p and q. Any path satisfying this condition is called a *t-spanner path* from p to q.

Since any t-spanner must be connected, it must have at least $n - 1$ edges. Therefore, the goal is to construct t-spanners having $O(n)$ edges, where the constant factor depends on t and d. In this paper, we consider two other quantitites that we want to optimize. First, we want to construct t-spanners in which each point of S has a *degree*—the sum of the *in-degree* and the *out-degree*—which is bounded by a constant that only depends on t and d. Clearly, this implies that the t-spanner contains $O(n)$ edges. Define the *weight* of a set of edges as the sum of the weights of all its elements. The *weight of a t-spanner* is the weight of its edge set. It is clear that the weight of any t-spanner is at least equal to the weight of a minimum spanning tree for S. Therefore, we want to construct a t-spanner whose weight is comparable to that of a minimum spanning tree. In order to estimate the weight of a t-spanner, Chandra et al.[1] introduced the *gap property*. Let $w \geq 0$. A set E of directed edges satisfies the *w-gap property* if for any two edges (p, q) and (r, s) in E, we have $|pr| > w \cdot \min(|pq|, |rs|)$ and $|qs| > w \cdot \min(|pq|, |rs|)$, i.e., the sources and sinks of any two edges are separated by at least w times the weight of the shorter edge.

Lemma 1 (Chandra et al.[1]) *Let E be a set of directed edges that satisfies the w-gap property. If $w \geq 0$, then no two edges share a source, and no two edges share a sink. Further, if $w > 0$, then the weight of E is at most equal to $O((1/w) \log n)$ times the weight of a minimum spanning tree for S.*

If p and q are points in $\mathbb{R}^d$, then we denote the angle between the vectors $\vec{0p}$ and $\vec{0q}$, which is a real number in the interval $[0 : \pi]$, denoted by $angle(p, q)$. The following lemma enables us to prove that a graph is a t-spanner. Its proof is closely related to the proof of Lemma 4.1 in Chandra et al.[1]. The proof is given in the full paper. It gives sufficient conditions for a graph to be a spanner. Intuitively the lemma says that a graph is a spanner if for any edge e missing from the graph there is a similarly-directed edge e' close by (relative to the length of e') with length not much greater than e.

Lemma 2 *Let t, θ and w be real numbers such that $0 < \theta < \pi/4$, $0 \le w < (\cos\theta - \sin\theta)/2$ and $t \ge 1/(\cos\theta - \sin\theta - 2w)$. Let S be a set of points in $\mathbb{R}^d$ and let $G = (S, E)$ be a directed graph such that the following holds. For any two points p and q of S, (i) there is an edge $(r, s) \in E$, such that $\text{angle}(q - p, s - r) \le \theta$, $|rs| \le |pq|/\cos\theta$ and $|pr| \le w|rs|$, or (ii) there is an edge $(r, s) \in E$, such that $\text{angle}(p - q, r - s) \le \theta$, $|rs| \le |pq|/\cos\theta$ and $|qs| \le w|rs|$. Then the graph G is a t-spanner for S.*

3 A greedy algorithm

In this section, we give a simple greedy algorithm for computing a spanner with bounded degree and low weight. In later sections, we modify this algorithm such that it can be implemented efficiently.

Let S be a set of n points in $\mathbb{R}^d$. The following algorithm $gap\_greedy(S, \theta, w)$ constructs a spanner for S. If $w > 0$, then the edges of this spanner can be partitioned into a constant number of subsets, such that within each subset the w-gap property holds. This will guarantee that the spanner has bounded degree and low weight.

The algorithm considers all ordered pairs (p, q) of points in increasing order of their distances. The edge (p, q) is added to the graph iff there is no edge (r, s) in the current graph such that (p, q) and (r, s) have roughly the same direction and the sources p and r are close to each other, or (q, p) and (s, r) have roughly the same direction and the sources q and s are close to each other.

> **Algorithm** $gap\_greedy(S, \theta, w)$
> (* S is a set of n points in $\mathbb{R}^d$, $0 < \theta < \pi/4$, $0 \le w < (\cos\theta - \sin\theta)/2$ *)
> **begin**
>   sort the $2\binom{n}{2}$ ordered pairs of points according to their distances (ties are broken arbitrarily) and store them in a list L;
> $E := \emptyset$;
>   **for all** ordered pairs $(p, q) \in L$ (* visit pairs in sorted order *)
>   **do** $add := true$;
>   **for each** edge $(r, s) \in E$
>   **do if** $\text{angle}(q - p, s - r) \le \theta$
>   **then** $add := add \wedge (|pr| > w|rs|)$
>   **fi**;
>   **if** $\text{angle}(p - q, r - s) \le \theta$
>   **then** $add := add \wedge (|qs| > w|rs|)$
>   **fi**
>   **od**;
>   **if** $add = true$ **then** $E := E \cup \{(p, q)\}$ **fi**
>   **od**;
>   output the set E
> **end**

Lemma 3 *Algorithm $gap\_greedy(S, \theta, w)$ computes a t-spanner for $t = 1/(\cos\theta - \sin\theta - 2w)$. If $w \ge 0$, then this spanner has degree at most $O((c/\theta)^{d-1})$, for a suitable constant c. Further, if $w > 0$, then the weight of this spanner is bounded by $O((c/\theta)^{d-1}(1/w)\log n)$ times the weight of a minimum spanning tree for S.*

Since algorithm gap_greedy inspects all pairs (p, q) of points explicitly, its running time is $\Omega(n^2)$. In the next section, we modify the algorithm. As we will see, the modified version can be implemented such that its running time is bounded by $O(n \log^d n)$.

4 Towards an efficient implementation

We start by introducing the notion of cones. A *(simplicial) cone* is the intersection of d halfspaces in $\mathbb{R}^d$. The intersection of the hyperplanes that bound these halfspaces is called the *apex* of the cone. We always assume that a cone is closed and that its apex is a point. In the plane, a cone having its apex at the point p is a wedge bounded by two rays emanating from p that make an angle at most equal to π.

Let C be any cone in $\mathbb{R}^d$ having its apex at the point p. The *angular diameter* of C is defined as the maximum value of $angle(q - p, r - p)$, where q and r range over all points of $C \cap \mathbb{R}^d$. For $d = 2$, this is exactly the angle between the two rays that form the boundary of C.

Let θ be a fixed real number such that $0 < \theta < \pi/4$. Let C be a collection of cones such that (i) each cone has its apex at the origin, (ii) each cone has angular diameter at most θ, (iii) all cones cover $\mathbb{R}^d$.

In [17], it is shown how such a collection C, consisting of $O((c/\theta)^{d-1})$ cones for a suitable constant c, can be obtained. In the plane and for $\theta = \pi/k$, we just rotate the positive x-axis over angles of $i \cdot \theta$, $0 \le i < 2k$. This gives $2k$ rays. Each wedge between two successive rays defines one cone of C.

For each cone $C \in C$, let l_C be a fixed ray that emanates from the origin and that is contained in C.

After having introduced the terminology, we can modify algorithm *gap_greedy*. There are three major modifications. Consider again the formal description of the algorithm. First, we replace the condition "$angle(q - p, s - r) \le \theta$" by "$q - p$ and $s - r$ are contained in the same cone of C". Clearly, the latter condition implies the first one. Second, we replace the condition "$|pr| > w|rs|$" by "$|pr|_\infty > (w/\sqrt{d})|rs|$", i.e., for the pair p, r, we switch from the Euclidean metric to the L_∞-metric. Note that all points r for which $|pr|_\infty \le \delta$ are contained in the d-dimensional axes-parallel cube centered at p having sides of length 2δ. Using range trees, we can find such points r efficiently. (Finding all points r such that $|pr| \le \delta'$ takes much more time.) Third, instead of inspecting all pairs in increasing order of their distances, we inspect them in order of their *approximate* distances, to be defined below. As we will see, in this way we do not have to inspect all pairs explicitly.

Let C be any cone of C and let p and q be two points in $\mathbb{R}^d$. Let $C_p := C + p := \{x + p : x \in C\}$, i.e., C_p is the cone obtained by translating C such that its apex is at p. Similarly, let $l_{C,p} := l_C + p$. Then we define

$$\delta_C(p, q) := \begin{cases} \text{Euclidean distance between } p \text{ and} \\ \text{the orthogonal projection of } q \text{ onto } l_{C,p} & \text{if } q \in C_p \\ \infty & \text{if } q \notin C_p. \end{cases}$$

Lemma 4 *Let p and q be points in $\mathbb{R}^d$. If $q \in C_p$, then $|pq| \cos \theta \le \delta_C(p, q) \le |pq|$.*

Now we can give the modified algorithm. For each fixed cone C, we compute a set E_C of edges (p, q) such that $q - p \in C$. The union of all these sets will form the edge set of our final spanner.

Consider a cone C. We find the pair (r, s) of distinct points for which $\delta_C(r, s)$ is minimal and add the edge (r, s) to E_C. Having added the edge (r, s), we do not want to add edges (p, q) such that $q - p \in C$ and the distance between p and r is small. That is, after having added (r, s), all points p that are "close" to r should not occur as sources of edges that are added later. Similarly, after having added the edge (r, s), all points q that are "close" to s should not occur as sinks of edges that are added later. That is, the addition of the edge (r, s) causes certain points to become "forbidden" as a source or a sink. In the next iteration, we find the pair (r', s') of non-forbidden points for which $\delta_C(r', s')$ is minimal and proceed in the same way. The formal algorithm is as follows:

```
Algorithm gap_greedy'(S, θ, w)
(* S is a set of n points in IR^d, 0 < θ < π/4, 0 ≤ w < (cos θ − sin θ)/2 *)
begin
for each cone C
do for each r ∈ S and s ∈ S do dist(r, s) := δ_C(r, s) od;
    E_C := ∅;
    while there are r ≠ s such that dist(r, s) < ∞
    do choose r ≠ s such that dist(r, s) is minimal;
        E_C := E_C ∪ {(r, s)};
        for each p ∈ S such that |pr|_∞ ≤ (w/√d)|rs|
        do for each q ∈ S do dist(p, q) := ∞ od
        od;
        for each q ∈ S such that |qs|_∞ ≤ (w/√d)|rs|
        do for each p ∈ S do dist(p, q) := ∞ od
        od
    od
od;
output the set E := ∪_C E_C
end
```

Lemma 5 *Algorithm gap_greedy'(S, θ, w) computes a t-spanner for $t = 1/(\cos\theta - \sin\theta - 2w)$. If $w \geq 0$, then this spanner has degree at most $O((c/\theta)^{d-1})$, for a suitable constant c. Further, if $w > 0$, then the weight of this spanner is bounded by $O((c/\theta)^{d-1}(1/w)\log n)$ times the weight of a minimum spanning tree for S.*

Proof: Consider the set E of edges that is computed by the algorithm. Let (p, q) be any ordered pair of points of S. If $(p, q) \in E$, then the conditions of Lemma 2 hold. So, assume that (p, q) is not contained in E. Let C be a cone such that $q \in C_p$. Consider the iteration during which the edge set E_C is constructed. At the start of this iteration, $dist(p, q)$ has a finite value. Since the edge (p, q) is not added to E_C, the value of $dist(p, q)$ changes to ∞ during some iteration of the while-loop. Let (r, s) be the edge that is added to E_C during that iteration. At the start of it, we have $dist(r, s) \leq dist(p, q) < \infty$, $dist(r, s) = \delta_C(r, s)$ and $dist(p, q) = \delta_C(p, q)$. Moreover, we have $|pr|_\infty \leq (w/\sqrt{d})|rs|$ or $|qs|_\infty \leq (w/\sqrt{d})|rs|$. We only consider the case where $|pr|_\infty \leq (w/\sqrt{d})|rs|$. Then, $|pr| \leq \sqrt{d} \cdot |pr|_\infty \leq w|rs|$. Since $s - r$ and $q - p$ are both contained in C, we have $angle(q - p, s - r) \leq \theta$. By Lemma 4, we have $|rs| \leq \delta_C(r, s)/\cos\theta$ and $\delta_C(p, q) \leq |pq|$. Since $\delta_C(r, s) \leq \delta_C(p, q)$, we conclude that $|rs| \leq |pq|/\cos\theta$. Hence, condition 1. of Lemma 2 holds for the pair (p, q).

We have shown that for each pair (p, q) of points one of the conditions of Lemma 2 is satisfied. This proves that the graph (S, E) is a t-spanner.

Consider any cone C. We will prove that the edges of E_C satisfy the $(w/\sqrt{d})$-gap property. Then, the claims about the degree and the weight follow from Lemma 1.

Consider any two edges (p, q) and (r, s) of E_C. Assume w.l.o.g. that (r, s) was added to E_C before (p, q). Then we must have $|pr|_\infty > (w/\sqrt{d})|rs|$ and $|qs|_\infty > (w/\sqrt{d})|rs|$. (Otherwise, the algorithm would have set $dist(p, q) := \infty$. Therefore, the pair (p, q) would never have been chosen as a pair with minimal and finite $dist(\cdot, \cdot)$-value and, hence, the edge (p, q) would never have been added to E_C.) But this implies that $|pr| \geq |pr|_\infty > (w/\sqrt{d})|rs| \geq (w/\sqrt{d}) \cdot \min(|pq|, |rs|)$, and $|qs| \geq |qs|_\infty > (w/\sqrt{d})|rs| \geq (w/\sqrt{d}) \cdot \min(|pq|, |rs|)$, i.e., the $(w/\sqrt{d})$-gap property holds. ∎

5 An efficient implementation

In this section, we show how to implement algorithm *gap_greedy'* such that its running time is bounded by $O(n \log^d n)$. The main idea is to use range trees (see [10]) for maintaining the minimal value $dist(r, s)$ for all "non-forbidden" points r and s. The technique is related to the ones in [6, 15] for maintaining the closest pair or k-point cluster in a dynamically changing set of points.

Let C be any cone of $\mathcal{C}$. Recall that C is the intersection of d halfspaces. Let $h_1, h_2, \ldots, h_d$ be the hyperplanes that bound these halfspaces, and let $H_1, H_2, \ldots, H_d$ be lines through the origin such that H_i is orthogonal to h_i, $1 \leq i \leq d$. We give the line H_i a direction such that the cone C is "above" h_i. Let L be the line that contains the ray l_C. We give L the same direction as l_C.

Let p be any point in $\mathbb{R}^d$. We write the coordinates of p w.r.t. the standard coordinate axes as $p_1, p_2, \ldots, p_d$. For $1 \leq i \leq d$, we denote by p'_i the signed Euclidean distance between the origin and the orthogonal projection of p onto H_i, where the sign is positive or negative according to whether this projection is to the "right" or "left" of the origin. Similarly, p'_{d+1} denotes the signed Euclidean distance between the origin and the orthogonal projection of p onto L.

In this way, we can write the cone C as $C = \{x \in \mathbb{R}^d : x'_i \geq 0, 1 \leq i \leq d\}$. For $p \in \mathbb{R}^d$, we can write the translated cone C_p with apex p as $C_p = \{x \in \mathbb{R}^d : x'_i \geq p'_i, 1 \leq i \leq d\}$. We define $-C_p := -C + p := \{-x + p : x \in C\}$. Then we have $-C_p = \{x \in \mathbb{R}^d : x'_i \leq p'_i, 1 \leq i \leq d\}$. If $q \in C_p$, then we have $\delta_C(p, q) = q'_{d+1} - p'_{d+1}$.

Let S be a set of n points in $\mathbb{R}^d$. During our algorithm we will maintain a data structure having the form of a $(d+1)$-layered range tree. This data structure depends on the cone C. We describe it in detail.

There is a balanced binary search tree storing the points of S in its leaves, sorted by their p'_1-coordinates. (Points with equal p'_1-coordinates are stored in lexicographical order.) Let v be any node of this tree and let S_v be the subset of S that is stored in the subtree of v. Then v contains a pointer to the root of a balanced binary search tree storing the points of S_v in its leaves, sorted by their p'_2-coordinates. (Points with equal p'_2-coordinates are stored such that the points $(p'_2, \ldots, p'_d)$ are in lexicographical order.) Any node w of this tree contains a pointer to the root of a balanced binary search tree storing the points of w's subtree in its leaves, sorted by their p'_3-coordinates, etc. At the d-th layer, there is a balanced binary search tree storing a subset of S in its leaves, sorted by their p'_d-coordinates. The binary tree that stores points sorted by their p'_i-coordinates is called a *layer-i tree*.

Before we can define the last layer of the data structure, we introduce some notation. Let u be any node of a layer-d tree. We inductively define a sequence $u_d, u_{d-1}, \ldots, u_1$ of nodes such that u_i belongs to a layer-i tree: Define $u_d = u$. Given u_i, walk to the root r of its layer-i tree. Then u_{i-1} is the node of the layer-$(i-1)$ tree that contains a pointer ro r.

For $1 \leq i \leq d$, let x'_{ui} be the maximal p'_i-coordinate that is stored in the left subtree of node u_i. Let x_u be the point with coordinates $x'_{u1}, x'_{u2}, \ldots, x'_{ud}$. (Note that these coordinates are w.r.t. the "axes" $H_1, H_2, \ldots, H_d$. In general, x_u is not a point of S.)

Now we can define the $(d+1)$-st layer of the data structure. Consider again any node u of a layer-d tree. Let S_{ud} be the subset of S that is stored in the subtree of u. Consider the point x_u. Let $S^+_{u,d+1}$ be a subset of $\{p \in S_{ud} : p'_i \geq x'_{ui}, 1 \leq i \leq d\}$ and let $S^-_{u,d+1}$ be a subset of $\{p \in S_{ud} : p'_i \leq x'_{ui}, 1 \leq i \leq d\}$. (The algorithm determines the sets $S^+_{u,d+1}$ and $S^-_{u,d+1}$. For the description of the data structure, we assume that they are any subsets.) Note that all points of $S^+_{u,d+1}$ and $S^-_{u,d+1}$ are contained in the cones C_{x_u} and $-C_{x_u}$, respectively.

Node u of the layer-d tree contains pointers to

1. a list $L^+_{u,d+1}$ storing the points of $S^+_{u,d+1}$, sorted by their p'_{d+1}-coordinates,

2. a list $L^-_{u,d+1}$ storing the points of $S^-_{u,d+1}$, sorted by their p'_{d+1}-coordinates,

3. a variable $\eta_{d+1}(u)$ having value $\eta_{d+1}(u) = \min\{\delta_C(p,q) : p \in S^-_{u,d+1}, q \in S^+_{u,d+1}\}$,

4. and, in case, $\eta_{d+1}(u) < \infty$, a pair of points that realizes $\eta_{d+1}(u)$.

These two lists are called layer-$(d+1)$ lists. If $S^-_{u,d+1}$ or $S^+_{u,d+1}$ is empty, then $\eta_{d+1}(u) = \infty$. (In particular, this is true if u is a leaf.) Otherwise, we have $\eta_{d+1}(u) = \delta_C(p,q) = q'_{d+1} - p'_{d+1}$, where p and q are the maximal and minimal elements that are stored in the lists $L^-_{u,d+1}$ and $L^+_{u,d+1}$, respectively.

During our algorithm, the layer-i trees for $1 \leq i \leq d$ do not change, except for certain η-variables that are defined below. For each node u of a layer-d tree, the corresponding layer-$(d+1)$ lists initially store the sets $\{p \in S_{ud} : p'_i \geq x'_{ui}, 1 \leq i \leq d\}$ and $\{p \in S_{ud} : p'_i \leq x'_{ui}, 1 \leq i \leq d\}$. During the algorithm, elements will be deleted from these lists.

In order to speed up searching during the algorithm, we store all points of S in a dictionary. With each point p, we store

1. a list of pointers to the positions of the occurrences of p in all lists $L^+_{u,d+1}$, and

2. a list of pointers to the positions of the occurrences of p in all lists $L^-_{u,d+1}$.

We are almost done with the description of the data structure. We saw that for each layer-$(d+1)$ structure there is a corresponding η_{d+1}-value. Let $1 \leq i \leq d$ and let v be any node of a layer-i tree. If v is a leaf then v stores a variable $\eta_i(v)$ having value ∞. If v is not a leaf, then let v_l and v_r be the left and right sons of v, respectively. Also, let $\eta_{i+1}(v)$ be the variable that is stored with the layer-$(i+1)$ structure that corresponds to v. Then node v stores a variable $\eta_i(v)$ having value

$$\eta_i(v) = \min(\eta_i(v_l), \eta_i(v_r), \eta_{i+1}(v)), \tag{1}$$

and, in case $\eta_i(v) < \infty$, a pair of points that realizes $\eta_i(v)$.

This concludes the description of our $(d+1)$-layered data structure. Recall that the entire structure depends on the cone C.

Let q be any point of S. We can delete q from all lists $L^+_{u,d+1}$ in which it occurs and update the entire data structure, as follows: Search for q in the dictionary, and follow the pointers to the positions of all occurrences of q in the lists $L^+_{u,d+1}$. For each such u, do the following:

1. Delete q from $L^+_{u,d+1}$. If the list $L^-_{u,d+1}$ is empty, then we are done. Otherwise, let p be the maximal element of $L^-_{u,d+1}$. Go to 2.

2. If q was not the minimal element of $L^+_{u,d+1}$, then we are done. If q was the only element in its list, then we set $\eta_{d+1}(u) := \infty$. Otherwise, if q was not the only element in its list, then let r be the new minimal element of $L^+_{u,d+1}$. Then, set $\eta_{d+1}(u) := \delta_C(p,r) = r'_{d+1} - p'_{d+1}$, and store the pair (p,r).

Now, all layer-$(d+1)$ structures are updated correctly. To update the rest of the data structure, we do the following: We search for q in the layer-1 tree. For each node on the path, we search for q in the corresponding layer-2 tree, etc., until we have located q in all layer-d trees that contain this point. Then we walk back along all these paths. During the walk, we update the values $\eta_i(\cdot)$ according to (1).

It is easy to see that the entire operation can be performed in time $O(\log^d n)$. In a completely symmetric way, we can delete a point p from all lists $L^-_{u,d+1}$ and update the entire data structure.

Now we can give the efficient implementation of algorithm gap_greedy'. As before, we consider all cones separately. If C is the current cone, then we maintain besides the above $(d+1)$-layered data structure two d-layered range trees storing subsets of S according to their standard coordinates $p_1, p_2, \ldots, p_d$. Recall that such a range tree can be used to find all points that are contained in a d-dimensional rectangle having sides that are parallel to the standard axes. Here is a complete description of the algorithm.

> **Algorithm** $gap\_greedy''(S, \theta, w)$
> (* S is a set of n points in $\mathbb{R}^d$, $0 < \theta < \pi/4$, $0 \le w < (\cos\theta - \sin\theta)/2$ *)
> **begin**
> **for each** cone C
> **do** store the points of S in the $(d+1)$-layered data structure T defined above;
>   the two layer-$(d+1)$ lists of each node u of each layer-d tree of T store
>   the sets $S_{u,d+1}^- = \{p \in S_{ud} : p_i' \le x_{ui}', 1 \le i \le d\}$ and
> $S_{u,d+1}^+ = \{p \in S_{ud} : p_i' \ge x_{ui}', 1 \le i \le d\}$;
>   store the points of S in two d-layered range trees RT_{source} and RT_{sink}
>   according to their standard coordinates;
> $E_C := \emptyset$;
> $\eta :=$ value stored with the root of the layer-1 tree of T;
>   **while** $\eta < \infty$
>   **do let** (r, s) be a pair such that $\eta = \delta_C(r, s)$;
> $E_C := E_C \cup \{(r, s)\}$;
>   **for each** $p \in RT_{source}$ such that $|pr|_\infty \le (w/\sqrt{d})|rs|$
>   **do** delete p from RT_{source};
>   delete p from all lists $L_{u,d+1}^-$, and update T and
> η as described in the text
>   **od**;
>   **for each** $q \in RT_{sink}$ such that $|qs|_\infty \le (w/\sqrt{d})|rs|$
>   **do** update RT_{sink}, T and η similarly **od**
>   **od**
> **od**;
> output the set $E := \bigcup_C E_C$
> **end**

Lemma 6 Consider the iteration for the cone C. During the execution of this iteration, if $\eta < \infty$, then $\eta = \min\{\delta_C(p, q) : p \in RT_{source}, q \in RT_{sink}, p \ne q\}$.

Proof: Since all η_i-variables, $1 \le i \le d+1$, either have value ∞ or $\delta_C(p, q)$ for some $p \in RT_{source}$ and $q \in RT_{sink}$, it is clear that

$$\eta \ge \min\{\delta_C(p, q) : p \in RT_{source}, q \in RT_{sink}, p \ne q\}. \tag{2}$$

If RT_{source} or RT_{sink} is empty, then $\eta = \infty$, which is a contradiction to our assumption that $\eta < \infty$. Hence, both these structures are non-empty. Let $r \in RT_{source}$ and $s \in RT_{sink}$ such that $\delta_C(r, s) = \min\{\delta_C(p, q) : p \in RT_{source}, q \in RT_{sink}, p \ne q\}$. If we can show that there is a node u in some layer-d tree of T such that $\eta_{d+1}(u) = \delta_C(r, s)$, then we must have $\eta \le \min\{\delta_C(p, q) : p \in RT_{source}, q \in RT_{sink}, p \ne q\}$. This will prove the lemma.

Consider the layer-1 tree of T. Let u_1 be the highest node in this binary tree such that r and s are contained in different subtrees of u_1. Let $1 < i \le d$ and assume that $u_1, u_2, \ldots, u_{i-1}$ have been defined already, and that u_{i-1} is a node of a layer-$(i-1)$ tree. Then, let u_i be the highest node in the layer-i tree that corresponds to u_{i-1} such that r

and s are contained in different subtrees of u_i. In this way, we get a sequence of nodes $u_1, u_2, \ldots, u_d$ such that (i) u_1 is a node of the layer-1 tree of T, (ii) u_i is a node of the layer-i tree that corresponds to u_{i-1}, $1 < i \leq d$, (iii) r and s are contained in different subtrees of u_i, $1 \leq i \leq d$. We claim that $\eta_{d+1}(u_d) = \delta_C(r, s)$, which will complete the proof.

Let $u = u_d$ and consider the point x_u as defined in the description of T. Since $\eta < \infty$, (2) implies that $\delta_C(r, s) < \infty$. Hence $s \in C_r$. This shows that $s'_i \geq r'_i$ for $1 \leq i \leq d$. Since r and s are in different subtrees of u_i, we know that x'_{u_i} separates the coordinates r'_i and s'_i. Therefore, we must have $r'_i \leq x'_{u_i} \leq s'_i$ for $1 \leq i \leq d$. Since $r \in RT_{source}$ and $s \in RT_{sink}$, it follows that r and s are contained in the lists $L^-_{u,d+1}$ and $L^+_{u,d+1}$, respectively. But then, since $\delta_C(r, s)$ is minimal, we must have $\eta_{d+1}(u) = \delta_C(r, s)$. ∎

In the full paper, we show that algorithms *gap_greedy'* and *gap_greedy''* compute the same graph (S, E). We proved in Lemma 5 that *gap_greedy'* always produces a t-spanner of bounded degree and, if $w > 0$, its weight is at most $O(\log n)$ times the weight of a minimum spanning tree for S. Hence, the same is true for algorithm *gap_greedy''*.

We analyze the complexity of our algorithm. Consider one cone C. The $(d+1)$-layered structure T has size $O(n \log^d n)$ and can be built in time $O(n \log^d n)$. The structures RT_{source} and RT_{sink} have size $O(n \log^{d-1} n)$ and can be built in time $O(n \log^{d-1} n)$. By applying dynamic fractional cascading ([9]) and observing that we only delete points, their amortized deletion time is bounded by $O(\log^{d-1} n)$, and their query time is bounded by $O(\log^{d-1} n)$ plus the number of reported points. Since each point of S is reported in at most one query for each RT-structure, the total query time is bounded by $O(n \log^{d-1} n)$. Consider one point p of S. It is deleted at most once from RT_{source}, taking $O(\log^{d-1} n)$ amortized time. If it is deleted from RT_{source}, then we delete p from all lists $L^-_{u,d+1}$ and update T and η. We saw already that this takes $O(\log^d n)$ time.

Hence for each point p of S, we spend $O(\log^d n)$ time for updating RT_{source} and T. The same bound holds for updating RT_{sink} and T. It follows that the entire algorithm has running time $O(n \log^d n)$. This proves:

Theorem 1 *Let t, θ and w be real numbers such that $0 < \theta < \pi/4$, $0 \leq w < (\cos \theta - \sin \theta)/2$ and $t \geq 1/(\cos \theta - \sin \theta - 2w)$. Let S be a set of n points in $\mathbb{R}^d$. In $O((c/\theta)^{d-1} n \log^d n)$ time and using $O((c/\theta)^{d-1} n + n \log^d n)$ space, algorithm gap_greedy''(S, θ, w) computes a t-spanner for S such that each point of S has degree at most $O((c/\theta)^{d-1})$, for some suitable constant c. If $w > 0$, then the weight of this t-spanner is at most $O((c/\theta)^{d-1}(1/w) \log n)$ times the weight of a minimum spanning tree for S.*

Corollary 1 *Let t and θ be real numbers such that $0 < \theta < \pi/4$ and $t \geq 1/(\cos \theta - \sin \theta)$. Let S be a set of n points in $\mathbb{R}^d$. In $O((c/\theta)^{d-1} n \log^d n)$ time and using $O((c/\theta)^{d-1} n + n \log^d n)$ space, we can compute a t-spanner for S such that each point of S has degree at most $O((c/\theta)^{d-1})$ and the weight of this t-spanner is at most a constant times the weight of a minimum spanning tree for S.*

Proof: This follows from results in [3, 4, 5] that were mentioned in Section 1. ∎

6 Application to distance enumeration

Salowe ([11, 14]) has suggested the use of Dijkstra's algorithm with bounded degree spanners for *interdistance enumeration*. Let S be a set of n points in $\mathbb{R}^d$ and let k be an integer between 1 and $\binom{n}{2}$. Then we want to enumerate the k smallest distances, sorted in non-decreasing order. The value of k may or may not be known in advance.

In Section 6.1, we show that we can use *any* bounded degree spanner to enumerate the k smallest interpoint distances *approximately* in $O(n + k \log k)$ time, not including the time to construct the spanner. In Section 6.2, we show how to improve the time bound for exact enumeration to $O(n+k\log k)$ by exploiting special properties of the bounded degree spanner constructed in this paper.

6.1 Approximate interdistance enumeration

Let $G = (S, E)$ be any t-spanner for S having bounded degree. Throughout this section, it is more convenient to consider a spanner as an *undirected* graph. Therefore, we assume that all edges of E are undirected. Let p and q be two points of S. The *weight* of this pair is defined as the Euclidean distance between p and q, and its *pseudo-weight* is defined as the Euclidean length of a shortest path in G between p and q.

The algorithm for approximate distance enumeration is similar to that of Dickerson et al.[7]. We initialize a priority queue with all pairs of points corresponding to the edges of G, with priority given by the pseudo-weight of the pair. In each iteration, we extract the pair p, q with smallest priority and report it together with its weight. For each edge (q, r) of G, we compute the priority of the pair p, r as the sum of the priority of the pair p, q and the weight of the edge (q, r). We insert the pair p, r into the priority queue if it has not already been reported and if it is not already in the queue with a smaller priority. We do the symmetrical thing with all edges (p, s) of G.

It is easy to see that this algorithm is running Dijkstra's shortest path algorithm simultaneously from all the points of S and that the pairs are reported in order of non-decreasing pseudo-weight. Our claim is that this implies that the pairs are reported approximately in order of non-decreasing weight. We make this precise in the following lemma. (Its proof is given in the full paper.)

Lemma 7 *Consider the t-spanner $G = (S, E)$. Arrange all pairs of points in order of non-decreasing weight and assign an index to each pair based on its rank in this sequence. Let w_i and w_i' denote the weight and pseudo-weight of the pair with index i, respectively. Let π be a permutation of the pairs that orders them on the basis of non-decreasing pseudo-weight, i.e., $w_{\pi(1)}' \leq w_{\pi(2)}' \leq w_{\pi(3)}' \leq \ldots$ Then for any i, $1 \leq i \leq \binom{n}{2}$, $w_i/t \leq w_{\pi(i)} \leq tw_i$ and $w_i \leq w_{\pi(i)}' \leq tw_i$.*

We estimate the running time of the algorithm. Assume that k is known in advance. To improve the efficiency of the priority queue, we maintain only k pairs in it. The time to initialize the priority queue is $O(n)$. Since the spanner G has bounded degree, the queue is updated $O(k)$ times. Each operation on the priority queue takes $O(\log k)$ time. Therefore, the total running time is bounded by $O(n + k\log k)$. If k is not known in advance, then the same bound can be obtained by applying the "doubling-method".

6.2 Improved solution for exact interdistance enumeration

We can achieve a better running time for exact enumeration by using the bounded degree spanner that is constructed by algorithm $gap\_greedy''(S, \theta, w)$ for $0 < \theta < \pi/4$ and $w = 0$. To enumerate the k exact closest pairs, we run the same algorithm as in Section 6.1, with one change: The priority of a pair of points is given by its weight.

The running time of this algorithm is clearly the same as that of Section 6.1: it is bounded by $O(n + k\log k)$. We give an inductive proof that the algorithm outputs the k closest pairs in order of non-decreasing weight.

Consider the closest pair p, q in S. Since p and q are connected by an edge in the spanner, this pair is put into the priority queue in the initialization step. Hence, it is the first pair to be reported.

Let $1 < m \leq k$, and assume that the $m - 1$ closest pairs have been reported by the algorithm. Let p, q be the m-th closest pair in S. We show that this pair is the next one to be reported. If p and q are connected by an edge in the spanner, then we are done, because then this pair was put into the queue in the initialization step. Hence, now this pair has smallest priority in the queue, and it will be reported.

Assume that p and q are not connected by an edge. Then (i) there is a point $s \in S$ such that (p, s) is an edge and $|sq| < |pq|$, or (ii) there is a point $r \in S$ such that (q, r) is an edge and $|pr| < |pq|$. Assume first that (i) holds. Then s, q must be one of the $m - 1$ closest pairs. At the moment when this pair was reported, the algorithm inserted the pair p, q into the queue. Hence, after $m - 1$ pairs have been reported, the pair p, q has minimal priority in the queue. Hence, it is the next pair to be reported. Case (ii) can be treated similarly.

References

[1] B. Chandra, G. Das, G. Narasimhan and J. Soares. *New sparseness results on graph spanners.* Proc. 8th ACM Sympos. Comput. Geom., 1992, pp. 192–201.

[2] G. Das and P.J. Heffernan. *Constructing degree-3 spanners with other sparseness properties.* Proc. 4th Annual Intern. Symp. on Algorithms, Lecture Notes in Computer Science, Vol. 762, Springer-Verlag, Berlin, 1993, pp. 11–20.

[3] G. Das, P. Heffernan and G. Narasimhan. *Optimally sparse spanners in 3-dimensional Euclidean space.* Proc. 9th Annu. ACM Sympos. Comput. Geom., 1993, pp. 53–62.

[4] G. Das and G. Narasimhan. *A fast algorithm for constructing sparse Euclidean spanners.* Proc. 10th Annu. ACM Sympos. Comput. Geom., 1994.

[5] G. Das, G. Narasimhan and J. Salowe. *Properties of Steiner minimum trees with applications to small weight Euclidean graphs.* Manuscript, 1994.

[6] A. Datta, H.P. Lenhof, C. Schwarz and M. Smid. *Static and dynamic algorithms for k-point clustering problems.* Proc. 3rd WADS, Lecture Notes in Computer Science, Vol. 709, Springer-Verlag, Berlin, 1993, pp. 265-276.

[7] M.T. Dickerson, R.L. Drysdale and J.R. Sack. *Simple algorithms for enumerating interpoint distances and finding k nearest neighbors.* Internat. J. Comput. Geom. Appl. 2 (1992), pp. 221–239.

[8] H.-P. Lenhof and M. Smid. *Enumerating the k closest pairs optimally.* Proc. 33rd Annu. IEEE Sympos. Found. Comput. Sci., 1992, pp. 380–386.

[9] K. Mehlhorn and S. Näher. *Dynamic fractional cascading.* Algorithmica 5 (1990), pp. 215-241.

[10] F.P. Preparata and M.I. Shamos. *Computational Geometry, an Introduction.* Springer-Verlag, New York, 1985.

[11] J.S. Salowe. *Constructing multidimensional spanner graphs.* Internat. J. Comput. Geom. Appl. 1 (1991), pp. 99–107.

[12] J.S. Salowe. *Enumerating interdistances in space.* Internat. J. Comput. Geom. Appl. 2 (1992), pp. 49–59.

[13] J.S. Salowe. *On Euclidean spanner graphs with small degree.* Proc. 8th Annu. ACM Sympos. Comput. Geom., 1992, pp. 186–191.

[14] J.S. Salowe. Personal communication, 1994.

[15] M. Smid. *Maintaining the minimal distance of a point set in polylogarithmic time.* Discrete Comput. Geom. 7 (1992), pp. 415-431.

[16] P.M. Vaidya. *A sparse graph almost as good as the complete graph on points in K dimensions.* Discrete Comput. Geom. 6 (1991), pp. 369–381.

[17] A.C. Yao. *On constructing minimum spanning trees in k-dimensional spaces and related problems.* SIAM J. Comput. 11 (1982), pp. 721-736.

Approaching the 5/4 – Approximation for Rectilinear Steiner Trees

Piotr Berman[1], Ulrich Fößmeier[2], Marek Karpinski[3], Michael Kaufmann[2], Alexander Zelikovsky[4]

[1] Computer Science Department, Pennsylvania State University,
University Park, PA 16802, USA,
[2] Wilhelm-Schickard-Institut für Informatik, Universität Tübingen,
Sand 13, 72076 Tübingen, Germany,
[3] Department of Computer Science, Universität Bonn,
Römerstraße 164, 53117 Bonn, Germany,
[4] Institute of Mathematics, Akademiei 5, Kishinev, 277028, Moldova,

Abstract. The rectilinear Steiner tree problem requires to find a shortest tree connecting a given set of terminal points in the plane with rectilinear distance. We show that the performance ratios of Zelikovsky's[17] heuristic is between 1.3 and 1.3125 (before it was only bounded from above by 1.375), while the performance ratio of the heuristic of Berman and Ramaiyer[1] is at most 1.271 (while the previous bound was 1.347). Moreover, we provide $O(n \cdot \log^2 n)$-time algorithms that satisfy these performance ratios.

1 Introduction

Consider a metric space with distance function d. For any set of *terminal* points S one can efficiently find $MST(S)$, a minimum spanning tree of S. Let $mst(S, d)$ be the cost of this tree in metric d. A Steiner tree is a spanning tree of a superset of the terminal points (the extra points are called Steiner points). It was already observed by Pierre Fermat that the cost of a Steiner tree of S may be smaller than $mst(S, d)$. Thus it is natural to look for the Steiner minimum tree, that is, for the least cost Steiner tree. However, finding such a tree is NP-hard for almost all interesting metrics, like Euclidean, rectilinear, Hamming distance, shortest-path distance in a graph etc. [7, 12]. Because these problems have many applications, they were subject of extensive research [3, 7, 8, 9, 15, 11].

In the last two decades many approximation algorithms for finding Steiner minimum trees appeared. The quality of an approximation algorithm is measured by its performance ratio: an upper bound of the ratio between the achieved length and the optimal length.

In the *rectilinear* metric, the distance between two points is the sum of the differences of their x- and y-coordinates. The rectilinear Steiner tree problem (RSP) got recently new importance in the development of techniques for VLSI routing [13, 14].

The most obvious heuristic for the Steiner tree problem approximates a Steiner minimum tree of S with $MST(S)$. While in all metric spaces the performance ratio of this heuristic is at most 2 [16], Hwang [10] proved that in the rectilinear plane the performance ratio of this heuristic equals exactly 1.5.

Zelikovsky [17] and Berman/Ramaiyer [1] gave two better heuristics for RSP. Now we give a more precise analysis of the performance ratio of these heuristics. Our results are the following:

1. Zelikovsky's algorithm has a performance ratio between 1.3 and 1.3125.
2. The Berman/Ramaiyer algorithm has a performance ratio of at most $\frac{61}{48} \approx 1.271$.
3. The now best approximation of factor 1.271 can be found in time $O(n \log^2 n)$ considerably improving the previous approximation [1] not only in quality, but also in efficiency. The previous time bound was $O(n^{3.5})$.

In the next section we provide a synopsis of the two approaches of Zelikovsky and Berman/Ramaiyer. In Sections 3, 4 and 5, we derive new estimates for the performance ratios. In particular, in Section 3 we introduce a new technical tool to study those ratios, so called *double covers*. Section 4 provides upper and lower bounds for approximations based on 3-restricted Steiner trees. Finally in Section 5, we improve the approximation further by considering 4-restricted trees.

Section 6 and 7 show how to obtain these approximation efficiently. We prove that for a heuristic that builds a 4-restricted tree it suffices to consider only linearly many quadruples of terminal points and demonstrate how to find those quadruples in time $O(n \log^2 n)$. This improves the running time of the approximation algorithm by a factor of $O(n^2)$ to $O(n^{1.5})$. Section 8 describes a slightly worse heuristic which comes arbitrarily close to the new performance ratio and which runs in time $O(n \log^2 n)$. We conclude with some final remarks and directions for further research.

Note that many proofs in the text are omitted; the details can be found in [2].

2   Berman/ Ramaiyer's and Zelikovsky's Heuristics

A Steiner tree T of a set of terminals S is *full* if every internal node of T is a Steiner point, i.e., not a terminal. If T is not full, it can be decomposed into full Steiner trees for subsets of terminals that overlap only at leaves. Such subtrees are called *full Steiner components* of T [8]. T is called *k-restricted* if every full component of T has at most k terminals. The length of the shortest $k-$restricted Steiner tree of S is denoted by $t_k = t_k(S)$, and $s = t_\infty$ denotes the length of the optimal Steiner tree.

So, $t_2(S)$ is the length of the min. spanning tree of S. $t_2 \le \frac{3}{2}s$ [10], $t_3 \le \frac{5}{4}s$ [17] and generally $t_k \le \frac{2k-1}{2k-2}s$ [1]. These bounds are tight for $k = 2, 3$ but in general $t_k \le \frac{2k}{2k-1}s$ for $k \ge 4$. The main idea of the new heuristics is to obtain good $k-$restricted Steiner trees and to show how they approximate a Steiner minimum tree.

The method described here can be applied with an arbitrary metric d. Without loss of generality, we may assume that the metric d on the set of terminals S is the shortest-path distance for the weighted edges D connecting S. This way, MST(d) is the minimum spanning tree of the graph $< S, D >$, we denote this tree with MST(D), and its cost with mst(D). If we increase the set of edges D by some extra edges, say forming a set E, the shortest-path distance will decrease; MST($D \cup E$) is the minimum spanning tree for the modified metric.

Let z be a triple of terminals. Let $T(z)$ be the Steiner minimum tree of z, $d(z)$ is the cost of $T(z)$ and $Z(z)$ is a spanning tree of z consisting of zero-cost edges.

If we decide to use $T(z)$ as a part of that tree, the remaining part can be computed optimally as MST($D \cup Z(z)$), from which we remove zero-cost edges of $Z(z)$. The improvement of the tree cost due to this decision is the *gain* of z, denoted $g(z, D)$. It is easy to see that $g(z, D) = \text{mst}(D)-\text{mst}(D \cup Z(z))-d(z)$. Now, if we have already decided to use some set of triples, so that the zero-cost edges of their $Z(z)$'s form set Z, the gain of a subsequent triple z_0 can be expressed as $g(z_0, D \cup Z) = \text{mst}(S, D \cup Z)-\text{mst}(S, D \cup Z \cup Z(z_0))-d(z_0)$.

In Zelikovsky's greedy approach (GA), Z is initially empty. In an iteration step, we choose a triple z that maximizes the gain $g(z, D \cup Z)$. If this gain is positive, we use $T(z)$ and replace Z with $Z + Z(z)$; otherwise we exit the loop. At the end, we remove zero-cost edges from MST$(D \cup Z)$ and replace them with the chosen $T(z)$'s. The output of this heuristic has length of at most $\frac{t_2 + t_3}{2}$ [17].

Before we describe the Berman/Ramaiyer heuristic (BR) [1], we have to look closer at the way how to obtain MST$(D \cup Z(z))$ from $M = $ MST(D). Say that $Z(z) = \{e_1, e_2\}$. When e_1 is inserted, the longest edge $H(e_1, D)$ in the path joining the ends of e_1 with cost $h(e_1, D)$ is removed from M. Then we do the same with e_2.

The idea of BR is to make the initial choices (performed in the *Evaluation Phase*) tentative, and to check later (in the *Selection Phase*) for better alternatives.

Evaluation Phase. Initially, $M = $ MST(D). For every triple z considered, find $g = g(z, M)$. If $g \leq 0$, z is simply discarded. Otherwise we do the following for every edge e of some spanning tree of z: find $e' = H(e, M)$ and $c = h(e, M)$, make the cost of e equal to $c - g$, replace in M edge e' with e, put e in a set B_{new} and e' in B_{old}. Once this spanning tree of z is processed, we place the tuple $< z, B_{new}, B_{old} >$ on a Stack (for the future inspection in the second phase). Repeat this while there are triples with positive gain. For later analysis, we define t_3^* to be the length of M at this point, continue the process with quadruples and get t_4^* as the final length of M.

Selection Phase. We initialize $D = M$. Then we repeatedly pop $< z, B_{new}, B_{old} >$ from the *Stack*, and insert B_{old} to D. If $B_{new} \subseteq$ MST(D), then such z is a 'good' one (i.e. we use $T(z)$ in the final output), otherwise we remove all edges of B_{new} from D.

All 'good' quadruples and triples with the rest of MST-edges form the output Steiner tree of the Berman/Ramaiyer heuristic. Its length is at most

$$\frac{t_2}{2} + \frac{t_3^*}{6} + \frac{t_4^*}{3}. \tag{1}$$

We call this version of BR, that considers triples and quadruples, BR4, to distinguish from the shorter version, BR3, that considers only triples. It is easy to see that any estimate for BR3 holds for GA. The length of the output of BR3 is at most

$$\frac{t_2}{2} + \frac{t_3^*}{2} \tag{2}$$

How can we bound the values (1) and (2)? Obviously $t_3^* \leq t_3$ and $t_4^* \leq t_4$. So (1) and (2) is at most $\frac{11}{8}s$ and $\frac{97}{72}s$, respectively [1]. In the next section we develop tools that provide bounds for the linear combinations of t_2, t_3^* and t_4^* that are far better than the linear combinations of the bounds above. For example, while bounds $t_2 \leq \frac{3}{2}s$, $t_3 \leq \frac{5}{4}s$ and $t_4 \leq \frac{8}{7}s$ are tight, we will show that $t_2 + t_3^* \leq \frac{21}{16}s$ and $t_2 + t_4 \leq \frac{5}{2}s$.

3 Double Covers and Spanning Trees

Let C be a union of pairs $C2$ of terminals (edges) and of triples $C3$ of terminals respectively. We say that a set of edges E is *implied* by C if it contains all edges of $C2$ and for each triple $z \in C3$ it contains two distinct edges contained in z. Both $C3$ and E are multisets, where some elements may belong "twice". We say that C is a *double cover* of the set of terminals S if every set of edges implied by C is a multiset union of two spanning trees of S (i.e., if an edge belongs to both trees, it has to belong to E twice). The total length of C is $d(C) = \sum_{x \in C} d(x)$, where $d(x)$ denotes the length of the Steiner minimum tree of x.

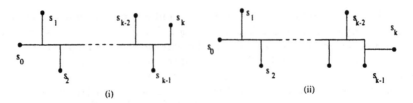

Fig. 1. Two shapes of a full Steiner component

Lemma 1 *If C is a double cover of S, then $d(C) \geq 2t_3^*$.* [2]

Our next tool is a pair of sufficiently short spanning trees. Hwang [10] proved that there is a Steiner minimum tree where every full component has one of the shapes shown in Fig. 1. Let $a_1, \ldots, a_k$ and $b_0 = 0, b_1, \ldots, b_k$ be the lengths of horizontal and vertical lines of a full Steiner component K with terminals $s_0, \ldots, s_k$. The horizontal lines form a *spine* of K. Moreover, in case (i) $b_k < b_{k-2}$ holds. In case (ii) assume that $b_k = 0$. Consider the sequences $b_0, b_1, b_3, \ldots, b_{2i+1}, \ldots$ and $b_0, b_2, \ldots, b_{2i}, \ldots$. Let

$$b_{h(0)} = b_0, b_{h(1)}, \ldots, b_{h(p+1)} = b_k \tag{3}$$

be the sequence of local minima of these sequences, i.e. $b_{h(j)-2} \geq b_{h(j)} < b_{h(j)+2}$. If $h(p) = k - 1$, we exclude the member $b_{h(p)}$ from (3). For the case of $h(j+1) = h(j) + 1$, $(j = 1, \ldots, p-1)$, we exclude arbitrarily either $b_{h(j+1)}$ or $b_{h(j)}$. So, we get $h(j+1) - h(j) \geq 3$. The elements of the refined sequence (3) are called *hooks*. Further we assume that a full Steiner tree K nontrivially contains at least 4 terminals $(k \geq 4)$. A *Steiner segment* K_j is a part of a full Steiner component bounded by two sequential hook terminals $s_{h(j)}, s_{h(j+1)}$. So two neighbouring Steiner segments have a common hook. K_j contains the two *furthest* terminals below and above the spine. Denote the index of the first of them by $f(j)$ and the last by $l(j)$, $(f(j) < l(j))$.

Lemma 2 *Let K_j be a Steiner segment. For the terminals $s_{h(j)}, \ldots, s_{h(j+1)}$ there are two spanning trees Top_j and Bot_j such that*

$$d(Top_j) + d(Bot_j) = 3d(K_j) - b_{h(j)} - b_{h(j+1)} - Rest_j; \tag{4}$$

$Rest_j$ sums the lengths of the thin drawn Steiner tree edges of both trees in Fig. 2.

Proof. Equation (4) can be easily drawn from Fig. 2. The Steiner segment edges are partitioned into three parts: the lengths of thick lines are counted twice, thin lines are counted once, and some hooks (dashed) do not appear at all.

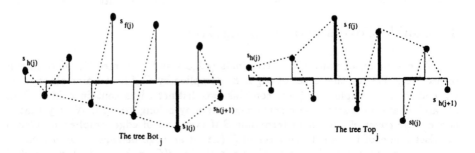

Fig. 2. Two spanning trees

Lemma 3 *For terminals $s_0, \ldots, s_k$, there are two spanning trees Top and Bot s.t.*

$$d(Top) + d(Bot) \leq 3d(K) - \sum_{j=1}^{p} b_{h(j)} - \sum_{j=1}^{p} Rest_j. \qquad (5)$$

Proof. Since $d(K) = \sum_{i=0}^{p} d(K_j) - \sum_{j=1}^{p} b_{h(j)}$ (5) is satisfied. $\Diamond$

Corollary 1 *(Hwang [10]) For any instance of RSP, $2t_2 \leq 3s$.* $\Diamond$

4 A Performance Ratio of BR3 and GA

Theorem 1 *For any instance of the rectilinear Steiner tree problem,*

$$6t_2 + 8t_3^* \leq 18s. \qquad (6)$$

Proof. (Sketch.) It is sufficient to prove inequality (6) for a full Steiner tree K. We will use three pairs of spanning trees *Top* and *Bot* and construct four double covers C^α, $\alpha = 1, \ldots, 4$ such that

$$3d(Top) + 3d(Bot) + \sum_{\alpha=1}^{4} d(C^\alpha) \leq 18s. \qquad (7)$$

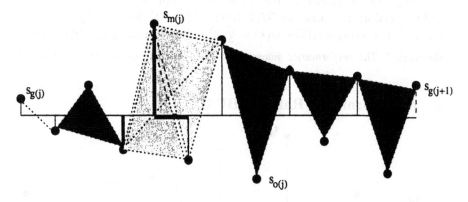

Fig. 3. The double cover for a segment B_j

At first we partition K into some other segments $B_j, j = 0, \ldots, r$, each a union of some Steiner segments. In this partition we must save the right hook of each segment. The first segment B_1 of K is bounded by $s_{g(0)} = s_0$ and $s_{g(1)}$ which is the first minimum at the same side of the spine as $s_{m(0)} = s_{f(0)}$, i.e. $s_{m(1)} \geq \cdots \geq s_{g(1)-2} \geq s_{g(1)} < s_{g(1)+2}$. Similarly, let $s_{m(2)}$ be the first furthest terminal after $s_{g(1)}$. Then B_2 is bounded by $s_{g(1)}$ and $s_{g(2)}$ which is the first minimum at the same side of the spine as $s_{m(2)}$, and so on. Only the last time, if the first minimum at the same side of the spine has index $k - 1$, we bound the last segment B_r by s_k instead of s_{k-1}. So $g(r + 1) = b_k$. Note that $g(j + 1) - m(j)$ is even for $j = 0, \ldots, r$.

If a set of triples and pairs C has a cut terminal and both parts are double covers of their terminal sets, then C is a double cover of the whole terminal set. Therefore, we may construct double covers C_j for each segment B_j separately. Moreover, a set of three triples is a double cover of four terminals. Each C_j consists of a set of three triples for some four terminals, a set of doubled triples (i.e. pairs of the same triple) and may be a double edge $(s_{g(j)}, s_{g(j)+1})$ (if $m(j) - g(j)$ is even).

The double cover C_1 is shown on Fig. 3. Dark triples are doubled and light triples cover four terminals; thick Steiner tree edges are those which lengths participate in $d(C_1)$ three times, thin Steiner tree edges participate in $d(C_1)$ only twice and the dashed hook does not participate in $d(C_1)$ at all. The double cover C_2 differs only in light triples: they have the common end in $s_{m(j)-1}$ instead of $s_{m(j)}$. Similarly for every B_j we construct the next pair of double covers around the opposite furthest terminal, called $s_{o(j)}$ of a Steiner segment which begins at the same terminal as B_j.

The main point of the proof is that the sets of thick edges for different double covers intersect in edges defining the term $Rest_j$ in (5). $\Diamond$

Corollary 2 *For any instance of RSP, $8t_2 + 8t_3^* \leq 21s$.* $\Diamond$

Corollary 3 *The performance guarantee of GA and BR3 is at most $\frac{21}{16} = 1.3125$.* $\Diamond$

To get a lower bound for the performance ratio, we construct a series of instances of the rectilinear Steiner tree problem for which the approximation ratio comes arbitrarily close to 1.3. The terminals are placed on the following coordinates:
$$a_i = (4^i, 0), \ i = 0, \ldots, k; \ b_i = (2 \cdot 4^{i-1}, -4^{i-1}), \ c_i = (3 \cdot 4^{i-1}, 4^{i-1}), \ i = 1, \ldots, k;$$
$$a_i' = (0, 4^i), \ i = 0, \ldots, k; \ b_i' = (-4^{i-1}, 2 \cdot 4^{i-1}), \ c_i' = (4^{i-1}, 3 \cdot 4^{i-1}), \ i = 1, \ldots, k.$$
The length of the Steiner minimum tree $s(k)$ is $\frac{10}{3}(4^k - 1) + 2$, $t_2(k) = 14\frac{4^k-1}{3} + 2$ and the total gain obtained by GA is $G(k) = \frac{1}{3}(4^k - 1)$. So, the length of the output tree of GA is $Gr(k) = t_2(k) - G(k) = \frac{13}{3}(4^k - 1) + 2$ and $\lim_{k \to \infty} Gr(k)/s(k) = 1.3$.

Remark 1 *The performance guarantee of GA and BR3 cannot be less than 1.3.*

5 A Performance Ratio of BR4

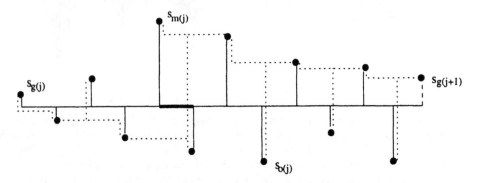

Fig. 4. The 4-restricted tree F for a segment B_j

Theorem 2 *For any instance of RSP, $2t_2 + 2t_4 \leq 5s$.*

Proof.(Sketch.) It is sufficient to prove Theorem 2 for a full Steiner component K. Similarly to the proof of Theorem 1, we construct two 4-restricted Steiner trees F

and L such that $d(Top) + d(Bot) + d(F) + d(L) \leq 5d(K)$. We use the same partition of K into the segments $B_j, j = 0, \ldots, r$. Recall that $g(j+1) - m(j)$ is even and the set of terminals $\{s_{m(j)}, j = 1, \ldots, r\}$ is a subset of the set $\{s_{f(j)}, j = 0, \ldots, p\}$ of the first furthest terminals in Steiner segments.

The 4-restricted Steiner tree F_j is constructed for each segment B_j (Fig. 4). It consists of a quadruple for $s_{m(j)}$, triples and a possible edge at the left end. Dotted lines denote the edges of F_j and, as above, thick Steiner tree edges are counted twice and thin Steiner tree edges only once in the length of F_j. We are confident that the right (dashed) hook is saved. Similarly, the tree L_j has a quadruple for $s_{o(j)}$.

The crucial point of the proof is that the sets of thick edges for both trees F and L belong to $Rest_j$ from (5).◊

Let *quadruple* mean a Steiner minimum tree for a set of four terminals. From Fig. 1 we know that there are only two different shapes for quadruples: *normal* (i) and *cross* (ii). Denote the terminals of the quadruple with u, v, w and z from the left to the right. We call a cross quadruple *long*, if $a2 \geq 2\min\{a1, a3, a4, a5\}$, (*short* otherwise), where $a2$ is the rectilinear distance between the Steiner points and $a1, a3, a4, a5$ are the distances between the terminals and their adjacent Steiner points.

Lemma 4 *If a Steiner minimum tree for four points is a long cross quadruple, then*

$$2t_2 + 2t_3 \leq 5s.$$

Proof. Let $a_1 = \min\{a1, a3, a4, a5\}$, then a Steiner tree for the set $\{u, w, z\}$ together with the edge (u, v) have the length $s + a1 \leq t_3$. Note, that $t_2 \leq s + a1 + a4$ and $t_2 \leq s + a3 + a5$. Therefore, $2t_3 + 2t_2 = 2s + 2a1 + 2s + a1 + a3 + a4 + a5 \leq 5s$. ◊

Let t'_4 be the length of the shortest 4-restricted Steiner tree without any long cross quadruples. Note that in the proof of Theorem 2 the quadruples of 4-restricted trees are normal (Fig. 4). Therefore, a cross quadruple may appear only as a full component. Thus, bound (1), Theorem 2, Corollary 2, and Lemma 4 imply

Theorem 3 *BR4 without long cross quadruples has a performance ratio of* $61/48 \approx$ 1.271.◊

6 How Many Quadruples Have to be Considered?

To keep the time complexity small we have to restrict the number of candidates where we look for our triples and quadruples. We show that it is sufficient to regard only a linear number of each triples and quadruples. In [5] this fact is shown for triples and there is given an algorithm to construct these triples in time $O(n \cdot \log^2 n)$. So it remains to show that a linear number of quadruples is enough for our algorithm.

For a point p, p.x and p.y denote its x- and y-coordinate respectively. At first we only consider normal quadruples. The segment containing the Steiner points is called *Steiner chain*, the unique terminal on the Steiner chain is the *root* of the quadruple. A quadruple is called a *left rooted quadruple*, if the Steiner chain is a horizontal line and the root is its left end. Otherwise we call a quadruple top-, right- or bottom rooted. So we have eight kinds of normal stars: For each possible root position there are two possibilities at which side of the Steiner chain is only one point (w) and at which side are the other two (v,z). Since left- and right rooted quadruples have the same length it suffices to consider only left rooted quadruples, for which the condition

$v.x - u.x < z.x - w.x$ holds. For a left rooted quadruple, v is called the *top point*, w the *bottom point* and z the *right point*. Also note that $z.y < v.y$ (cf. Fig. 1).

A quadruple is called a *tree quadruple* if the following condition holds: An MST for the point set $V \cup \{c_1, c_2\}$ where c_1 and c_2 are the Steiner points of the quadruple, contains the edges $(u, c_1), (v, c_1), (c_1, c_2), (w, c_2)$ and (z, c_2). From results in [1] and [5] we know that we only have to consider tree quadruples where the rectangles defined by the point sets $\{u,v\}$, $\{v,w\}$ and $\{w,z\}$ do not contain any other points.

Lemma 5 *For a given left root u and a top point v there is at most one bottom point w. If the bottom point w is also given, there are at most two right points z.* ◊

For a point v the union of the triangles T_1, T_2 and T_3 is called *butterfly(v)*, where these triangles are defined in Fig. 5. Triangle T_4 is the *right wing* of the butterfly.

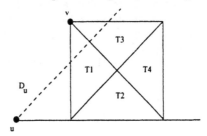

Fig. 5: The butterfly of the top point v

Lemma 6 *There is a 4-restricted Steiner minimum tree T where at each used quadruple the top point has an empty butterfly.*

Proof. Let Δ_1 be $T_1 \cup T_2$, $\Delta_2 = T_1 \cup T_3$. Δ_1 has to be empty, because otherwise (v, c_1) could not be part of an MST of $V \cup \{c_1, c_2\}$; so let p be a point in $\Delta_2 \backslash \Delta_1$ (i.e. T_3); let $x \in \{u, v, w, z\}$ be the first point on the path from p to the quadruple in T. If $x \in \{u, w, z\}$ we can add the edge (v, p), delete the edge (v, c_1) and get a 4-restricted Steiner tree with smaller cost, a contradiction to the optimality of T. So $x = v$: Case(i): $z.x < v.x$: Then $dist(p, z) < dist(v, c_1)$, because z does not lie in Δ_1. So we get a better 4-restricted Steiner tree by adding the edge (p, z) and deleting (v, c_1). Case(ii): $z.x > v.x$: Here we add a vertical edge from p down to the Steiner chain, delete the edge (v, c_1) and again we get a better 4-restricted Steiner tree. ◊

From Lemma 5 follows that for given u and v the number of tree quadruples is at most four. So we try to bound the number of top points with empty butterflies. Let D_u be the 45°-diagonal through u with positive gradient (Fig. 5).

Lemma 7 *For a given u there is at most one top point v with empty butterfly at the left side of D_u.*

Proof. Let v and v' two such points with $v.x < v'.x$. Then v' has to lie below v, because the rectangle defined by u and v' has to be empty. But then v' lies in the butterfly of v, so this butterfly is not empty. ◊

So we can restrict our search for other top points to the region at the right side of D_u. Let v be the leftmost candidate, i.e. v is at the right side of D_u, the rectangle defined by u and v, and the butterfly of v are empty and v is the leftmost point with these characteristics. The tree star condition for quadruples implies that all other candidates for top points have to be in the right wing of the butterfly of v.

Lemma 8 *There are at most two possible top points in the right wing of the butterfly of v.*

So there are at most four top points for the left root u: one at the left of D_u, and at its right v, and two claimed in Lemma 8. For a given pair (u, v) the bottom point is unique and we have at most two right points (Lemma 5), so we have eight tree quadruples with root u of normal shape and since there are eight possibilities for the state of the quadruple, the number of normal quadruples is at most $8 \times 8n = 64n$.

Lemma 9 *At most $64n$ normal quadruples are necessary for the approximation claimed in Theorem 3.*

The number of necessary cross quadruples can be bounded by $32n$ using similar techniques and the tree star property [2].

Theorem 4 *At most $96n$ quadruples are necessary for the approximation claimed in Theorem 3.*

7 Computation of Quadruples

To compute the normal quadruples we need a fast method to find top points v with empty butterfly. But the size of the butterfly of a top point depends of its left root, so checking the butterflies of all points for every root would require quadratic time. Therefore we use the following construction: For every point v, let p_v be the highest point (maximal y-coordinate) in the region at the right of the vertical line through v and at the left of the 45°-diagonal through v with negative gradient. The horizontal line through p_v together with this region defines the triangle Δ_1. Let q_v be the point with smallest rectilinear distance to v lying right below v (i.e. $v.x < q_v.x$ and $v.y > q_v.y$). Then Δ_2, the triangle defined by the vertical and horizontal lines through v and the 45°-diagonal through q_v with positive gradient is empty and it is maximal among all such triangles. The length of the leg of the smaller triangle denotes the size of the largest empty butterfly of v.

Now we draw for every point v a vertical line segment, starting at v in the bottom direction, with the length of the largest empty butterfly of v. Regarding a left root u, we only have to compute intersections of the horizontal line through u with such vertical line segments; the corresponding points are those having an empty butterfly with respect to u.

With a similar construction we can handle empty rectangles. Following the proof of Lemma 8, we can find the corresponding four top point candidates for a given left root u. The implementation of these operations can be performed in a quite standard way in time $O(n \log^2 n)$ using priority search trees to find the maximal empty butterflies and segment trees to compute the candidates for top points.

The necessary cross quadruples can be found in time $O(n \log^2 n)$ [2]. After computing the triples and quadruples, we can apply the algorithm of Berman/Ramaiyer for 4-restricted Steiner trees and get an approximation as stated in Theorem 3. The operations of the algorithm itself are mainly update-operations on a minimum spanning tree, which can be performed in time $O(\sqrt{n})$ each [6]. We conclude:

Theorem 5 *For any rectilinear Steiner tree problem an 1.271-approximation can be found in time $O(n^{3/2})$.*

8 A Faster Approximation

In this section we give a generalization of the fast parameterized version of BR presented in [5]. The main idea is, not to update the whole configuration after each step, but to sort triples and later quadruples according to their gain, and to evaluate those stars according to this sorting, even if it changed meanwhile. Stars are considered to be good, if their gain lies close to the optimal gain (in the range of $m/(m+1)$). We refuse to take triples (quadruples) with smaller gains, or if they would destroy planarity or if they contain artificial edges.

We use the following notations: g_i is the gain of a star (triple or quadruple) at the beginning of a phase, a_i its actual gain at the time of its treatment and a_i' its actual gain without using artificial edges, that means edges which were created in the same phase. B_{old} and B_{new} are the sets of edges from Section 2. Importantly, we have several possibilities to connect the terminals of a quadruple, i.e. e_1 and e_2 are not strictly fixed. We choose these new edges such that the structure remains planar.

The main difference to the algorithm described in [5] is that we now have to run a Selection Phase which we could avoid in the case of considering only triples. For the running time the only problem is maintaining an MST in the Selection Phase. But since the structure always stays planar we can use the data structure introduced in [4] that allows maintaining an MST in time $O(\log n)$ per step. So we run BR4 with an involved version of the Evaluation Phase:

Phase 1: (triple insertion)
$\quad E := \{d(u,v) : u,v \in S\}; M := \text{MST}(E);$
$\quad$ Compute the triples τ_i^3 and store them;
$\quad$ repeat $r \cdot m \cdot \log n$ times:
$\quad\quad$ Sort the triples due to their gains in decreasing order;
$\quad\quad$ compute j: $g_j \geq g_1 \frac{m}{m+1}, g_{j+1} < g_1 \frac{m}{m+1}$;
$\quad\quad$ for $i := 1$ to j do
$\quad\quad\quad$ if $a_i' \geq g_1 \frac{m}{m+1}$ and τ_i^3 is planar
$\quad\quad\quad$ then $M := M \backslash B_{old}(\tau_i^3) \cup B_{new}(\tau_i^3);$
$\quad\quad\quad\quad$ push $(\tau_i^3, B_{old}(\tau_i^3), B_{new}(\tau_i^3))$ on a stack;
Phase 2: (quadruple insertion)
$\quad$ Repeat Phase 1 for quadruples τ_i^4;
$\quad$ all edges in M at the beginning of Phase 2 are said to be original for this phase.

The rest of this section is devoted to the proof of

Theorem 6 *The algorithm computes in time $O(r \cdot m \cdot n \log^2 n)$ a Steiner tree for a given set S of points, $|S| = n$. Its approximation ratio is at most $\frac{61}{48} + \frac{7}{4m} + \frac{1}{(2 \cdot 32^{r-2})}$.*

Now we have to count how much we lose compared to the analysis of section 5 by

a) finishing the Evaluation Phases after $O(\log n)$ rounds, i.e. although there might still be some stars with positive gain.
b) ignoring stars which would use artificial edges.
c) ignoring non-planar stars.

Phase 1 is the same as in the algorithm in [5], so we only analyse Phase 2.

The loss due to a) and b) is easily counted: With the same arguments as in the proofs of Lemma 7 and 8 in [5] we can state that the total loss here can be bounded by the fraction $\frac{1}{96^{r-2}} + \frac{3}{m}$ of the total gain. The constant 3 comes from the fact that every quadruple ignored because of artificial edges would decrease the total cost by three times its gain by decreasing the length of three edges.

Let *tuple* mean a triple or a quadruple. The planarity restriction requires to ignore every quadruple that would cross an original edge or a tuple already lying on the stack. Two tuples 'cross' or 'intersect' each other if there is a crossing between an artificial edge created by one tuple and an artificial edge created earlier by the other, and the tuples have at most one common point. If a quadruple and a triple have more than one point in common, we can avoid intersections by placing the new edges in such a way that they do not cross artificial edges of the triple. This can be done by never adding the 'diagonals' of the quadruple which are the only edges that may cross. Two quadruples having more than one point in common cause no problems, because the second one always uses an artificial edge created by the other.

Lemma 10 *There is a 4-restricted Steiner tree without intersections.*

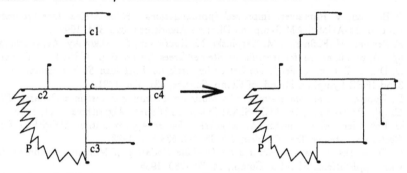

Fig. 6. Proof of Lemma 10: Triples are better than crossing quadruples

So the gain of all the quadruples ignored because of non-planarity can be bounded by the gains of all triples after the triple insertion phase.

Corollary 2 *After the quadruple insertion phase all quadruples with positive gain that were refused because of non-planarity together have a gain of at most $\frac{1}{32^{r-2}} + \frac{4}{m}$.*

Note that the planarity test can be done in time $O(\log n)$ by maintaining a planar map of the actual edge set where the edges incident to the same point are ordered in lists according to their direction. Since every pair of edges that ever appear in the course of the algorithm has no crossing, we can maintain an MST in the Selection Phase in time $O(\log n)$ per step [4]. Thus the time bound of Theorem 6 is proved.

The total gain that could be realized is

$$\left(\frac{t_2}{2} - \frac{t_3^*}{6} - \frac{t_4}{3}\right)\left(1 - \left(\frac{1}{32^{r-2}} + \frac{4}{m} + \frac{1}{96^{r-2}} + \frac{3}{m}\right)\right)$$

and the performance ratio is at least

$$\frac{61}{48} + \left(\frac{t_2}{2} - \frac{t_3^*}{6} - \frac{t_4}{3}\right)\left(\frac{7}{m} + \frac{2}{32^{r-2}}\right) \le \frac{61}{48} + \frac{7}{4m} + \frac{1}{2 \cdot 32^{r-2}}.$$

Corollary 3 *We can achieve in time $O(n \log^3 n)$ an approximation ratio $61/48 + \log\log n / \log n$ and in time $O(n \log^2 n)$ a ratio $61/48 + \varepsilon$ for any $\varepsilon > 0$.*

9 Conclusion

In this paper refined analysis of the new approximation algorithms for rectilinear Steiner trees of Zelikovsky and Berman/Ramaiyer were presented. The introduction of new techniques for estimating the length of different classes of Steiner trees enables to considerably improve the best known approximation factors. Somewhat surprisingly, the same time bounds of $O(n \log^2 n)$ can be achieved as before [5], although we have to use much more involved methods.

This paper almost closes the first approach to improve the approximation ratio for rectilinear Steiner trees, bringing the ratio down from 3/2 [10], via 11/8 [17, 1] now to 61/48, which is very close to 5/4. The next task is to jump below the 5/4-barrier.

References

1. P. Berman, V. Ramaiyer. *Improved approximations for the Steiner tree problem*. In Proc. of 3d ACM-SIAM Symp. on Discrete Algorithms, 325–334, 1992.
2. P. Berman, U. Fößmeier, M. Karpinski, M. Kaufmann, A. Zelikovsky. *Approaching the 5/4–Approximations for Rectilinear Steiner Trees*. Techn. Rep. WSI-94-6, Tübingen.
3. D. Du, Y. Zhang, Q. Feng. *On better heuristic for Euclidean Steiner minimum trees*. 32nd IEEE Symp. on Found. of Comp. Science, 431–439, 1991.
4. D. Eppstein, et al. *Maintaining of a Minimum Spanning Forest in a Dynamic Planar Graph*. Proceedings, 1st ACM-SIAM Symp. on Discrete Algorithms, 1-11, 1990.
5. U. Fößmeier, M. Kaufmann, A. Zelikovsky. *Faster Approximation Algorithms for the Rectilinear Steiner Tree Problem*. LNCS 762, 533-542, 1993.
6. G. Frederickson. *Data Structures for On-Line Updating of Minimum Spanning Trees, with Applications*. SIAM J. Comp., 14, 781-789, 1985.
7. M. R. Garey, D. S. Johnson. *The Rectilinear Steiner Problem is NP-Complete*. SIAM J. Appl. Math., 32, 826-834, 1977.
8. E. N. Gilbert, H. O. Pollak. *Steiner Minimal Trees*. SIAM Appl. Math., 16, 1-29, 1968.
9. M. Hanan. *On Steiner's Problem with Rectilinear Distance*. SIAM J. Appl. Math.,14, 255-265, 1966.
10. F. K. Hwang. *On Steiner Minimal Trees with Rectilinear Distance*. SIAM J. Appl. Math., 30, 104-114, 1976.
11. F. K. Hwang, D. S. Richards, P. Winter. *The Steiner Tree Problem*. Annals of Disc. Math. 53, 1992.
12. R. M. Karp. *Reducibility among combinatorial problems*. In Miller and Thatcher (eds.), Complexity of Computer Computations, Plenum Press, 85–103, 1972.
13. B.Korte, H.J.Prömel, A.Steger. *Steiner Trees in VLSI-Layouts*. In Korte et al.: Paths, Flows and VLSI-Layout, Springer, 1990.
14. Th. Lengauer. *Combinatorial Algorithms for Integrated Circuit Layout*. John Wiley, 1990.
15. D. Richards. *Fast Heuristic Algorithms for Rectilinear Steiner Trees*. Algorithmica, 4, 191-207, 1989.
16. H. Takahashi, A. Matsuyama. *An approximate solution for the Steiner problem in graphs*. Math. Japonica, 24: 573–577, 1980.
17. A. Z. Zelikovsky. *An 11/8-approximation Algorithm for the Steiner Problem on Networks with Rectilinear Distance*. In Coll. Math. Soc. J. Bolyai 60: 733-745, 1992.
18. A.Z. Zelikovsky. *A Faster Approximation Algorithm for the Steiner Tree Problem in Graphs*. Inf. Process. Lett. 46: 79-83, 1993.

Membership in Constant Time and Minimum Space*

Andrej Brodnik and J. Ian Munro

Department of Computer Science
University of Waterloo
Waterloo, Ontario, Canada
N2L 3G1
{ABrodnik,IMunro}@UWaterloo.CA

Abstract. We investigate the problem of storing a subset of the elements of a bounded universe so that searches can be performed in constant time and the space used is within a constant factor of the minimum required. Initially we focus on the static version of this problem and conclude with an enhancement that permits insertions and deletions.

1 Introduction

Given a universal set $\mathcal{M} = \{0, \ldots, M - 1\}$ and any subset $\mathcal{N} = \{e_1, \ldots, e_N\}$ the membership problem is to determine whether given query element in $\mathcal{M}$ is an element of $\mathcal{N}$. There are two standard approaches to solve this problem: to list all elements of $\mathcal{N}$ (e.g. in a hash table) or to list all the answers (e.g. a bit map of size M). When N is small the former approach comes close to the information theoretic lower bound on the number of bits needed to represent an arbitrary subset of the given size (i.e. a function of both N and M, $\left\lceil \lg \binom{M}{N} \right\rceil$). Similarly, when N is large (say αM) the later approach is near optimal in space utilization. This paper focuses on the middle ground where both of these natural approaches require space nonlinear in the lower bound. The techique we introduce has the feature that not only is the space used within a constant factor of the lower bound and the query time is constant, but also that both of these constants are modest.

The basic approach is to use either perfect hashing or a bit map whenever they achieve the optimum space bound, and otherwise to partition the universe into subranges of equal size. We discover that, with care, after a couple of iterations of this splitting, the subranges are small enough so that succinct indices into a single table of all possible configurations of these small ranges (*table of small ranges*) permit the encoding in the minimal space bound. Our technique is, then, an example of what we call *word-size truncated recursion* (cf. [11, 12]). Namely, the recursion needs only to continue down to a level of "small enough" subproblems, at which point indexing into a table of all solutions suffices. This happens to a large extent because at this level a single word in

---

* This research was supported in part by the Natural Science and Engineering Research Council of Canada under grant number A-8237 and the Information Technology Research Centre of Ontario.

the machine model is big enough to encode a complete solution to each of these small problems.

Our paper is organized as follows: in the next section we introduce the notation, some definitions and a short overview of a relevant literature. In Sect. 3 we present a complete static solution, which is followed by a solution of the dynamic problem in Sect. 4. In Sect. 5 we present a pair of examples of usage of our static structure and end the paper with final conclusions and remarks.

2 Notation, Definitions and Background

As noted above, our universal set, $\mathcal{M}$, is assumed to be $\{0, \ldots, M - 1\}$, hence of size M. On the other hand, $\mathcal{N}$ is an arbitrary subset of $\mathcal{M}$ of cardinality N. Since solving the membership problem for $\mathcal{N}$ trivially gives a solution for $\overline{\mathcal{N}}$ we assume $0 \leq N \leq \frac{M}{2}$.

The membership problem is to answer a query whether a given element $x \in \mathcal{M}$ is in $\mathcal{N}$. The problem has the obvious dynamic extension which also includes the operations insert and delete.

We use the standard *cell probe model* (cf. [6, 10, 15]) with word size $\lceil \lg M \rceil$ bits.[2] In other words, one register can be used to represent a single element from the entire domain $\mathcal{M}$, specify all elements which are present in a set of a size $\lg M$ elements, refer to some portion of the data structure, or have some other role that is a ($\lceil \lg M \rceil$ bit) blend of these. For convenience we measure the space usage in bits (rather than in words). Furthermore, we assume that the machine can perform usual operations on words in one unit of time.

The information theoretic lower bound on the number of bits needed to represent an arbitrary set of size N from the universe of size M is

$$B \equiv \left\lceil \lg \binom{M}{N} \right\rceil . \tag{1}$$

Using Stirling's approximation and defining $r = \frac{M}{N}$, this formula can be replaced by the approximation

$$B \approx N \lg \frac{M}{N} = N \lg r . \tag{2}$$

The error in this estimate is less than $N \lg e$ and so (2) can be used for B throughout this paper.

2.1 A Short Stroll Through the Literature

The searching problem is one of central problems of computer science with an extensive literature. In this section we point out particular work that is essential to ours. There are three aspects which are addressed: the static case, the dynamic case, and the information theoretic tradeoffs. In the first two cases it has always been assumed that there is at least enough space to list all keys that are present (or to use a bit map). However, we are dealing with the situation in which we can not always afford that much space.

---

[2] We use a notation $\lg x$ to describe a binary logarithm $\log_2 x$.

We start with the static case, and, in particular, with the implicit data structures in Yao's sense. His notion of implicit structure was that one has room to store only N data items, although these were not constrained to be those in the "logical" table. In other words, by the term implicit structure he meant a structure which uses only N registers ($N \lg M$ bits) and no additional storage.

For such data structures Yao ([15]) showed that in a bounded universe there always exists some subset for which any implicit data structure requires at least logarithmic search time. Later Fiat and Naor ([7]) put a better bound on the size of the subset N. They proved that for this awkward case, N has to be larger than a quantity which lies somewhere between $\lg^{(O(1))} M$ and $(\lg M)^\epsilon$ $(0 \le \epsilon < 1)$.[3] In particular, they proved that there exists an implicit scheme for $N = \Omega((\log M)^\epsilon)$.

However, adding some storage changes the situation. For example, Yao ([15]) showed that there exists a constant time solution for $N \approx M$ or $N \le \frac{1}{4}\sqrt{\lg M}$, while Tarjan and Yao ([14]) presented a constant time solution using $O(N \lg M)$ bits of space for $N = O(M^\epsilon)$ and $0 < \epsilon \le 1$. Fredman, Komlós and Szemerédi ([9]) closed the gap developing a constant time algorithm which uses a data structure of size $N \lg M$ bits and $O(N\sqrt{\log N} + \log^{(2)} M)$ additional bits of storage, for any N and M. Fiat, Naor, Schmidt and Siegel in [8] decreased the number of additional bits to $\lceil \lg N \rceil + \left\lceil \log^{(2)} M \right\rceil + O(1)$. Moreover, combining their result with Fiat and Naor's ([7]) construction of an implicit search scheme for $N = \Omega((\log M)^p)$ we get a scheme which uses less than $(1 + p) \left\lceil \log^{(2)} M \right\rceil + O(1)$ additional bits of memory.

Moving to the dynamic case, Dietzfelbinger et al. ([2]) proved an $\Omega(\log N)$ worst case lower bound for a class of realistic hashing schemes. In the same paper they also presented a scheme which, using results of [9] and a standard doubling technique, achieved a constant amortized expected time per operation with a high probability. However, the worst case time per operation (non-amortized) was $\Omega(N)$. Later Dietzfelbinger and Meyer auf der Heide in [3] upgraded the scheme and achieved a constant worst case time per operation with a high probability. A similar result was also obtained by Dietzfelbinger, Gil, Matias and Pippenger in [1].

Finally we turn to some information theoretic references. Elias ([4]) addressed a more general version of a static membership problem which involved several different types of queries. For these queries he discussed a tradeoff between the size of the data structure and the average number of bit probes required to answer the queries. In particular, for the set membership problem he described a data structure of a size $N \lg \frac{M}{N} + O(N)$ – that is, using (2), $B + o(B)$ – bits which required an average of $(1 + \epsilon) \lg N + 2$ bit probes to answer a query. However, in the worst case, it was still necessary to probe N bits. Elias and Flower in [5] further generalized the notion of a query into a database. They defined the set of data and a set of queries, and in such a general setting they studied the relation between the size of the data structure and the number of bits probed, given the set of all possible queries. The same arrangement was later more rigorously studied by Miltersen in [13].

---

[3] $\log^{(p)} x$ denotes the iterated logarithm, namely the logarithm function applied p times to x.

3 A Solution for the Static Case

As mentioned earlier, if more than half the elements are present we can solve the complementary problem. In the solutions below it is implicit that if any subrange (bucket) has more than half its elements present we use a single bit flag to indicate that we are solving the complementary problem. The static solution is presented in two parts. In the first part we describe a solution for $N \leq \frac{M}{\log_\lambda M}$, that is for $r \geq \log_\lambda M$, while in the second part we deal with the case when $\frac{M}{\log_\lambda M} \leq N \leq \frac{M}{2}$, that is for $\log_\lambda M \geq r \geq \frac{1}{2}$. In both cases λ is some predefined constant greater than 1. We refer to the first case as *sparse* and to the second as *dense*.

Each of these cases has one extreme situation. The first is when $N \leq M^{1-\epsilon}$ for some $0 < \epsilon \leq 1$. In this case we need $\Omega(N \log M)$ bits which is enough to list all the keys. Moreover, this is also enough space to use a perfect hashing function in any of its flavours (cf. [7, 8, 9]) and so retain a constant response time in the worst case. The other extreme is when N is very large, that is $N \geq \alpha M$ for some $0 < \alpha \leq \frac{1}{2}$. The number of bits necessary in this case is $\Omega(M)$ and hence we can use a bit map of size M to represent the subset.

3.1 Indexing – Solution for $N \leq \log_\lambda M$

In this section we focus on case $N \leq \frac{M}{\log_\lambda M}$ for some $\lambda > 1$. The top end of this range, that is $N = \frac{M}{\log_\lambda M}$, typifies the case in which both simple approaches require too much space as $B = \Theta(N \log^{(2)} M)$. Furthermore, this solution suggests the first iteration of our general approach in the $N \geq \frac{M}{\log_\lambda M}$ case.

Lemma 1. *If $N \leq \frac{M}{\log_\lambda M}$ (i.e. $M \geq r \geq \log_\lambda M$) for some $\lambda > 1$, then there is an algorithm which answers a membership query in constant time using $O(B)$ bits of space for a data structure.*

Proof. The idea is to split a universe $\mathcal{M}$ into p buckets, where p is to be determined later and will be as large as possible. The data falling into individual buckets is then organized using perfect hashing. The buckets are contiguous ranges of equal sizes, so that a key $x \in \mathcal{M}$ falls into bucket $\lfloor \frac{x}{p} \rfloor$ and to reach individual buckets, we index through an array of pointers.

To consider this in detail, let us assume that we split the universe into p buckets. We build an index of pointers to individual buckets, where each pointer occupies $\lceil \lg M \rceil$ bits. Hence, the total size of the index is $p \cdot \lceil \lg M \rceil$ bits.

We store all elements that fall in the same bucket in a perfect hash table ([7, 8, 9]) for that bucket. Since the ranges of all buckets are equal, the space required for these hash tables is $\beta = \left\lceil \lg \frac{M}{p} \right\rceil$ bits per element, and so, to describe all elements in all buckets we require only $N \cdot \beta$ bits.

If we allocate B bits for the index and about the same for the buckets, we achieve the desired space bound. Moreover, this structure also supports constant time queries.

It remains to show that $O(B)$ bits are enough to store elements in the buckets' perfect hash tables. From the limit on the size of the index and using (2) we get

$$p = \left\lfloor \frac{B}{\lceil \lg M \rceil} \right\rfloor = \Theta(\frac{M \log r}{r \log M})$$

buckets, each of which has a range $\frac{M}{p} = \Theta(\frac{r \log M}{\log r})$. The range of a bucket defines the number of bits used by each element in a hash table $\beta = \Theta(\log^{(2)} M + \log r - \log^{(2)} r)$, which, in turn, gives the total size of all hash tables $\Theta(N(\log^{(2)} M + \log r - \log^{(2)} r))$. Moreover, this order matches information theoretic lower bound in (2) for $r \geq \log_\lambda M$. $\square$

The cases covered so far are summarized in Table 1. The first column of the table holds ranges of N as an increasing function of M. The second column contains approximations of B on the top end of the range in the first column and was obtained using (2), while the third column defines the data structure which should be used for N in the range defined in the first column. In the table we can find solutions for small or large N, while as yet we have no solution for

$$\frac{M}{\log_\lambda M} \leq N \leq \alpha M \leq \frac{M}{2}$$
$$\log_\lambda M \geq r \geq \frac{1}{\alpha} \geq 2 .$$

(3)

Next we will assume that r and N lie in this awkward range, that is that subset at hand is dense.

Table 1. Ranges of N and structures used in them to represent subset $\mathcal{N}$.

range of N	B	structure
0	1	single bit
1 to c	$\Theta(\log M)$	unordered list
c to $M^{1-\epsilon}$	$\Theta(N \log M)$	hash table
$M^{1-\epsilon}$ to $\frac{M}{\log_\lambda M}$	$\Theta(N \log^{(2)} M)$	indexing
$\frac{M}{\log_\lambda M}$ to αM	$\Theta(N \log r)$	?
αM to $\frac{1}{2} M$	$\Theta(M)$	bit map

3.2 A Complete Solution

We return to Table 1 and in particular to subsets $\mathcal{N}$ whose sizes lie in the range given in (3). For such dense $\mathcal{N}$ we apply the technique of Lemma 1, that is split the universe $\mathcal{M}$ into equal-range buckets (using at most B bits for the pointers). However, this time the ranges of buckets remain too big to use hash tables, and therefore we apply the splitting scheme again. In particular, we treat each bucket as a new, separate, but smaller universe and if the number of elements falling into the bucket falls in the range defined by (3) (with respect to the size of its smaller universe) we can recursively split it.

Such a straightforward strategy leads, in the worst case, to an $O(\log^* M)$ level structure and therefore to an $O(\log^* M)$ search time and to an $O((\log^* M) \cdot B)$ space requirement. However, we observe that at each level the number of buckets with the same range increases and, furthermore, their number is becoming so large that not all of them can be different. Therefore we could build a table of all possible subsets of universes of size up to a certain threshold. That is, we could build a *table of small ranges* and replace buckets in the main structure by pointers (indices) into the table (see [11, 12] for example). We refer to this technique as *word-size truncated recursion*. In our structure the truncation occurs after two splittings. In the rest of this section we give a detailed description of the structure and its analysis.

Returning to the problem at hand, on the first split we partition the universe into $p = \frac{M}{(\log_\lambda M)^2}$ buckets each of which has a range $M_1 = \frac{M}{p} = (\log_\lambda M)^2$. This gives three kinds of buckets at the second level: sparse, dense and very dense ones. For sparse buckets we apply the solution from Subsect. 3.1 and for very dense ones with more than the fraction α of their elements present we use a bit map. For the remaining dense buckets (they have a number of elements within the range defined in (3)) we re-apply the splitting. However, this time the number of buckets is $p_1 = \frac{M_1}{(\log_\lambda M_1)^2}$ so that each of these smaller buckets has range $M_2 = \frac{M_1}{p_1} = 4(\log_\lambda^{(2)} M)^2$.

At this point we build the table of small ranges, which consists of a bit map representations of all possible subsets chosen from the universe of size $M_2 = \sigma$, and replace buckets in the main structure with "indices" (of varying sizes) into the table. We order the table first according to the number of elements in the subset and then lexicographically. If we store, as the representation of a pointer to the table of small ranges, a record consisting of two fields: the number of elements in the bucket, ν (this takes $\lceil \lg \sigma \rceil$ bits), and the lexicographic order of the bucket in question among all buckets containing ν elements, η (from (1) this is $\lceil \lg \binom{\sigma}{\nu} \rceil$ bits), then we can compute the actual position of the corresponding bit map of the bucket using

$$\sum_{i=1}^{\nu-1} \binom{\sigma}{i} + \eta - 1 .\tag{4}$$

The sum is found by table lookup and so a search is performed in constant time.

This concludes the description of our data structure. Obviously, the structure allows constant time membership queries, but it remains to be seen how much space it occupies. In the analysis we are interested only in dense subsets, that is in the range defined in (3), as otherwise we use structure presented in Subsect. 3.1.

First we analyze the main structure and in order to bind later results together we will need the following lemma:

Lemma 2. *Suppose we are given a subset of N elements from the universe M and B defined as in (1). If this universe is partitioned into p buckets of sizes M_i containing N_i elements (now, using (1), $B_i = \lceil \lg \binom{M_i}{N_i} \rceil$) then $B \geq \sum_{i=1}^{p} B_i - p$.*

Proof sketch. We observe that $B_i < \lg \binom{M_i}{N_i} + 1$ and that $\prod_{i=1}^p \binom{M_i}{N_i} \leq \binom{M}{N}$ where $\sum_{i=1}^p M_i = M$ and $\sum_{i=1}^p N_i = N$. □

The main structure itself is analyzed from the top to the bottom. The first level index consists of p pointers each of which is of size $\beta = \lg M$ bits. Therefore the size of that complete index is

$$p \cdot \beta = \frac{M}{(\log_\lambda M)^2} \cdot \lg M = o(\frac{M}{\log M}(\log^{(2)} M)) = o(B) \qquad (5)$$

for the range defined in (3).

On the second level we have three kinds of buckets: sparse, dense, and very dense. For the sparse buckets we use solution presented in Subsect. 3.1 and for the very dense we use a bit map. Both of these structures guarantee space requirements within a constant factor of the information theoretic bound on the number of bits. If the same also holds for the dense buckets, then, using Lemma 2 and (5), the complete main structure uses $O(B)$ bits. Note, that we can apply Lemma 2 freely because the number of buckets p is, by (5), $o(B)$.

It remains to see how large the second level dense buckets can be. For this purpose we first consider the size of bottom level pointers (indices) into the table of small ranges. As mentioned, the pointers are records consisting of two fields, where the first field occupies $\lceil \lg \sigma \rceil$ bits and the second field the information theoretic bound on the number of bits required, B_ν, as defined in (1). Since $B_\nu \geq \lceil \lg \sigma \rceil$ the complete pointer[4] takes at most twice the information theoretic bound on the number of bits, B_ν. On the other hand, the size of an index is bounded using an expression similar to (3). Subsequently, this, together with Lemma 2, also limits the size of dense second level buckets to be within a constant factor of the information theoretic necessary number of bits. This, in turn, limits the size of a complete main structure to $O(B)$ bits.

It remains to compute the size of the table of small ranges. There are 2^σ entries in the table and each of the entries is σ bits wide, where $\sigma = 4(\log_\lambda^{(2)} M)^2$. This gives us the total size of the table

$$T_\sigma = \sigma \cdot 2^\sigma = o(B) \ . \qquad (6)$$

Moreover, this also limits the size of the whole structure to $O(B)$ bits and, hence, proves the theorem:

Theorem 3. *There is an algorithm which solves the static membership problem in $O(1)$ time using data structure of size $O(B)$ bits.*

Let us note that the constants in order notation are relatively small: 7 for time − 2 per level in recursion and a final probe in the table of small ranges − and certainly less than 2 for space. For fine tuning of the structure one can adjust the constants ϵ, λ, and α mentioned in Table 1.

---

[4] Note, that the size of a pointer depends on the number of elements that fall into the bucket, and the more elements there are in the bucket the bigger the pointer is.

4 A Dynamic Version

The solution presented in the previous section deals with a fixed set. It is natural to ask how we can incorporate updates while maintaining the time and space bound.

Our dynamic solution retains the space bound of $O(B)$ bits, but the time per operation of $O(1)$ is an expected bound with a high probability. The solution uses the standard technique of "doubling". The core of this method is to maintain, at any given time, the most appropriate data structure and when the situation changes, to switch to some other better structure. Moreover, the new structure is not built when a certain threshold is reached, but *smoothly* through a sufficient number of preceding operations. Since the additional work due to build-up adds only a constant amount of work per operation, this retains our $O(1)$ time bound.

It remains to be shown that we can apply the doubling technique to our static structure. The structure employs a number of different substructures each of which supports the smooth version of doubling technique including hashing (cf. dynamic perfect hashing [1, 3]) and construction of table of small ranges (when appropriate extend the table by doubling the range).

However, there is a slight memory management problem with indexing. Namely, when the size of a bucket remains the same, but the size of one of its sub-buckets grows to the level that it has to be doubled, we have to find space for the outgrown sub-bucket inside the big bucket.[5] The obvious solution is to allocate more space when the big bucket is created and then copy the outgrown sub-bucket into the first sufficiently big empty slot. Since the copying can be done using a smooth version of doubling technique it does not affect our time and space bounds.

Still, we can construct a case when this technique requires initial allocation of more than $O(B)$ bits of space for the big bucket. To avoid such a situation we use a background daemon that squeezes sub-bucket together. Roughly, in time proportional to the size of a big bucket it goes through all sub-buckets squeezes out all empty slots and compresses sub-buckets together. It is not hard to see that the daemon does not require more than constant additional time per operation, which leaves total time per operation at $O(1)$ worst case with the high probability and total space at $O(B)$ bits.

5 Two "Natural" Examples

There are many situations in dealing with a subset chosen from the bounded universe, in which the size of the subset is relatively big but not big enough to use a bit map representation. Here we mention only two such examples. The first one is a set of all primes smaller then some number M, for which it is known that there are approximately $\frac{M}{\ln M}$ such primes. We pretend that the set of primes is random and that we are to store primes in some kind of structure in order to allow quick answering on a query whether given number is prime. Clearly, we could use some kind of compression (e.g. in a bit map representation we could leave out all even numbers, sieve more carefully etc.), but for the purpose of this example we will not do so.

---

[5] This happens when, for example, elements from one sub-bucket are shuffled into another sub-bucket.

The second example we consider is population of Canada and Social Insurance Numbers (S.I.N.'s) allocated to each individual person. Canada has approximately 28 million people and each person has a 9 digit Social Insurance Number. One may want to determine whether or not a given number is allocated. This query is in fact a membership query in the universe of size $M = 10^9$ with a subset of size $N = 28 \cdot 10^6$.

In both of these examples N is approximately $\frac{M}{\log_\lambda M}$ and we can use the method presented in Subsect. 3.1 directly. Furthermore, no special features of data are used which makes our space calculations slightly pessimistic. Still, as is typically the case in applying complex data structures to real world data, some tuning is very helpful. In the original version we allocated B bits of memory for the index, but in our examples we allocated ζB bits for some constant ζ.

The perfect function used for the buckets is the one described in [8]. Using an argument similar to that of Lemma 2, we observe that the worst case distribution occurs when all buckets are equally full and, therefore, we can assume that in each bucket there are $\frac{N}{p}$ elements.

In total we use ζB bits of memory for the index, while for buckets we use $N \lg \frac{M}{p}$ bits to list all elements, $p \lg \frac{N}{p}$ bits to describe the number of elements in individual buckets, and $\lg^{(2)} \frac{M}{p}$ bits to describe the range of buckets. Table 2 gives the sizes of data structures for both examples for the standard approaches using a hash function or bit map, and the size of our structure for $\zeta = 0.29$. The table also includes the size of the universe, the size of the subset we are dealing with, and the information theoretic lower bound on the number of necessary bits which is computed using (2).

Table 2. Space usage for subsets of primes and SINs for different data structures.

Example	M	N	B	ours	hash	bitmap
primes	$1.0\,2^{32}$	$1.4\,2^{27}$	$1.6\,2^{29}$	$1.9\,2^{30}$	$1.4\,2^{32}$	$1.0\,2^{32}$
SINs	$1.9\,2^{29}$	$1.7\,2^{24}$	$1.1\,2^{27}$	$1.2\,2^{28}$	$1.6\,2^{29}$	$1.8\,2^{29}$

6 Discussion and Conclusions

In this paper we have presented a solution to a static membership problem. Our solution answers queries in constant time and uses space within a small constant factor of the minimum required by the information theoretic lower bound.

The data structure consists of three major sub-structures which are used in different ranges depending on the ratio of the size of the subset N to universe M. When the size of the subset is at most $M^{1-\epsilon}$ for some fixed $0 \le \epsilon < 1$ we use perfect hash tables; when the size of the subset is at least αM for some fixed $\alpha \le \frac{1}{2}$ we use bit map; and when the size of the subset is in the range between these two values we use recursive splitting. The depth of the recursion is bounded by the use of *word-size truncation* and in our case it is 2. We also addressed the dynamic problem and proposed the solution based on a standard doubling technique.

Acknowledgments

We thank Martin Dietzfelbinger for several very helpful comments including pointing out the relevance of non-oblivious hashing to our work. We also thank to anonymous referee for a suggestion of a direct proof of Lemma 2.

References

1. M. Dietzfelbinger, J. Gil, Y. Matias, and N. Pippenger. Polynomial hash functions are reliable. In *Proceedings 19th International Colloquium on Automata, Languages and Programming*, volume 623 of *Lecture Notes in Computer Science*, pages 235–246. Springer-Verlag, 1992.
2. M. Dietzfelbinger, A. Karlin, K. Mehlhorn, F. Meyer auf der Heide, H. Rohnert, and R.E. Tarjan. Dynamic perfect hashing: Upper and lower bounds. In *29th IEEE Symposium on Foundations of Computer Science*, pages 524–531, 1988.
3. M. Dietzfelbinger and F. Meyer auf der Heide. A new universal class of hash functions and dynamic hashing in real time. In *Proceedings 17th International Colloquium on Automata, Languages and Programming*, volume 443 of *Lecture Notes in Computer Science*, pages 6–19. Springer-Verlag, 1990.
4. P. Elias. Efficient storage retrieval by content and address of static files. *Journal of the ACM*, 21(2):246–260, April 1974.
5. P. Elias and R.A. Flower. The complexity of some simple retrieval problems. *Journal of the ACM*, 22(3):367–379, July 1975.
6. P. van Emde Boas. Machine models and simulations. In J. van Leeuwen, editor, *Handbook of Theoretical Computer Science*, volume A: Algorithms and Complexity, chapter 1, pages 1 – 66. Elsevier, Amsterdam, Holland, 1990.
7. A. Fiat and M. Naor. Implicit $O(1)$ probe search. *SIAM Journal on Computing*, 22(1):1 – 10, 1993.
8. A. Fiat, M. Naor, J.P. Schmidt, and A. Siegel. Nonoblivious hashing. *Journal of the ACM*, 39(4):764–782, October 1992.
9. M.L. Fredman, J. Komlós, and E. Szemerédi. Storing a sparse table with $O(1)$ worst case access time. *Journal of the ACM*, 31(3):538–544, July 1984.
10. M.L. Fredman and M.E. Saks. The cell probe complexity of dynamic data structures. In *21st ACM Symposium on Theory of Computing*, pages 345–354, Seattle, Washington, 1989.
11. H.N. Gabow and R.E. Tarjan. A linear-time algorithm for a special case of disjoint set union. *Journal of Computer and System Sciences*, 30:209–221, 1985.
12. T. Hagerup, K. Mehlhorn, and J.I. Munro. Optimal algorithms for generating discrete random variables with changing distributions. In *Proceedings 20th International Colloquium on Automata, Languages and Programming*, volume 700 of *Lecture Notes in Computer Science*, pages 253 – 264. Springer-Verlag, 1993.
13. P.B. Miltersen. The bit probe complexity measure revisited. In *Proceedings 10th Symposium on Theoretical Aspects of Computer Science*, volume 665 of *Lecture Notes in Computer Science*, pages 662–671. Springer-Verlag, 1993.
14. R.E. Tarjan and A.C. Yao. Storing a sparse table. *Communications of the ACM*, 22(11):606–611, November 1979.
15. A.C.-C. Yao. Should tables be sorted? *Journal of the ACM*, 28(3):614–628, July 1981.

Faster Searching in Tries and Quadtrees— An Analysis of Level Compression

Arne Andersson and Stefan Nilsson

Department of Computer Science, Lund University,
Box 118, S-221 00 Lund, Sweden

Abstract. We analyze the behavior of the level-compressed trie, LC-trie, a compact version of the standard trie data structure. Based on this analysis, we argue that level compression improves the performance of both tries and quadtrees considerably in many practical situations. In particular, we show that LC-tries can be of great use for string searching in compressed text.

Both tries and quadtrees are extensively used and much effort has been spent obtaining detailed analyses. Since the LC-trie performs significantly better than standard tries, for a large class of common distributions, while still being easy to implement, we believe that the LC-trie is a strong candidate for inclusion in the standard repertoire of basic data structures.

1 Introduction

A fundamental, well known, and very well studied technique for storing and retrieving data is to use *tries* [9, 10, 14]. In its original form, the trie is a data structure where a set of strings from an alphabet containing m symbols is stored in a natural way in an m-ary tree where each string corresponds to a unique path. This data structure is used in many different settings, including string matching, approximate string matching, compression schemes, and genetic sequences. The results presented in this paper also apply to *quadtrees* [17], which are widely used in computer graphics, image processing, geographic information systems, and robotics.

The expected average depth of a trie containing n independent random strings is $\Theta(\log n)$ [5]. A common method to decrease the size of a trie is to use a *path compression* method known as Patricia compression. A path compressed trie is often called a *Patricia tree* [10]. However, this compression technique does not give an asymptotic improvement. The expected average depth is still $\Theta(\log n)$ [13, 14].

We recently proposed another compression method [3]. The main idea is that the i highest complete levels of a trie can be replaced—without losing any relevant information—by a single node of degree m^i, the replacement being made top-down. This data structure is called a *level-compressed trie*, or *LC-trie* and it performs much better than a conventional trie. The expected average depth of an LC-trie is $\Theta(\log^* n)$ for uniformly distributed data.[1] We also want to point out that for a reasonable machine model the time spent at each node in an LC-trie will be constant and hence the decrease in average depth reflects a real improvement in search time in practical applications.

The trie has been the subject of thorough theoretic analysis [5, 14, 15, 16, 19]; an extensive list of references can be found in Handbook of theoretical computer science [19]. Using simple analytic methods we show how level compression improves the search time in a trie for a wide variety of common statistical models.

When analyzing the behavior of trie structures used for text retrieval, it is common to assume that the input consists of independent random strings from a Bernoulli-type process [7, 8]. For this kind of input, the expected search cost in an LC-trie will be $O(\log \log n)$ which is significantly better than $\Theta(\log n)$, achieved by the conventional trie. Another frequently used model, particularly pertinent when studying quadtrees, is independent random variables with common density. We show that for a large class of densities, the expected search cost of a level-compressed trie structure is $\Theta(\log^* n)$. Once again, this complexity is significantly better than $\Theta(\log n)$, achieved by a conventional trie or quadtree.

The LC-trie does not only have a good asymptotic behavior, it is also easy to implement, and it can be used in a wide range of different applications. We therefore consider the LC-trie to be a strong candidate for inclusion in the standard repertoire of basic data structures.

2 Preliminary definitions

To keep the notation simple we will first and foremost consider binary strings and binary trie structures. The generalization to m-ary strings is straightforward. We say that a string v of length i is the i-prefix of a string u if there is a string w such that $u = vw$. The string w is called the i-suffix of u. To make things rigorous we give the following definition of a (binary) trie:

Definition 1. A trie *containing n strings is*

if $n = 0$: an empty leaf;
if $n = 1$: a leaf containing the string;
if $n > 1$: an internal node of degree 2. For each character b there is a child, which is a trie, containing the 1-suffixes of all strings starting with b.

Note that the strings are assumed to be prefix-free, i.e. no string is a prefix of another string. In particular this implies that there can't be any duplicate strings. However, if all strings to be stored in the trie are unique, it is easy to ensure that the strings are prefix-free by using a special string terminator. Instead of using a special terminator we can append the string "100..." to the end of each string.

---

[1] The function $\log^* n$ is the iterated logarithm function, which is defined as follows. $\log^* 1 = 1$. For any $n > 1$, $\log^* n = 1 + \log^*(\lceil \log n \rceil)$.

Path compression, or Patricia compression, is the most common method to decrease the size of a trie. A trie where all internal nodes with only one child has been removed is called a *Patricia tree.* At each internal node of the Patricia tree an index is used to indicate the character used for branching at this node. Observe that it is not possible, in general, to retrieve the full strings from a Patricia tree; some information is lost when internal nodes are removed. Therefore it is often necessary to use an auxiliary data structure to store the complete strings. As pointed out above the Patricia tree does not offer an asymptotic improvement in search time as compared to the plain trie. In practice, however, it is an important size reduction technique [4].

Another natural way to decrease the search cost in a trie is to use more than one digit for branching. In this way we obtain a *multi-digit trie* as defined below.

Definition 2. *A* multi-digit *trie containing n strings is*

> *if $n = 0$: an empty leaf;*
> *if $n = 1$: a leaf containing the string;*
> *if $n > 1$: an internal node of degree 2^i, $i \geq 1$. For each possible i-prefix v there is a child, which is a multi-digit trie, containing the i-suffixes of all strings starting with v.*

The next step is to chose the outdegrees efficiently. On the one hand we want the outdegree to be as large as possible, on the other hand we do not want to introduce a large number of superfluous leaves.

Definition 3. *A* level-compressed trie, LC-trie, *is a multi-digit trie with the following properties:*

> *– the degree of the root is 2^i, where i is the smallest number such that at least one of the children becomes a leaf;*
> *– each child is an LC-trie.*

The LC-trie may be viewed as a binary trie where the i highest complete levels are replaced by a single node of degree 2^i; the replacement being made top-down. Path compression and level compression can be combined. We will consider the case where we first apply level compression and then path compression to the trie.

Definition 4. *A path compressed LC-trie is a trie to which first level compression and then path compression has been applied.*

This is a straightforward procedure; each node in the LC-trie that only has one child is removed and an index is used at each node to indicate the bit used for branching.

Let $X_1, \ldots, X_n$ be infinite binary strings. We will study the trie formed by the smallest prefixes of $X_1, \ldots, X_n$ that are pairwise different. Let D_{ni} be the depth of X_i in this trie. The average depth is defined as

$$A_n = \frac{1}{n} \sum_{i=1}^{n} D_{ni}.$$

This parameter is proportional to the average successful search time. Let

$$A_n^{\text{trie}}, \ A_n^{\text{LC}}, \ \text{and} \ A_n^{\text{path-LC}}$$

be the average depth in a trie, an LC-trie, and a path compressed LC-trie, respectively.

3 Independent strings from a Bernoulli-type process

Since tries are typically used·for storing textual data, it is natural to consider input from a Bernoulli-type process. This model can be considered as a first approximation of textual data, where we disregard all dependencies between letters and only consider the different probabilities of the letters. The behavior of a trie under this model has been thoroughly studied [15, 16]. In this section we show that the LC-trie performs significantly better. The expected average depth of an LC-trie is shown to be $O(\log \log n)$, which should be compared to $\Theta(\log n)$ for a trie.

Consider independent binary strings from a Bernoulli-type process, i.e. each string $v = (v_i)_{i \geq 1}$ is a sequence of independent identical random variables with values in the alphabet $S = \{0, 1\}$, $P(v_i = s) = p(s) > 0$, $s \in S$, $i \geq 1$.

Theorem 5. *For an independent random sample from a Bernoulli-type process, with character probabilities $p(s)$ not all equal we have $E(A_n^{LC}) = \Theta(\log \log n)$. If the probabilities are equal $E(A_n^{LC}) = \Theta(\log^* n)$.*

Proof. If all characters have the same probability the strings will be uniformly distributed and hence $E(A_n^{LC}) = \Theta(\log^* n)$ as was shown in our original paper [3].

Now consider the case where $\alpha = \min_{s \in S} p(s) < 1/|S|$. Let $k = -\log n / \log \alpha$, and let q be a number such that $0 < q < 1$. Without loss of generality we can assume that $p(0) = \alpha$. We will give a lower bound for the probability $P_{\text{all}}(n)$ that all possible strings of length qk will occur at least once as a prefix among the n strings. It can be shown that $P_{\text{all}}(n)$ is larger than the probability that the string consisting of qk zeroes occurs at least 2^{qk} times among the prefixes of the strings. In fact, to prove this we may assume without loss of generality that the probabilities of the prefixes are equal. Now consider all possible sequences of length n of prefixes of length qk. We just need to check that the number of different sequences that contain at least 2^{qk} prefixes consisting of only zeroes is not greater than the number of sequences that contain at least one copy of each possible prefix. This is a straightforward exercise in combinatorics; the details are omitted. Let A be the number of prefixes consisting of qk zeroes among the n input strings.

$$P_{\text{all}}(n) \geq P(A \geq 2^{qk}) = P(A \geq 2^{-q \frac{\log n}{\log \alpha}}) = P(A \geq n^{-(1 + q(\frac{1}{\log \alpha} - 1))} n^{1-q}).$$

A obeys the binomial distribution $B(n, \alpha^{qk}) = B(n, n^{-q})$. And in particular $E(A) = n^{1-q}$. Using the following Chernoff bound

$$P(A \geq (1 - \epsilon)E(A)) \geq 1 - e^{-\epsilon^2 E(A)/2}$$

we see that $P_{\text{all}}(n) \to 1$ as $n \to \infty$ if $1 + q(\frac{1}{\log \alpha} - 1) > 0$, or equivalently $q < \frac{\log \alpha}{\log \alpha - 1}$. In particular, $P_{\text{all}}(n) \to 1$ for all α such that $0 < \alpha < 1/2$, if $q < 1/2$. Clearly $E(A_n^{\text{LC}}) \leq E(A_n^{\text{trie}})$, but $E(A_n^{\text{trie}}) = O(\log n)$ [16] and hence we get the following recursion inequality for $D(n) = E(A_n^{\text{LC}})$.

$$D(n) \leq 1 + P_{\text{all}}(n) \cdot D(n/2^{qk}) + (1 - P_{\text{all}}(n)) \cdot O(\log n)$$

It is easy to check that for $q = 1/4$, $(1 - P_{\text{all}}(n)) \log n \to 0$ as $n \to \infty$ and hence

$$D(n) \leq O(1) + D(n/2^{qk}) = O(1) + D(n^{1+q/\log \alpha}).$$

But $-1/4 \leq q/\log \alpha < 0$ and hence $E(A_n^{\text{LC}}) = O(\log \log n)$.

The lower bound can be established in a similar way. Choose α as above and let Q be a number such that $1 < Q < -\log \alpha$. Denote by $P_{\text{none}}(n)$ the probability that there will be no string with a prefix consisting of Qk zeroes among the n strings. Assuming that $min_{s \in S} p(s) < 1/|S|$ we make the following two observations. $P_{\text{none}}(n) = (1 - \alpha^{Qk})^n = (1 - \frac{1}{n^Q})^n \to 1$ as $n \to \infty$ and $n/2^{Qk} = n^{1+Q/\log \alpha}$, where $-1 < Q/\log \alpha < 0$. We have the following inequality

$$D(n) \geq 1 + (1 - P_{\text{none}}(n)) \cdot 0 + P_{\text{none}}(n) \cdot D\left(\frac{n}{2^{Qk}}\right)$$

and hence $E(A_n^{\text{LC}}) = \Omega(\log \log n)$ in this case.

Observe that the assumption that the probability of each character is positive is necessary. For example, consider a ternary alphabet a, b, c where $P(a) = 0$. No level compression will take place in this case since every node in the trie will have at most two children.

3.1 Compressed strings

Pattern matching in compressed text documents is a new and interesting problem that has been investigated by Amir et al. [1, 2, 6]. In this section we argue that the LC-trie could be particularly useful for this kind of applications. We show that under certain assumptions the LC-trie actually behaves better when the input strings have been compressed. Finally, we indicate how this fact could be used in a database application.

Most modern model-based compression schemes use either Huffman coding [12] och Arithmetic coding [20] to code a message with respect to a statistical model of the text. Arithmetic coding is the more efficient of the two. The only drawback being that it is slightly more difficult to implement.

In arithmetic coding a message is represented by a subinterval of $[0, 1)$. Successive characters of the text reduce the size of the interval according to the character probability generated by the statistical model. Consider the following example. The text contains three different characters a, b, and c with fixed probabilities 0.2, 0.3, and 0.5, respectively. We want to encode the text cab. To each of the characters we assign an interval whose size is proportional to the probabilities. In this example the intervals $[0, 0.2)$, $[0.2, 0.5)$, and $[0.5, 1]$ are assigned to the characters a, b, and c, respectively. After the first character (c) has been processed the interval is reduced to $[0.5, 1)$, the second character (a) reduces the interval to $[0.5, 0.6)$, and after the third character (b) we get $[0.52, 0.55)$.

Theorem 6. *For an independent random sample from a Bernoulli-type process where each string has been arithmetically coded, $E(A_n^{LC}) = O(\log^* n)$.*

Proof. The strings from the Bernoulli-type process will be uniformly distributed after having been arithmetically coded and hence the result follows immediately from Theorem 4. In fact, consider any subinterval I of $[0, 1)$ and an arithmetically coded infinite string s from a Bernoulli-type process. It follows from the definition of arithmetic coding that $P(s \in I)$ equals the size of I.

Huffman coding also has the effect of smoothing the input, but to a lesser degree.

Theorem 7. *For an independent random sample from a Bernoulli-type process where each string has been Huffmann-coded, $E(A_n^{LC}) = O(\log \log n)$.*

Proof. In this case the distribution will not be uniform and the bits will not be independent. The probability of a bit being either 0 or 1 depends on the preceding bits. However, we can make the following observation. At each node r of the Huffman tree there will be a positive probability $P_r(0)$ that the next character occurring will be 0. In fact, $P_r(0) \geq \min_{s \in S} P(s)$, since $P_r(0)$ is the sum of the weights of the leaves in the left subtree of r. Similarly, $P_r(1) \geq \min_{s \in S} P(s)$. The theorem can now be proved using the same technique as in the proof of Theorem 5.

As a possible application we mention a large dictionary where each entry, starting with its keyword, is coded separately by arithmetic coding. This dictionary, together with an LC-trie representing the coded keywords, would require less space that the uncompressed document and the LC-trie would support fast location of keywords. When a keyword is found, the appropriate article could be decoded separately and presented to the user.

4   Independent strings with common density

Independent random strings with common density is perhaps the most frequently used statistical model and for many applications this model will be more natural than a Bernoulli-type process. For example, geometric data can often be accurately modelled in this way and, as pointed out above, both path compression and level compression can be applied not only to tries by also to quadtrees, a data structure that is frequently used in applications that handle geometric data.

Assume that $X_1, \ldots, X_n$ are the binary representations of independent random variables with common density f on $[0, 1)$. The behavior of binary tries under this model has been analyzed by Devroye [5]. If the density $f \in L^2$, i.e. $\int_{x=0}^{1} f^2(x)\, dx < \infty$, then $E(A_n^{\text{trie}}) = O(\log n)$ else $E(A_n^{\text{trie}}) = \infty$ for all $n \geq 2$. Even for LC-tries the case $f \notin L^2$ is intractable, as seen by the following observation.

Observation 8. *For an independent random sample of size $n \geq 2$ taken from a distribution with density $f(x) \notin L^2$, $0 \leq x \leq 1$, $E(A_n^{LC}) = \infty$.*

Proof. We prove this using a simple relation between LC-tries and tries. In a trie, consider an element x with depth $n + h$. On the path of x there are at least h nodes that have an empty child. Hence, in the LC-trie containing the same set, the depth of x will be at least h. From this follows that the average depth of the elements in an LC-trie is not less than the average depth in a trie minus n. The result now follows since the expected depth of an element in a trie is infinite if $f \notin L^2$.

We also note that the depth of an LC-trie will never be greater than that of a conventional trie. In fact, for a large number of densities the expected average depth will be much smaller as will be shown in the following theorems.

Theorem 9. *For an independent random sample taken from a distribution with density $f(x) \in L^2$, $0 < \alpha \leq f(x) \leq \beta$, $0 \leq x \leq 1$, $E(A_n^{LC}) = \Theta(\log^* n)$.*

Proof. We first consider the case $\alpha = \beta = 1$.

We choose the indices in such a way that $X_1 < \cdots < X_n$, and write $X_0 = 0$ and $X_{n+1} = 1$ and consider the random variable

$$M_n = \max_{0 \leq i \leq n} (X_{i+1} - X_i),$$

the maximal spacing between adjacent elements in the sample. We will use the following lemma by Slud [18]:

Lemma 10. *If δ is a positive constant, then*

$$P(|nM_n / \ln n - 1| > \delta) = O(n^{-\delta}).$$

First we determine an upper bound. We will show that the degree at the root is $\Omega(n / \log n)$ with high probability. This also holds true at the lower levels of the trie, since the elements in a subtree are a random sample of independently chosen points in an interval of some length 2^{-j}; thus, the analysis applies with due alteration of details.

Let x be an element and let T_x denote the number of elements that are stored in the same subtrie as x. Choose the integer i such that $\frac{2 \ln n}{n} < 2^{-i} \leq \frac{4 \ln n}{n}$ and split the interval $[0, 1)$ into subintervals of size 2^{-i}. Let I denote the subinterval into which x falls. Let y be any of the other $n - 1$ elements and let p be the probability that y falls into I. Then, $p = 2^{-i} \leq \frac{4 \ln n}{n}$.

In the expected case, the effect of choosing i like this will be that $M_n \leq 2^{-i}$. Therefore every subinterval of size 2^{1-i} will contain at least two elements and hence the number of children of the root will be at least 2^i. This implies that the elements that are stored in the same subtrie as x will be a subset of the elements in I. From this we will deduce that for some constant C, $P(T_x > C \ln n)$ is small. This is done by studying the random variables M_n and n_I, where n_I denotes the number of elements (not counting x) falling into I.

Clearly, n_I obeys the binomial distribution $B(n - 1, p)$. Let $C = 2e$. We use a Chernoff bound [11] to get a tail estimate for the binomial function.

$$P(n_I > C \ln n) \leq e^{-(n-1)p} \left(\frac{e(n-1)p}{C \ln n} \right)^{C \ln n} = O(n^{-1}) \tag{1}$$

If $M_n \leq 2^{-i}$ then, as explained above, $T_x \leq n_I$. In particular

$$M_n \leq 2^{-i} \text{ and } n_I \leq C \ln n \Rightarrow T_x \leq C \ln n$$

and hence

$$P(M_n \leq 2^{-i} \text{ and } n_I \leq C \ln n) \leq P(T_x \leq C \ln n). \tag{2}$$

From Lemma 10 we have

$$P\left(M_n > \frac{\ln n}{n} (1 + \delta) \right) = O(n^{-\delta}).$$

Hence

$$P(M_n > 2^{-i}) = P\left(M_n > \frac{\ln n}{n} \left(1 + \frac{n2^{-i}}{\ln n} - 1 \right) \right) = O(n^{-\varepsilon}), \tag{3}$$

where $\varepsilon = \frac{n2^{-i}}{\ln n} - 1 \geq 1$.

Combining these results, we get

$$\begin{aligned} P(T_x > C \ln n) &\leq P(\neg(M_n \leq 2^{-i} \text{ and } n_I \leq C \ln n)) \\ &\leq P(M_n > 2^{-i}) + P(n_I > C \ln n) \\ &= O(n^{-1}). \end{aligned} \tag{4}$$

The depth of an element in an LC-trie is never larger than that of the corresponding binary trie. It has been shown by Devroye [5] that the expected average depth of a binary trie is $O(\log n)$. Therefore, we can use the estimate $O(\log n)$ for the expected depth of an element in an LC-trie when $T_x > C \ln n$.

We inductively assert that the expected depth $E(A_n^{LC})$ can be bounded by

$$E(A_n^{LC}) \leq D \log^* n \tag{5}$$

for some constant D. This is obviously true for $n = 1$. For $n \geq 2$ there is a constant D' such that

$$E(A_n^{\text{LC}}) \leq 1 + \sum_{m=1}^{C \ln n} P(T_x = m) \cdot E(A_m^{\text{LC}}) + P(T_x > C \ln n) \cdot O(\log n)$$

$$\leq 1 + \sum_{m=1}^{C \ln n} P(T_x = m) \cdot D \log^*(C \ln n) + O(1)$$

$$\leq D(\log^*(\log n)) + D'. \tag{6}$$

It can be arranged that $D' \leq D$. Then the last expression can be bounded from above by $D(\log^* n - 1) + D' \leq D \log^* n$ and hence $E(A_n^{\text{LC}}) = O(\log^* n)$.

The lower bound can be established in a similar way. By choosing the integer i such that $\frac{\ln n}{4n} < 2^{-i} \leq \frac{\ln n}{2n}$ we can ensure that $P(T_x \leq c \ln n) \cdot E(A_{c \ln n}^{\text{LC}}) = o(1)$ for some positive constant c and the result follows by an induction argument similar to the the one in the upper bound analysis.

Now consider the case $0 < \alpha < \beta < \infty$. We make the following observations:

- Let Δ be an interval of size $|\Delta|$ and let n_Δ be the expected number of elements falling into Δ. Then $n_\Delta \in \text{Bin}(n, p)$, where $p \leq \beta |\Delta|$.
- Let M_n^f be the maximum spacing for data taken from f and let M_n^u be the maximum spacing for data from the uniform distribution over the interval $[0, 1)$, then $P(M_n^f > x) \leq P(M_n^u > \alpha x)$. This can be shown in the following way: Consider the density $g(x) = \alpha$, $0 \leq x \leq 1/\alpha$. Then, $P(M_n^g > x) = P(M_n^u > \alpha x)$. The fact now follows since the domain of f is a subinterval of the domain of g and $f(x) \geq g(x)$ when $x \in [0, 1)$.

Using the first observation we see that Eq. 1 holds if C is chosen large enough. If we make the interval I larger by a factor of $\frac{1}{\alpha}$, i. e. we choose i such that $\frac{\ln n}{\alpha n} < 2^{-i} \leq \frac{2 \ln n}{\alpha n}$, the second observation gives

$$P(M_n^f > 2^{-i}) \leq P(M_n^u > \alpha 2^{-i})$$

$$= P\left(M_n^u > \frac{\ln n}{n}\left(1 + \frac{\alpha n 2^{-i}}{\ln n} - 1\right)\right)$$

$$= O(n^{-\varepsilon}),$$

where $\varepsilon = \frac{\alpha n 2^{-i}}{\ln n} - 1 > 0$. Hence, equation 6 still holds if we choose C large enough ($C = 2\beta e/\alpha$ will do) and we conclude that $E(A_n^{\text{LC}}) = O(\log^* n)$ in this case also.

The lower bound can be established using the same technique.

We now show that if path compression is applied to the LC-trie, the class of densities for which the expected search time is $O(\log^* n)$ can be further extended. Using this technique, we can efficiently handle a class of densities that is zero on a finite number of intervals as proved in the following lemma and theorem.

Lemma 11. *Given a constant r, $0 \leq r \leq 1$, and a density $f(x) \in L^2$,*

$$\begin{cases} 0 < \alpha \leq f(x) \leq \beta, & 0 \leq x \leq r \\ f(x) = 0 & otherwise. \end{cases}$$

For an independent random sample taken from the distribution with density $f(x)$, $0 \leq x \leq 1$, $E(A_n^{path\text{-}LC}) = O(\log^* n)$.

Proof. We say that a trie *covers* an interval if all elements that fall in this interval is stored in that trie. Let L be an LC-trie containing n elements from the distribution in the lemma. For the purpose of this proof we construct a trie T containing the same elements as L in the following way: If the interval covered by T contains r the root is made binary even if we could afford a higher degree. Otherwise, we choose the maximal branching factor according to Definition 3. This construction is repeated in the subtries.

It is not hard to show [3] that the external path length of T will not be smaller than the external path length of L. Hence, we will be done if we can prove that the expected average depth of T is $O(\log^* n)$.

Since we use Patricia compression the root of T has two nonempty subtries T_{left} and T_{right}. If r is contained in any of the intervals covered by these two subtries, it has to be the interval covered by T_{right}. Thus, we have that

1. In the interval covered by T_{left}, f satisfies $0 < \alpha \leq f(x) \leq \beta$. From Theorem 9 follows that this subtrie can be represented by an LC-trie with an expected average depth of $O(\log^* n)$. The expected number of elements in this interval is at least $\frac{\alpha}{2}n$. In particular, the number of elements is more than $\frac{\alpha}{4}n$ with very high probability.
2. In the interval covered by T_{right} we can repeat the analysis recursively.

This gives us the following recursion inequality for $D(n) = E(A_n^{path\text{-}LC})$:

$$\begin{cases} D(n) \leq 2 + \left(1 - \frac{\alpha}{4}\right) D\left(n\left(1 - \frac{\alpha}{4}\right)\right) + \frac{\alpha}{4}O(\log^* n) \\ D(1) = 1 \end{cases}$$

This equation is of the form

$$\begin{cases} T(n) \leq a + bT(bn) + c\log^* n \\ T(1) = 1 \end{cases}$$

where a, b, and c are positive constants, $b < 1$. (The case when $b < 0$ is irrelevant, since in this case the lemma follows immediately). This equation is easily solved:

$$T(n) < (a + c\log^* n)(1 + b + b^2 + b^3 + \ldots) = O(\log^* n).$$

Theorem 12. *Let $\delta_1, \ldots, \delta_k$ be a number of disjoint subintervals of the unit interval, such that the length of the shortest interval is positive. Let $f(x) \in L^2$ be a function such that*

$$\begin{cases} f(x) = 0, & x \in \bigcup_{1 \leq i \leq k} \delta_i \\ 0 < \alpha \leq f(x) \leq \beta, & otherwise. \end{cases}$$

If an independent random sample taken from the distribution with density $f(x)$, $0 \leq x \leq 1$, $E(A_n^{path\text{-}LC}) = O(\log^* n)$.

Proof. Let $|\delta|$ be the length of the smallest interval. At each level we have a branching factor of at least two. Therefore, the function f restricted to the interval covered by a subtrie at level $\left\lceil \log \frac{1}{|\delta|} \right\rceil$ will be of the form described in Lemma 11. Thus, the expected average depth of an element will be at most $\left\lceil \log \frac{1}{|\delta|} \right\rceil + O(\log^* n)$.

5   Conclusions

We note that the methods of path compression and level compression can be used to reduce the search time in two of the most prominent data structures, the trie and the quadtree.

The trie is perhaps most widely used in text processing applications, such as string matching, approximate string matching, and compression schemes. But it has also been used in algorithms for genetic sequences. The Bernoulli-type process is often a good first approximation for text and genetic sequences and since Theorem 3 shows that we get an asymptotic reduction in search time for data from this model, we conjecture that the LC-trie should give a significant improvement in many of these settings.

We also observe that the assumptions of Theorem 12 are likely to be met in many geometrical applications and hence the methods of path compression and level compression should give significant improvements in many applications in computer graphics, image processing, geographic information systems, and robotics, since quadtrees are often an integral part of these systems.

Furthermore, the LC-trie seems to behave even better for compressed strings as shown in Theorems 6 and 7. Consequently the LC-trie should be of interest in the recently initiated study of search algorithms for compressed text.

References

1. A. Amir, G. Benson, and M. Farach. Efficient pattern matching with scaling. *Journal of Algorithms*, 13(1), 1992.
2. A. Amir, G. Benson, and M. Farach. Let sleeping files lie: pattern matching in z-compressed files. In *Proc. of 5th Symposium on Discrete Algorithms*, 1994.
3. A. Andersson and S. Nilsson. Improved behaviour of tries by adaptive branching. *Information Processing Letters*, 46:295–300, 1993.
4. A. Andersson and S. Nilsson. Efficient implementation of suffix trees. *Technical Report, Dept. of Computer Science, Lund University*, 1994.
5. L. Devroye. A note on the average depth of tries. *Computing*, 28:367–371, 1982.
6. T. Eilam-Tsoreff and U. Vishkin. Matching patterns in a string subject to multi-linear transformation. In *Proc. of the International Workshop on Sequences, Combinatorics, Compression, Security and Transmission, Salerno, Italy*, June, 1988.

7. Ph. Flajolet. On the performace evaluation of extendible hashing and trie searching. *Acta Informatica*, 20:345–369, 1983.

8. Ph. Flajolet, M. Régnier, and D. Sotteau. Algebraic methods for trie statistics. *Ann. Discrete Math.*, 25:145–188, 1985.

9. E. H. Fredkin. Trie memory. *Communications of the ACM*, 3:490–500, 1960.

10. G. H. Gonnet and R. Baeza-Yates. *Handbook of Algorithms and Data Structures*. Addison-Wesley, 1991.

11. T. Hagerup and C. Rüb. A guided tour of chernoff bounds. *Information Processing Letters*, 33(6):305–308, 1990.

12. D. A. Huffman. A method for the construction of minimum redundancy codes. In *Proc. IRE*, volume 40, pages 1098–1101, 1952.

13. P. Kirschenhofer and H. Prodinger. Some further results on digital search trees. In *Proc. 13th ICALP*, pages 177–185. Springer Verlag, 1986. Lecture Notes in Computer Science vol. 26.

14. D. E. Knuth. *The Art of Computer Programming, Volume 3: Sorting and Searching*. Addison-Wesley, Reading, Massachusetts, 1973.

15. B. Pittel. Asymptotical growth of a class of random trees. *The Annals of Probability*, 13(2):414–427, 1985.

16. B. Pittel. Paths in a random digital tree: limiting distributions. *Advances in Applied Probability*, 18:139–155, 1986.

17. H. Samet. The quadtree and related hierarchical data structures. *Computing Surveys*, 16(2):187–260, 1984.

18. E. Slud. Entropy and maximal spacings for random partitions. *Zeitschrift für Wahrscheinlichkeitstheorie und verwandte Gebiete*, 41:341–352, 1978.

19. J. van Leeuwen, editor. *Handbook of Theoretical Computer Science, vol. A*. Elsevier Science Publishers B.V., 1990. ISBN 0-444-88071-2.

20. I. H. Witten, R. M. Neal, and J. G. Cleary. Arithmetic coding for data compression. *Communications of the ACM*, 30(6):520–540K, 1987.

The Analysis of a Hashing Scheme by the Diagonal Poisson Transform *

(Extended Abstract)

Patricio V. Poblete[1], Alfredo Viola[2], J. Ian Munro[2]

[1] Department of Computer Science, University of Chile, Casilla 2777, Santiago, Chile.
E-mail: ppoblete@dcc.uchile.cl
[2] Department of Computer Science, University of Waterloo, Waterloo, Ont. N2L 3G1, Canada.
E-mail: aviola@uwaterloo.ca, imunro@uwaterloo.ca

Abstract. We present an analysis of the effect of the last-come-first-served heuristic on a linear probing hash table. We study the behavior of successful searches, assuming searches for all elements of the table are equally likely. It is known that the Robin Hood heuristic achieves minimum variance over all linear probing algorithms. We show that the last-come-first-served heuristic achieves this minimum up to lower order terms.

An accurate analysis of this algorithm is made by introducing a new transform which we call the diagonal Poisson transform as it resembles the Poisson transform. We present important properties of this transform, as well as its application to solve some classes of recurrences. We feel this is the main contribution of the paper.

1 Introduction

The simplest collision resolution scheme for open addressing hash tables is *linear probing*, which uses the cyclic probe sequence

$$h(K), h(K) + 1, \ldots m - 1, 0, 1, \ldots, h(K) - 1$$

assuming the table slots are numbered from 0 to $m - 1$. Linear probing works reasonably well for tables that are not too full, but as the load factor increases, its performance deteriorates rapidly.

If A_n denotes the number of probes in a successful search in a hash table of n elements (assuming all elements in the table are equally likely to be searched), and if we assume that the hash function h takes all the values in $0 \ldots m - 1$ with equal probabilities, then we know from [11]:

$$\mathbf{E}[A_n] = \frac{1}{2}(1 + Q_0(m, n - 1)) \tag{1}$$

$$\mathbf{V}[A_n] = \frac{1}{3}Q_2(m, n - 1) - \frac{1}{4}Q_0(m, n - 1)^2 - \frac{1}{12} \tag{2}$$

---

* This research was supported in part by the Natural Sciences and Engineering Research Council of Canada under grant No. A8237, the Information Technology Research Centre of Ontario, and FONDECYT(Chile) under grant 1940271.

Part of this work was done while the first author was on sabbatical at the University of Waterloo.

where the functions $Q_i(m, n)$ are a generalization of the Ramanujan's Q-function. For a table with $n = \alpha m$ elements, and fixed $\alpha < 1$ and $n, m \to \infty$, these quantities depend (essentially) only on α:

$$E[A_{\alpha m}] = \frac{1}{2}\left(1 + \frac{1}{1 - \alpha}\right) - \frac{1}{2(1 - \alpha)^3 m} + O\left(\frac{1}{m^2}\right)$$

$$V[A_{\alpha m}] = \frac{1}{3(1 - \alpha)^3} - \frac{1}{4(1 - \alpha)^2} - \frac{1}{12} - \frac{1 + 3\alpha}{2(1 - \alpha^5)m} + O\left(\frac{1}{m^2}\right)$$

For a full table, these approximations are useless, but the properties of the Q functions can be used to obtain the following expressions:

$$E[A_m] = \frac{\sqrt{2\pi m}}{4} + \frac{1}{3} + \frac{1}{48}\sqrt{\frac{2\pi}{m}} + O\left(\frac{1}{m}\right)$$

$$V[A_m] = \frac{\sqrt{2\pi m^3}}{12} + \left(\frac{1}{9} - \frac{1}{8}\right)m + \frac{13\sqrt{2\pi m}}{144} - \frac{47}{405} - \frac{\pi}{48} + O\left(\frac{1}{\sqrt{m}}\right)$$

It is clear from these expressions that not only is the expected search time high, but also the variances are quite large, and therefore the expected value is not a very reliable predictor for the actual running time of a successful search.

Operating primarily in the context of double hashing, several authors [3, 2, 7] observed that a collision could be resolved in favor of *any* of the keys involved, and used this additional degree of freedom to decrease the expected search time in the table. Celis *et al* [6] were the first to observe that collisions could be resolved having *variance reduction* as a goal. They defined the Robin Hood heuristic, where each collision occurring on each insertion is resolved in favor of the key that is farthest away from its home location. Later, Poblete and Munro [14] defined the last-come-first-served heuristic, where collisions are resolved in favor of the incoming key, and others are moved ahead one position in their probe sequences. In both cases, the reduction of the variance can be used to speed up searches by replacing the standard search algorithm by a "mean-centered" one that first searches in the vicinity of where we would expect the element to have "drifted" to, rather than its initial probe location.

It was shown in [5] that the Robin Hood linear probing algorithm achieves minimum variance over all linear probing algorithms. This variance, for a full table, is $\theta(m)$, instead of the $\theta(m^{3/2})$ of the standard algorithm. They derived the following expressions for the variance of the successful search time:

$$V[A_n] = \frac{1}{2}Q_1(m, n - 1) - \frac{1}{4}Q_0(m, n - 1)^2 - \frac{1}{6}Q_0(m, n - 1) + \frac{1}{6}\frac{n - 1}{m} - \frac{1}{12}$$

$$V[A_{\alpha m}] = \frac{1}{4(1 - \alpha)^2} - \frac{1}{6(1 - \alpha)} - \frac{1}{12} + \frac{\alpha}{6} - \frac{1}{6m} - \frac{1 + 2\alpha}{3(1 - \alpha^4)m} + O\left(\frac{1}{m^2}\right)$$

$$V[A_m] = \frac{4 - \pi}{8}m + \frac{1}{9} - \frac{\pi}{48} + \frac{1}{135}\sqrt{\frac{2\pi}{m}} + O\left(\frac{1}{m^2}\right) \tag{3}$$

In this paper we study the effect of the LCFS heuristic on the linear probing scheme. We prove that this algorithm also achieves this optimal variance up to lower order terms. Section 2 introduces the necessary background to understand the mathematical tools we use. Section 3 presents the derivation of the main recurrence to solve. The succeeding sections (Sections 4-6) are the core of the paper, and introduce the new diagonal Poisson transform and its application to solve a general class of recurrences. Finally Sections 7 and 8 present the analysis of the variance of successful searches and the conclusions of our work.

2 Preliminaries and the Poisson Transform

The Q functions are defined by $Q_r(m, n) = \sum_{i \geq 0} (i, r) \frac{n^{\underline{i}}}{m^i}$ where $n^{\underline{i}} = n!/(n-i)!$ and $(i, j) = \binom{i+j}{j}$. In [12] a more general class of Q functions is presented, and several properties are proved. In [10] we find the proof of this asymptotic expansion, that is used in our analysis:

$$Q_0(m, m-1) = \frac{\sqrt{2\pi m}}{2} - \frac{1}{3} + \frac{1}{24}\sqrt{\frac{2\pi}{m}} - \frac{4}{135m} + O\left(\frac{1}{m^{3/2}}\right) \quad (4)$$

We also have $Q_1(m, m-1) = m$.

Following the analysis of the standard linear probing algorithm given in [11], we use the function $\hat{f}(m, n)$ to denote the number of ways to create a table of size m, with n elements inserted in such a way that the last location of the table is empty. Note that the probability of the last location being empty is $(1 - n/m)$. As there are m^n possible arragements, then we see that $\hat{f}(m, n) = m^{n-1}(m - n)$.

If we have a function $F(x_1, \ldots, x_n, z)$ we use the following operators:

$$\mathbf{U}_z F(x_1, \ldots, x_n, z) = F(x_1, \ldots, x_n, 1) \qquad \text{(unit)}$$

$$\mathbf{D}_z^k F(x_1, \ldots, x_n, z) = \frac{\partial^k F(x_1, \ldots, x_n, z)}{\partial z^k} \qquad \text{(derivative)}$$

If X is an integer valued random variable, we denote $p_i = Prob[X = i], i = 1 \ldots n$. In our case, X is the number of probes for a successful search in a table of size m where n elements have been inserted. The generating function for the probability distribution p_i is defined by $P_{m,n}(z) = \sum_{i \geq 0} p_i z^i$. We use the following well known properties of generating functions:

$$E[X] = \mathbf{U}_z \mathbf{D}_z P_{m,n}(z) \quad (5)$$

$$V[X] = \mathbf{U}_z \mathbf{D}_z^2 P_{m,n}(z) + E[X] - E[X]^2 \quad (6)$$

We define an almost full table of size m, as a hash table of size m with $m - 1$ elements inserted, in such a way that the last location of the table is empty.

The Poisson transform is discussed in detail in [8]. The key notions that we will need are outlined as follows: Let $f(m, n)$ be an expected value computed using a model of n objects randomly distributed among m locations. Then, the *Poisson transform* of $f(m, n)$ is defined as

$$\mathcal{P}_m[f(m, n); x] = \tilde{f}_m(x) = e^{-mx} \sum_{n=0}^{\infty} f(m, n) \frac{(mx)^n}{n!} \quad (7)$$

The main theorem presented in [8] is

Theorem 1. *If* $\mathcal{P}_m[f(m,n); x] = \sum_{i \geq 0} a_i x^i$ *is the Poisson transform of f(m,n) then*
$$f(m, n) = \sum_{i=0}^{\infty} a_i \frac{n^{\underline{i}}}{m^{\underline{i}}}.$$

3 Analysis of Last-Come-First-Served Linear Probing Hashing

Consider a hash table of size m, with $n + 1$ elements inserted using the last-come-first-served linear probing algorithm. We will consider a randomly chosen element as a "tagged" one, and denote it by •. We define $P_{m,n}(z)$ as the probability generating function for the cost of searching for this tagged element.

Without loss of generality, we may assume that after inserting the first n elements, the hash table is as shown in Fig. 1, and that as a result of the insertion of the $(n + 1)^{st}$ element, the last location of the table is filled. We may see the table as a concatenation of two tables of sizes $m - i - 2$ and $i + 2$ with $n - i - 1$ and $i + 1$ elements respectively. We may also assume that • belongs to the last cluster of the hash table. Consider now the insertion of the last element. With probability $1/(n + 1)$, this element is •, and so its cost is 1 (generating function z). With probability $n/(n + 1)$ the new element is not •. If we assume this insertion does not force • to move, then we have the recurrence:

$$P_{m,n}(z) = \frac{z}{n+1} + \frac{n}{n+1} P_{m,n-1}(z) \tag{8}$$

We must, of course, include a correction term to account for this shortcomming. As we can see in Fig. 1, the last insertion increments the cost of searching for • when it maps into any of the first $\ell + 1$ positions of the last cluster.

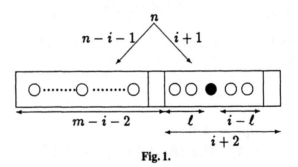

Fig. 1.

We now introduce two auxiliary functions. Given a table of size $\ell + r + 2$, we define $F_{\ell,r}(z)$ as the generating function for the number of ways of inserting $\ell + r + 1$ elements in the table, where one element is tagged (•), such that the rightmost location is empty, and such that there are ℓ elements to the left of • and r elements to its right. Fig. 2 helps to understand this definition. It is easy to see that if we insert a new element in any of the first $\ell + 1$ locations of the table, the cost of • increases by one. By the definition of $F_{\ell,r}(z)$ we know that $F_{\ell,r}(1) = \hat{f}(\ell + r + 2, \ell + r + 1) = (\ell + r + 2)^{\ell+r}$, where $\hat{f}(m, n)$ is defined in Section 2.

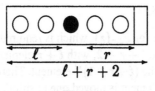

Fig. 2.

We define $C_i(z)$ as the generating function for the number of ways of inserting $i+1$ elements into a table of size $i+2$, where one element is tagged ($\bullet$), and such that the rightmost location is empty, with z keeping track of the cost of $\bullet$. As $\bullet$ may be any of the $i+1$ elements inserted

$$C_i(z) = \sum_{\substack{\ell+r=i \\ \ell,r\geq 0}} F_{\ell,r}(z) \tag{9}$$

Equation (9) and the value of $F_{\ell,r}(1)$ imply that $C_i(1) = (i+2)^i(i+1)$. The function $C_i(z)/C_i(1)$ is the probability generating function for the cost of a successful search for $\bullet$ in an almost full table of size $i+2$. Therefore, by (1) we have $\mathbf{U}_z\mathbf{D}_zC_i(z)/C_i(1) = \frac{1}{2}(1+Q_0(i+2,i))$.

We now have the tools to find the correction term $T_{m,n}(z)$.

$$T_{m,n}(z) = \frac{z-1}{m^n(n+1)}\sum_{i=0}^{n-1}\binom{n}{i+1}\hat{f}(m-i-2,n-i-2)\sum_{\ell+r=i}(\ell+1)F_{\ell,r}(z)$$

So, we have the following recurrence for $P_{m,n}(z)$:

$$P_{m,n}(z) = \frac{z}{n+1} + \frac{n}{n+1}P_{m,n-1}(z) + T_{m,n}(z) \tag{10}$$

If we solve recurrence (10), and define $G_i(z) = \sum_{\ell+r=i}(\ell+1)F_{\ell,r}(z)$, we find

$$P_{m,n}(z) = z + \frac{z-1}{m^n(n+1)}\sum_{0\leq i\leq n-1}\binom{n+1}{i+2}(m-i-2)^{n-i-1}G_i(z) \tag{11}$$

Following the ideas presented in [8] we will find the Poisson transform $\tilde{P}_m(x,z)$ of $P_{m,n}(z)$. So, we first obtain an accurate analysis under a Poisson-filling model, and then by the use of the Poisson transform convert $\tilde{P}_m(x,z)$ back to $P_{m,n}(z)$. If we use the definition of the Poisson transform we obtain

$$\tilde{P}_m(x,z) = z + (z-1)\sum_{i\geq 0}e^{-(i+2)x}\frac{x^{i+1}}{(i+2)!}G_i(z) \tag{12}$$

Now, we have to find a recurrence for $G_i(z)$, and try to solve it. Note that in this case we are working with almost full tables of size $i+2$. Using the value of $F_{\ell,r}(1)$ and the definition of $G_i(z)$ we may easily check that for $z=1$

$$\mathbf{U}_zG_i(z) = \frac{(i+1)(i+2)^{i+1}}{2} \tag{13}$$

3.1 A Recurrence for $G_i(z)$

We first present a recurrence for $F_{\ell,r}(z)$, which is required to derive the recurrence we need. We have a table of size $\ell + r + 2$, with $\ell + r$ elements inserted, and want to see what happens when we add the $(\ell+r+1)^{st}$ element. There are four cases as described in Fig. 3. When the tagged element is moved one position, then the label z of the arrow shows that we need z as a factor in the recurrence.

Case a) is the insertion of the tagged element. In this case case the generating function is z times the number of ways of generating a table of size $\ell + r + 2$ with $\ell + r$ elements, in such a way that the last cluster is of size k. For a fixed k, this factor is $\binom{\ell+r}{k}(k+1)^{k-1}(\ell+r-k+1)^{\ell+r-k-1}$. As k ranges from 0 to r,

$$F_{\ell,r}(z) \leftarrow z \sum_{0 \leq k \leq r} \binom{\ell+r}{k}(k+1)^{k-1}(\ell+r-k+1)^{\ell+r-k-1} \qquad (14)$$

For the last three cases, we assume that the inserted element is not the tagged one.

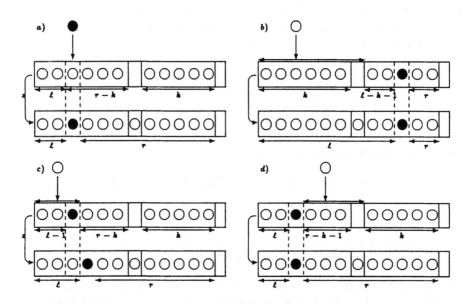

Fig. 3.

Case b) is the insertion of an element in the cluster that precedes the one that has $\bullet$. The cost of searching for $\bullet$ does not increase. As k ranges from 0 to $\ell - 1$,

$$F_{\ell,r}(z) \leftarrow \sum_{0 \leq k \leq \ell-1} \binom{\ell+r}{k}(k+1)^{k-1}F_{\ell-k-1,r}(z)(k+1) \qquad (15)$$

Case c) is the insertion of an element to the left of the tagged element. Now, the cost of searching for it increases by 1, and therefore we multiply by z. We have ℓ positions where the element may hash. Following a similar analysis as for the previous cases we have

$$F_{\ell,r}(z) \leftarrow \ell z \sum_{0 \le k \le r} \binom{\ell+r}{k}(k+1)^{k-1}F_{\ell-1,r-k}(z) \tag{16}$$

Case d) is the insertion of an element to the right of $\bullet$. Again, in this case the cost of searching for $\bullet$ does not increase. We have $r - k$ positions where the element may hash. Therefore,

$$F_{\ell,r}(z) \leftarrow \sum_{0 \le k \le r-1} \binom{\ell+r}{k}(k+1)^{k-1}F_{\ell,r-k-1}(z)(r-k) \tag{17}$$

From (14),(15),(16) and (17) we can derive a recurrence for $F_{\ell,r}(z)$. Then, if we use the definition of $G_i(z)$ and $C_i(z)$ we arrive at the following recurrence for $G_i(z)$:

$$G_i(z) = \sum_k \binom{i}{k}(k+1)^{k-1}\left(z\frac{(i-k+1)^{i-k}(i-k+2)}{2} + (i+3)G_{i-k-1}(z)\right.$$

$$\left. + (k+1)^2 C_{i-k-1}(z) + (z-1)\sum_{\ell+r=i-k-1}(\ell+1)(\ell+2)F_{\ell,r}(z)\right) \tag{18}$$

Later we will require the value of $\mathbf{U}_z\mathbf{D}_zG_i(z)$. If we differentiate each side of (18) and evaluate at $z = 1$, we have the following recurrence

$$\mathbf{U}_z\mathbf{D}_zG_i(z) = \frac{(i+2)^{i+2}}{2} + \frac{(i+2)^i}{6} - \frac{7(i+2)^{i+1}}{12}Q_0(i+2,i)$$

$$+ (i+3)\sum_{k\ge0}\binom{i}{k}(k+1)^{k-1}\mathbf{U}_z\mathbf{D}_zG_{i-k-1}(z) \tag{19}$$

In order to solve the above recurrence we introduce our *diagonal Poisson transform*.

4   The Diagonal Poisson Transform

We define $\mathcal{D}_c[f(n); x] = \hat{f}_c(x)$, the *diagonal Poisson transform* of $f(n)$ as

$$\mathcal{D}_c[f(n); x] = (1-x)\sum_{n\ge0}e^{-(n+c)x}\frac{((n+c)x)^n}{n!}f(n) \tag{20}$$

The name diagonal Poisson transform comes from the similarity to the Poisson transform. If we consider an infinite matrix where the rows represent the values of m and the columns represent the values of n, we may easily see the relationship. The Poisson transform has m fixed, while n varies from 0 to infinity; hence, it follows a row of this matrix. The diagonal Poisson transform, has the property that $m - n = c$, where c is a constant. Therefore, it follows a principal diagonal of the matrix.

The other main difference is the factor $(1 - x)$, that does not appear in the Poisson transform. We include this factor, because, with this adjustment, some of the most important properties of the diagonal Poisson transform are the same as for the Poisson transform when $m = n + c$. However, while in [8] an interpretation of the Poisson transform is presented, we were unable to find a similar interpretation for the diagonal Poisson transform. We can show a number of useful properties of this transform, including the following:

$$\mathcal{D}_c[\alpha f(n) + \beta g(n); x] = \alpha \dot{f}_c(x) + \beta \dot{g}_c(x) \quad \alpha, \beta \text{ constants} \tag{21}$$

$$\mathcal{D}_c\left[\frac{f(n)}{n+1}; x\right] = \frac{e^{-(c-1)x}(1-x)}{x} \int_0^x e^{(c-1)t} \mathcal{D}_c[f(n); t] dt \tag{22}$$

$$(1 + (c-1)x)\frac{\mathcal{D}_c[f(n); x]}{1-x} + x\frac{\partial \frac{\mathcal{D}_c[f(n); x]}{1-x}}{\partial x} = \mathcal{D}_c[(n+1)f(n); x] \tag{23}$$

$$\frac{\partial\left(x^c \frac{\mathcal{D}_c[f(n); x]}{1-x}\right)}{\partial x} = x^{c-1}\mathcal{D}_c[(n+c)f(n); x] \tag{24}$$

$$\mathcal{D}_c[1; x] = 1 \tag{25}$$

$$\mathcal{D}_c[Q_r(n+c, n); x] = \frac{1}{(1-x)^{r+1}} \tag{26}$$

$$\mathcal{D}_c\left[\frac{n^{\underline{k}}}{(n+c)^k}; x\right] = x^k \tag{27}$$

Our main theorem follows easily from these facts.
Inversion Theorem. *If $\mathcal{D}_c[f(n); x] = \sum_{k\geq 0} a_k x^k$ is the diagonal Poisson transform of $f(n)$ then $f(n) = \sum_{k\geq 0} a_k \frac{n^{\underline{k}}}{(n+c)^k}$*

Proof. By (21) and (27) we know

$$\mathcal{D}_c\left[\sum_{k\geq 0} a_k \frac{n^{\underline{k}}}{(n+c)^k}; x\right] = \sum_{k\geq 0} a_k \mathcal{D}_c\left[\frac{n^{\underline{k}}}{(n+c)^k}; x\right] = \sum_{k\geq 0} a_k x^k = \mathcal{D}_c[f(n); x]\square$$

A trivial but useful corollary of the Inversion Theorem is the following inversion formula

Corollary 2. $\frac{(-1)^n}{n!}(n+c)\sum_{k\geq 0}(-1)^k \binom{n}{k}(k+c)^{n-1}b_k = a_n \Leftrightarrow b_n = \sum_{k\geq 0} a_k \frac{n^{\underline{k}}}{(n+c)^k}$ $\square$

A useful consequence of the Inversion Theorem is:
Transfer Theorem. *Let $\tilde{a}_m(x) = \mathcal{P}_m[f(m, n); x]$ and $\dot{b}_c(x) = \mathcal{D}_c[f(n+c, n); x]$ then $\tilde{a}_m(x) = \dot{b}_c(x)$ if and only if $\tilde{a}_m(x)$ does not depend on m.*

Proof. The necessity condition is trivial: if $\tilde{a}_m(x)$ depends on m, then it cannot be equal to $\dot{b}_c(x)$, because the latter function does not depend on m.

Now suppose $\tilde{a}_m(x) = \tilde{a}(x)$. We may also suppose that $\tilde{a}(x) = \sum_{k\geq 0} a_k x^k$ and $\dot{b}_c(x) = \sum_{k\geq 0} b_k x^k$. Then by Theorem 1 and the Inversion Theorem we know $f(m, n) = \sum_{i\geq 0} a_i n^{\underline{i}}/m^i$ and $f(n+c, n) = \sum_{i\geq 0} b_i n^{\underline{i}}/(n+c)^i$. Then, if we substitute $m = n+c$ in the first sum we obtain $f(n+c, n) = \sum_{i\geq 0} a_i n^{\underline{i}}/(n+c)^i$. Therefore, we have two expansions for $f(n+c, n)$. Both expansions are rational functions in n with the same denominator. Hence, the numerators should be equal. As both numerators are polynomials in n, their coefficients should be equal. Therefore, $a_i = b_i$ for $i \geq 0$. As a consequence we have $\tilde{a}(x) = \tilde{b}_c(x) = \tilde{b}(x)$. $\qquad \square$

As a very nice consequence of Theorem 1, the Inversion Theorem, and the Transfer Theorem, we have:

Corollary 3. *The set of all functions $f(m, n)$ that verify the conditions of the Transfer Theorem are $\sum_{k \geq 0} a_k \frac{n^k}{m^k}$ where a_k does not depend on m.* $\qquad \square$

Let $\tilde{a}(x) = \mathcal{P}_m[f(m, n); x]$ and $\tilde{b}(x) = \mathcal{D}_c[f(n + c, n); x]$, and suppose $\tilde{a}(x) = \tilde{b}(x)$. If we consider the Taylor expansion of $e^{mx} \tilde{a}(x)$ and $e^{mx} \tilde{b}(x)$, then the coefficients of x^n from both expansions should be equal. As a consequence we have the following equation

$$\sum_{k=0}^{n} \frac{m^k}{k!} f(m, k) = \frac{1}{n!} \sum_{k=0}^{n} \binom{n}{k} (k + c)^k (m - c - k)^{n-k} f(k + c, k) \qquad (28)$$

Hence, Corollary 3 gives the complete solution to (28).

5 Verification of Known Results

In this section we rewrite (12) as a function of $\mathcal{D}_2[g_i(z); x]$ and then verify that $\mathbf{E}[A_{n+1}] = \frac{1}{2}(1 + Q_0(m, n))$. If we denote $\mathring{g}_2(x, z) = \mathcal{D}_2[g_i(z); x]$, where $g_i(z) = \frac{G_i(z)}{(i+2)^i(i+1)}$, then

$$\frac{\partial(x \tilde{P}_m(x, z))}{\partial x} = z + (z - 1)x \mathring{g}_2(x, z)$$

Therefore we derive the equation

$$\tilde{P}_m(x, z) = \frac{1}{x} \int_0^x (z + t(z - 1)\mathring{g}_2(t, z)) \, dt = z + \frac{z - 1}{x} \int_0^x t \mathring{g}_2(t, z) dt \qquad (29)$$

Taking derivatives with respect to z we obtain

$$\mathbf{U}_z \mathbf{D}_z \tilde{P}_m(x, z) = 1 + \frac{1}{x} \int_0^x t \mathbf{U}_z \mathring{g}_2(t, z) dt \qquad (30)$$

$$\mathbf{U}_z \mathbf{D}_z^2 \tilde{P}_m(x, z) = \frac{2}{x} \int_0^x t \mathbf{U}_z \mathbf{D}_z \mathring{g}_2(t, z) dt \qquad (31)$$

As we know by (13) that $\mathbf{U}_x g_i(z) = (i + 2)/2$, then $\mathbf{U}_x \mathring{g}_2(x, z) = \mathbf{U}_x \mathcal{D}_2[g_i(z); x] = \mathcal{D}_2\left[\frac{i+2}{2}; x\right]$. By (24) we know $\mathcal{D}_2\left[\frac{(i+2)}{2}; x\right] = \frac{1}{2}\left(\frac{1}{(1-x)^2} + \frac{1}{(1-x)}\right)$. Therefore, if we substitute into (30) and integrate, we find that $\mathbf{U}_z \mathbf{D}_z \tilde{P}_m(x, z) = \frac{1}{2}\left(1 + \frac{1}{1-x}\right)$.

As $1/(1 - x)$ is the Poisson transform of $Q_0(m, n)$, we have given an alternative proof to that of [11].

6 Solving Recurrences with the Diagonal Poisson Transform

In (29), we wrote $\tilde{P}_m(x, z)$ as a function of $\mathcal{D}_2[g_i(z); x]$, and in (30) and (31) we found $\mathbf{U}_z \mathbf{D}_z \tilde{P}_m(x, z)$ and $\mathbf{U}_z \mathbf{D}_z^2 \tilde{P}_m(x, z)$ as functions of $\mathbf{U}_z \dot{g}_2(x, z)$ and $\mathbf{U}_z \mathbf{D}_z \dot{g}_2(x, z)$. However, we still do not know the value of $\mathbf{U}_z \mathbf{D}_z \dot{g}_2(x, z)$. In this section, by applying the diagonal Poisson transform to both sides of (19), we find $\mathbf{U}_z \mathbf{D}_z \dot{g}_2(x, z)$ and $\mathbf{U}_z \mathbf{D}_z^2 \tilde{P}_m(x, z)$. Finally, we find $\mathbf{U}_z \mathbf{D}_z^2 P_{m,n}(z)$. Hence, we may now use formula (6) to find the variance of the number of probes for a successful search.

We require a solution to a recurrence of the form:

$$H_i = B_i + \sum_{k \geq 0} \binom{i}{k}(k+1)^{k+p}(i+d)H_{i-k-1} \tag{32}$$

Writing $h_i = \frac{H_i}{(i+c)^i(i+1)}$ and $b_i = \frac{B_i}{(i+c)^i(i+1)}$ we are to solve

$$h_i = b_i + \sum_{k \geq 0} \binom{i}{k}(k+1)^{k+p}\frac{i+d}{(i+c)^i(i+1)}(i-k+c-1)^{i-k-1}(i-k)h_{i-k-1}$$

$$= b_i + \left(1 + \frac{d-1}{i+1}\right)a_i \tag{33}$$

where a_i denotes the factor that multiplies $\frac{i+d}{i+1}$. Applying the diagonal Poisson transform to both sides of (33) we get

$$\dot{h}_c(x) = \dot{b}_c(x) + \mathcal{D}_c[a_i; x] + (d-1)\,\mathcal{D}_c\left[\frac{a_i}{i+1}; x\right]$$

If we find the values of $\mathcal{D}_c[a_i; x]$ and $\mathcal{D}_c[\frac{a_i}{i+1}; x]$, we arrive at the following integral equation:

$$\dot{h}_c(x) = \dot{b}_c(x) + s_p(x)\dot{h}_c(x) + \frac{(d-1)e^{-(c-1)x}(1-x)}{x}\int_0^x e^{(c-1)t}s_p(t)\dot{h}_c(t)dt$$

where $s_p(x) = x\sum_{n \geq 0}\left\{{n+p+1 \atop n+1}\right\}x^n$, and $\left\{{n \atop k}\right\}$, as in [9], denotes the Stirling numbers of the second kind. Solving the integral equation and using (23), we have the following solution

$$\dot{h}_c(x) = \frac{(1-x)e^{(d-c)x}}{x^d(1-s_p(x))}e^{(d-1)A(x)}\int_0^x x^{d-1}e^{(c-d)t}e^{-(d-1)A(t)}\mathcal{D}_c[(i+1)b_i; t]dt$$

where $A(x) = \int_{t=1}^x (1-t)/(t(1-s_p(t))dt$.

We are now able to solve (19), a special case of (32). In our case $c = 2$, $d = 3$ and $p = -1$. As $p = -1$ then $s_p(x) = x$. Therefore, the general solution simplifies to

$$\dot{h}_2(x) = \frac{e^x}{x}\int_0^x e^{-t}\mathcal{D}_2[(i+1)b_i; t]dt \tag{34}$$

In (31) $\dot{h}_2(x) = \mathbf{U}_z \mathbf{D}_z \dot{g}_2(x, z)$. Applying (34) to (31):

$$\mathbf{U}_z \mathbf{D}_z^2 \tilde{P}_m(x, z) = \frac{2}{x} \int_0^x (e^{x-t} - 1) \, \mathcal{D}_2[(i+1)b_i; t] dt \tag{35}$$

In (19) we have $(i+1)b_i = \frac{(i+2)^2}{2} + \frac{1}{6} - \frac{7(i+2)}{12} Q_0(i+2, i)$. Using properties (24),(25) and (26) for $c = 2$, we arrive at the final result:

$$\mathbf{U}_z \mathbf{D}_z^2 \tilde{P}_m(x, z) = \frac{2}{x} \int_0^x (e^{x-t} - 1) \left(\frac{3}{2(1-t)^4} - \frac{2}{3(1-t)^3} + \frac{1}{6} \right) dt$$

$$= \frac{1}{3(1-x)} + \frac{1}{2(1-x)^2} - \frac{1}{3} - \frac{1}{3x}(e^x - 1) - \frac{e^{x-1}}{6x} \int_{1-x}^1 \frac{e^t}{t} dt \tag{36}$$

Now, we apply the inversion formulae presented in [8] to find $\mathbf{U}_z \mathbf{D}_z^2 P_{m,n}(z)$. As the Poisson transform is linear, we need only to find the inverse of each summand of (36).

Lemma 4. $\mathbf{U}_z \mathbf{D}_z^2 P_{m,n}(z) = \frac{1}{3} Q_0(m, n) + \frac{1}{2} Q_1(m, n) - \frac{1}{3} - \frac{m+1}{3(n+1)} \left(\frac{m+1}{m} \right)^n + \frac{m}{3(n+1)}$

$-\frac{1}{6} \left(\frac{m+1}{m} \right)^n \frac{1}{n+1} \sum_{k=0}^n \left(\frac{m}{m+1} \right)^k Q_0(m, k)$ ☐

7 Analysis of the Variance

As a consequence of (6) and Lemma 4 we have the following

Theorem 5. $\mathbf{V}[A_{n+1}] = \frac{1}{2} Q_1(m, n) - \frac{1}{4} Q_0^2(m, n) + \frac{1}{3} Q_0(m, n) - \frac{m+1}{3(n+1)} \left(\frac{m+1}{m} \right)^n$

$-\frac{1}{12} + \frac{m}{3(n+1)} - \frac{1}{6} \left(\frac{m+1}{m} \right)^n \frac{1}{n+1} \sum_{k=0}^n \left(\frac{m}{m+1} \right)^k Q_0(m, k)$ ☐

Using the approximation theorem presented in [13], we have the following result for a table with $n = \alpha m$ elements, for fixed $0 \le \alpha < 1$ and $n, m \to \infty$

Theorem 6. $\mathbf{V}[A_{\alpha m}] = \frac{1}{4(1-\alpha)^2} + \frac{1}{3(1-\alpha)} - \frac{1}{3\alpha}(e^\alpha - 1) - \frac{e^{\alpha-1}}{6\alpha} \int_{1-\alpha}^1 \frac{e^t}{t} dt - \frac{1}{12} + O\left(\frac{1}{m} \right)$ ☐

Studying the asymptotic behavior of the variance for a full table ($n = m - 1$), we obtain:

Theorem 7. $\mathbf{V}[A_m] = \frac{4-\pi}{8} m + \frac{\sqrt{2\pi m}}{4} - \frac{1}{12} H_m + \left(\frac{1}{9} - \frac{\ln(2)}{12} - \frac{\pi}{48} - \frac{Ei(1)}{6} - \frac{e}{3} + \frac{\gamma}{6} \right)$

$+ \frac{181}{2160} \sqrt{\frac{2\pi}{m}} + o\left(\frac{1}{\sqrt{m}} \right)$ ☐

where $Ei(x)$ is the exponential integral function (See [1]). Comparing with (3), we have shown that for a full table, the last-come-first-served heuristic on a linear probing hash table achieves the optimal variance for the distribution of successful searches, up to lower order terms.

8 Conclusions and Future Work

We have proved that the LCFS heuristic on a linear probing hash table achieves the optimal variance for the distribution of successful searches, up to lower order terms. In the development of the proof, we introduced the diagonal Poisson transform that was very helpful to solve the main recurrence. We have presented different classes of recurrences that can be nicely solved using this transform. We would like to find other classes of problems that can be solved naturally with the use of the diagonal Poisson transform.

Acknowledgements

We thank Bruce Richmond for suggesting several important ideas leading to the proof of Theorem 7. We also thank to the creators of Maple [4]. Most of the formulae we have derived in this paper have been checked with this system.

References

1. M. Abramowitz and I.A. Stegun. *Handbook of Mathematical Functions*. Dover Publications, New York, 1972.
2. O. Amble and D.E. Knuth. Ordered hash tables. *Computer Journal*, 17(2):135–142, 1974.
3. R.P. Brent. Reducing the retrieval time of scatter storage techniques. *C.ACM*, 16(2):105–109, 1973.
4. B.W.Char, K.O.Geddes, G.H.Gonnet, B.L.Leong, M.B.Monagan, and S.M.Watt. *MAPLE V Reference Manual*. Springer-Verlag, 1991.
5. S. Carlsson, J.I. Munro, and P.V. Poblete. On linear probing hashing. Unpublished Manuscript.
6. P. Celis, P.-A. Larson, and J.I. Munro. Robin Hood hashing. In *26th IEEE Sympusium on the Foundations of Computer Science*, pages 281–288, 1985.
7. G.H. Gonnet and J.I. Munro. Efficient ordering of hash tables. *SIAM Journal on Computing*, 8(3):463–478, 1979.
8. G.H. Gonnet and J.I. Munro. The analysis of linear probing sort by the use of a new mathematical transform. *Journal of Algorithms*, 5:451–470, 1984.
9. R.L. Graham, D.E. Knuth, and O.Patashnik. *Concrete Mathematics*. Addison-Wesley, 1989.
10. D.E. Knuth. *The Art of Computer Programming: Fundamental Algorithms*, volume 1. Addison-Wesley, 1973.
11. D.E. Knuth. *The Art of Computer Programming: Sorting and Searching*, volume 3. Addison-Wesley, 1973.
12. D.E. Knuth and A. Schönhage. The expected linearity of a simple equivalence algorithm. *Theoretical Computer Science*, 6:281–315, 1978.
13. P.V. Poblete. Approximating functions by their poisson transform. *Information Processing Letters*, 23:127–130, 1986.
14. P.V. Poblete and J.I. Munro. Last-come-first-served hashing. *Journal of Algorithms*, 10:228–248, 1989.

Some Lower Bounds for Comparison-Based Algorithms

Svante Carlsson   Jingsen Chen

Department of Computer Science, Luleå University, S–971 87 Luleå, Sweden

Abstract. Any comparison-based algorithm for solving a given problem can be viewed as a partial order production. The productions of some particular partial orders, such as sorting and selection, have received much attention in the past decades. As to general partial orders, very little is known about the inherent complexity of their productions. This paper investigates how different sequences of comparisons affect the complexity of the production.

We first disprove a folk theorem stating that there always exists an optimal algorithm for producing a partial order that involves the maximum number of disjoint comparisons between singleton elements. We also present some algorithmic properties of partial orders, which demand that any optimal production algorithm must make the all possible disjoint comparisons. Moreover, we show how to and when one can save comparisons in producing a given partial order by constructing many isomorphic copies of the partial order simultaneously or by providing extra elements. Furthermore, the techniques obtained and the results presented have been successful applied to investigate the problem of constructing data structures, resulting in efficient algorithms and tight lower bounds.

1 Introduction

A large number of problems in algorithm theory involve relative ordering between elements. One of the frame-works for determining the inherent complexity of the problems is the so-called partial order production, initiated by Schönhage [12]. For example, the traditional sorting is to produce a linear order on the set of singleton elements. The selection problem can be thought of as producing a partial order from which the kth smallest element can be deduced without further comparisons. In general, any comparison-based algorithm can be viewed from the perspective of partial order productions. Hence, the attempt to find optimal comparison-based algorithms becomes an investigation of the partial order productions.

Given a set, S, of elements drawn from some ordered domain, the *partial order production* problem is to produce a partial order $\mathcal{P}$ on S which is initially configured as some partial order $\mathcal{Q}$. Any comparison-based algorithm $\mathcal{A}$ for the production can be represented by a binary decision tree $\mathcal{T}$. We say that the algorithm $\mathcal{A}$ *produces* $\mathcal{P}$ from $\mathcal{Q}$ if $\mathcal{P}$ is a *suborder* of every *end-poset* of $\mathcal{T}$, which is the partial order at each leaf of $\mathcal{T}$ consistent with all comparisons made on the internal path from the root to the leaf. That is, one can determine $\mathcal{P}$

from each end-poset of T with no further comparisons. Denote the *minimax complexity* of producing $\mathcal{P}$ starting from Q by $C(Q \to \mathcal{P})$. We just write $C(\mathcal{P})$ instead of $C(Q \to \mathcal{P})$ when Q is an antichain. See [1,12–14] for the definitions.

For a given partial order, the main interest is to answer the question of how many comparisons are necessary and sufficient to produce it. Apart from some work done in [1,12,14], very little is known about the inherent complexity of the production problem. In the general case it seems very hard, or impossible, to determine the *exact* production cost. Even for concrete partial orders, there are numerous open problems. For example, the worst-case number of comparisons required to find the median of an n-element set is between $2n$ and $3n$ [3,13].

Folk Theorems on Productions

Perhaps as a result of the difficulty of determining the inherent complexity of partial order production, intuitions concerning the complexity and algorithmic aspects of the production have been proposed. One of these "folk theorems", known as "$\lfloor \frac{n}{2} \rfloor$ singleton-comparisons", states that for any connected partial order on n elements, there always exists an optimal production algorithm that starting by making $\lfloor \frac{n}{2} \rfloor$ disjoint comparisons between singletons [1,4]. This "folklore" has permeated into the study of comparison-based problems and no counter-example to this conjecture was previously known. Surprisingly, we now know that this conjecture is not true in general, by presenting a class of partial orders for which no optimal production algorithm performs maximum number of disjoint comparisons. Moreover, we examine algorithmic properties of partial orders with respect to sequences of comparisons made and give conditions under which the "$\lfloor \frac{n}{2} \rfloor$ singleton-comparisons" are unavoidable in order to produce partial order optimally. Furthermore, we are able to use these properties to obtain improved lower bounds for specific partial orders.

Mass Productions and Size-Sensitivity

Given a partial order $\mathcal{P}$ on n elements, one finds that producing k partial orders on different element-sets of size n all isomorphic to $\mathcal{P}$, called the *mass production* of $\mathcal{P}$, does not necessarily require k times the cost to produce a single $\mathcal{P}$ on n elements. The original motivation for studying the mass production of partial orders is derived from the selection problem [13]. There are a few partial orders for which the mass production costs less than that of producing the copies of the partial order separately. For example, Schönhage [12] gave some examples of such partial orders (called *non-additive* partial orders). However, no non-trivial result for mass production was known. In Section 3, we shall demonstrate interesting properties concerning mass productions and related problems. The characterizations obtained also enable us to exhibit an infinite family of partial orders from which we can benefit through mass productions. We also apply the mass production technique to heap-like structures, resulting in efficient algorithms.

The existence of non-additive partial orders is hardly surprising. What is surprising is perhaps the fact that there is a special class of partial orders, called *size-sensitive* partial orders, for which the production cost decreases if some extra elements are available. Similar to non-additive partial orders, examples of size-sensitive partial orders are very few. Paterson [12] provided a size-sensitive partial order on 7 elements. The study of size-sensitivity is important in understanding the intrinsic complexity of partial order production. However, we are not aware of any other example of such partial orders and of any work of characterizing them. By demonstrating algorithmic properties of size-sensitive partial orders, we successfully identify a large class of partial orders as size-sensitive.

2 Behavior of Optimal Production Algorithms

It has often been suggested, although not been proved, that there always exists an optimal algorithm, for producing any connected poset of size n, that involves $\lfloor \frac{n}{2} \rfloor$ disjoint comparisons between singleton elements [1,4]. The results known for specific partial orders [2,4,11] provide justification for this intuition. Moreover, when optimally producing partial orders with special properties, we might expect that making as many comparisons between singletons as possible would generate maximum ordering information. However, this is not always the case. We will show, surprisingly, that there exists posets for which this conjecture is not true. Consider posets $\mathcal{P}_k$ on $n = 2^k + 2$ $(k \geq 2)$ elements that connects two binomial trees of size 2^{k-1} plus two elements that hang below the roots of the binomial trees, respectively. The simplest instance of $\mathcal{P}_k$ (i.e., $k = 2$) is shown in Figure 1. Let the *sum*, $\mathcal{P} + \mathcal{Q}$, of partial orders $\mathcal{P}$ and $\mathcal{Q}$ on disjoint sets S_1 and S_2, respectively, be a partial order on the union of the sets S_1 and S_2 such that $x \prec y$ in $\mathcal{P} + \mathcal{Q}$ if either $x \prec y$ in $\mathcal{P}$ for $x, y \in S_1$ or $x \prec y$ in $\mathcal{Q}$ for $x, y \in S_2$.

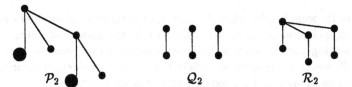

$$\mathcal{P}_2 \qquad\qquad \mathcal{Q}_2 \qquad\qquad \mathcal{R}_2$$

FIGURE 1: A counter example of the singleton-comparison conjecture.

Theorem 1. *Let $\mathcal{B}_m^{(i)}$ be a binomial tree of size 2^m with root $r_m^{(i)}$ $(i = 1, 2)$ and*

$$\mathcal{P}_k = \mathcal{B}_{k-1}^{(1)} + \mathcal{B}_{k-1}^{(2)} + \{x\} + \{y\} + \left\{ r_{k-1}^{(1)} \prec r_{k-1}^{(2)} \right\} + \left\{ r_{k-1}^{(1)} \prec x \right\} + \left\{ r_{k-1}^{(2)} \prec y \right\}.$$

Then, any comparison-based algorithm for producing the poset $\mathcal{P}_k$ that makes $\lfloor \frac{n}{2} \rfloor$ disjoint comparisons between singletons is not optimal.

Proof. It is easy to verify that $\mathcal{C}(\mathcal{P}_k) = n - 1 = 2^k + 1$. Observe first that partial order $\mathcal{B}'_m = \mathcal{B}_m + \{z\} + \{r_m \prec z\}$ can be produced in 2^m comparisons, where $\mathcal{B}_m$ is a binomial tree of size 2^m with root r_m. This is done simply by first building $\mathcal{B}_m$ (Cost: $2^m - 1$ comparisons) and then comparing r_m with z (Cost: 1 comparison). Finally, the partial order $\mathcal{P}_k$ can be produced by first constructing two partial orders $\mathcal{S}_1$ and $\mathcal{S}_2$ that are isomorphic to $\mathcal{B}'_{k-1}$ and then comparing their roots r_{s_1} and r_{s_2}. That is, either $\mathcal{P}_k = \mathcal{S}_1 + \mathcal{S}_2 + \{r_{s_1} \prec r_{s_2}\}$ or symmetrically $\mathcal{P}_k = \mathcal{S}_2 + \mathcal{S}_1 + \{r_{s_2} \prec r_{s_1}\}$. The total cost to produce $\mathcal{P}_k$ is $2^{k-1} + 2^{k-1} + 1 = n - 1$ comparisons, which is optimal since $\mathcal{P}_k$ is connected.

We shall show below that if we perform $\lfloor \frac{n}{2} \rfloor$ $(= 2^{k-1}+1)$ comparisons between the singletons while producing $\mathcal{P}_k$, then $n - 1$ comparisons are not enough in order to solve the problem. First, notice that $\lfloor \frac{n}{2} \rfloor$ $(2^{k-1}+1)$ singleton-comparisons will build a partial order $\mathcal{Q}_k$ that consists of $\lfloor \frac{n}{2} \rfloor$ ordered pairs of elements from an order-theoretic point of view; as shown in Figure 1 for $k = 2$. We can think, without loss of generality, that an algorithm starts with $\lfloor \frac{n}{2} \rfloor$ $(2^{k-1} + 1)$ comparisons between singletons, since otherwise the algorithm can rearrange the order of making comparisons between singletons and achieve the same complexity. Now, to produce $\mathcal{P}_k$ from the partial order $\mathcal{Q}_k$, we need $2^{k-1} + 1$ additional comparisons. In fact, 2^{k-1} comparisons between the minimal elements of $\mathcal{Q}_k$ are necessary to find the minimum in the set. As to these comparisons between the minima, an adversary is established that selects the element with a larger *down-set* (i.e., the set of all elements that is greater than the given element) to be the winner of a comparison. Under this adversary, the best poset that can be produced by those 2^{k-1} comparisons between the minimal elements is the one of the form $\mathcal{R}_k = \mathcal{B}_k + \mathcal{B}_2 + \{r_k \prec r_2\}$, where again $\mathcal{B}_i$ denotes a binomial tree of size 2^i with root r_i; see Figure 1 for the case when $k = 2$. Notice that $\mathcal{P}_k$ cannot be deduced from $\mathcal{R}_k$ without further comparisons. Hence, at least one more comparison is needed to convert the resulting poset $\mathcal{R}_k$ into $\mathcal{P}_k$. The total cost for this algorithm will be $2^{k-1} + 1 + 2^{k-1} + 1 = 2^k + 2 = n$ comparisons. Therefore, the method that compares $\lfloor \frac{n}{2} \rfloor$ disjoint pairs of singletons is more expensive for constructing $\mathcal{P}_k$ than its inherent complexity.

The above theorem implies that the information theoretical production method, which always chooses the next comparison to cut the number of linear extensions of the resulting partial order in half, or as nearly so as possible, is not always optimal. However, there are cases where the optimal production algorithms have to perform as many disjoint singleton-comparisons as possible.

Proposition 2. *To be able to produce any n-element connected poset $\mathcal{P}$ in $\lceil \frac{3}{2}n \rceil - \|Max(\mathcal{P})\| - \|Min(\mathcal{P})\|$ comparisons, we must perform $\lfloor \frac{n}{2} \rfloor$ comparisons between singletons during the course of the production, where $Max(\mathcal{P})$ and $Min(\mathcal{P})$ are the sets of all maximal and minimal elements of $\mathcal{P}$, respectively.*

Proof. A rigorous analysis of the adversary in [1], which is used to prove that $\mathcal{C}(\mathcal{P}) \geq \lceil \frac{3}{2}n \rceil - \|Max(\mathcal{P})\| - \|Min(\mathcal{P})\|$ for an n-element poset $\mathcal{P}$, will lead to our claim. Suppose that the algorithm is up to compare x with y and $\mathcal{Q}$ is the partial order that has been generated by the algorithm just before the comparison $(x : y)$. The strategy for the adversary in [1] is to select the outcome $x \prec y$ if $x \in Min(\mathcal{Q})$ and $y \notin Min(\mathcal{Q})$, or if $x \notin Max(\mathcal{Q})$ and $y \in Max(\mathcal{Q})$; and make arbitrary choices consistent with $\mathcal{Q}$ in all other cases. With this adversary, the total number $\|Max(\mathcal{Q})\| + \|Min(\mathcal{Q})\|$ is reduced by at most 1 after each comparison made, unless two singletons are being compared in which case this number drops by 2. Initially, Q is an antichain of size n and $\|\mathrm{Max}(Q)\| + \|\mathrm{Min}(Q)\| = 2n$. When $\mathcal{P}$ has been produced, the poset Q clearly contains $\mathcal{P}$ as a suborder. Thus $\|\mathrm{Max}(Q)\| + \|\mathrm{Min}(Q)\| \le \|\mathrm{Max}(\mathcal{P})\| + \|\mathrm{Min}(\mathcal{P})\|$. Suppose that an optimal algorithm performs k comparisons between singletons and m other types of comparisons. Then, $2n - 2k - m \le \|\mathrm{Max}(Q)\| + \|\mathrm{Min}(Q)\| \le \|\mathrm{Max}(\mathcal{P})\| + \|\mathrm{Min}(\mathcal{P})\|$. Consequently,

$$C(\mathcal{P}) = k + m \ge 2n - 2k - \|\mathrm{Max}(\mathcal{P})\| - \|\mathrm{Min}(\mathcal{P})\| + k > \lceil \tfrac{3}{2} n \rceil - \|\mathrm{Max}(\mathcal{P})\| - \|\mathrm{Min}(\mathcal{P})\|$$

and hence $k \ge \lfloor \tfrac{n}{2} \rfloor$. This proves the proposition.

One of the cases to which the proposition has been applied successfully is when $\|\mathrm{Max}(\mathcal{P})\| = \|\mathrm{Min}(\mathcal{P})\| = 1$.

Corollary 3. *Any algorithm $\mathcal{A}$ for finding the minimum and maximum elements always makes $\lfloor \tfrac{n}{2} \rfloor$ comparisons between singletons, $\lceil \tfrac{n}{2} \rceil - 1$ comparisons between the minimal elements, and $\lceil \tfrac{n}{2} \rceil - 1$ comparisons between the maxima.*

We have successful applied the optimal algorithmic behaviors incorporated in adversaries to investigate the problem of constructing heap-like structures [7].

Proposition 4. *In the worst case,*

1. *building a min-max heap on 10 elements costs exactly 13 comparisons;*
2. *building a deap on 6 elements costs exactly 8 comparisons; and*
3. *building a heap on 7 elements costs exactly 8 comparisons.*

The combination of different adversary arguments we adapted proves to be advantageous and establishes a new method of deriving lower bounds. Previously, only enumerating or exhaustive searching methods were available.

3 Additive Partial Orders and Mass Productions

Despite of the difficulty of determining the inherent complexity of partial order production, it is possible that partial orders with special decomposition properties may admit efficient ways of estimating their production costs. For a partial order representable as the sum of disjoint posets, namely $\mathcal{P} = \mathcal{P}_1 + \mathcal{P}_2 + \cdots + \mathcal{P}_k$, it has been observed that

$$C(\mathcal{P}_1 + \mathcal{P}_2 + \cdots + \mathcal{P}_k) \le C(\mathcal{P}_1) + C(\mathcal{P}_2) + \cdots + C(\mathcal{P}_k) \qquad (1)$$

In fact, $\mathcal{P}$ can be produced by constructing each $\mathcal{P}_i$ separately. Posets for which the equality in (1) holds are interesting due to the fact that the production complexity of such posets can be studied by investigating the complexity of producing simpler and smaller connected components. Such posets do exist. For example, $\mathcal{P} = \mathcal{P}_1 + \mathcal{P}_2 + \cdots + \mathcal{P}_k$, where $C(\mathcal{P}_i) = \log \frac{(\|\mathcal{P}_i\|)!}{\ell(\mathcal{P}_i)}$ for all $1 \le i \le k$, where $\ell(\mathcal{P}_i)$ denotes the number of linear extensions of the partial order $\mathcal{P}_i$. One should be aware that the inequality (1) can be strict for special cases. Consider two 6-element posets $\mathcal{P}$ and $\mathcal{Q}$ shown in Figure 2. It is easy to verify that $\mathcal{C}(\mathcal{P}) = \mathcal{C}(\mathcal{Q}) = 6$. Interestingly, if we produce $\mathcal{P}$ and $\mathcal{Q}$ simultaneously, then the total cost will be at most 11 comparisons. In fact, after 4 comparisons which are made to create two inverse binomial trees of sizes 4 and 2, respectively, we compare elements x with y (see Figure 2). Either poset $\mathcal{P}$ or $\mathcal{Q}$ can be produced according to whether $x \prec y$ or not. The other one can then be produced with 6 comparisons more. Therefore, $\mathcal{C}(\mathcal{P} + \mathcal{Q}) \leq 11$. Notice that $\ell(\mathcal{P}) = 19$ and $\ell(\mathcal{Q}) = 26$. Hence, we have from the information-theoretic lower bound [12] that $\mathcal{C}(\mathcal{P} + \mathcal{Q}) \geq \lceil \log \frac{12!}{\ell(\mathcal{P}+\mathcal{Q})} \rceil = 11$. Consequently, $\mathcal{C}(\mathcal{P} + \mathcal{Q}) = 11$.

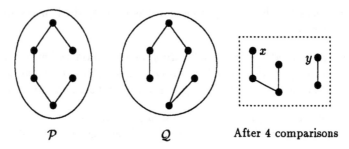

$\mathcal{P}$ $\mathcal{Q}$ After 4 comparisons

FIGURE 2: Posets with additivity.

An important special case occurs when producing simultaneously many isomorphic copies of a partial order, called the *mass production* of the partial order. There exist partial orders whose mass production can cost less than that of successively building the copies of the partial order separately, as will be shown shortly. We will refer to such a partial order as a *non-additive partial order* (as opposed to an *additive* partial order). More precisely, a partial order $\mathcal{P}$ is non-additive if there exists an integer $k \geq 2$ such that

$$C(k \cdot \mathcal{P}) \triangleq \mathcal{C}(\overbrace{\mathcal{P} + \cdots + \mathcal{P}}^{k \text{ times}}) < \overbrace{\mathcal{C}(\mathcal{P}) + \cdots + \mathcal{C}(\mathcal{P})}^{k \text{ times}}$$

The benefits of the mass production of a partial order have been noted and applied to develop the fastest known median-finding algorithm [13]. See the proposition below for some examples of non-additive partial orders.

Proposition 5. $\mathcal{C}(\mathcal{P} + \mathcal{P}) = \mathcal{C}(\mathcal{P}) + \mathcal{C}(\mathcal{P}) - 1$ if $\mathcal{P}$ is

1. a partial order on a set on $2^k + 1$ $(k \geq 2)$ elements that computes the second smallest element;
2. a partial order of form $\{x\} \times \mathcal{B}$, where x is the minimum and $\mathcal{B}$ is a binomial tree of size 2^k; or
3. a partial order on a set on $2^k + 1$ $(k \geq 2)$ elements that determines the minimum and maximum elements.

A special case of this proposition that corresponds to finding the second smallest element among five elements was described in [12]. The partial orders in the proposition above enlarge the class of non-additive partial orders on a *large* number of elements. Previously, the only partial order on a large set that is recognized to be non-additive is the partial order corresponding to the median-finding problem. While the computation of bonus from the mass production of non-additive partial orders is important in understanding the inherent complexity of partial order production and in designing efficient production algorithms, the posets having the additivity property are also interesting. The major advantage of the posets with additivity is that their production complexity can simply be examined through their connected components. In this way, we only need to deal with the subproblems that are smaller and easier than the original one. The recognition problem of the posets possessing the additivity property seems very hard in general. Only a few results are given for special classes of posets [2,4]. The theorem below establishes sufficient conditions for the additivity property.

Theorem 6. *Denote the cardinality of k disjoint posets $\mathcal{P}_1, \mathcal{P}_2, \cdots, \mathcal{P}_k$ respectively by $n_1, n_2, \cdots, n_k$. Then $C(\mathcal{P}_1 + \mathcal{P}_2 + \cdots + \mathcal{P}_k) = C(\mathcal{P}_1) + C(\mathcal{P}_2) + \cdots + C(\mathcal{P}_k)$ provided that one of the following conditions holds:*

1. *$C(\mathcal{P}_i) = \frac{3}{2}n_i - \|Max(\mathcal{P}_i)\| - \|Min(\mathcal{P}_i)\|$ all $i = 1, 2, \cdots, k$; or*
2. *$C(\mathcal{P}_i) = n_i - comp(\mathcal{P}_i)$ for all $i = 1, 2, \cdots, k$, where $comp(\mathcal{P}_i)$ denotes the number of connected components of $\mathcal{P}_i$.*

The claim can easily be verified by the facts that $C(\mathcal{P}) \geq \lceil \frac{3}{2}n \rceil - \|Max(\mathcal{P})\| - \|Min(\mathcal{P})\|$ and $C(\mathcal{P}) \geq n - comp(\mathcal{P})$ [1] for any n-element poset $\mathcal{P}$. The above theorem generalize the result in [4] which solves the case when $\|Max(\mathcal{P}_i)\| = \|Min(\mathcal{P}_i)\| = comp(\mathcal{P}_i) = 1$. Define the *asymptotic mass-production cost* of a poset $\mathcal{P}$ as $\widetilde{C}(\mathcal{P}) \overset{\triangle}{=} \lim_{k \to \infty} \frac{C(k \cdot \mathcal{P})}{k}$. Hence, $\widetilde{C}(\mathcal{P}) \leq C(\mathcal{P})$. From the proof of Theorem 6, we know that $C(k \cdot \mathcal{P}) \geq k \cdot \left(\frac{3}{2}n - \|Max(\mathcal{P})\| - \|Min(\mathcal{P})\| \right)$ and $C(k \cdot \mathcal{P}) \geq k \cdot (n - comp(\mathcal{P}))$. Therefore,

Corollary 7. *For any partial order $\mathcal{P}$ on n element, we have*

$$\widetilde{C}(\mathcal{P}) \geq \frac{3}{2}n - \|Max(\mathcal{P})\| - \|Min(\mathcal{P})\| \quad and \quad \widetilde{C}(\mathcal{P}) \geq n - comp(\mathcal{P}).$$

There are only six posets whose asymptotic costs are previously known [12]. Moreover, the computations of asymptotic costs of partial orders are very extensive. Unlike $C(\mathcal{P})$, the cost $\widetilde{C}(\mathcal{P})$ cannot be determined by an enumeration of finite many cases. With Corollary 7, we can successfully demonstrate the asymptotic costs for a large class of posets. The class of partial orders that satisfy the first condition in the theorem above generalized the partial orders of minimum & maximum finding which been studied in [4]. Furthermore, we have also successfully applied the mass production technique to the problem of constructing heap-like structures, resulting in efficient algorithms. For example,

Proposition 8. *In the worst case,*

- *four heaps of size 7 can be built in at most 31 comparisons, and*
- *two deaps of size 6 can be constructed optimally with 15 comparisons.*

4 Size-Sensitive Posets

We illustrate in this section a family of partial orders, size-sensitive partial orders, whose production complexity can be strictly decreased by providing extra elements. It may come as something of a surprise that such partial orders exist. The (only) example known previously of this kind was provided in [12,13], which shows that the extra element can reduce the cost to produce the poset $\mathcal{P}$ in Figure 3. First, it is clear that $\mathcal{C}(\mathcal{Q}) \geq \lceil \log{(7 \times 2 \times 4 \times 2)} \rceil = 7$. To show that $\mathcal{C}(\mathcal{Q}) \leq 7$, we first create a binomial tree $\mathcal{B}$ of size 8 and then "forget" the ordering relation between the root and x. More precisely, we produce a poset $\mathcal{B}$ such that $\mathcal{B}$ contains $\mathcal{Q}$ as a suborder. Since $\mathcal{Q}$ can be deduced from $\mathcal{B}$ and thus $\mathcal{P}$ from $\mathcal{Q}$ without any further comparison, we say that $\mathcal{P}$ is producible in 7 comparisons if we have one extra element available. On the other hand, $\mathcal{C}(\mathcal{P}) = 8$, which was shown by enumerating all the cases [12,13]. Besides its order theoretical appeal, the study of the advantage of extra elements over partial order production was stemmed primarily from its relation to selection problem [10,12,13].

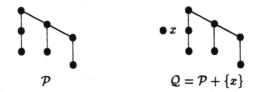

$\mathcal{P}$ $\mathcal{Q} = \mathcal{P} + \{x\}$

FIGURE 3: An example of size-sensitive posets.

Formally, a poset $\mathcal{P}$ is *size-sensitive* if and only if there exists some integer k such that $\mathcal{C}(\mathcal{P} + \{x_1\} + \{x_2\} + \cdots + \{x_k\}) < \mathcal{C}(\mathcal{P})$, where $x_i \notin \mathcal{P}$ and $x_i \neq x_j$, for all $1 \leq i, j \leq k$ and $i \neq j$. Of course, from this definition, any size-sensitive poset is also non-additive. We sometimes refer to $\mathcal{C}_k(\mathcal{P})$ as the cost to produce the poset $\mathcal{P} + \{x_1\} + \{x_2\} + \cdots + \{x_k\}$. Of course, it is always that $\mathcal{C}_k(\mathcal{P}) \leq \mathcal{C}(\mathcal{P})$ for any $k \geq 1$. The discovery of the poset shown in Figure 3 sparked an interest in specializing posets that have the size-sensitivity. Despite many efforts, it has not been possible to identify size-sensitive posets of large sizes. With the assistance of the example illustrated in Figure 3, we can now design a class of posets such that adding one singleton element to the posets can help in producing them, as shown in the following proposition. An interesting aspect of the proposition below is that its proof does not rely on enumeration.

Proposition 9. *Let $\mathcal{P}$ be a partial order on $n = 2^k - 1$ $(k \geq 3)$ elements whose Hasse diagram is a minimum element connecting to $k-1$ disjoint binomial trees of sizes $2, 4, \cdots, 2^{k-1}$. Then $\mathcal{C}(\mathcal{P}) = n+1$ and $\mathcal{C}(\mathcal{P} + \{x'\}) = n$ for some element $x' \notin \mathcal{P}$. Moreover, for $m \geq 2$, we have $\mathcal{C}(m \cdot \mathcal{P}) \leq m \cdot n + 1 < m(n+1) = m \cdot \mathcal{C}(\mathcal{P})$.*

Proof. The proof of $\mathcal{C}(\mathcal{P} + \{x'\}) = n$ and of $\mathcal{C}(\mathcal{P}) \leq n + 1$ are not difficult, which are omitted due to the space limit. We now proceed to show that $\mathcal{C}(\mathcal{P}) \geq n + 1$. Notice that $\ell(\mathcal{P}) = \frac{(n+1)!}{n \cdot 2^n}$ and hence $\mathcal{C}(\mathcal{P}) \geq \lceil \log \frac{n!}{\ell(\mathcal{P})} \rceil = n$.

We shall demonstrate that the assumption that $\mathcal{C}(\mathcal{P}) = n$ will lead to a contradiction and hence $\mathcal{C}(\mathcal{P}) \geq n + 1$. To this end, let us denote $\mathcal{P}$ by $\{r\} \times \mathcal{P}'$, where r is the root of $\mathcal{P}$, $\mathcal{P}' = \mathcal{B}_2 + \mathcal{B}_4 + \cdots + \mathcal{B}_{2^k-1}$, and $\mathcal{B}_i$ is the binomial tree of size i in $\mathcal{P}$. It is known that the root r must win at least $\lceil \log n \rceil = k$ comparisons ([11], p.210). These k comparisons do not facilitate the productions of the posets $\mathcal{B}_2, \mathcal{B}_4, \cdots, \mathcal{B}_{2^k-1}$. The reason is that $\mathcal{C}(\mathcal{B}_i) = \|\mathcal{B}_i\| - 1$ and each of these comparisons must only involves elements in $\mathcal{B}_i$ in order to obtain a connected poset $\mathcal{B}_i$. Moreover, the $\lceil \log n \rceil$ comparisons involving r do not generate any order relation between elements in $\mathcal{P}'$. To see this, let us employ a simple adversary that replies $x \prec y$ for a comparison $(x : y)$ whenever $d[x] \triangleq \|D[x]\| \geq \|D[y]\| \triangleq d[y]$, where $D[x]$ and $D[y]$ are the set of all elements which are greater than or equal to x and y known so far, respectively. Since r will always win its comparisons according to the adversary, no order relation between elements in $\mathcal{P}'$ can be derived by these comparisons. Notice that $\mathcal{C}(\mathcal{P}') = \sum_{i=1}^{k-1} \mathcal{C}(\mathcal{B}_{2^i}) = \sum_{i=1}^{k-1} (2^i - 1) = 2^k - 1 - k = n - \lceil \log n \rceil$.

Moreover, since we are only allowed to use at most $n - \lceil \log n \rceil$ comparisons for constructing $\mathcal{P}'$ due to the assumption of $\mathcal{C}(\mathcal{P}) = n$, no element of $\mathcal{B}_i$ is allowed to be compared with any of the elements in $\mathcal{P} - \mathcal{B}_i$ for any i. Again, this is because that $\mathcal{C}(\mathcal{B}_i) = \|\mathcal{B}_i\| - 1$ and each of these comparisons must only involves elements in $\mathcal{B}_i$ in order to obtain a connected poset $\mathcal{B}_i$. Furthermore, every comparison performed on $\mathcal{B}_i$ must involve two elements in $\mathcal{B}_i$ whose downsets have the same size. Otherwise, we cannot build the binomial tree $\mathcal{B}_i$ in $i - 1$ comparisons. Analogously, one should note that no comparison between two elements that belong to different components of $\mathcal{B}_i$ is allowed since only $2^i - 1$ comparisons are available for producing $\mathcal{B}_i$ for $i = 2, 4, \cdots, 2^{k-1}$.

Now, we are ready to set up an adversary that enables us to show that n comparisons are not enough for the production of $\mathcal{P}$. To do this, consider any comparison between two previously unrelated elements x and y. Let $u[x]$ and $u[y]$ be the number of all elements which are smaller than or equal to x and y known so far, respectively. The adversary will choose the relation $x \prec y$ according to the first rule in the following it can apply:

1. $d[x] = d[y] = 1$, $u[x] = 1$, and $u[y] \geq 2$;
2. $d[x] > d[y]$;
3. $d[x] = d[y]$, $D[x] \cap D[y] = \emptyset$, and $u[y] = 2$;
4. $d[x] = d[y]$ and there exists at least one element $z \in D[x] \cap D[y]$ that was already in $D[y]$ while being put into $D[x]$.

Remark that this adversary can be regarded as a fine version of the previous one. Since we have only k comparisons, denoted by $C_1, C_2, \cdots, C_k$, available for the root r and since any comparison can only at most double the size of the down-set by the adversary computations, it follows that $d[r] \leq 2^k$ (more precisely, $\leq 2^k - 1$ since we have only $2^k - 1$ elements totally) after the kth comparison. In order to ensure that $d[r] = 2^k - 1$ (i.e., to guarantee that r is the minimum element of $\mathcal{P}$), the following conditions must hold (otherwise, by the adversary rules 2, 3, and 4, $d[r] < 2^k - 1$ after k comparisons):

a) Before the $(i+1)$th comparison C_i, $d[r] = 2^i$;
b) The $(i+1)$th comparison C_i must compare r with an element x of $d[x] = 2^i$, and $D[r] \cap D[x] = \emptyset$ (except for the kth comparison).

Notice that to have $d[x] = 2^i$, the element x must be the smaller element in at least i comparisons. Suppose that the comparison C_k compares r with x. From the condition b) above, we know that either $d[x] = 2^{k-1}$ or $d[x] = 2^{k-1} - 1$. If $d[x] = 2^{k-1} - 1$, then x must have already been the winner of at least k comparisons. However, as seen above, every comparison performed on $\mathcal{P}'$ must involve two elements in the same binomial tree $\mathcal{B}_i$ and the size of their down-sets are the same. Therefore, $d[x] = 2^{k-1}$, which contradicts to the assumption that $d[x] = 2^{k-1} - 1$.

If $d[x] = 2^{k-1}$, then $D[r] \cap D[x] \neq \emptyset$ because we have only $2^k - 1$ elements. Moreover, we also know that $\|D[r] \cap D[x]\| = 1$ since otherwise $d[r] \leq 2^k - 2$. Let $D[r] \cap D[x] = \{z\}$. Notice that r must be the winner of the comparison C_k. Hence, z must already be in $D[x]$ while it was put into $D[r]$ according to Adversary Rule 4. Suppose that at the time t, $z \in D_t[x]$. In order to put z into $D[r]$, we can only perform a comparison either $(r : z)$ or $(r' : z)$, where $r' \in D_t(r)$.

If the comparison is made between r and z, then $d_t[r] = 1$ due to Conditions a) and b) above. However, in this case, z will be the winner of the comparison according to Adversary Rule 1. This is a contradiction.

If the comparison is made between r' and z (where $r' \in D_t(r)$), then $d_t[r'] = 1$ because comparisons on $\mathcal{P}'$ must compare elements whose down-sets have the same size. Since now $d_t[r'] = d_t[z] = 1$, we know that $D[r'] \cap D[z] = \emptyset$. Remember that r' is the winner against z. Hence, Adversary Rule 3 can apply to this case. Before the comparison, we know that $u[r'] \geq 2$ because $r \prec r'$. If $u[r'] = 2$, then r' will be the loser of the comparison according to Adversary Rule 3, which causes a contradiction. On the other hand, suppose that $u[r'] > 2$. But this means that we have made a comparison on $\mathcal{P}'$ that compares two elements whose down-sets have different sizes, which results in a contradiction as well. In other words, we cannot establish all the ordering relations in $\mathcal{P}$ by only doing n comparisons, and this completes the proof of the first part of the proposition.

We are now left only to demonstrate that $\mathcal{C}(m \cdot \mathcal{P}) \leq m \cdot n + 1$. Notice that by taking advantage of the extra element when producing $\mathcal{P}$, we can save $m - 1$ comparisons for building the first $m - 1$ copies of $\mathcal{P}$ instead of producing each of them separately. Recall that producing one single copy of $\mathcal{P}$ costs $n + 1$ comparisons. Therefore, $\mathcal{C}(m \cdot \mathcal{P}) \leq (n + 1) \cdot m - (m - 1) = m \cdot n + 1$.

The partial orders described in this proposition enriches our knowledge of size-sensitive partial orders on *large* element-sets. We shall see, in the proposition below, that if the cost to produce a poset equals to some general lower bounds, then extra elements cannot provide any advantage for the partial order production. More precisely, we claim that:

Proposition 10. *For any poset $\mathcal{P}$ of size n, one of the following conditions will guarantee that $\mathcal{C}(\mathcal{P} + \{x\}) = \mathcal{C}(\mathcal{P})$ for $x \notin \mathcal{P}$:*

1. $C(\mathcal{P}) = \left\lceil \log \frac{n!}{l(\mathcal{P})} \right\rceil$;
2. $C(\mathcal{P}) = n - comp(\mathcal{P})$; or
3. $C(\mathcal{P}) = \lceil \frac{3}{2}n \rceil - \|Max(\mathcal{P})\| - \|Min(\mathcal{P})\|$.

Proof. We only need to prove that $C(\mathcal{P} + \{x\}) \geq C(\mathcal{P})$. The first two conditions can easily be confirmed by the information-theoretic and the connectivity-based lower bound, respectively. To see that the condition 3 holds, suppose that we have performed at most $C(\mathcal{P}) - 1$ comparisons between $n + 1$ elements. From the adversary presented in [1] (Proposition 3.5 (ii)), we know that a poset $\mathcal{P}'$ can be produced with these comparisons, where the number of both the minimum and maximum elements in $\mathcal{P}'$ is at least $\|Max(\mathcal{P})\| + \|Min(\mathcal{P})\| + 2$. We now show that $\mathcal{P}' - \{x'\} \not\succeq \mathcal{P}$ for any $x' \in P'$. If $x' \in Sing(\mathcal{P}')$ $(\neq \emptyset)$, then x' has not been involved in any comparison during the production of $\mathcal{P}'$. Thus, $C(\mathcal{P}' - \{x'\}) = C(\mathcal{P}') = C(\mathcal{P}) - 1$, which means that $\mathcal{P}' - \{x'\} \not\succeq \mathcal{P}$. On the other hand, if x' is not a singleton element of $\mathcal{P}'$, then the number of both the minimum and maximum elements in $\mathcal{P}' - \{x'\}$ is at least $\|Max(\mathcal{P})\| + \|Min(\mathcal{P})\| + 1$, which also implies that $\mathcal{P}$ is not a suborder of $\mathcal{P}' - \{x'\}$. Therefore, to produce a poset of size $n + 1$ of which $\mathcal{P}$ is a subposet, $C(\mathcal{P})$ comparisons are necessary.

Proposition 10 gives simple algorithmic characterizations of some partial orders whose production cost cannot be reduced by providing extra elements. In general, no matter how many extra elements are available, one can always ignore the extra elements and produce only the connected poset $\mathcal{P}$. Namely, $C_m(\mathcal{P})$ is a non-increasing function of m; that is, $C_m(\mathcal{P}) \geq C_{m+1}(\mathcal{P})$ for all $m \geq 0$ ($C_0(\mathcal{P}) \equiv C(\mathcal{P})$). Repeating the similar argument to Proposition 10 yields:

Proposition 11. *Let $\mathcal{Q}_n$ is a poset on n elements and $f(\mathcal{Q}_n, m)$ a non-decreasing function, depending on m, in the sense that $f(\mathcal{Q}_n, m) \leq f(\mathcal{Q}_n, m + 1)$ and $f(\mathcal{Q}'_n, m) \leq f(\mathcal{Q}''_n, m)$ whenever $\mathcal{Q}'_n$ is a suborder of $\mathcal{Q}''_n$. Then, for any poset $\mathcal{P}_n$ of size n and an antichain $\mathcal{A}_m$ of size m, if $C(\mathcal{P}_n) \geq f(\mathcal{P}_n, n)$ then*

$$C_m(\mathcal{P}_n) = C(\mathcal{P}_n + \mathcal{A}_m) \geq f(\mathcal{P}_n, n)$$

for all $m \geq 0$. Further, if there exists some integer m such that $C_{m+1}(\mathcal{P}_n) < C_m(\mathcal{P}_n)$, then $C(\mathcal{P}_n) \geq C_m(\mathcal{P}_n) > f(\mathcal{P}_n, n)$.

From Proposition 11, by letting $f(\mathcal{P}_n, n)$ be one of $\lceil \log \frac{n!}{l(\mathcal{P}_n)} \rceil$, $n - comp(\mathcal{P})$, and $\lceil \frac{3}{2}n \rceil - \|Max(\mathcal{P}_n)\| - \|Min(\mathcal{P}_n)\|$, we immediately have

Corollary 12. *For any connected poset $\mathcal{P}$ of size n and for all $m \geq 0$, $C_m(\mathcal{P}) \geq \left\lceil \log \frac{n!}{l(\mathcal{P})} \right\rceil$, $C_m(\mathcal{P}) \geq n - comp(\mathcal{P})$, and $C_m(\mathcal{P}) \geq \lceil \frac{3}{2}n \rceil - \|Max(\mathcal{P})\| - \|Min(\mathcal{P})\|$.*

The fact that $C_m(\mathcal{P}) \geq \lceil \log \frac{n!}{l(\mathcal{P})} \rceil$ was originally observed by Schönhage [12]. While Proposition 10 claims that if the cost to produce a poset $\mathcal{P}$ matches some known general lower bounds, then the production complexity cannot be decreased by providing one extra element, Corollary 12 confirms that even if we have more than one extra elements available, it does not help for the problem either, under the same situation. Using a connectivity argument, we know from Proposition 11 that

Corollary 13. *For any connected poset $\mathcal{P}$ of size n, there exists some integer $0 \leq m \leq \mathcal{C}(\mathcal{P}) - n$ such that*

$$\mathcal{C}(\mathcal{P}) \stackrel{\Delta}{=} \mathcal{C}_0(\mathcal{P}) \geq \mathcal{C}_1(\mathcal{P}) \geq \cdots \geq \mathcal{C}_m(\mathcal{P}) = \mathcal{C}_{m+1}(\mathcal{P}) = \cdots = \mathcal{C}_\infty(\mathcal{P}) \geq n + m - 1,$$

where $\mathcal{C}_\infty(\mathcal{P}) \stackrel{\Delta}{=} \lim_{k \to \infty} \mathcal{C}_k(\mathcal{P})$.

We end this section with a remark about the relationship between size-sensitive posets and non-additive posets. Clearly, any size-sensitive poset is also non-additive. However, not all non-additive posets have the size-sensitivity property. For example, consider the poset $\mathcal{P}$ whose Hasse diagram is a binomial tree of size four hanging below a minimum node. A simple calculation shows that $\mathcal{C}(\mathcal{P}) = \lceil \log \frac{n!}{l(\mathcal{P})} \rceil = 6$. Hence, $\mathcal{P}$ is not a size-sensitive poset from Proposition 10. However, two isomorphic copies of $\mathcal{P}$ can be built simultaneously in at most 11 comparisons. As a further example, let $\mathcal{Q}$ be the poset of finding both the maximum and the minimum among five elements. Then, $\mathcal{Q}$ has not the size-sensitivity since $\mathcal{C}(\mathcal{Q}) = 6 = \lceil \frac{3}{2} \cdot 5 \rceil - 2$, while producing two isomorphic copies of $\mathcal{Q}$ will save at least one comparison.

References

1. M. Aigner: Producing posets. *Discrete Mathematics* **35** (1) (1981), 1-15.
2. M. D. Atkinson: The complexity of orders. In: *Proceedings of the NATO Advanced Study Institute on Algorithms and Order* (1989), 195-230.
3. S. W. Bent and J. W. John: Finding the median requires $2n$ comparisons. In: *Proce. 17th Annu. ACM Symp. Theo. Comput.* (1985), 213-216.
4. J. C. Culberson and G. J. E. Rawlins: On the comparison cost of partial orders. *Technical Report*, Department of Computer Science, Indiana University, 1988.
5. F. Fussenegger and H. N. Gabow: A counting approach to lower bounds for selection problems. *Journal of the ACM* **26** (2) (1979), 227-238.
6. A. Hadian and M. Sobel: Selecting the tth largest using binary errorless comparisons. *Colloquia Mathematica Societatis János Bolyai* **4** (1970), 585-599.
7. E. Horowitz and S. Sahni: *Fundamentals of Data Structures in Pascal*. Third Edition. Computer Science Press, New York, 1990.
8. L. Hyafil: Bounds for selection. *SIAM Journal on Computing* **5** (1) (1976), 109-114.
9. J. W. John: A new lower bound for the set-partitioning problem. *SIAM Journal on Computing* **17** (4) (1988), 640-647.
10. D. G. Kirkpatrick: A unified lower bound for selection and set partitioning problems. *Journal of the ACM* **28** (1) (1981), 150-165.
11. D. E. Knuth: *The Art of Computer Programming. Vol. 3: Sorting and Searching*. Addison-Wesley, Reading, Massachusetts, 1973.
12. A. Schönhage: The production of partial orders. *Astérisque* **38-39** (1976), 229-246.
13. A. Schönhage, M. Paterson, and N. Pippenger: Finding the median. *Journal of Computer and System Sciences* **13** (2) (1976), 184-199.
14. A. C.-C. Yao: On the complexity of partial order productions. *SIAM Journal on Computing* **18** (4) (1989), 679-689.
15. C. K. Yap: New upper bounds for selection. *Comm. ACM* **19** (9) (1976), 501-508.

An Efficient Algorithm for Edge-Ranking Trees

Xiao Zhou and Takao Nishizeki *

Department of System Information Sciences
Graduate School of Information Sciences
Tohoku University, Sendai 980-77, Japan

Abstract. An edge-ranking of an undirected graph G is a labeling of the edges of G with integers such that all paths between two edges with the same label i contain an edge with label $j > i$. The problem of finding an edge-ranking of G using a minimum number of ranks has applications in scheduling the manufacture of complex multi-part products; it is equivalent to finding the minimum height edge separator tree. Deogun and Peng and independently de la Torre *et al.* have given polynomial-time algorithms which find an edge-ranking of trees T using a minimum number of ranks in time $O(n^3)$ and $O(n^3 \log n)$ respectively, where n is the number of nodes in T. This paper presents a more efficient and simple algorithm, which finds an edge-ranking of trees using a minimum number of ranks in $O(n^2)$ time.

1 Introduction

An *edge-ranking* of a graph G is a labeling of edges of G with positive integers such that all paths between two edges with the same label i contain an edge with label $j > i$ [DP90, IRV91]. The integer label of an edge is called the *rank* of the edge. The minimum number of ranks needed for an edge-ranking of G is called the *edge-ranking number* and denoted by $\chi'_r(G)$. An edge-ranking of G using $\chi'_r(G)$ ranks is called an *optimal edge-ranking* of G. The *edge-ranking problem* is to find an optimal edge-ranking of a given graph G. The problem seems to be NP-complete in general [dlTGS93]. The problem has applications in scheduling the manufacture of complex multi-part products; it is equivalent to finding the minimum height edge separator trees of G. The edge-ranking of trees, in particular, can be used to model the parallel assembly for a product from its components in a quite natural manner [BDJ+94, dlTGS93, IRV88a, IRV88b, IRV91]. Fig. 1 depicts an optimal edge-ranking of a tree using four ranks.

An *edge-coloring* of a graph G is to color all the edges of G so that no two adjacent edges are colored with the same color. By the definition of edge-rankings, two edges which are incident to the same vertex cannot have the same rank, and hence any edge-ranking of a graph is also an edge-coloring. This implies that the maximum vertex degree of a graph G is a lower bound on the edge-ranking number $\chi'_r(G)$ of G [DP90, IRV91]. Furthermore it is known that $\log_2 n \le \chi'_r(G)$ for any connected graph G of n vertices [IRV91].

---

* E-mail:(zhou|nishi)@ecei.tohoku.ac.jp

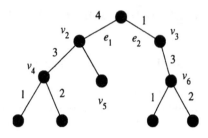

Fig. 1. An optimal edge-ranking of a tree T.

The node-ranking problem of a graph G is defined similarly [IRV88a, KMS88]. A *node-ranking* of G is a labeling of vertices of G with integers such that all paths between two vertices with the same label i contain a vertex with label $j > i$. The *node-ranking problem* is to find a node-ranking of a given graph G using the minimum number of ranks. The node-ranking problem is NP-complete in general [BDJ+94, Pot88]. On the other hand Schaffer has presented a linear algorithm to solve the node-ranking problem for trees [Sch89]. Very recently Bodlaénder *et al.* have given a polynomial time algorithm to solve the node-ranking problem for graphs with treewidth at most k for fixed k [BDJ+94].

Deogun and Peng [DP90] have given the first polynomial-time algorithm to solve the edge-ranking problem for trees, which takes $O(n^3)$ time. Their algorithm consists of the following two steps: (1) transform the edge-ranking problem for trees into the equivalent node-ranking problem for line-graphs of trees which they call "clique-trees"; and (2) find a node-ranking of clique-trees with a minimum number of ranks. Recently de la Torre *et al.* have rediscovered a polynomial-time algorithm which solves the edge-ranking problem for trees by means of a complicated greedy procedure, but the time complexity is $O(n^3 \log n)$ [dlTGS93].

In this paper we give a more efficient and simple algorithm which solves the edge-ranking problem for trees in $O(n^2)$ time. Our algorithm improves the known algorithms [DP90, dlTGS93] in time-complexity and simplicity. We directly solve the edge-ranking problem for trees without transforming it to the node-ranking problem for cliques trees or using a complicated greedy procedure. More precisely, we find an optimal edge-ranking of a given tree T by means of "bottom-up tree computation" on T: we find an optimal edge-ranking of a subtree rooted at an internal node of T from those of the subtrees rooted at its children.

2 Preliminaries

In this section we define some terms and present easy observations. Let $T = (V, E)$ denote a tree with node set V and edge set E. We often denote by $V(T)$

and $E(T)$ the node set and the edge set of T, respectively. We denote by n the number of nodes in T. T is a "free tree," but we regard T as a "rooted tree" for convenience sake: an arbitrary node of tree T is designated as the *root* of T. We will use notions as: root, node, internal node, child and leaf in their usual meaning. An edge joining nodes u and v is denoted by (u, v). The *degree* of node v is denoted by $d(v)$. A *sum* $T + T'$ of two trees T and T' is a graph $(V(T) \cup V(T'), E(T) \cup E(T'))$.

The definition of an edge-ranking immediately implies the following lemma [DP90, IRV91].

Lemma 1. *Any edge-ranking of a connected graph labels exactly one edge with the largest rank.*

Let φ be an edge-ranking of a tree T. The number of ranks used by φ is denoted by $\#\varphi$. One may assume w.l.o.g. that φ uses the ranks in $[1, \#\varphi]$. An edge e of T and rank $\varphi(e)$ are *visible* (from the root with respect to φ) if all the edges (except e) in the path from the root to e have ranks smaller than $\varphi(e)$. Define $L(\varphi)$ as

$$L(\varphi) = \{\varphi(e) \mid e \in E \text{ is visible}\},$$

and call $L(\varphi)$ the *list of an edge-ranking* φ of the rooted tree T. Obviously $\#\varphi \in L(\varphi)$ and $\varphi(e) \in L(\varphi)$ for every edge e incident to the root. The ranks in the list $L(\varphi)$ are sorted in decreasing order. Thus the edge-ranking φ in Fig. 1 has the list $L(\varphi) = \{4, 3, 1\}$.

We define the lexicographical order $\prec$ on the set of decreasing sequences (lists) of positive integers as follows: let $A = \{a_1, \cdots, a_p\}$ and $B = \{b_1, \cdots, b_q\}$ be two sets (lists) of positive integers such that $a_1 > \cdots > a_p$ and $b_1 > \cdots > b_q$, then $A \prec B$ if and only if there exists an integer i such that

(a) $a_j = b_j$ for all $1 \le j < i$, and
(b) either $a_i < b_i$ or $p < i \le q$.

We write $A \preceq B$ if $A = B$ or $A \prec B$. For integers α and β, $\alpha \le \beta$, we denote by $[\alpha, \beta]$ the set of integers between α and β, that is, $[\alpha, \beta] = \{\alpha, \alpha + 1, \cdots, \beta\}$. If $\alpha > \beta$, then $[\alpha, \beta] = \phi$. Obviously the following (1)–(3) hold for any integer $\alpha \ge 1$:

$$\text{if } A \preceq B, \text{ then } A - [1, \alpha] \preceq B - [1, \alpha]; \tag{1}$$

$$\text{if } A \preceq B \text{ and } A - [1, \alpha] = B - [1, \alpha], \text{ then } A \cap [1, \alpha] \preceq B \cap [1, \alpha]; \text{ and} \tag{2}$$

$$\text{if } A \prec B \text{ and } A - [1, \alpha] = B - [1, \alpha], \text{ then } A \cap [1, \alpha] \prec B \cap [1, \alpha]. \tag{3}$$

An edge-ranking φ of T is *critical* if $L(\varphi) \preceq L(\eta)$ for any edge-ranking η of T. The optimal edge-ranking depicted in Fig. 1 is indeed critical. The list of a critical edge-ranking of T is called the *critical list of tree T* and denoted by $L^*(T)$. Thus $L^*(T)$ corresponds to an equivalent class of optimal edge-rankings of T. Since any critical edge-ranking is optimal, it suffices to find a critical edge-ranking of T.

The subtree of T rooted at node v is denoted by $T(v)$. Let $e = (u, v)$ be an edge in T such that v is a child of u. Then the tree obtained from $T(v)$ by adding

e is called the *subtree of T rooted at edge e* and is denoted by $T(e)$. We denote by $T - T(e)$ the tree obtained from T by deleting all edges in $T(e)$ and all nodes in $T(e)$ except u.

For an edge-ranking φ of tree T and a subtree T' of T, we denote by $\varphi|T'$ a restriction of φ to $E(T')$: let $\varphi' = \varphi|T'$, then $\varphi'(e) = \varphi(e)$ for $e \in E(T')$. By Lemma 1 T has exactly one edge e with the largest rank $\#\varphi$. Denote by $T(\#\varphi)$ the subtree $T(e)$ rooted at the edge e. Let $T' = T - T(\#\varphi)$, then

$$L(\varphi) = \{\#\varphi\} \cup L(\varphi|T'). \tag{4}$$

If φ is a critical edge-ranking of T, then by (2) and (4) $\varphi|T'$ is a critical edge-ranking of T'. Let T'' be the tree obtained from $T(e)$ by deleting e. If φ is an optimal edge-ranking, then

$$\chi'_r(T'), \chi'_r(T'') < \#\varphi = \chi'_r(T). \tag{5}$$

3 Optimal Edge-Ranking

The main result of this paper is the following theorem.

Theorem 2. *An optimal edge-ranking of a tree T can be found in $O(n^2)$ time where n is the number of nodes in T.*

In the remaining of this section we give an $O(n^2)$ algorithm for finding a critical edge-ranking of a tree T. Our algorithm uses the technique of "bottom-up tree computation." For each node u of a tree T, we construct a critical edge-ranking of $T(u)$ from those of u's children. This can be done in time $O(nd(u))$, and hence the total running time is $O(n^2)$, as shown later.

We first have the following lemma.

Lemma 3. *If a tree T has two edge-disjoint subtrees T_a and T_b with $\chi'_r(T_a) = \chi'_r(T_b) = c$, then $c + 1 \leq \chi'_r(T)$.*

Proof. Clearly $c \leq \chi'_r(T)$. Suppose that $\chi'_r(T) = c$. Then by Lemma 1 T has exactly one edge e with rank c in an optimal edge-ranking of T. Remove the edge e from T, and let T' and T'' be the resulting two trees. Then by (5) $\chi'_r(T')$, $\chi'_r(T'') < c$. Since T_a and T_b are edge-disjoint, either T_a or T_b, say T_a, does not contain e. Then T_a is a subgraph of either T' or T'', say T', and hence $\chi'_r(T_a) \leq \chi'_r(T') < c$, contrary to the assumption $\chi'_r(T_a) = c$. $\mathcal{Q.E.D.}$

Let $v_1, \cdots, v_p$ be children of an internal node u of T, let $e_i = (u, v_i)$, $1 \leq i \leq p$, and let $T^p = T(e_1) + \cdots + T(e_p)$. Our idea is that a critical edge-ranking of $T(e_i)$ can be constructed from a critical edge-ranking of $T(v_i)$ and that a critical edge-ranking of T^p can be constructed from those of $T(e_1), \cdots, T(e_p)$. We have the following two lemmas.

Lemma 4. *If φ is an edge-ranking of T^p, then*

(a) $L(\varphi) = \bigcup_{1 \leq i \leq p} L(\varphi|T(e_i))$, and

(b) sets $L(\varphi|T(e_i))$, $1 \leq i \leq p$, are pairwise disjoint.

Conversely, if φ_i, $1 \leq i \leq p$, are edge-rankings of $T(e_i)$ and sets $L(\varphi_i)$ are pairwise disjoint, then the labeling φ of $E(T^p)$ extended from φ_i, $1 \leq i \leq p$, is an edge-ranking of T^p and $L(\varphi) = \bigcup_{1 \leq i \leq p} L(\varphi_i)$.

Proof. immediate from the definition of an edge-ranking and its list. $\mathcal{Q}.\mathcal{E}.\mathcal{D}.$

Lemma 5. *Let* $m = \max\{\chi_r'(T^{p-1}), \chi_r'(T(e_p))\}$, *then* $\chi_r'(T^p) = m$ *or* $m + 1$.

Proof. Clearly $m \leq \chi_r'(T^p)$. Therefore it suffices to prove that $\chi_r'(T^p) \leq m + 1$. Let φ and ψ be critical edge-rankings of T^{p-1} and $T(e_p)$, respectively. Extend φ and ψ to an edge-labeling η of T^p as follows:

$$\eta(e) = \begin{cases} \varphi(e) & \text{if } e \in E(T^{p-1}), \\ \psi(e) & \text{if } e \in E(T(e_p)) - \{e_p\}, \text{ and} \\ m+1 & \text{if } e = e_p. \end{cases}$$

Clearly η is an edge-ranking of T^p and $\#\eta \leq m + 1$. Therefore $\chi_r'(T^p) \leq m + 1$.
$\mathcal{Q}.\mathcal{E}.\mathcal{D}.$

The following Lemma 6 gives a necessary and sufficient condition for $\chi_r'(T^p) = m$.

Lemma 6. *Let* $m = \max\{\chi_r'(T^{p-1}), \chi_r'(T(e_p))\}$. *Then* $\chi_r'(T^p) = m$ *if and only if the following* (a) *or* (b) *holds:*

(a) $L^*(T^{p-1}) \cap L^*(T(e_p)) = \phi$; *or*

(b) *there exists* $\beta \in [\alpha + 1, m]$ *such that* $\beta \notin L^*(T^{p-1}) \cup L^*(T(e_p))$ *where* α *is the maximum integer in* $L^*(T^{p-1}) \cap L^*(T(e_p))$.

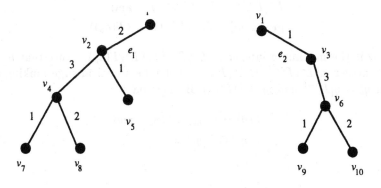

Fig. 2. Critical edge-rankings of $T(e_1)$ and $T(e_2)$.

Before presenting a proof, we illustrate the conditions (a) and (b) for the tree T in Fig. 1. Fig. 2 depicts critical edge-rankings of $T(e_1)$ and $T(e_2)$. Let $p = 2$, then $T^{p-1} = T(e_1)$, $\chi_r'(T^{p-1}) = 3$, $\chi_r'(T(e_p)) = 3$, $m = 3$, $L^*(T^{p-1}) = \{3, 2\}$, $L^*(T(e_p)) = \{3, 1\}$, $L^*(T^{p-1}) \cap L^*(T(e_p)) = \{3\}$, $\alpha = 3$, and $[\alpha + 1, m] = \phi$. Thus neither (a) nor (b) holds.

Proof of Lemma 6. $\impliedby$ Let φ and ψ be critical edge-rankings of T^{p-1} and $T(e_p)$, respectively. Then $L(\varphi) = L^*(T^{p-1})$ and $L(\psi) = L^*(T(e_p))$. From φ and ψ we construct an edge-ranking η of T^p with $\#\eta = m$.

Suppose first that (a) holds, that is, $L(\varphi) \cap L(\psi) = \phi$. Then by Lemma 4

$$\eta(e) = \begin{cases} \varphi(e) & \text{if } e \in E(T^{p-1}), \text{ and} \\ \psi(e) & \text{if } e \in E(T(e_p)) \end{cases}$$

is an edge-ranking of T^p and $\#\eta = m$.

Suppose next that (b) holds. Let ψ' be an edge-ranking of $T(e_p)$ such that

$$\psi'(e) = \begin{cases} \beta & \text{if } e = e_p; \\ \psi(e) & \text{otherwise.} \end{cases}$$

Then β is visible, but none of the ranks in $L(\psi) \cap [1, \beta]$ is visible with respect to ψ' in $T(e_p)$. Hence $L(\psi') = (L(\psi) - [1, \beta]) \cup \{\beta\}$. Since $L(\varphi) \cap L(\psi) \subseteq [1, \beta - 1]$,

$$L(\varphi) \cap L(\psi') = L(\varphi) \cap L(\psi) - [1, \beta] = \phi.$$

Hence by Lemma 4

$$\eta(e) = \begin{cases} \varphi(e) & \text{if } e \in E(T^{p-1}), \text{ and} \\ \psi'(e) & \text{if } e \in E(T(e_p)) \end{cases}$$

is an edge-ranking of T^p with $\#\eta \le m$.

$\implies$ Suppose that $\chi_r'(T^p) = m$ but neither (a) nor (b) holds:

$$L^*(T^{p-1}) \cap L^*(T(e_p)) \ne \phi \text{ and} \tag{6}$$
$$[\alpha + 1, m] \subseteq L^*(T^{p-1}) \cup L^*(T(e_p)) \tag{7}$$

where α is the maximum integer in $L^*(T^{p-1}) \cap L^*(T(e_p))$. Note that m is the largest integer in $L^*(T^{p-1}) \cup L^*(T(e_p))$. Let η be a critical edge-ranking of T^p, and let $\eta_a = \eta|T^{p-1}$ and $\eta_b = \eta|T(e_p)$. By Lemma 4

$$L(\eta) = L(\eta_a) \cup L(\eta_b), \text{ and} \tag{8}$$
$$L(\eta_a) \cap L(\eta_b) = \phi. \tag{9}$$

Clearly

$$L(\eta_a) \succeq L^*(T^{p-1}), \text{ and} \tag{10}$$
$$L(\eta_b) \succeq L^*(T(e_p)). \tag{11}$$

By (1) $L(\eta_a) - [1, \alpha] \succeq L^*(T^{p-1}) - [1, \alpha]$ and $L(\eta_b) - [1, \alpha] \succeq L^*(T(e_p)) - [1, \alpha]$. We then claim

$$L(\eta_a) - [1, \alpha] = L^*(T^{p-1}) - [1, \alpha], \quad \text{and} \tag{12}$$
$$L(\eta_b) - [1, \alpha] = L^*(T(e_p)) - [1, \alpha]. \tag{13}$$

If (12) or (13) does not hold, that is, $L(\eta_a) - [1, \alpha] \succ L^*(T^{p-1}) - [1, \alpha]$ or $L(\eta_b) - [1, \alpha] \succ L^*(T(e_p)) - [1, \alpha]$, then

$$
\begin{aligned}
L(\eta) - [1, \alpha] &= (L(\eta_a) - [1, \alpha]) \cup (L(\eta_b) - [1, \alpha]) &&(\text{ by } (8)) \\
&\succ (L^*(T^{p-1}) - [1, \alpha]) \cup (L^*(T(e_p)) - [1, \alpha]) &&(\text{ by } (9)) \\
&= L^*(T^{p-1}) \cup L^*(T(e_p)) - [1, \alpha] \\
&= [\alpha + 1, m]. &&(\text{ by } (7))
\end{aligned}
$$

Thus $L(\eta) \succeq \{m + 1\}$ and hence $\#\eta \geq m + 1$, contrary to the supposition $\chi'_r(T^p) = m$. Thus we have verified (12) and (13).

By (2) and (10) – (13) we have

$$L(\eta_a) \cap [1, \alpha] \succeq L^*(T^{p-1}) \cap [1, \alpha] \succeq \{\alpha\}, \quad \text{and} \tag{14}$$
$$L(\eta_b) \cap [1, \alpha] \succeq L^*(T(e_p)) \cap [1, \alpha] \succeq \{\alpha\} \tag{15}$$

since $\alpha \in L^*(T^{p-1}) \cap L^*(T(e_p))$. However, by (9) we have either $\alpha \notin L(\eta_a)$ or $\alpha \notin L(\eta_b)$ and hence either $L(\eta_a) \cap [1, \alpha] \prec \{\alpha\}$ or $L(\eta_b) \cap [1, \alpha] \prec \{\alpha\}$, contrary to (14) or (15). $\mathcal{Q.E.D.}$

In order to find a critical edge-ranking of T^p from those of T^{p-1} and $T(e_p)$, we need the following four lemmas.

Lemma 7. *Let v_1, v_2 and v_3 be children of an internal node u, and let $e_i = (u, v_i)$, $i = 1, 2, 3$. If $L^*(T(e_1)) \succeq L^*(T(e_2))$, then $L^*(T(e_1) + T(e_3)) \succeq L^*(T(e_2) + T(e_3))$.*

Lemma 8. *Let v be a child of an internal node u, and let $e_1 = (u, v)$. Let φ be a critical edge-ranking of $T(v)$, and let $\gamma = \min\{l \geq 1 \mid l \notin L(\varphi)\}$. Then*

$$
\eta(e) = \begin{cases} \varphi(e) & \text{if } e \in E(T(v)), \text{ and} \\ \gamma & \text{if } e = e_1 \end{cases}
$$

is a critical edge-ranking of $T(e_1)$. Given φ, one can find such η in $O(n)$ time.

Proof. One can easily observe that η is an edge-ranking of $T(e_1)$ and

$$L(\eta) = (L(\varphi) - [1, \gamma - 1]) \cup \{\gamma\}. \tag{16}$$

Let η' be any critical edge-ranking of $T(e_1)$. Then

$$L(\eta') = (L(\eta'|T(v)) - [1, \eta'(e_1) - 1]) \cup \{\eta'(e_1)\}. \tag{17}$$

Furthermore

$$\eta'(e_1) \notin L(\eta'|T(v)); \tag{18}$$

otherwise η' would not be an edge-ranking of $T(e_1)$. By (17) and (18) we have

$$L^*(T(e_1)) = L(\eta') \succ L(\eta'|T(v)) \succeq L(\varphi).$$

On the other hand, since $[1, \gamma - 1] \subseteq L(\varphi)$ and $\gamma \notin L(\varphi)$, by (16) $L(\eta)$ is the lexicographically smallest one among all the lists $\succ L(\varphi)$. Thus we have $L^*(T(e_1)) \succeq L(\eta)$. $\qquad$ Q.E.D.

Lemma 9. *Let v_1 and v_2 be children of an internal node u, and let $e_1 = (u, v_1)$ and $e_2 = (u, v_2)$. Let φ and ψ be critical edge-rankings of $T_a = T(e_1)$ and $T_b = T(e_2)$, respectively. Let $T_a'' = T_a(\#\varphi)$ and $T_a' = T_a - T_a''$. Let $m = \max\{\#\varphi, \#\psi\}$. If $\chi_r'(T^2) = m$, $m = \#\varphi$, and η' is a critical edge-ranking of $T_a' + T_b$, then*

$$\eta(e) = \begin{cases} \varphi(e) & \text{if } e \in E(T_a''), \text{ and} \\ \eta'(e) & \text{if } e \in E(T_a' + T_b) \end{cases}$$

is a critical edge-ranking of T^2. (Remember $T^2 = T(e_1) + T(e_2) = T_a + T_b$.)

Proof. We first claim that $\#\eta' \leq m - 1$, that is, $\chi_r'(T_a' + T_b) \leq m - 1$. Since $\chi_r'(T^2) = \chi_r'(T_a) = m$ and T_a and T_b are edge-disjoint subtrees of T^2, by Lemma 3 we have

$$\chi_r'(T_b) \leq m - 1. \tag{19}$$

By (5) we have

$$\chi_r'(T_a') < m = \chi_r'(T_a). \tag{20}$$

Suppose for a contradiction that $\chi_r'(T_a' + T_b) = m$. Then by (19), (20) and Lemma 6 $L^*(T_a') \cap L^*(T_b) \neq \phi$ and $[\alpha + 1, m - 1] \subseteq L^*(T_a') \cup L^*(T_b)$ where α is the maximum integer in $L^*(T_a') \cap L^*(T_b)$. Since $L^*(T_a) = \{m\} \cup L^*(T_a')$ by (4), we have $\alpha \in L^*(T_a) \cap L^*(T_b)$ and $[\alpha + 1, m] \subseteq L^*(T_a) \cup L^*(T_b)$. Hence by Lemma 6 $\chi_r'(T^2) = m + 1$, a contradiction. Thus we have verified $\#\eta' \leq m - 1$.

Since $\#\eta' \leq m - 1$, one can easily observe that η constructed from η' and φ is an edge-ranking of T^2 and $L(\eta) = \{m\} \cup L^*(T_a' + T_b)$.

We finally prove that η is a critical edge-ranking of T^2, that is, $L(\eta) \preceq L(\xi)$ for any critical edge-ranking ξ of T^2. Since $\chi_r'(T^2) = m$, $m = \#\xi \in L(\xi)$. Since $L(\varphi)$ is a critical list of T_a, $L(\xi|T_a) \succeq L(\varphi) \succeq \{m\}$ and hence $m \in L(\xi|T_a)$. Since $T_a' = T_a - T_a''$, by (4)

$$L(\varphi) = \{m\} \cup L(\varphi|T_a') \succeq \{m\} \cup L^*(T_a').$$

Let $\xi' = \xi|T_a$ and $T_a''' = T_a - T_a(\#\xi') = T_a - T_a(\#\xi|T_a)$. Let ρ be a critical edge-ranking of T_a''', and let

$$\mu(e) = \begin{cases} \xi(e) & \text{if } e \in E(T_a(\#\xi')), \\ \rho(e) & \text{if } e \in E(T_a'''). \end{cases}$$

Then μ is an edge-ranking of T_a, and $L(\mu) = \{m\} \cup L(\rho)$. Since φ is a critical edge-ranking of T_a, $L(\varphi) \preceq L(\mu)$. We therefore have

$$\{m\} \cup L^*(T_a') \preceq L(\varphi) \preceq L(\mu) = \{m\} \cup L(\rho) = \{m\} \cup L^*(T_a'''),$$

and consequently $L^*(T'_a) \preceq L^*(T'''_a)$ by (2). Therefore by Lemma 7

$$L^*(T'_a + T_b) \preceq L^*(T'''_a + T_b). \tag{21}$$

Clearly

$$T'''_a + T_b = T^2 - T_a(\#\xi'). \tag{22}$$

By (21) and (22) we have

$$L^*(T'_a + T_b) \preceq L^*(T^2 - T_a(\#\xi')).$$

Thus we have

$$L(\eta) = \{m\} \cup L^*(T'_a + T_b) \preceq \{m\} \cup L^*(T^2 - T_a(\#\xi')) = L(\xi).$$

$$Q.\mathcal{E}.\mathcal{D}.$$

Thus, since $\varphi|T'_a$ is a critical edge-ranking of T'_a, the problem of constructing a critical edge-ranking η of $T^2 = T_a + T_b$ from critical edge-rankings φ and ψ can be reduced to the problem of constructing a critical edge-ranking η' of a smaller tree $T'_a + T_b = T^2 - T''_a$ from critical edge-rankings $\varphi|T'_a$ and ψ when $\chi'_r(T^2) = m$.

Lemma 10. *Let v_1 and v_2 be children of an internal node u, and let $e_1 = (u, v_1)$ and $e_2 = (u, v_2)$. Let φ and ψ be critical edge-rankings of $T(e_1)$ and $T(e_2)$, respectively. Let $m = \max\{\#\varphi, \#\psi\}$. If $\chi'_r(T^2) = m + 1$ and $L(\varphi) \preceq L(\psi)$, then*

$$\eta(e) = \begin{cases} \varphi(e) & \text{if } e \in E(T(e_1)), \\ \psi(e) & \text{if } e \in E(T(e_2)) - \{e_2\}, \text{ and} \\ m + 1 & \text{if } e = e_2 \end{cases}$$

is a critical edge-ranking of T^2.

We have the following two lemmas which imply Lemmas 9 and 10. We omit the proofs since they can be proved similarly as Lemmas 9 and 10.

Lemma 11. *Let $v_1, \cdots, v_p$ be children of an internal node u, and let $e_i = (u, v_i)$ for each i. Let φ and ψ be critical edge-rankings of $T_a = T^{p-1}$ and $T_b = T(e_p)$, respectively. Let $m = \max\{\#\varphi, \#\psi\}$, and suppose that $\chi'_r(T^p) = m$.*
(a) If $L(\varphi) \succeq L(\psi)$, then

$$\eta(e) = \begin{cases} \varphi(e) & \text{if } e \in E(T''_a), \text{ and} \\ \eta'(e) & \text{if } e \in E(T'_a + T_b) \end{cases}$$

is a critical edge-ranking of T^p, where $T''_a = T_a(\#\varphi)$, $T'_a = T_a - T''_a$ and η' is a critical edge-ranking of $T'_a + T_b$.
(b) If $L(\psi) \succ L(\varphi)$, then

$$\eta(e) = \begin{cases} \psi(e) & \text{if } e \in E(T''_b), \text{ and} \\ \eta'(e) & \text{if } e \in E(T_a + T'_b) \end{cases}$$

is a critical edge-ranking of T^p, where $T''_b = T_b(\#\psi)$, $T'_b = T_b - T''_b$ and η' is a critical edge-ranking of $T_a + T'_b$.

Lemma 12. *Let $v_1, \cdots, v_p$ be children of an internal node u, and let $e_i = (u, v_i)$ for each i. Let φ and ψ be critical edge-rankings of $T_a = T^{p-1}$ and $T_b = T(e_p)$, respectively. Let $m = \max\{\#\varphi, \#\psi\}$, and suppose that $\chi'_r(T^p) = m + 1$. Let $L(\varphi|T(e_j))$ be the lexicographically largest list among $L(\varphi|T(e_i))$, $1 \le i \le p - 1$.*
  (a) *If $L(\varphi|T(e_j)) \prec L(\psi)$, then*

$$\eta(e) = \begin{cases} \varphi(e) & \text{if } e \in E(T_a), \\ \psi(e) & \text{if } e \in E(T_b) - \{e_p\}, \text{ and} \\ m+1 & \text{if } e = e_p \end{cases}$$

is a critical edge-ranking of T^p.
  (b) *If $L(\varphi|T(e_j)) \succeq L(\psi)$, and η' is a critical edge-ranking of $T'_a = T^p - T(e_j)$, then*

$$\eta(e) = \begin{cases} \varphi(e) & \text{if } e \in E(T(e_j)) - \{e_j\}, \\ \eta'(e) & \text{if } e \in E(T'_a + T_b), \text{ and} \\ m+1 & \text{if } e = e_j \end{cases}$$

is a critical edge-ranking of T^p.

By Lemmas 8, 11 and 12 above we have the following recursive algorithm to find a critical edge-ranking of a subtree $T(u)$ rooted at a node u.

Procedure Edge-Ranking($T(u)$);
  begin
1   **if** u is a leaf **then return** a trivial edge-ranking: $\phi \to \phi$
2   **else**
3   **begin**
4   let $v_1, \cdots, v_p$ be the children of u, and let $e_i = (u, v_i)$ for all i, $1 \le i \le p$;
5   **for** $i := 1$ **to** p **do** Edge-Ranking($T(v_i)$);
6   **for** $i := 1$ **to** p **do**
          find a critical edge-ranking of $T(e_i)$
          from a critical edge-ranking of $T(v_i)$ by Lemma 8;
7   **for** $i := 2$ **to** p **do**
8   find a critical edge-ranking of T^i
          from those of T^{i-1} and $T(e_i)$ by Lemmas 6, 11 and 12;
9   **return** a critical edge-ranking of $T(u)$ $\{T(u) = T^p\}$
10   **end**
  end.

Clearly one can find a critical edge-ranking of a tree T by calling **Procedure** Edge-Ranking($T(r)$) for the root r of T. Therefore it suffices to verify the time-complexity of the algorithm. As we show the detail later, one execution of line 8 takes in $O(n)$ time. Furthermore line 8 is executed at most $d(u) - 1$ times in the for-loop at line 7. Therefore the for-loop at line 7 can be done in $O(d(u)n)$ time. Since **Procedure** Edge-Ranking is recursively called for all nodes, the total running time is

$$O\left(\sum_{u \in V(T)} d(u)n\right) = O(n^2).$$

This completes the proof of Theorem 3.1.

We finally give an implementation of line 8 of Edge-Ranking, which finds a critical edge-ranking η of T^i from a critical edge-ranking φ of T^{i-1} and a critical edge-ranking ψ of $T(e_i)$.

Procedure Line-8(φ, ψ, η);
begin

1    $L := \phi$; $\{ L = L(\eta) \}$

2   let $T_a := T^{i-1}$ and $T_b := T(e_i)$;

3   let $L_a := L(\varphi)$ and $L_b := L(\psi)$;

4   **while** $L_a \cap L_b \neq \phi$ **do begin**

5        $\alpha := \max\{l \in L_a \cap L_b\}$; $\{\alpha$ does not increase $\}$

6        $\beta := \min\{l \geq \alpha + 1 \mid l \notin L_a \cup L_b\}$; $\{ \beta$ decreases $\}$

7   **if** $L_a \cup L_b - [1, \beta] \neq \phi$ **then**
              $\{ \chi_r'(T_a + T_b) = m$ where $m = \max\{\chi_r'(T_a), \chi_r'(T_b)\}$
              since (b) in Lemma 6 holds $\}$

8   **for** each label $l \in L_a \cup L_b - [1, \beta]$ in decreasing order **do**

9   **if** $l \in L_a$ **then begin** $\{$ cf. Lemma 11(a) $\}$

10              $T_a'' := T_a(\#(\varphi|T_a))$; $\{ T_a''$ is the subtree (of T_a) rooted at
                      the edge e with the largest label $l = \varphi(e)$ $\}$

11              $\eta|T_a'' := \varphi|T_a''$; $\{ \eta|T_a''$ has been decided $\}$

12              $T_a := T_a - T_a''$; $\{$ prune T_a'' $\}$

13              $L_a := L_a - \{l\}$; $\{ L_a = L(\varphi|T_a) \}$

14              $L := L \cup \{l\}$

15   **end**

16   **else begin** $\{ l \in L_b$, cf. Lemma 11(b) $\}$

17              $T_b'' := T_b(\#(\psi|T_b))$;

18              $\eta|T_b'' := \psi|T_b''$; $\{ \eta|T_b''$ has been decided $\}$

19              $T_b := T_b - T_b''$; $\{$ prune T_b'' $\}$

20              $L_b := L_b - \{l\}$; $\{ L_b = L(\psi|T_b) \}$

21              $L := L \cup \{l\}$

22   **end**;
                $\{ \beta = m + 1$ and $\chi_r'(T_a + T_b) = m + 1 \}$

23   let e_j be the edge with the lexicographically largest list $L(\varphi|T_a(e_j))$
           among the edges e_j, $1 \leq j \leq i - 1$, remaining in T_a;

24   **if** $L(\varphi|T_a(e_j)) \succ L_b$ **then begin** $\{$ cf. Lemma 12 $\}$

25          $\eta|(T_a(e_j) - e_j) := \varphi|(T_a(e_j) - e_j)$;

26          $\eta(e_j) := \beta$; $\{ \eta|T_a(e_j)$ has been decided $\}$

27           $L_a := L_a - L(\varphi|T_a(e_j))$;

28           $T_a := T_a - T_a(e_j)$; $\{$ prune $T_a(e_j)$ $\}$

29           $L := L \cup \{\beta\}$;

30   **end**

31   **else begin** $\{ L(\varphi|T_a(e_j)) \preceq L_b \}$

32          $\eta|(T_b - e_i) := \psi|(T_b - e_i)$;

33          $\eta(e_i) := \beta$; $\{ \eta|T_b$ has been decided $\}$

34          $L_b := \phi$;

```
35        let T_b be empty tree;
36        L := L ∪ {β}
37     end
38    end;
```

$$39 \qquad\qquad\qquad\qquad \{\ L_a \cap L_b = \phi,\ \text{cf. Lemma 11}\ \}$$

$$40 \quad \eta|T_a := \varphi|T_a; \qquad\qquad\quad \{\ \eta|T_a \text{ has been decided}\ \}$$

$$41 \quad \eta|T_b := \psi|T_b; \qquad\qquad\quad \{\ \eta|T_b \text{ has been decided}\ \}$$

$$42 \quad L := L \cup L_a \cup L_b \qquad\qquad \{\ \eta \text{ has been decided}\ \}$$

```
   end;
```

Whenever lines 5 and 6 are executed, α does not increase and β always decreases. Furthermore $\beta > \alpha$. Thus lines 5–37 in the **while**-loop are executed at most n times. Noting this fact, one can easily observe that the book-keeping operations takes $O(n)$ time in total. Thus **Procedure** Line-8 takes $O(n)$ time.

Remark: A more sophisticated implementation leads to an $O(n \log n)$ algorithm. We omit the detail for the sake of space.

References

[BDJ+94] H. Bodlaender, J.S. Deogun, K. Jansen, T. Kloks, D. Kratsch, H. Müller, and Zs. Tuza. Ranking of graphs. In *Proc. of International Workshop on Graph-Theoretic Concepts in Computer Science*, Herrsching, Bavaria, Germany, 1994.

[dlTGS93] P. de la Torre, R. Greenlaw, and A. A. Schäffer. Optimal ranking of trees in polynomial time. In *Proc. of the 4th Ann. ACM-SIAM Symp. on Discrete Algorithms*, pp. 138–144, Austin, Texas, 1993.

[DP90] J. S. Deogun and Y. P. Peng. Edge ranking of trees. *Congressus Numerantium*, 79, pp.19–28, 1990.

[IRV88a] A. V. Iyer, H. D. Ratliff, and G. Vijayan. Optimal node ranking of trees. *Information Processing Letters*, 28, pp.225–229, 1988.

[IRV88b] A.V. Iyer, H.D. Ratliff, and G. Vijayan. Parallel assembly of modular products - an analysis. Technical Report PDRC, Technical Report 88-06, Georgia Intitute of Technology, 1988.

[IRV91] A. V. Iyer, H. D. Ratliff, and G. Vijayan. On an edge-ranking problem of trees and graphs. *Discrete Applied Mathematics*, 30, pp.43–52, 1991.

[KMS88] M. Katchalski, W. McCuaig, and S. Seager. Ordered colourings. manuscript, *University of Waterloo*, 1988.

[Pot88] A. Pothen. The complexity of optimal elimination trees. Technical Report CS-88-13, Pennsylvania State University, U.S.A., 1988.

[Sch89] A. A. Schaffer. Optimal node ranking of trees in linear time. *Information Processing Letters*, 33(2), pp.91–99, 1989.

Edge–Disjoint (s, t)–Paths in Undirected Planar Graphs in Linear Time*

Karsten Weihe

Technische Universität Berlin, Sekr. MA 6-1, Str. d. 17. Juni. 136, 10623 Berlin, Germany, *e-mail:* weihe@math.tu-berlin.de .

Abstract. We consider the following problem. Let $G = (V, E)$ be an undirected planar graph and let $s, t \in V$, $s \neq t$. The problem is to find a set of pairwise edge–disjoint paths, each connecting s with t, with maximum cardinality. In other words, the problem is to find a maximum unit flow from s to t. The fastest algorithm in the literature has running time $\mathcal{O}(|V| \log |V|)$. In this paper now, we give a linear time algorithm.

1 Introduction

Let $G = (V, E)$ be an undirected, planar graph and let $s, t \in V$, $s \neq t$. The problem is to find a maximum number of paths connecting s with t such that no two paths have an edge of G in common (edge–disjoint *Menger problem*). This problem is of some interest in connection with VLSI design and in the layout of traffic and communication networks, since there the underlying networks are often undirected and planar.

The best worst–case bound obtained so far is based on network flows. In [Re], Reif gives an algorithm that finds a minimum (s, t)–cut in undirected, planar graphs with respect to *arbitrary* positive capacities. This algorithm consists in a reduction to $\mathcal{O}(\log n)$ shortest path computations in the dual graph and, hence, requires $\mathcal{O}(n \log^2 n)$ time, where $n = |V|$. Hassin and Johnson showed how to make use of this idea to solve, within the same asymptotic running time, the maximum flow problem itself in undirected, planar graphs with arbitrary positive capacities [HJ]. Based on Frederickson's decomposition technique, both algorithms can be implemented so as to run in $\mathcal{O}(n \log n)$ time. Frederickson's technique is applied to solve a single shortest path problem in linear time even for arbitrary capacities (with $\mathcal{O}(n \log n)$ preprocessing time in addition). Therefore, in the unit case, which is the special case we consider here, we cannot do better by replacing Dijkstra's algorithm with a simple breadth–first search.

In this paper now, another approach is introduced, which yields an $\mathcal{O}(n)$ algorithm. Instead of G itself, the algorithm works on the directed, symmetric graph $G^{\rightarrow} = (V, E^{\rightarrow})$, which arises from G by replacing each undirected edge $\{v, w\} \in E$ with the two corresponding directed arcs (v, w) and (w, v). It suffices to find a solution for $G^{\rightarrow}$, since this immediately yields a solution for G as well: An edge $\{v, w\} \in E$ belongs to the solution for G if and only if exactly one of (v, w) and (w, v) belongs to the solution for $G^{\rightarrow}$. See Figs. 11 and 12.

---

* This work was supported by the Technische Universität Berlin under grant FIP 1/3.

The basic idea is the following. First we determine a circulation in $G^\rightarrow$ such that the residual graph of this circulation contains no clockwise cycle. This is crucial for the core procedure, which constructs a maximum number of edge–disjoint (s, t)–paths in a certain class of directed, planar graphs: All cycles have the same orientation and the indegree of each vertex equals its outdegree. The core procedure is applied to the residual graph of that circulation, and the result is then easily transformed into a maximum solution for $G^\rightarrow$ itself.

Very recently, the *vertex–disjoint* version of this problem could be solved in linear time as well [RWW]. That algorithm is based on similar ideas, but runs fairly differently. In particular, completely different strategies are applied to avoid clockwise cycles. The correctness proof for the vertex–disjoint case is *much* more complicated. In view of the result presented now, it seems that the general approach behind these algorithms conforms to the edge–disjoint case much better than to the vertex–disjoint case.

The paper is organized as follows. In Section 2, all necessary terminology is introduced. Then, in Section 3, the algorithm itself is given. It will turn out that the overall running time is dominated by the time required to perform a certain sequence of *union–find* operations. Using a technique developed by Gabow and Tarjan [GT], these operations can, too, be implemented so as to run in linear time. Finally, in Section 4, correctness of the algorithm is proved.

2 Preliminaries

Let $G^\rightarrow = (V, E^\rightarrow)$ be the directed, symmetric graph that arises from G when we replace each undirected edge $\{v, w\} \in E$ with the two corresponding directed arcs (v, w) and (w, v). See Figs. 1 and 2. From now on, we will restrict attention to the problem of finding a maximum *unit flow* F from s to t in $G^\rightarrow$. That is, F is a $0/1$–weighting of all arcs of $G^\rightarrow$ such that

$$\sum_{w:(v,w)\in E^\rightarrow} F_{(v,w)} = \sum_{w:(w,v)\in E^\rightarrow} F_{(w,v)}$$

for $v \in V \setminus \{s, t\}$ (*flow conservation conditions*) and the *total flow value*

$$\sum_{w:(s,w)\in E^\rightarrow} F_{(s,w)} - \sum_{w:(w,s)\in E^\rightarrow} F_{(w,s)}$$

is maximum. (See [AMO1, AMO2] for an introduction to and a comprehensive survey of network flows, respectively.) Obviously, a maximum collection of edge–disjoint (s, t)–paths in G can be constructed from any maximum unit (s, t)–flow in $G^\rightarrow$ in linear time.

A *circulation* c is an (s, t)–flow with total flow value 0. In other words, *all* vertices in V satisfy the flow conservation conditions with respect to c. If a circulation c is a unit flow, this means that c decomposes into edge–disjoint cycles. Henceforth, let c be a unit circulation in $G^\rightarrow$. See Fig. 3.

The *residual graph* with respect to c, denoted by $G_c^\rightarrow = (V, E_c^\rightarrow)$, is defined as follows. For each arc $(v, w) \in E^\rightarrow$ with $c_{(v,w)} = 0$, we have $(v, w) \in E_c^\rightarrow$, too. On the other hand, for each arc $(v, w) \in E^{\rightarrow}$ with $c_{(v,w)} = 1$, we have $(w, v) \in E_c^{\rightarrow}$. (This is not the usual definition of residual networks, but in case of unit capacities, it is completely equivalent.) In particular, $G_c^{\rightarrow}$ contains double arcs in general. Whenever we refer to a double arc (v, w), we mean exactly either of the copies. If $(v, w) \in E_c^{\rightarrow}$ appears twice, one copy corresponds to $(v, w) \in E^{\rightarrow}$, and the other one, to $(w, v) \in E^{\rightarrow}$. Note that for $v \in V$, the number of arcs of $E_c^{\rightarrow}$ pointing to v equals the number of arcs of $E_c^{\rightarrow}$ leaving v.

There is a natural one-to-one correspondence between unit (s, t)-flows in $G^{\rightarrow}$ and in $G_c^{\rightarrow}$, namely, a flow F in $G^{\rightarrow}$ and a flow f in $G_c^{\rightarrow}$ correspond to each other if and only if we have $F_{(v,w)} = f_{(v,w)}$ for $c_{(v,w)} = 0$ and $F_{(v,w)} = 1 - f_{(w,v)}$ for $c_{(v,w)} = 1$. In other words, the flow value is switched if and only if the arc itself is switched. Corresponding flows have the same total flow value. Therefore, it suffices to find a maximum flow in $G_c^{\rightarrow}$, and exactly this is the core of the algorithm.

As mentioned in the Introduction, we are interested in determining a unit circulation c in $G^{\rightarrow}$ such that $G_c^{\rightarrow}$ contains no clockwise cycles. See Figs. 3 and 4. The residual graph $G_c^{\rightarrow}$ of c may be characterized as follows. For each pair (v, w), $(w, v) \in E^{\rightarrow}$, consider the faces $\mathcal{F}_1$ and $\mathcal{F}_2$ of $G^{\rightarrow}$ that are neighbored to this pair. Precisely, v, $\mathcal{F}_1$, w, and $\mathcal{F}_2$ appear in the counterclockwise order $v \prec \mathcal{F}_1 \prec w \prec \mathcal{F}_2$ around the pair (v, w), (w, v). Then the distances of $\mathcal{F}_1$ and $\mathcal{F}_2$ from the outer face, that is the minimum number of faces in between, differ from each other by at most 1. If they are equal, we have (v, w), $(w, v) \in E_c^{\rightarrow}$, too. Else if $\mathcal{F}_1$ is closer to the outer face than $\mathcal{F}_2$, we have two copies of (v, w) in $E_c^{\rightarrow}$. Else we have two copies of (w, v) in $E_c^{\rightarrow}$. See Fig. 4. Obviously, the distances of all faces from the outer face can be computed in linear time by applying a breadth–first search to the dual graph of $G^{\rightarrow}$, with the outer face being the root. From that, the circulation c and the corresponding residual graph $G_c^{\rightarrow}$ are easily computed in linear time as well. This justifies our focus on the problem of finding a maximum unit (s, t)-flow in $G_c^{\rightarrow}$.

The *adjacency list* of $v \in V$ consists of all arcs in $E_c^{\rightarrow}$ that are incident to v, that is, leaving v or pointing to v. We assume that the adjacency lists of all vertices are sorted according to a fixed embedding of $G_c^{\rightarrow}$ in the plane. In particular, this defines a fixed ordering of both copies of a double arc. Such an ordering of all adjacency lists can always be obtained in linear time, see [HT], for example. W.l.o.g., t is incident to the outer face of $G_c^{\rightarrow}$, as shown in Fig. 1.

3 The Algorithm

We are now going to give an informal explanation of the algorithm. A *formal* description is given in Table 1, and an example, in Figs. 1–12.

As mentioned in the last section, it suffices to construct a maximum unit (s, t)-flow f in $G_c^{\rightarrow}$. In the beginning, all arcs $a \in E_c^{\rightarrow}$ carry zero flow, $f_a = 0$. Let $a_1, \ldots, a_k$ be the arcs of $G_c^{\rightarrow}$ leaving s, in arbitrary order. The procedure to determine a maximum flow f is essentially a loop over $1, \ldots, k$. In the i-th iteration of this loop, we determine a directed path in $G_c^{\rightarrow}$ and switch, on–line, the flow values of all arcs on this path from 0 to 1. This path starts with a_i, ends either at s or at t, and, clearly, contains only arcs which still carry zero flow. In particular, after each iteration, f is a unit (s, t)-flow with nonnegative total flow value. We never need to correct the flow value of any arc, once we have changed it from 0 to 1.

To determine the path in the i-th iteration, say, we start a *right-first search* with a_i. This is a depth-first search where in each step the rightmost possibility of going forward is chosen. More precisely, if v and a are the leading vertex and the leading arc of the current search path, respectively, then the next arc to be added is the counterclockwise next arc after a in the adjacency list of v that leaves v and carries zero flow. (Recall that all adjacency lists are sorted according to a fixed embedding in the plane.)

Since the number of arcs in $G_c^\rightarrow$ pointing to $v \in V$ equals the number of arcs in $G_c^\rightarrow$ leaving v, there always remains a possibility of going forward from v, as long as we need one to proceed with. Hence, this right-first search must finally reach s or t without performing a backtrack step in the meantime. (We allow that the path is non-simple.) This already implies that the total number of forward steps, taken over *all* iterations of the overall loop, is linear, because in each single forward step the flow value of one arc is set from 0 to 1, and no arc with flow value 1 is touched later on again. Therefore, for the linear worst-case bound it suffices to show that each forward step can be done in (amortized) constant time. The rest of the informal description is devoted to this particular task.

Single forward step in amortized constant time. Consider a single step of the i-th right-first search, where $v \in V \setminus \{s,t\}$ is the leading vertex and $(u,v) \in E_c^\rightarrow$ is the leading arc of the search path. Then we have to determine the counterclockwise next arc after (u,v) in the adjacency list of v which leaves v and still carries zero flow. In order to determine this arc efficiently in each step, we in turn maintain a (dynamically changing) arc set $S_{(v,w)}$ for each arc (v,w) with $f_{(v,w)} = 0$. The set $S_{(v,w)}$ consists of all arcs $(u,v) \in E_c^\rightarrow$ such that (v,w) is the counterclockwise next arc after (u,v) in the adjacency list of v which leaves v and still carries zero flow. Thus, for fixed v, the sets $S_{(v,\cdot)}$ partition the arcs pointing to v into disjoint intervals. More precisely, at each stage this is the coarsest partition of the arcs pointing to v such that any two subsequent partition sets are separated by an arc that leaves v and carries zero flow.

Each set $S_{(v,w)}$ keeps the corresponding arc (v,w) "in mind." Hence, finding the next arc for going forward amounts to finding the set to which the leading arc of the search path belongs. So let $(u,v) \in S_{(v,w)}$ be the leading arc of the search path at some stage. Let a be the counterclockwise next arc after (v,w) in the adjacency list of v. If a leaves v, then a must still carry zero flow, because anyway, (v,w) must have been considered before a for going forward from v. Then $S_{(v,w)}$ must from now on keep a "in mind" instead of (v,w). On the other hand, if a points to v, $S_{(v,w)}$ must from now on store the arc (v,x) with $a \in S_{(v,x)}$. In other words, the arc sets $S_{(v,w)}$ and $S_{(v,x)}$ have to be united, unless $S_{(v,w)} = S_{(v,x)}$ already.

Therefore, maintaining and using these arc sets amounts to one *union-find* problem for each vertex. These union-find problems are of a very restricted shape. In fact, the sets to be maintained for vertex v are disjoint intervals in the cyclic list of all arcs pointing to v. Moreover, only intervals which are neighbored in this cyclic list are subject to union operations. Gabow and Tarjan have given a linear time technique for union-find problems like this, but only for the case that the underlying list is not *cyclic* but *linear* [GT]. (In general, this technique applies to *tree* structured

Table 1. Formal description of the algorithm

1. Construct $G^{\rightarrow} = (V, E^{\rightarrow})$ from G;
2. construct the circulation c and the residual graph $G_c^{\rightarrow} = (V, E_c^{\rightarrow})$;
3. initialize the union-find structures of all vertices: each inclusion-maximal interval of arcs pointing to $v \in V$ in $G_c^{\rightarrow}$ forms one initial set in the union-find structure of v;
4. FOR $a \in E_c^{\rightarrow}$ DO $f_a := 0$;
5. let $a_1, \ldots, a_k \in E_c^{\rightarrow}$ be the arcs of $G_c^{\rightarrow}$ leaving s;
6. FOR $i := 1$ TO k DO $right\_first\_search\ (i)$;
7. FOR $(v, w) \in E^{\rightarrow}$ DO
   IF $c_{(v,w)} = 0$ THEN $F_{(v,w)} := f_{(v,w)}$
   ELSE $F_{(v,w)} := 1 - f_{(w,v)}$;
8. construct from F a collection of edge-disjoint (s, t)-paths in G;

PROCEDURE $right\_first\_search\ (i : Integer)$;

1. $f_{a_i} := 1$;
2. $a := a_i$;
3. WHILE $head\ (a) \notin \{s, t\}$ DO $go\_forward\ (a)$;

PROCEDURE $go\_forward\ (\text{VAR } a \in E_c^{\rightarrow})$;

1. $v := head\ (a)$;
2. $S := find\ (a)$;
3. $a := S.next$;
4. $f_a := 1$;
5. $a' :=$ the counterclockwise next arc after a in the adjacency list of v;
6. IF a' leaves v
   THEN $S.next := a'$
   ELSE $T := find\ (a')$;
       IF $S \neq T$
       THEN $U := union\ (S, T)$;
           $U.next := T.next$;

problems, not only to linear ones.) In [WW, We] it is, respectively, shown how to extend this technique to cyclic lists: Roughly speaking, in the beginning we break the cyclic list between two initial partition sets, which makes the list linear. All union operations except at most one are done according to Gabow and Tarjan's technique. The only possible exception is the case that the two sets around the breakpoint have to be united. In this case, we do *not* apply the union operation provided by Gabow and Tarjan's data structure, but keep simply "in mind" that, henceforth, these two sets are merely two halves of the same set.

This completes the informal description. An example is given in Figs. 1–12. We refer to Table 1 for further details. There the algorithm is more or less formulated in a *top-down* fashion, that is, each subroutine is defined only after its application is given. For each arc set S, $S.next$ is the arc kept "in mind" by S. The arc set to which arc a belongs is returned by the function call $find\,(a)$. Two arc sets S and T are united by the operation $union\,(S,T)$, provided S and T are neighbored intervals in the adjacency list of the same vertex. (For convenience, this includes the irregular "unions" mentioned above.) In particular, we require $S \neq T$, that is, at least two arc sets in that adjacency list have still "survived" all previous union operations. The name of the resulting, united, arc set is returned by function $union$. Finally, the function call $head\,(a)$ returns the vertex that arc $a \in E_c^{\rightarrow}$ points to.

The overall result of this section is summarized in the following theorem.

Theorem 1. *The algorithm in Table 1 constructs a unit (s,t)-flow f in $G_c^{\rightarrow}$ with nonnegative total flow value. In particular, it constructs a unit (s,t)-flow F in $G^{\rightarrow}$ with nonnegative total flow value and, hence, a (possibly empty) set of pairwise edge-disjoint (s,t)-paths in G. All this is done in linear time (if the technique of Gabow and Tarjan is applied to construct f).*

4 Correctness

Because of Theorem 1, it remains to show that the final flow f has maximum total flow value among all unit (s,t)-flows in $G_c^{\rightarrow}$.

By the famous Labeling Theorem of Ford and Fulkerson [FF], it suffices to show that there is no (simple) augmenting (s,t)-path in $G_c^{\rightarrow}$ with respect to f. That is, a path from s to t that may contain an arc a of $G_c^{\rightarrow}$ in forward *or* backward direction, but in forward direction only if $f_a = 0$, and in backward direction only if $f_a = 1$. Since all arcs $a_1, \ldots, a_k$ carry unit flow, such an augmenting (s,t)-path must start with an arc that points to s and carries unit flow. We will show that this is not possible. So suppose for a contradiction that there is an arc (w,s) where a simple augmenting (s,t)-path starts.

We will consider a certain sequence of arcs, $\alpha^1 = (v_0,v_1)$, $\alpha^2 = (v_1,v_2)$, $\alpha^3 = (v_2,v_3)$, up to $\alpha^l = (v_{l-1},v_l)$, with $f_{\alpha^j} = 1$ for all $j = 1,\ldots,l$. See Fig. 10. The arcs $\alpha^1,\ldots,\alpha^l$ are defined recursively "the other way round." More precisely, we have $v_l = s$ and $v_{l-1} = w$ and hence $\alpha^l = (w,s)$. For $j < l$, α^j is the clockwise next arc after $\alpha^{j+1} = (v_j,v_{j+1})$ in the adjacency list of v_j that points to v_j and carries positive flow. The arc α^1 is the first arc in this sequence such that either $v_0 = s$ or the arc to be chosen next already belongs to the sequence. Clearly, for $v_j \neq s$, we always find an arc with positive flow that points to v_j, since we reach v_j via an arc with positive flow that leaves v_j. Hence, this sequence of arcs is well defined (and finite, of course).

The idea is to show that there is $\mu \in \{1,\ldots,l\}$ such that the arcs $\alpha^1,\ldots,\alpha^\mu$ form a cycle C which excludes t but does not exclude s or w, and no augmenting path can leave C and enter its exterior except, possibly, at s, if s belongs to C. As a consequence, no *simple* augmenting path starting with (w,s) can reach t, because C blocks all these paths. See Fig. 10.

Now, if the sequence terminates with α^1 because of $v_0 = s$, the cycle C is defined as $C = (\alpha^1 - \alpha^2 - \cdots - \alpha^{l-1} - \alpha^l - \alpha^1)$, which means $\mu = l$. Else C is defined as $C = (\alpha^1 - \alpha^2 - \cdots - \alpha^{\mu-1} - \alpha^\mu - \alpha^1)$, where α^μ had been added next to the sequence, if we had not deliberately terminated the sequence with α^1. In other words, α^μ is the clockwise next arc after α^1 in the adjacency list of $v_0 = v_\mu$ which points to v_μ and carries positive flow. For convenience, we henceforth identify v_i with $v_{i \bmod \mu}$.

Notice that C may not be a simple cycle. In fact, in general we may have $v_i = v_j$ for $i \neq j$ and, thus, C may contain subcycles. But at least one can prove the following fact.

Lemma 2. *The cycle C does not cross itself, that is: There are no $i, j \in \{1, \ldots, \mu\}$, $i \neq j$, such that $v_i = v_j$ and the counterclockwise ordering of α^i, α^{i+1}, α^j, and α^{j+1} around v_i is $\alpha^i \prec \alpha^j \prec \alpha^{i+1} \prec \alpha^{j+1}$.*

Proof. In this case, α^j had been chosen as the next arc after α^{i+1} rather than α^i. $\square$

Corollary 3. *The interior and the exterior, the right side and the left side of C are well defined. Since $G_{\vec{C}}$ contains no clockwise cycle, C is a counterclockwise cycle. In other words, the exterior of C is its right side.*

Lemma 4. *Any arc that points to a vertex on C from the exterior of C carries zero flow, except, possibly, in one situation: s belongs to C and s is the head of the arc.*

Proof. Suppose a points to a vertex $v_j \neq s$ on C, is outside C, and carries positive flow. There may be several arcs of C pointing to v_j, but w.l.o.g. (v_{j-1}, v_j) is the clockwise next such arc after a in the adjacency list of v_j. As a is outside C, a points to C from the right side. Hence, the clockwise order of (v_j, v_{j+1}), a, and (v_{j-1}, v_j) around v_j is $(v_j, v_{j+1}) \prec a \prec (v_{j-1}, v_j)$. Therefore, a had become the next member of the sequence after (v_j, v_{j+1}) rather than (v_{j-1}, v_j). $\square$

Lemma 5. *Any arc that points from a vertex of C to the exterior of C carries positive flow.*

Proof. Suppose there is an arc a that points from v_j to the exterior of C and carries zero flow. Then we have $v_j \neq s$, because $f_{a_1} = f_{a_2} = \cdots = f_{a_k} = 1$.

Consider the counterclockwise next arc of C after a in the adjacency list of v_j. As the exterior of C is its right side, this arc leaves v_j, say (v_j, v_{j+1}). Now let (x, v_j) be the arc that was assigned flow value 1 immediately before (v_j, v_{j+1}) during the algorithm. Then the clockwise ordering of a, (v_{j-1}, v_j), (x, v_j), and (v_j, v_{j+1}) around v_j is $a \prec (v_{j-1}, v_j) \preceq (x, v_j) \prec (v_j, v_{j+1})$, where the placement of (x, v_j) follows from Lemma 4. But then, because of our right–first strategy, a had been preferred to (v_j, v_{j+1}) as the next arc after (x, v_j) to be assigned a positive flow value during the algorithm. $\square$

Lemma 6. *The exterior of C contains t, but neither s nor w.*

Proof. First note that t cannot lie on C, because no arc leaving t is given positive flow value during the algorithm. Recall that t is incident to the outer face of G. Therefore, t cannot belong to the interior of C either. Hence, t belongs to the exterior of C.

Now consider s and w. We make a case distinction. If s belongs to C, then we have $v_0 = v_\mu = s$, and all arcs α^j belong to C. In particular, $(w, s) = \alpha^l$ does, and w belongs to C, too.

On the other hand, assume that s does not belong to C. As the arcs of C have positive flow value, there must be a directed path p from s to C such that all arcs on p have positive flow value. If s belongs to the exterior of C, then the last arc of p points to C from outside, which contradicts Lemma 4. Therefore, s belongs to the interior of C, and since w is incident to s, w must belong either to C itself or to its interior. $\square$

Now we are in a position to prove correctness of the algorithm.

Theorem 7. *For $(w,s) \in E_c^{\rightarrow}$, no simple augmenting (s,t)-path with respect to the final flow f may start with (w,s). In particular, the final flow f is maximum.*

Proof. Suppose there is a simple augmenting (s,t)-path p starting with (w,s). By Lemma 6, p must leave C and enter its exterior at some vertex different from s. There are two possibilities for the first arc of p outside C: Either it points to C and carries positive flow, or it leaves C and carries zero flow. But this contradicts either Lemma 4 or Lemma 5. $\square$

References

[AMO1] R.K. Ahuja, T.L. Magnanti and J.B. Orlin: *Network Flows*. In: G.L. Nemhauser, A.H.G. Rinnooy Kan and M.J. Todd (eds.), *Handbooks in Operations Research and Management Science* 1. Elsevier Science Publishers, North-Holland, 1989.

[AMO2] R.K. Ahuja, T.L. Magnanti und J.B. Orlin: *Network Flows*. Prentice Hall, 1993.

[FF] L.R. Ford and D.R. Fulkerson (1962): *Flows in networks*. Princeton University Press.

[Fr] G.N. Frederickson (1987): *Fast algorithms for shortest paths in planar graphs, with applications*. SIAM J. Comput. 16, 1004-1022.

[GT] Gabow and Tarjan (1985): *A linear-time algorithm for a special case of disjoint set union*. J. Comp. System Sci. 30, 209-221.

[HJ] R. Hassin and D.B. Johnson (1985): *An $\mathcal{O}(n \log^2 n)$ algorithm for maximum flow in undirected planar networks*. SIAM J. Comput. 14, 612-624.

[HT] J. Hopcroft and R.E. Tarjan (1974): *Efficient planarity testing*. J. ACM 21, 549-568.

[Re] J.H. Reif (1983): *Minimum $s - t$ cut of a planar undirected network in $\mathcal{O}(\log^2(n))$ time*. SIAM J. Comput. 12, 71-81.

[RWW] H. Ripphausen-Lipa, D. Wagner and K. Weihe (1992): *The vertex-disjoint Menger problem in planar graphs*. Proc. 4th Ann. ACM-SIAM Symp. Discrete Algorithms (SODA '93), 112-119.

[WW] D. Wagner and K. Weihe (1992): *A linear-time algorithm for edge-disjoint paths in planar graphs*. Proc. 1st Europ. Symp. Algorithms (ESA '93), Lect. Notes Comp. Science 726, 384-395.

[We] K. Weihe (1993): *Multicommodity flows in even, planar networks*. Proc. 4th Ann. Symp. Algorithms and Comp. (ISAAC '93), Lect. Notes Comp. Science 762, 333-342.

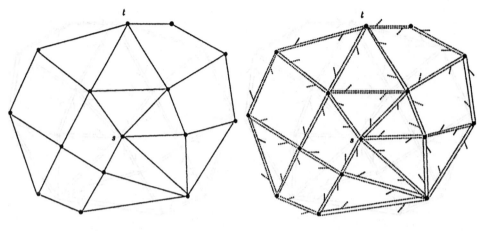

Fig. 1. An input instance $G = (V, E)$.

Fig. 2. The directed, symmetric graph $G^\rightarrow = (V, E^\rightarrow)$. The direction of an arc is indicated by a "half-arrow."

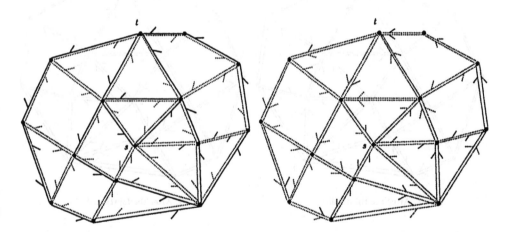

Fig. 3. The circulation c. Arcs with positive flow value are emphasized. The circulation c here consists of three nested cycles, and the i-th cycle from outside separates the faces with distance at least i from the other faces.

Fig. 4. The residual graph $G_c^\rightarrow = (V, E_c^\rightarrow)$. Two parallel arcs have the same orientation if and only if the face on the right side is closer to the outer face than the face on the left side. Note that $G^\rightarrow$ has no clockwise cycles.

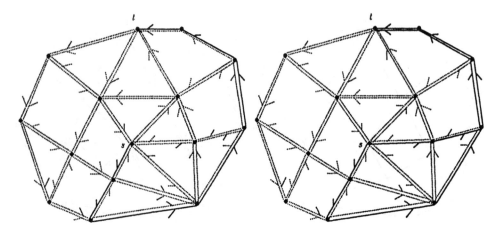

Fig. 5. After the first call to the procedure *right_first_search*.

Fig. 6. After the second call.

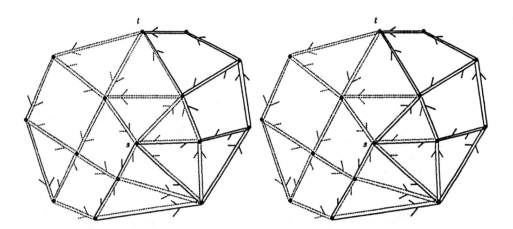

Fig. 7. After the third call.

Fig. 8. After the fourth call.

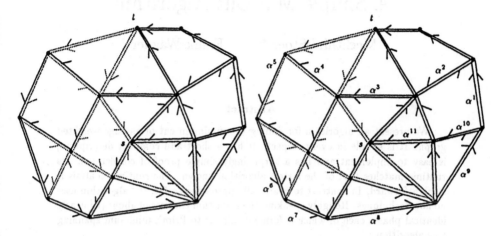

Fig. 9. The final flow f after the fifth, and last, call to *right_first_search*. The path found in this iteration ends with s and is highly pathological.

Fig. 10. The sequence of arcs constructed in Sect. 4 to prove maximality of the final flow f. We have $l = 11$ and $\mu = 9$. In particular, $\mathcal{C} = (\alpha^1 - \alpha^2 - \cdots - \alpha^9 - \alpha^1)$.

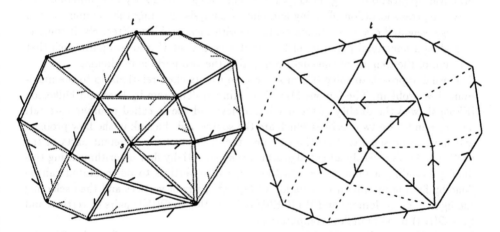

Fig. 11. The flow F in $G^{\rightarrow}$ corresponding to f.

Fig. 12. The collection of (s,t)-paths in G induced by F.

A Simple Min Cut Algorithm

Mechthild Stoer *   Frank Wagner †

Abstract

We present an algorithm for finding the minimum cut of an edge-weighted graph. It is simple in every respect. It has a short and compact description, is easy to implement and has a surprisingly simple proof of correctness. Its runtime matches that of the fastest algorithm known. The runtime analysis is straightforward. In contrast to nearly all approaches so far, the algorithm uses no flow techniques. Roughly speaking the algorithm consists of about $|V|$ nearly identical phases each of which is formally similar to Prim's minimum spanning tree algorithm.

1   Overview

Graph connectivity is one of the classical subjects in graph theory, and has many practical applications, e. g. in chip and circuit design, reliability of communication networks, transportation planning and cluster analysis. Finding the minimum cut of an edge-weighted graph is a fundamental algorithmical problem. Precisely, it consists in finding a nontrivial partition of the graph's vertex set V into two parts such that the sum of the weights of the edges connecting the two parts is minimum.

The usual approach to solve this problem is to use its close relationship to the maximum flow problem. The famous Max-Flow-Min-Cut-Theorem by Ford and Fulkerson [FF56] showed the duality of the maximum flow and the so-called minimum s-t cut. There, s and t are two vertices which are the source and the sink in the flow problem and have to be separated by the cut, i.e., they have to lie in different parts of the partition. Until recently all cut algorithms were essentially flow algorithms using this duality. Finding a minimum cut without specified vertices to be separated can be done by finding minimum s-t cuts for all $\mathcal{O}(|V|^2)$ pairs of vertices and then selecting the lightest one. Gomory and Hu [GH61] reduced this to $\mathcal{O}(|V|)$ pairs of vertices and thus $\mathcal{O}(|V|)$ maximum flow computations.

Recently Hao and Orlin [HO92] showed how to use the maximum flow algorithm by Goldberg and Tarjan [GT88], the fastest so far, in order to solve the minimum cut problem as fast as the maximum flow problem, i.e., in time $\mathcal{O}(|V||E|\log(|V|^2/|E|))$.

In the same year Nagamochi and Ibaraki [NI92a] published the first minimum cut algorithm that is not based on a flow algorithm, has the slightly better running time of $\mathcal{O}(|V||E|+|V|^2 \log |V|)$ but is still rather complicated. In the unweighted case they

---

*Norwegian Telecom Research, P. O. Box 83, N-2007 Kjeller, e-mail:mechthild.stoer@nta.no

†Institut für Informatik, Fachbereich Mathematik und Informatik, Freie Universität Berlin, Takustraße 9, 14195 Berlin-Dahlem, Germany, e-mail: wagner@math.fu-berlin.de

use a fast search technique to decompose a graph's edge set E into subsets $E_1, \ldots, E_\lambda$ such that the union of the first k E_i's is a k-edge-connected spanning subgraph of the given graph and has size at most $k|V|$. They simulate this approach in the weighted case. Their work is one of a small number of papers treating questions of graph connectivity by non-flow-based methods [NP89, NI92a, M93].

In this context we present in this paper a remarkably simple minimum cut algorithm with the optimal running time established in [NI92b]. We reduce the complexity of the algorithm of Nagamochi and Ibaraki by avoiding the unnecessary simulated decomposition of the edge set. This enables us to give a comparably straightforward proof of correctness avoiding e. g. the distinction between the unweighted, integer-, rational-, and real-weighted case.

2   The Algorithm

Throughout the paper we deal with an ordinary undirected graph G with vertex set V and edge set E. Every edge e has positive real weight $w(e)$.

In order to describe the idea of the algorithm we start by reminding the reader of Prim's minimum spanning tree algorithm:

$\text{MINIMUMSPANNINGTREE}(G, w, a)$
$A \leftarrow \{a\}$
$T \leftarrow \emptyset$
while $A \neq V$
    add to A the most loosely connected vertex
    add to T the connecting edge

A subset A of the graph's vertices grows starting with an arbitrary single vertex until A is equal to V. In each step the vertex outside of A *most loosely connected* with A is added. Formally, we add a vertex

$$z \notin A \text{ such that } w(a, z) = \min\{w(b, y) \mid b \in A, y \notin A, by \in E\}$$

where $w(b, y)$ is the weight of edge by and az is the *connecting* edge.

At the end the set of all the connecting edges then forms a minimum spanning tree. The simple minimum cut algorithm we describe here consists of $|V| - 1$ phases each of which is very similar to Prim's algorithm:

$\text{MINIMUMCUTPHASE}(G, w, a)$
$A \leftarrow \{a\}$
while $A \neq V$
    add to A the most tightly connected vertex
store the cut-of-the-phase and shrink G by merging the two vertices added last

A subset A of the graph's vertices grows starting with an arbitrary single vertex until A is equal to V. In each step the vertex outside of A *most tightly connected* with A is added. Formally, we add a vertex

$$z \notin A \text{ such that } w(A, z) = \max\{w(A, y) \mid y \notin A\}$$

where $w(A, y)$ is the sum of the weights of all the edges between A and y.

At the end of each such phase the two vertices added last are *merged*, i.e., the two vertices are replaced by a new vertex, and any edges from the two vertices to a remaining vertex are replaced by an edge weighted by the sum of the weights of the previous two edges. Edges joining the merged nodes are removed.

The cut of V that separates the vertex added last from the rest of the graph is called the *cut-of-the-phase*. The lightest of these cuts-of-the-phase is the result of the algorithm, the desired minimum cut.

MINIMUMCUT(G, w, a)
while $|V| > 1$
    MINIMUMCUTPHASE(G, w, a)
    if the cut-of-the-phase is lighter than the current minimum cut
        then store the cut-of-the-phase as the current minimum cut

Notice that the starting vertex a stays the same throughout the whole algorithm.

3 Correctness

The core of the proof of correctness is the following somewhat surprising lemma.

Lemma *Each cut-of-the-phase is a minimum s-t cut in the current graph, where s and t are the two vertices added last in the phase.*

Assuming, that the lemma holds, we can show by a simple case distinction, that the smallest of these cuts-of-the-phase is indeed the minimum cut we are looking for. This is done by induction on $|V|$. The case $|V| = 2$ is trivial. If $|V| \geq 3$, look at the first phase: If G has a minimum cut, that is at the same time a minimum s-t cut, then, according to the lemma, the cut-of-the phase is already a minimum cut. If not, then G has a minimum cut with s and t on the same side. Therefore, a minimum cut of G', the input graph of phase 2, that differs from G by the merging of s and t, is a minimum cut of G. Now, by induction, the lightest of the cuts-of-the-phases 2 to $|V| - 1$ is such a minimum cut of G'. Notice that the application of phases 2 to $|V| - 1$ to G' is the same as the application of the complete algorithm to G'. Thus we have shown the correctness of the whole algorithm.

Theorem *The smallest of the $|V| - 1$ cuts-of-the-phase considered during the algorithm is the minimum cut of the graph.*

Finally, we show that the lemma holds. The run of a MINIMUMCUTPHASE orders the vertices of the current graph linearly, starting with a and ending with s and t, according to their order of addition to A. Now we look at an arbitrary s-t cut C of the current graph and show, that it is at least as heavy as the cut-of-the-phase.

We call a vertex $v \neq a$ *active* (with respect to C) when v and the vertex added just before v are in the two different parts of C. Let $w(C)$ be the weight of C, A_v the set of all vertices added before v (excluding v), C_v the cut of $A_v \cup \{v\}$ induced by C, and $w(C_v)$ the weight of the induced cut, i.e., the sum of the weights of the edges going from one part of the induced partition to the other.

We show that for every active vertex v

$$w(A_v, v) \leq w(C_v)$$

by induction on the set of active vertices:
For the first active vertex the inequality is satisfied with equality. Let the inequality be true for all active vertices added up to the active vertex v, and let u be the next active vertex that is added. Then we have

$$w(A_u, u) = w(A_v, u) + w(A_u \setminus A_v, u) =: \alpha$$

Now, $w(A_v, u) \leq w(A_v, v)$ as v was chosen as the vertex most tightly connected with A_v. By induction $w(A_v, v) \leq w(C_v)$. All edges between $A_u \setminus A_v$ and u connect the different parts of C. Thus they contribute to $w(C_u)$ but not to $w(C_v)$. So

$$\alpha \leq w(C_v) + w(A_u \setminus A_v, u) \leq w(C_u)$$

As t is always an active vertex with respect to C we can conclude that $w(A_t, t) \leq w(C_t)$ which says exactly that the cut-of-the-phase is at most as heavy as C.

4 Running Time

As the running time of the algorithm MINIMUMCUT is essentially equal to the added running time of the $|V| - 1$ runs of MINIMUMCUTPHASE, which is called on graphs with decreasing number of vertices and edges, it suffices to show that a single MINI-MUMCUTPHASE needs at most $\mathcal{O}(|E| + |V| \log |V|)$ time yielding an overall running time of $\mathcal{O}(|V||E| + |V|^2 \log |V|)$.
The key to implementing a phase efficiently is to make it easy to select the next vertex to be added to the set A, the most tightly connected vertex. During execution of a phase, all vertices that are not in A reside in a priority queue based on a key field. The key of a vertex v is the sum of the weights of the edges connecting it to the current A, i.e., $w(A, v)$. Whenever a vertex v is added to A we have to perform an update of the queue. v has to be deleted from the queue, and the key of every vertex w not in A, connected to v has to be increased by the weight of the edge vw, if it exists. As this is done exactly once for every edge, overall we have to perform $|V|$ EXTRACTMAX and $|E|$ INCREASEKEY operations. Using Fibonacci heaps [FT87], we can perform an EXTRACTMAX operation in $\mathcal{O}(\log |V|)$ amortized time and a INCREASEKEY operation in $\mathcal{O}(1)$ amortized time.
Thus the time we need for this key step that dominates the rest of the phase, is $\mathcal{O}(|E| + |V| \log |V|)$.
Notice, that this runtime analysis is very similar to the analysis of Prim's minimum spanning tree algorithm.

Acknowledgement

The authors thank Dorothea Wagner for her helpful remarks.

An Example

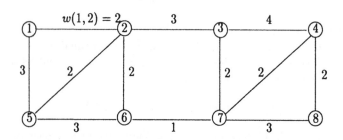

Figure 1: A graph $G = (V, E)$ with edge-weights.

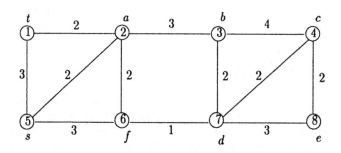

Figure 2: The graph after the first MINIMUMCUTPHASE(G, w, a), $a = 2$, and the induced ordering a, b, c, d, e, f, s, t of the vertices. The first *cut-of-the-phase* corresponds to the partition $\{1\}, \{2, 3, 4, 5, 6, 7, 8\}$ of V with weight $w = 5$.

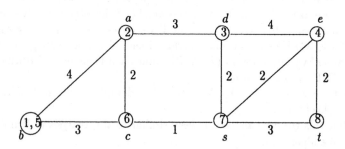

Figure 3: The graph after the second MINIMUMCUTPHASE(G, w, a), and the induced ordering a, b, c, d, e, s, t of the vertices. The second *cut-of-the-phase* corresponds to the partition $\{8\}, \{1, 2, 3, 4, 5, 6, 7\}$ of V with weight $w = 5$.

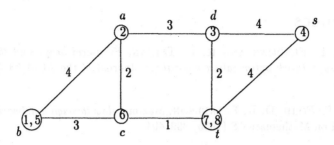

Figure 4: After the third MINIMUMCUTPHASE(G, w, a). The third *cut-of-the-phase* corresponds to the partition $\{7, 8\}, \{1, 2, 3, 4, 5, 6\}$ of V with weight $w = 7$.

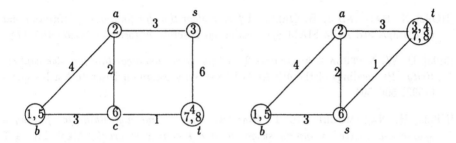

Figure 5: After the fourth and fifth MINIMUMCUTPHASE(G, w, a), respectively. The fourth *cut-of-the-phase* corresponds to the partition $\{4, 7, 8\}, \{1, 2, 3, 5, 6\}$. The fifth *cut-of-the-phase* corresponds to the partition $\{3, 4, 7, 8\}, \{1, 2, 5, 6\}$ with weight $w = 4$.

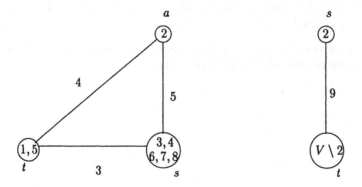

Figure 6: After the sixth and seventh MINIMUMCUTPHASE(G, w, a) respectively. The sixth *cut-of-the-phase* corresponds to the partition $\{1, 5\}, \{2, 3, 4, 6, 7, 8\}$ with weight $w = 7$. The last *cut-of-the-phase* corresponds to the partition $\{2\}, V \setminus \{2\}$; its weight is $w = 9$. The minimum cut of the graph G is the fifth *cut-of-the-phase* and the weight is $w = 4$.

References

[FT87] M. L. FREDMAN AND R. E. TARJAN, *Fibonacci heaps and their uses in improved network optimization algorithms*, Journal of the ACM **34** (1987) 596–615

[FF56] L. R. FORD, D. R. FULKERSON, *Maximal flow through a network*, Canadian Journal on Mathematics **8** (1956) 399-404

[GT88] A. V. GOLDBERG AND R. E. TARJAN, *A new approach to the maximum flow problem*, Journal of the ACM **35** (1988) 921-940

[GH61] R. E. GOMORY, *Multi-terminal network flows*, Journal of the SIAM **9** (1961) 551-570

[HO92] X. HAO AND J. B. ORLIN, *A faster algorithm for finding the minimum cut in a graph*, 3rd ACM-SIAM Symposium on Discrete Algorithms (1992) 165-174

[M93] D. W. MATULA *A linear time $2 + \epsilon$ approximation algorithm for edge connectivity* , Proceedings of the 4th ACM-SIAM Symposium on Discrete Mathematics (1993) 500-504

[NI92a] H. NAGAMOCHI AND T. IBARAKI, *Linear time algorithms for finding a sparse k-connected spanning subgraph of a k-connected graph*, Algorithmica **7** (1992) 583-596

[NI92b] H. NAGAMOCHI AND T. IBARAKI, *Computing edge-connectivity in multigraphs and capacitated graphs*, SIAM Journal on Discrete Mathematics **5** (1992) 54-66

[NP89] T. NISHIZEKI AND S. POLJAK, *Highly connected factors with a small number of edges*, Preprint (1989)

[P57] R. C. PRIM, *Shortest connection networks and some generalizations*, Bell System Technical Journal **36** (1957) 1389-1401

Approximation Algorithm on Multi-Way Maxcut Partitioning[*]

Jun Dong Cho[1], Salil Raje[2] and Majid Sarrafzadeh[2]

[1] Samsung Electronics, Buchun Kyunggi-Do, Korea
[2] Northwestern University, Evanston IL 60208, USA

Abstract. Given arbitrary positive weights associated with edges, the *maximum cut* problem is to find a cut of the maximum cardinality (or weight in general) that partitions the graph G into X and $\overline{X}$. Our maxcut approximation algorithm runs in $O(e + n)$ sequential time yielding a node-balanced maxcut with size at least $\lfloor (e + e/n)/2 \rfloor$, improving the time complexity of $O(e \log e)$ known before. Employing a height-balanced binary decomposition, an $O(e + n \log k)$ time algorithm is devised for the maxcut k-coloring problem which always finds a k-partition of vertices such that the number of bad edges (or "defected" edges with the same color on two of its end-points) does not exceed $\lceil (e/k)((n - 1)/n)^h \rceil$, where $h = \lceil \log_2 k \rceil$, thus improving both the time complexity $O(enk)$ and the bound $\lfloor e/k \rfloor$ known before. The bound on maxcut k-coloring is also extended to find an approximation bound for the maximum k-covering problem. The relative simplicity of the algorithms and their computational economy are both keys to their practical applications. The proposed algorithms have a number of applications, for example, in VLSI design. ...

1 Introduction

Given an undirected graph $G = (V, E)$, $|V| = n$ and $|E| = e$, a *cut* is a set of all edges of E with one end in a proper subset X of V, and the other end in $\overline{X} = V - X$. The maxcut problem is to maximize

$$CUT(X, \overline{X}) = \sum_{(i,j) \in E, i \in X, j \notin X} w_{i,j},$$

where $w_{i,j}$ is the weight on the edge connecting vertices i and j in V. It is well known that even simple maxcut (maximum cut with edge weights restricted to value 1) is NP-complete [18] on general graphs. For a history of the maxcut problem see [18, 33].

Another well-known combinatorial optimization problem is the *coloring* problem that assigns a color to each vertex where adjacent vertices are assigned dif-

---

[*] This work has been supported in part by the National Science Foundation under Grant MIP-9207267.

ferent colors[3]. A variation of it is the problem of finding a k-coloring of a given graph for a fixed number k [22, 25]. We call the problem *maxcut k-coloring* problem. The problem can be restated as follows. Given a fixed k, the problem finds k-color partitions $V_1, V_2, \cdots, V_k$ of V and aims to minimize the number of "bad" edges[4] in $V_i, 1 \leq i \leq k$. The proposed graph k-coloring algorithm is very robust and has many applications in VLSI design (e.g., constrained layer assignment [8, 23, 29]).

In this paper, we present a linear-time approximation algorithm on *maxcut* and *maxcut k-coloring* problems. Furthermore, extending the algorithm of maxcut k-coloring problem, we obtain an approximation bound on the maximum k-covering problem. We shall discuss the problem in details in Section 4. Table 1 gives a summary of results obtained by us for the three problems.

Problems	Before		Ours	
	time	ratio[5]	time	ratio
MC	$O(e \log e)^{\dagger}$	$\frac{1}{2} + \frac{1}{2n}^{\dagger}$	$O(e + n)$	$\frac{1}{2} + \frac{1}{2n}$
MWC	$O(en)^{\ddagger}$	$\frac{1}{2}^{\ddagger}$	$O(e + n)$	$\frac{w}{2}$
MkC	$O(enk)^{\ddagger}$	$\frac{1}{k}^{\ddagger}$	$O(e + n \log k)$	$\frac{1}{k}(\frac{n-1}{n})^{h}$ *
MWkC	$O(enk)^{\ddagger}$	$\frac{1}{k}^{\ddagger}$	$O(e + n \log k)$	$1/k$
MkCV	Unknown	Unknown	$O(e + n \log k)$	$\frac{2e+n-nk}{2k}$

MC: maxcut; MWC: maximum weighted cut; MkC: maxcut k-coloring;

MWkC: maxcut weighted k-coloring; MkCV: maximum k-covering (see Section 4 for the detailed definition).

$\dagger$: [24]. A linear time algorithm for the maxcut problem with a ratio of $1/2$ is given in [11].

$\ddagger$: [30].   *: $h = \lceil \log_2 k \rceil$.

Table 1. A comparison between the previous results and ours

---

[3] An optimal coloring of an arbitrary graph, i.e., the problem of determining the smallest number of colors used for an arbitrary graph, is a well-known NP-complete problem [16].

[4] The set of all *bad edges* B is defined as $B = \{a \in E \| a \in \bigcup_{i=1}^{k} V_i \times V_i\}$. That is, the bad edges is the total number of edges having the same color on both its incident vertices.

[5] Let $w_G = \sum_{i \in E} w_i$ where w_i is the weight on edge i in E. Let $C(\mathcal{A})$ be the cutsize generated by a maxcut algorithm $\mathcal{A}$ on a graph G. Then the performance ratio of the algorithm with respect to the number of edges (edge weights) is defined as $C(\mathcal{A})$ $/w_G$. Let $B(\mathcal{B})$ be the total weight on bad edges generated by a maxcut k-coloring algorithm $\mathcal{B}$ on a graph G, then the performance ratio of the algorithm with respect to the edge weights is defined as $B(\mathcal{B})$ $/w_G$.

Some of our proofs are omitted in the subsequent subsections for brevity. This paper is organized as follows. Section 2 presents an approximation algorithm for the maxcut problem. Section 3 devotes to generalize the algorithm into maxcut k-coloring problem. Section 4 is extended to find an approximation algorithm for the maximum k-covering problem. Finally, conclusion is presented in Section 5.

2 Maxcut

Recently, Hagin and Venkatesan [24] proposed an approximation algorithm running $O(e \log e)$ sequential time for the maxcut problem. Their algorithm guarantees that a matching M of size e/n is added to improve the bound developed by [30]. To the best of our knowledge, their algorithm yields the best approximation bound known today. In this section, we develop a so-called *Simmax* algorithm for the problem. The algorithm is remarkably simple and employs an incremental updating approach considering a vertex pair (or a vertex) at a time. The algorithm improves the computing time to $O(n + e)$ sequential time over the algorithm developed by [24] and finds the same performance ratio of $1/2 + 1/(2n)$.

2.1 Fundamental Algorithm

Our algorithm proceeds as shown in Algorithm $\mathcal{A}$. The given graph is represented by an adjacency list. Each header vertex $v_i, 1 \leq i \leq n$, in the list contains a list of vertices adjacent to itself denoted by N_i. Let $L_i = \{v_i\} \cup N_i$. We assign the vertices into two sets, V_1 and V_2, one vertex pair at a time such that $CUT(V_1, V_2)$ is maximum. The *connectivity* $C_1[v_i]$ or $C_2[v_i]$ to either of the sets is the number of edges (edge weights for the weighted case) connecting the vertex v_i with the set V_1 or V_2. The connectivity information is stored in a *connectivity table* and incrementally updated at each phase of the algorithm.

We first find a matching M of size e/n in $O(e + n)$ time (refer to [24]). Two sets V_1 and V_2, and two arrays $C_1[v_i]$ and $C_2[v_i]$, for all $v_i \in V$, are initialized to be empty. Scanning through the adjacency list once, we identify matching edges induced by header vertices v_i and a vertex $v_j \in N_i$, for all $i \in V$. While looping through each L_i, if there is a matching edge $(v_i, v_j) \in M$ in L_i (**Case 1**), then compute the costs of two cut candidates, $A = C_2[v_i] + C_1[v_j]$ and $B = C_2[v_j] + C_1[v_i]$. If $A \geq B$, then place v_i into set V_1 and v_j into set V_2; otherwise, do the opposite (called **VS-assignment**). In either case, both the matching edge (v_i, v_j) and the newly selected edges associated with the two cut candidates A or B will be added to the cut set. We call the vertex that has not yet been assigned to either of the sets during the scanning process a *"free" vertex*. Whereas, if there is no matching edge of M, but a free vertex in L_i (**Case 2**), we perform the VS-assignment as in Case 1. For the case where there is no free vertices in N_i (**Case 3**), the header vertex v_i (called a Case 3 vertex) is inserted into a queue to be paired with another Case 3 vertex later on.

After determining the VS-assignment for the vertex pair of Case 1 or Case 2, we scan through the list L_i once more to update the connectivities from S1 and S2 to all free vertices in L_i. We denote by $w(v_i, v_j)$ a weight on the edge connecting two vertices v_i and v_j. We denote by $w(v_i, v_j)$ the weight on an edge connecting v_i and v_j.

Finally, to deal with Case 3 vertices stored in the queue, we perform the VS-assignment as follows. While the queue is not empty, we remove one vertex pair from the queue at a time, and perform the set assignment for the vertex pair, followed by updating the connectivity.

If the number of vertices in the queue is odd, then, for the last vertex in the queue, we simply check if $C_1[v_i] > C_2[v_i]$. If so, we place v_i into set V_1; otherwise do the opposite.

We loop through each list L_i only twice (once for finding a matching edge or a free edge, and once for updating the connectivity), and the adjacency list once to compute $CUT(V_1, V_2)$ after finding two partitions, V_1 and V_2. Thus, the algorithm runs in $O(e + n)$ time. The resulting partition guarantees a vertex-balanced cut[6] such that $|X||\overline{X}|CUT(X, \overline{X})$ is maximal. The algorithm described above is called *Simmax-II* (because we are dealing with two vertices at a time).

Now, we have the following result.

Lemma 1 refer to Figure 1. *Simmax-II partitions a graph G into two sets V_1 and V_2 with a cut of size more than half the number of edges (or edge weights) in G.*

Proof: Consider an incremental update (using the connectivity check) of two sets in each phase. Let $G_1^i = (V_1^i, E_1^i)$ and $G_2^i = (V_2^i, E_2^i)$ be the two induced subgraphs in phase i. Let v_1^i and v_2^i be two vertices chosen to be considered in phase i. Initially, we have chosen two vertices v_1^1 and v_2^1 arbitrarily and assigned v_1 to V_1^1 and v_2 to V_2^1. Since $G_1^1 = (v_1, \emptyset)$, $G_2^1 = (v_2, \emptyset)$, $CUT(G^1) = 1$, and $E(G^1) = 1$, $CUT(G^1)/E(G^1) \geq 1/2$, thus the lemma holds. Inductively assume that G^i admits $CUT(G^i)$ of size at least $E^i/2$. Consider $G^{i+1} = (V^{i+1}, E^{i+1})$. Let $E(S^i) = A^i + B^i$ and $V(S^i) = \{v_1^i\} + \{v_2^i\}$. Then, we define $V^{i+1} = V^i + V(S^i)$, and $E^{i+1} = E^i + E(S^i)$. We have:

$$CUT(G^{i+1}) = CUT(G^i) + CUT(S^i) \tag{1}$$
$$\geq E^i/2 + \max(A^i, B^i) \geq E^i/2 + (A^i + B^i)/2 \tag{2}$$
$$\geq E^i/2 + E(S^i)/2 \geq E^{i+1}/2, \tag{3}$$

thus, satisfies the lemma. $\square$

Theorem 2. *There exists an $O(e + n)$ time sequential algorithm such that an n-vertex e-edge graph $G = (V, E)$ admits a cut with at least the fraction $1/2 + 1/(2n)$ of its total number of edges.*

Proof: In our algorithm, we first identified a matching of e/n edges from G in $O(e+n)$ time. In each phase of the algorithm, whenever a matching edge M^i is found, we increase the cut size by one. The matching edge was not involved in computing $CUT(S^i)$ of the subgraph S^i in phase i. Along with Lemma 1, any graph G admits a cut with at least $(e - e/n)/2 + e/n = (e + e/n)/2$ edges; $\square$

When the balanced condition is not of interest, we simply check for one vertex at a time during the VS-assignment; if $C_1[v_i] > C_2[v_i]$, then we place v_i into set V_1; otherwise do the opposite. We call the simpler variation *Simmax-I*. We can easily see that Lemma 1 still holds for this case so that the performance ratio is $1/2$.

---

[6] Balanced partitions are important for certain applications (e.g., layer assignment problem in VLSI layout to achieve uniform wiring density on two layers).

Algorithm $\mathcal{A}$: Simmax-II(G, G_1, G_2)
Input: $G = (V, E)$
Output: $G_1 = (V_1^{n/2}, E_1^{n/2})$, $G_2 = (V_2^{n/2}, E_2^{n/2})$ with $CUT(G_1, G_2) \geq w_G/2$
    begin-1
(1)   Find a matching M of size e/n;
(2)   for each $v_i \in V$
(3) $C_1[v_i] = 0$;
(4) $C_2[v_i] = 0$;
(5) $assign[v_i] = 0$;
(6)   Initialize two partition sets V_1 and V_2 as empty;
(7)   for each header $v_i \in V$
(8)   if $(\exists v_j \in N_{v_i}$ s.t. $(v_i, v_j) \in M$ and $assign[v_j] = 0)$ then
(9)   VS-assignment(v_i, v_j);
(10)   else if $(\exists v_j \in N_{v_i}$ s.t. $assign[v_j] = 0)$ then
(11)   VS-assignment(v_i, v_j);
(12)   else (* no free vertices in N_{v_i} *)
(13)   Inqueue(v_i);
(14)   Update-Connectivity(v_i);
(15)   while $(empty(Q) \neq \emptyset)$
(16) $v_1 = dequeue(Q)$;
(17) $v_2 = dequeue(Q)$;
(18)   VS-Assignment(v_1, v_2);
(19)   Update-Connectivity(v_1); Update-Connectivity(v_2);
(20)   Compute $CUT(V_1, V_2)$;
    end-1

Procedure $\mathcal{A}$-1: VS-assignment(v_1, v_2)
    begin-2
(1)   if $(v_2 \neq \emptyset)$
(2) $A = C_2[v_1] + C_1[v_2]$; $B = C_2[v_2] + C_1[v_1]$;
(3)   if $(A \geq B)$ then
(4) $V_1 = V_1 \cup \{v_1\}$; $V_2 = V_2 \cup \{v_2\}$;
(5) $assign[v_1] = 1$; $assign[v_2] = 2$;
(6)   else do the opposite.
(7)   else (* if $v_2 = \emptyset$ *)
(8)   if $(C_1[v_1] \geq C_2[v_1])$ then
(9) $V_1 = V_1 \cup \{v_1\}$;
(10)   else do the opposite.
    end-2

Procedure $\mathcal{A}$-2: Update-Connectivity(v_1)
    begin-3
(1)   for (each $v_k \in L_{v_1}$ s.t. $assign[v_k] = 0$)
(2)   if $(assign[v_1] = 1)$ then $C_1[v_k] = C_1[v_k] + w(v_1, v_k)$;
(3)   else if $(assign[v_1] = 2)$ then $C_2[v_k] = C_2[v_k] + w(v_1, v_k)$;
    end-3

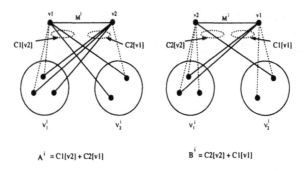

Fig. 1. Finding max (A^i, B^i) in phase i of Algorithm Simmax-II

2.2 Extensions

The bound achieved by the basic algorithm can be improved by the following heuristic (called Simmax-IV). That is, we compare four assignment candidates $A = C_2[v_i] + C_1[v_j]$, $B = C_2[v_j] + C_1[v_i]$, $C = C_1[v_j] + C_1[v_j]$, and $D = C_2[v_i] + C_2[v_i]$, instead of comparing only two assignment candidates A and B. If C or D is larger than A and B, then the bound on cut size will be improved. Applying the Simmax-IV to bipartite graphs, it is not difficult to see:

Theorem 3. *Both Simmax-I and Simmax-IV (not Simmax-II) are exact procedures to test if a graph is bipartite in $O(e + n)$ time.*

In general, consider δ vertices at the same time at each phase (Theorem 2 uses $\delta = 2$ considering only A and B). Then, computing a maximal cut at each stage takes 2^δ time which is the number of ways of arranging δ distinct vertices into two sets. Note that the storage requirement for the new algorithm increases to $O(n2^\delta)$ for constructing the connectivity tables. In all, time complexity of the extended algorithm is $O(2^\delta n/\delta + e)$. We call the general version of the Simmax algorithm *Simmax-X*. Notice that in this case a vertex-balanced partition may not be guaranteed.

3 Maxcut k-Coloring

In the maxcut k-coloring problem, we ask for a partition of the set of vertices V into k subsets $V_1, V_2, \cdots, V_k$ such that

$$\sum_{x \in \bigcup_{i=1}^{k} V_i \times V_i} weight(x)$$

is minimal. Observe that the problem becomes the maxcut problem when $k = 2$. In this section, we present an $O(e + n \log k)$ time approximation algorithm to color a graph using k-colors such that the number of bad edges does not exceed $(e/k)((n-1)/n)^h$, where $h = \lceil \log_2 k \rceil$.

3.1 Fundamental Algorithm

We employ a binary decomposition tree to partition the graph into two subgraphs recursively. At each step, we generate a maxcut between two subgraphs using our Simmax algorithm proposed in the previous section. The subgraphs obtained are further partitioned until we get a total of $k = 2^h$ subgraphs. Note that a subgraph is not partitioned further if it does not contain any (bad) edges. We call the generated binary tree a *maxcut color tree* (see Figure 2). An intuition behind the approach is based on the following fact.

Lemma 4. *The $k - 1$ maxcuts in a maxcut color tree define an optimal partition in the maxcut k-coloring problem.*

By maximizing the cut at each level of the top-down color tree, when we go down the tree, the number of bad edges will be gradually reduced. Now, based on a binary decomposition tree construction, a detailed implementation of a so-called *k-Coloring* (*k*-C) algorithm is shown in Algorithm $\mathcal{B}$.

Algorithm $\mathcal{B}$: k-Coloring
Input: $G = (V, E)$
Output: k partitions $V_1, V_2, \cdots, V_k$ such that the number of bad edges in G is minimal
begin
    $ncut = 1$; (* the number of cuts *)
    $ncol = 0$; (* the number of colors *)
    initqueue(Q); (* initialize a queue Q *)
    inqueue(Q, G); (* insert G into Q *)
    while (*ncut < k*)
    begin
        $g = $ dequeue(Q); (* remove an item from Q *)
        if ($\exists$ edges in g)
        begin
            Simmax-II(g, a, b);
            $ncut = ncut + 1$;
            inqueue(Q, a);
            inqueue(Q, b);
        end
        else $ncol = ncol + 1$; $C[ncol] = g$;
    end
    while (empty(Q) $\neq$ TRUE)
    begin
        $ncol = ncol + 1$;
        $C[ncol] = $ dequeue(Q);
    end
end

Theorem 5. *There exists an $O(e+n)$ time algorithm for the maxcut k-coloring problem such that the total number of edges in the k subgraphs is at most $(e/k)((n - 1)/n)^h$, where $h = (\lceil \log_2 k \rceil)$ is the height of the maxcut color tree.*

Proof: We know from Theorem 2 that a graph $G = (V, E)$ admits a cut such that the total number of edges in V_1 plus the total number of edges in V_2 ($V = V_1 \cup V_2$) are at most $|B_1| = e(1 - \lfloor (n+1)/2n \rfloor) = \lfloor \frac{e(G)(n-1)}{2n} \rfloor$.

Thus at level 2 of the maxcut color tree, the subgraphs V_1 and V_2 are further divided into V_a and V_b, V_c and V_d respectively, such that again from Theorem 2 we have:

$$|B_2| = e(V_a) + e(V_b) + e(V_c) + e(V_d)$$

$$\leq |B_1|(1 - \lfloor (n+1)/2n \rfloor) = \lceil e(\frac{n-1}{2n})^2 \rceil.$$

Similar partitions will give us the number of bad edges at level $h = \lceil \log k \rceil$,

$$|B_h| = e(V_1) + e(V_2) + \ldots + e(V_k)$$

$$\leq \lfloor e(\frac{n-1}{2n})^{\log k} \rfloor = \lceil \frac{e}{k}(\frac{n-1}{n})^h \rceil.$$

$\square$

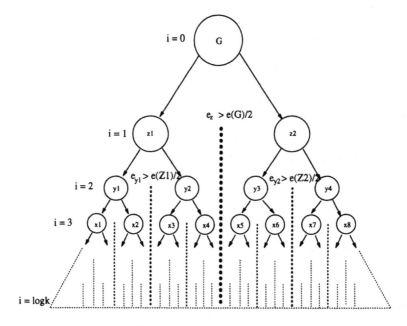

Fig. 2. A maxcut color tree

3.2 Extensions

Applying Algorithm B to bipartite graphs, we obtain:

Theorem 6. *The k-Coloring algorithm is an exact procedure for finding a set of k-color partitions for bipartite graphs (i.e., without bad edges).*

Let us consider the restricted case where all edge weights in a complete graph are required to be 1 (which we call simple complete graph).

Theorem 7. *Given a simple complete graph, there exists a linear time algorithm that produces a maximal k-color partition optimally when $k = 2^h$.*

4   Maximum k-Covering

Given an undirected graph $G = (V, E)$, a stable set (independent set, vertex packing) in G is a set of pairwise nonadjacent vertices, and a clique in G is a set of pairwise adjacent vertices. A clique in G is a stable set in $\overline{G}$, and vice versa. A vertex coloring of G is an assignment of colors to the vertices of G in such a way that adjacent vertices are assigned by different colors. Equivalently, a vertex coloring is a collection of stable sets (color classes) such that each vertex belongs to at least one color classes. A clique covering of G is a collection of cliques such that every vertex belongs to at least one clique. A clique covering of G is a vertex coloring of $\overline{G}$, and vice versa.

A well-known pair of combinatorial optimization problems asks for finding a maximum clique and a minimum vertex coloring in G or, equivalently, for finding a maximum stable set and a minimum clique covering of $\overline{G}$. Here we introduce a notion of maximum k-covering such that finding a maximum k-covering of G is equivalent to finding a maxcut k-coloring of $\overline{G}$. It is easy to see that the new problem is a generalization of mincut partitioning (that is, a multi-way mincut partitioning).

Theorem 8. *There exists an $O(e + n \log k)$ time approximation algorithm for maximum k-covering problem such that the total number of edges in the k subgraphs is at least $\lfloor \frac{2e + n - nk}{2k} \rfloor$.*

Proof: Let us consider applying Theorem 5 (but, based on Simmax-I) for finding the maxcut k-coloring of $\overline{G}$. Then, the total number of edges in the k subgraphs is at most $\lceil e(\overline{G})/2^h \rceil$, where $h = (\lceil \log_2 k \rceil)$. Thus, the maxcut, such that the number of edges that partition the graph $\overline{G}$ into k subgraphs is maximum, is at least $\lfloor e(\overline{G})(k-1)/k \rfloor$. Here $e(\overline{G}) = n(n-1)/2 - e(G)$. We have a maxcut k-coloring when each subgraph contains n/k vertices, i.e., the number of vertices are balanced over the k subgraphs. Thus, the mincut, such that the number of edges that partition the graph G into k subgraphs is minimum, is at most $\lceil ((\frac{n}{k})^2 \frac{k(k-1)}{2} - \frac{e(\overline{G})(k-1)}{k}) \rceil$. That is, the total number of edges in the k subgraphs of G is at least $(e(G) - \lceil \frac{(n+2e(G))(k-1)}{2k} \rceil = \lfloor \frac{2e(G)+n-nk}{2k} \rfloor)$. $\square$

5   Conclusion

In this paper we proposed new approximation algorithms on three combinatorial optimization problems: maxcut, maxcut k-coloring, and maximum k-covering. Not only are the algorithms simple and fast, but also, the performance bounds achieved in this paper are the best known (to the best of our knowledge). The devised k-Coloring algorithm has also been implemented in C on a SUN SPARC station IPC, under UNIX, but omitted for brevity. Experimental result showed that the algorithm is also of practical interest. The proposed graph k-coloring algorithm is very robust and has many applications in VLSI design (e.g., constrained layer assignment [8, 23, 29]).

References

1. D. Adolphson and T. C. Hu. "Optimal Linear Ordering,". *SIAM Journal on Applied Mathematics*, 25:403–423, 1973.

2. E. Auer, W. Schiele, and G. Sigl. "A New Linear Placement Algorithm for Cell Generation". In *International Conference on Computer-Aided Design*. IEEE, November 1991.

3. K. Chang and D. Du. "Efficient Algorithms for the Layer Assignment Problem". *IEEE Transactions on Computer Aided Design*, 6:67–78, 1987.

4. C. K. Cheng, X. Deng, Y. Z. Liao, and S. Z. Yao. "Symbolic Layout Compaction Under Conditional Design Rules". *IEEE Transactions on Computer Aided Design*, 11(4):475–486, 1992.

5. C. K. Cheng. "Linear Placement Algorithms and Application to VLSI Design". *Networks*, 17:439–464, 1987.

6. J. D. Cho, K. F. Liao, and M. Sarrafzadeh. "Multilayer Routing Algorithm for High Performance MCMs". In *Proceedings of the Fifth Annual IEEE International ASIC Conference and Exhibit*, pages 226–229, 1992.

7. W. T. Cheng, J. L. Lewandowski, and E. Wu. "Optimal Diagnostic Methods for Wiring Interconnects". *IEEE Transactions on Computer Aided Design*, 11(9):1161–1165, September 1992.

8. J. D. Cho, S. Raje, M. Sarrafzadeh, M. Sriram, and S. M. Kang. "A Multilayer Assignment Algorithm for Interference Minimization". In *Proc. ACM/SIGDA Physical Design Workshop, Lake Arrowhead, CA*, pages 63–67, 1993.

9. J. D. Cho, S. Raje, M. Sarrafzadeh, M. Sriram, and S. M. Kang. "Crosstalk Minimum Layer Assignment". In *Proc. IEEE Custom Integr. Circuits Conf., San Diego, CA*, pages 29.7.1–29.7.4, 1993.

10. J. D. Cho and M. Sarrafzadeh. "The Pin Redistribution Problem in Multichip Modules". In Proceedings of the Fourth Annual IEEE International ASIC Conference and Exhibit, pages p9-2.1 – p9-2.4. IEEE, September 1991.

11. L. H. Clark, F. Shahrokhi, and L. A. Székely. "A Linear Time Algorithm for Graph Partition Problems". *Information Processing Letters*, 42:19–24, 1992.

12. H. H. Chen and C. K. Wong. "Wiring and Crosstalk Avoidance in Multi-Chip Module Design". In *IEEE Custom Integrated Circuits Conference*. IEEE, May 1992.

13. W. Wei-Ming Dai. "Performance Driven Layout of Thin-film Substrates for Multichip Modules". In *Proceedings of Multichip Module Workshop*, pages 114–121. IEEE, 1990.

14. J. P. de Gyvez and C. Di. "IC Defect Sensitivity for Footprint-Type Spot Defects". *IEEE Transactions on Computer Aided Design*, 11(5):638–658, 1992.

15. P. Erdős, R. Faudee, J. Pach, and J. Spencer. "How to Make a Graph Bipartite". *Journal of Combinatorial Theory (b)*, 45:86–98, 1988.

16. M. R. Garey and D. S. Johnson. *Computers and Intractability: A Guide to the Theory of NP-completeness.* Freeman, 1979.

17. M. R. Garey, D. S. Johnson, and H. C. So. "An Application of Graph Coloring to Printed Circuit Testing". *IEEE Transactions on Circuits and Systems*, CAS-23:591–599, 1976.

18. M. R. Garey, D. S. Johnson, and L. Stockmeyer. "Some Simplified NP-Complete Graph Problems". *Theoretical Computer Science*, 1:237–267, 1976.

19. M. Grötschel, L. Lovasz, and A. Schrijver. "The Ellipsoid Method and Its Consequences in Combinatorial Optimixation". *Combinatorica*, 1:169–197, 1981.

20. M. C. Golumbic. *Algorithmic Graph Theory and Perfect Graph*, pages 80–103. New York: Academic, 1980.

21. M. M. Halldórsson. "A Still Better Performance Guarantee for Approximate Graph Coloring". *Information Processing Letters*, 45:19–23, 1993.

22. A. Hertz and D. de Werra. "Using Tabu Search Techniques for Graph Coloring". Computing, 39:345–351, 1987.

23. J. M. Ho, M. Sarrafzadeh, G. Vijayan, and C. K. Wong. "Layer Assignment for Multi-Chip Modules". *IEEE Transactions on Computer Aided Design*, CAD-9(12):1272–1277, December 1990.

24. D. J. Hagin and S. M. Venkatesan. "Approximation and Intractability Results for the Maximum Cut Problem and Its Variants". *IEEE Transactions on Computers*, 40(1):110–113, January 1991.

25. D. S. Johnson, C. R. Aragon, L. A. McGeoch, and C. Schevon. "Optimization by Simulated Annealing: An Experimental Evaluation, Part II (Graph Coloring and Number Partitioning)". Oper. Res., 39:378–406, 1991.

26. D. S. Johnson. "The NP-completeness Column: On Going Guide". J. Algorithms, 6:434–451, 1985.

27. D. S. Johnson. *Private communication.* Bell Lab. Murray Hill, NJ, October 1992.

28. R. Y. Pinter. "Optimal Layer Assignment for Interconnect". In *IEEE Int. Conf. Circuits and Computers*, pages 398–401, 1982.

29. M. Sriram and S. M. Kang. "Detailed Layer Assignment for MCM Routing". In *International Conference on Computer-Aided Design*, pages 386–389. IEEE, 1992.

30. P. M. B. Vitányi. "How Well Can A Graph Be n-Colored?". *Discrete Mathematics*, 34:69–80, 1981.

31. Y. C. Wei and C. K. Cheng. "Ratio-Cut Partitioning for Hierachical Designs". *IEEE Transactions on Computer Aided Design*, 40(7):911–921, July 1991.

32. M. Yannakakis. "Edge-Deletion Problems". *SIAM Journal on Computing*, 10:297–309, 1981.

33. M. Yannakakis. "On the Approximation of Maximum Satisfiability". In *Proc. 3rd Annual ACM-SIAM Symp. on Discrete Algorithms*, pages 1–8, 1992.

A Linear-Time Algorithm for Finding a Central Vertex of a Chordal Graph *

Victor Chepoi and Feodor Dragan

Department of Mathematics & Cybernetics,
Moldova State University, A. Mateevici str., 60,
Chişinău 277009, Moldova

Abstract. In a graph $G = (V, E)$, the eccentricity $e(v)$ of a vertex v is $\max\{d(v, u) : u \in V\}$. The center of a graph is the set of vertices with minimum eccentricity. A graph G is chordal if every cycle of length at least four has a chord. We present an algorithm which computes in linear time a central vertex of a chordal graph. The algorithm uses the metric properties of chordal graphs and Tarjan and Yannakakis linear–time test for graph chordality.

1 Introduction

All graphs in this paper are connected and simple, i.e. finite, undirected, loopless and without multiple edges. In a graph $G = (V, E)$ the *length* of a path from a vertex v to a vertex u is the number of edges in the path. The *distance* $d(u, v)$ from vertex u to vertex v is the length of a minimum length path from u to v and the *interval* $I(u, v)$ between these vertices is the set

$$I(u, v) = \{w \in V : d(u, v) = d(u, w) + d(w, v)\}.$$

The *eccentricity* $e(v)$ of a vertex v is the maximum distance from v to any vertex in G. Denote by $D(v)$ the set of all *farthest from v* vertices, i.e. $D(v) = \{w \in V : d(v, w) = e(v)\}$. The *radius* $r(G)$ is the minimum eccentricity of a vertex in G and the *diameter* $d(G)$ the maximum eccentricity. The *center* $C(G)$ of G is the subgraph induced by the set of all *central vertices* , i.e. vertices whose eccentricities are equal to $r(G)$. A *clique* of G is a set of pairwise adjacent vertices. A vertex v of G is called *simplicial* if its neighborhood is a clique.

A graph G is *chordal* (*triangulated*) if every cycle of length greater than three possesses a chord, i.e. an edge joining two nonconsecutive vertices on the cycle. Chordal graphs arise in the study of Gaussian elimination of sparse symmetric

*This work was partially supported by the VW–Stiftung Project No. I/69041
e-mail addresses: chepoi@university.moldova.su, dragan@university.moldova.su

matrices. They were first introduced by Hajnal & Suranyi [10] and then studied extensively by many people, see [6,9] for general results. Recently, the metric properties of chordal graphs were also investigated, see [1–7,13,15]. In particular, the inequality $2r(G) \geq d(G) \geq 2r(G) - 2$ was proven [1,3,4] and the centers of chordal graphs were characterized [4]. The class of chordal graphs contains trees, block graphs, maximal outerplanar graphs, k–trees, interval graphs and strongly chordal graphs.

In this paper we will present a linear time algorithm for finding a central vertex of a chordal graph. Note that for general graphs with n vertices and m edges the upper bound on the time complexity of this problem is $O(nm)$ and the lower bound is $\Omega(m)$. Hence the presented algorithm is optimal. Linear time algorithms for finding the central vertices are known for trees [11,12], 2–trees and maximal outerplanar graphs [8], strongly chordal graphs [5] and interval graphs [14].

The key idea of our algorithm is that with a few applications of breadth first search, it is possible to find two vertices y, z such that the distance $d(y, z)$ is at most two less than the diameter of the chordal graph G. Intuitively, it makes sense to look for a central vertex in the vicinity of the middle of shortest y, z–paths. We either find a central vertex of G in this set or we replace the pair y, z by a new pair of vertices at distance one step closer to the diameter of G.

2 Metric Properties of Chordal Graphs

In this section we present the metric properties of chordal graphs, used in our algorithms.

Lemma 1 [1] *For any two vertices x and y of a chordal graph G and integer $k \leq d(x, y)$ the set $L(x, k, y) = \{z \in I(x, y) : d(x, z) = k\}$ is a clique.*

Lemma 2 [1] *In a chordal graph G, if C is a clique and x is a vertex not in C such that $d(x, y) = k$ is a constant for all $y \in C$ then there exists a vertex $x^* \in \cap\{I(x, y) : y \in C\}$ adjacent to all vertices from C.*

Such a vertex x^* we will call a *gate* of a vertex x in C.

Lemma 3 [3] *If $x, y, u, v \in V$ are distinct vertices of a chordal graph G, such that $x \in I(u, y)$, $y \in I(x, v)$ and $d(x, y) = 1$ then $d(u, v) \geq d(u, x) + d(y, v)$. The equality holds if and only if there is a vertex $w \in I(x, v) \cap I(u, y)$ adjacent to x and y.*

For any subset $M \subset V$ and any vertex v we denote by

$$Pr(v, M) = \{y \in M : d(v, y) = d(v, M)\}$$

the *metric projection* of v on M (recall that $d(v, M) = \min\{d(v, z) : z \in M\}$).

Lemma 4 *In a chordal graph G, for any clique C and any adjacent vertices $u, v \notin C$ the metric projections $Pr(u, C)$ and $Pr(v, C)$ are comparable, i.e. either $Pr(u, C) \subseteq Pr(v, C)$ or $Pr(v, C) \subseteq Pr(u, C)$.*

Proof Since vertices u and v are adjacent $\mid d(u, C) - d(v, C) \mid \le 1$. Without loss of generality assume that $d(u, C) \ge d(v, C)$. If $d(u, C) = d(v, C) + 1$ then for any vertex $w \in Pr(v, C)$ we have $d(u, w) \le d(u, C)$ and therefore $w \in Pr(u, C)$, i.e. $Pr(v, C) \subseteq Pr(u, C)$. Now assume that $d(u, C) = d(v, C)$, but the sets $Pr(u, C)$ and $Pr(v, C)$ are incomparable. Then we find the vertices

$$x \in Pr(v, C) \backslash Pr(u, C), y \in Pr(u, C) \backslash Pr(v, C).$$

Since $x \in I(v, y)$ and $y \in I(u, x)$, by Lemma 3 we have

$$d(u, v) \ge d(u, y) + d(x, v) = 2d(u, C) \ge 2,$$

thus yielding a contradiction. $\square$

Lemma 5 *For any graph G if $d(x, y) = d(G) = 2r(G)$ then* $C(G) \subset L(x, r(G), y).$

Lemma 6 *For any vertex v of a chordal graph G and any vertex u that is farthest from v we have $e(u) \ge 2r(G) - 3$.*

Proof Assume the contrary and among the vertices which fail our assertion choose a vertex v with minimal eccentricity. Let $u \in D(v)$ be a vertex for which $e(u) < 2r(G) - 3$. From our assumption we deduce that for any vertex $x \in L(v, 1, u)$ we have $u \notin D(x)$, i.e. $e(x) \ge e(v)$. If $e(x) > e(v)$ for some vertex $x \in L(v, 1, u)$ then $v \in I(x, y)$ for any vertex $y \in D(x)$. By Lemma 3

$$d(u, y) \ge e(v) - 1 + e(x) - 1 \ge 2e(v) - 1 \ge 2r(G) - 1.$$

Hence all vertices from $L(v, 1, u)$ have the same eccentricity $e(v)$. Now, if for some vertex $x \in L(v, 1, u)$ there is a vertex $z \in D(x) \backslash D(v)$, then $v \in I(z, x)$ and by Lemma 3

$$d(u, z) \ge e(v) - 1 + e(x) - 1 \ge 2r(G) - 2.$$

Finally assume that $\cup \{D(x) : x \in L(v, 1, y)\} \subset D(v)$. Let x^* be a vertex from $L(v, 1, u)$ having a minimum number of farthest vertices and let $y^* \in D(x^*)$. Since $d(v, y^*) = d(x^*, y^*)$ by Lemma 2 there is a vertex $x \in I(v, y^*) \cap I(x^*, y^*)$ adjacent to v and x^*. If $x \notin I(v, u)$ then $x^* \in I(x, u)$ and again, by Lemma 3 $d(u, y^*) \ge 2e(v) - 2 \ge 2r(G) - 2$. Therefore $x \in I(v, u)$. From our choice of the vertex x^* there exists a vertex $y \in D(x) \backslash D(x^*)$. Since the vertices x and x^* are adjacent and equidistant from u, according to Lemma 2 there is a vertex $w \in I(x, u) \cap I(x^*, u)$ adjacent to x and x^*. Since $x^* \in I(x, y)$ and $x \in I(x^*, y^*)$ $d(w, y^*) \ge e(v) - 1$ and $d(w, y) \ge e(v) - 1$.

If $x \in I(w, y^*)$ or $x^* \in I(w, y)$ then by Lemma 3 at least one of the following inequalities holds

$$d(u, y^*) \ge d(x, y^*) + d(w, u) = 2e(v) - 3 \ge 2r(G) - 3,$$

$$d(u, y) \ge d(x^*, y) + d(w, u) = 2e(v) - 3 \ge 2r(G) - 3.$$

So $d(w, y^*) = d(x, y^*)$ and $d(w, y) = d(x^*, y)$. Again, by Lemma 2 there exist vertices $s \in I(x, y^*) \cap I(w, y^*)$ and $t \in I(x^*, y) \cap I(w, y)$ adjacent to x, w

and x^*, w correspondingly. Note that $w \in I(s, u) \cup I(t, u)$, otherwise $s, t, w \in L(v, 2, u)$ and by Lemma 1 the vertices s and t must be adjacent. The obtained cycle (t, x^*, x, s, t) has one of the chords (x, t) or (x^*, s), thus violating that $x \in I(x^*, y^*)$ and $x^* \in I(x, y)$. Without loss of generality assume that $w \in I(t, u)$. By Lemma 3 $d(u, y) \geq d(u, w) + d(t, y) \geq 2r(G) - 4$ and equality holds if and only if there exists a vertex $z' \in I(w, y) \cap I(t, u)$ adjacent to w and t. We distinguish two cases.

Case 1 $d(s, u) = d(w, u)$.

By Lemma 2 there is a common neighbour $u'' \in I(s, u) \cap I(w, u)$ of vertices s and w; see Fig. 1. Since vertices w and z' are equidistant from u they have a common neighbour $u' \in I(w, u) \cap I(z', u)$. Since $u', u'' \in I(w, u)$ the vertices u' and u'' are adjacent or coincide. In any case we obtain the cycle $(x^*, x, s, u'', u', z', t, x^*)$; see Fig. 1. From the previous conditions on vertices v, x, x^*, y, y^*, u we deduce that vertices x and x^* do not have a common neighbour among the vertices of this cycle. Therefore there is no triangulation of this cycle in which the edge (x, x^*) belongs to some triangle, contradicting the chordality of the graph G.

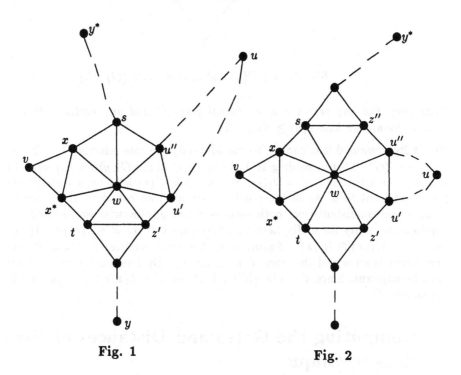

Fig. 1   **Fig. 2**

Case 2 $d(s, u) = d(w, u) + 1$.

By Lemma 3 $d(u, y^*) \geq d(y^*, s) + d(w, u) \geq 2r(G) - 4$ and equality holds only if there is a vertex $z'' \in I(w, y^*) \cap I(s, u)$ adjacent to w and s; see Fig.

2. As in the first case, since $d(z',u) = d(w,u) = d(z'',u)$, there exist vertices $u' \in I(w,u) \cap I(z',u)$ and $u'' \in I(w,u) \cap I(z'',u)$ adjacent to z', w and z, w respectively. Then u' and u'' either are adjacent or coincide. In any case we obtain the cycle $(x^*, x, s, z'', u'', u', z', t, x^*)$. As in the first case there is no common neighbour of x and x^* among the vertices of this cycle, yielding a contradiction.

So the assumption that for some vertex $u \in D(v)$ we have $e(u) < 2r(G) - 3$ leads to a contradiction. $\square$

The following example (see Fig. 3) show the sharpness of the inequality $e(u) \geq 2r(G) - 3$.

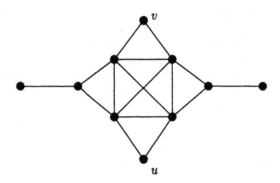

Fig. 3 $(u \in D(v)$ and $e(u) = 3 = 2r(G) - 3)$

Corollary *For any vertex v of a chordal graph G and any vertex u that is farthest from v we have $e(u) \geq d(G) - 2$.*

Proof By Lemma 6 it is enough to consider only the case when $d(G) = 2r(G)$ and $v \in C(G)$. Then according to Lemma 5 $v \in L(x, r(G), y)$ for any vertices $x, y \in V$ such that $d(x, y) = d(G)$. For any $u \in D(v)$ let t be a vertex from $I(v, u)$ adjacent to v. As in the proof of Lemma 6 we can show that vertices v and t are equidistant from both vertices x and y, otherwise we received the required inequality for u. By Lemma 2 there are vertices $s' \in I(v, y) \cap I(t, y)$ and $s'' \in I(v, x) \cap I(t, x)$. Again, as in Lemma 6 vertices v, s' and s'' are equidistant from u and therefore $s', s'' \in I(v, u)$. By Lemma 1 these vertices must be adjacent. Since $s' \in L(x, r(G) + 1, y)$ and $s'' \in L(x, r(G) - 1, y)$ this is impossible. $\square$

3 Computing the Gates and Distances of Vertices to Clique

In this section we present a linear–time algorithm for computing the distances from all vertices to any fixed clique C of a chordal graph G. For any vertex $v \in V$ besides the distance $dist(v) = d(v, C)$ we find some gate

$$gate(v) \in \cap\{I(v, u) : u \in Pr(v, C)\},$$

and also the size *num(v)* of this projection $Pr(v, C)$. Note that the existence of such a vertex *gate(v)* follows from Lemma 2. The algorithm is based on a modification of a *maximum cardinality search*; see Tarjan and Yannakakis [16]. Their linear–time test for graph chordality numbers the vertices of a graph from n to 1. A vertex adjacent to the largest number of previously numbered vertices will be the next selected vertex. In our algorithm we process the vertices of the graph beginning with the vertices of the clique C. As in [16] we maintain an array of sets *set(i)* for $0 \leq i \leq m - 1$. We store in *set(i)* all unnumbered vertices adjacent to exactly i numbered vertices. Initially *set(0)* contains all the vertices. After steps (1)–(13) any vertex $w \notin C$ adjacent to some vertex of the clique C is included in *set(size(w))*, where *size(w)* is the number of vertices from C adjacent to w, all other vertices stay in *set(0)*. Further we maintain the largest index j such that *set(j)* is nonempty. Then we remove a vertex v from *set(j)* and number it. For each unnumbered vertex w adjacent to v, we move w from the set containing it, say *set(i)*, to *set(i+1)* (steps (16)–(21)). Among the numbered vertices adjacent to v we find a vertex z with minimal *dist(z)*= $d(z, C)$ and if there are several such vertices a vertex z with maximal *num(z)* is chosen. In other words this is a vertex with the largest projection on the clique C (steps (22)–(23)). Finally for the vertex v define

$$dist(v) = dist(z) + 1, num(v) = num(z), gate(v) = gate(z)$$

(steps (24)–(26)). The correctness of these steps follows from Lemma 4. Note that v is a simplicial vertex of a subgraph induced by all numbered vertices, i.e. the set $C(v)$ of numbered vertices adjacent to v is a clique. According to Lemma 4 the projections of any two vertices from $C(v)$ are comparable. Hence in $C(v)$ there is a vertex, whose projection covers any other projection on C. Obviously, any vertex with maximal projection, in particular the vertex z, has this property. Since the subgraph induced by numbered vertices is a distance–preserving subgraph of a graph G, *dist(v)* computed in steps (25) or (26) is a distance from v to the clique C. After all this we add one to j and while *set(j)* is empty we update the value of j. As in the maximum cardinality search we represent each set by a doubly linked list of vertices and maintain for each vertex v the index of the set containing it, denoted by *size(v)*. Since the complexity of the updates of j is bounded by the total number of times j is incremented and the access to a set element requires a constant time, the whole complexity of the algorithm is $O(m)$.

Procedure *distance_to_clique*

Input: A chordal graph and its clique C;
Output: For all vertices of the graph the pairs *dist(v)* and *gate(v)*;

begin
(1)   for $i \in \{0, 1, \ldots, n - 1\}$ do $set(i) := \emptyset$;
(2)   for $v \in$ *vertices* **do**
(3)   **if** $(v \in C)$ **then**

(4)   **begin** $size(v) := -1; dist(v) := 0; num(v) := 1; gate(v) = \hat{\ }$ **end**;
           else

(5)   **begin** $size(v) := 0;$ **add** v **to** $set(0)$ **end**;

(6) $j := 0; k := 0;$

(7) **for** $v \in C$ **do begin**

(8)   **for** $(v, w) \in E$ such that $size(w) \geq 0$ **do**
        begin

(9)   **delete** w **from** $set(size(w))$;

(10) $size(w) := size(w) + 1; k := 1; num(w) := size(w);$

(11)   **add** w **to** $set(size(w))$
      end;

(12)   **if** $(k = 1)$ **then**

(13)   **begin** $j := j + 1; k := 0$ **end**;
            end;

(14) $i := | C | + 1;$ (* $| C |$ is the number of vertices in clique C *)

(15) **while** $(i \leq n)$ **do begin**

(16) $v :=$ **delete any from** $set(j); size(v) := -1;$

(17) $d := \infty; nm := 0;$

(18)   **for** $(v, w) \in E$ **do**

(19)   **if** $(size(w) \geq 0)$ **then begin**

(20)   **delete** w **from** $set(size(w)); size(w) := size(w) + 1;$

(21)   **add** w **to** $set(size(w))$
           end;
     else (* $size(w) = -1$ *)

(22)   **if** $(dist(w) < d$ **or** $(dist(w) = d$ **and** $num(w) > nm))$ **then**

(23)   **begin** $z := w; d := dist(w); nm := num(w)$ **end**;

(24)   **if** $(dist(z) = 0)$ **then**

(25)   **begin** $dist(v) := 1; gate(v) := v$ **end**;
    else
    begin

(26) $dist(v) := dist(z) + 1; num(v) := num(z); gate(v) = gate(z)$
    end;

(27) $i := i + 1; j := j + 1;$

(28)   **while** $(j \geq 0$ **and** $set(j) = \emptyset)$ **do** $j := j - 1;$
           end;

end;

4 Computing the Radius and a Central Vertex of a Chordal Graph

The algorithm presented below for finding a central vertex is based on the properties of chordal graphs stated in Lemmas 3, 5 and 6.

Procedure *central_vertex*

Input: A chordal graph G;
Output: A central vertex and the radius of G;

begin
(1) $x :=$ **any vertex of graph;**
(2) $d := 0;$
(3) $y :=$ **the farthest vertex from** x;
(4) $delta := d(x, y);$
(5)   **while** ($d < delta$) **do begin**
(6) $z := y; d := delta$;
(7) $y :=$ **the farthest vertex from** z;
(8) $delta := d(y, z)$
                             end;
(9) $C :=$ **the set** $L(y, \lfloor delta/2 \rfloor, z)$;
(10)   **call the procedure** *distance_to_clique* **for the clique** C;
(11) $R := \max\{ dist(u) : u \in vertices\};$
(12)   **for** $v \in vertices$ **do** $num(v) := 0;$
(13) $param := 0;$
(14)   **for** $v \in vertices$ **do**
(15)   **if** $(dist(v) = R)$ **then**
(16)   **begin** $param := param + 1; num(gate(v)) := num(gate(v)) + 1$ **end;**
(17)   **for** $v \in C$ **do** $size(v) := 0;$
(18)   **for** $v \in C$ **do**
(19)   **for** $(v, w) \in E$ **do** $size(v) := size(v) + num(w);$
(20) $c :=$ **vertex in** C **with maximal** $size();$
(21)   **if** $(size(c) = param)$ **then** (* $e(c) = R$ *)
(22)   **begin** c **is central vertex of** $G; r := R;$ **stop end;**
      else (* $size(c) < param$ and so $e(c) = R + 1$ *)
(23)   **if** (($delta$ **is even**) **and** ($R = delta /2$)) **then**
(24)   **begin** c **is a central vertex of** $G; r := R + 1;$ **stop end;**
       else
(25)   **begin** $x :=$ **the farthest vertex from** $c; d := delta$;
(26)   **goto step** (3)
          end;
end;

Theorem *The central_vertex algorithm correctly finds a central vertex of a chordal graph G in time $O(m)$.*

Proof We begin with the description of the algorithm. First we find a pair of mutually farthest vertices (steps (1)–(8)). To find these vertices, we choose an arbitrary vertex x, find a farthest vertex y from x, and then find a farthest vertex z from y. Put $delta = d(y, z)$. By Corollary $delta \geq d(G) - 2$. If $e(z) > e(y)$ then we add one to $delta$ and choose the farthest vertex from z. Denote this vertex by y and repeat the same operations until $delta = e(y) = e(z)$. Since $delta \leq d(G)$ and initially $delta \geq d(G) - 2$ there are at most two improvements of the value of $delta$. Hence the procedure of finding farthest vertices and future improvement of this pair requires a total computational effort of $O(m)$.

In step (9) in time $O(m)$ we find the clique

$$C = L(y, \lfloor delta/2 \rfloor, z),$$

where y and z is the pair of mutually farthest vertices computed in steps (3)–(8). In order to do this, by the breadth first search algorithm we compute the distances from y and z to all other vertices of G and select whose vertices $v \in V$ for which the equalities $d(v,y) = \lfloor delta/2 \rfloor$ and $d(v,z) = d(y,z) - \lfloor delta/2 \rfloor$ hold. At the following step using the procedure *distance_to_clique* we compute $dist(v)$ and $gate(v)$ for the clique C and all vertices $v \in V$. Using the list $dist()$ we compute the maximal distance R from vertices of G to C (step (11)). Put $D^* = \{v \in V : d(v,C) = R\}$. At the steps (12)–(16) for each vertex from the list $gate()$ we compute $num(w)$: the number of vertices v from D^* for which $gate(v)=w$. These steps can be done in $O(n)$ time. At the following step for each vertex $v \in C$ find $size(v)$, which is the number of vertices from D^* whose projections on C contain vertex v (this operation takes $O(m)$ time). Further we choose a vertex $c \in C$ with maximal $size(c)$, i.e. a vertex of C which belongs to projections of a maximum number of vertices from D^*. As we will show below either c is a central vertex of G or for any vertex $x \in D(c)$ $e(x) > delta$ and we improve the value of *delta*. Since initially $delta \geq d(G) - 2$ there are at most two returns from step (26) to the step (3). All steps of the algorithm require linear time and therefore the algorithm computes a central vertex of G in $O(m)$ time.

We now finally come to proving the correctness of our algorithm.

Claim 1 If *delta* is even ($delta = 2k$) then $R \leq k + 1$.

Proof of Claim 1 Assume the contrary and let $u \in D^*$, i.e. $d(u,C) = R \geq k+2$. Denote by v some vertex from metric projection of u on C and by w some vertex of the interval $I(u,v)$ adjacent to v. Then either $w \in I(v,y) \cup I(v,z)$ or $w \notin I(y,z)$. In the first case, if, say $w \in I(v,y)$, then by Lemma 3 $d(z,u) \geq d(z,v) + d(w,u) = k + R - 1 > delta$, contradicting our assumption that y and z are mutually farthest vertices. Now assume that $w \notin I(y,z)$. Then $v \in I(w,y) \cup I(w,z)$. Therefore if, say $v \in I(y,w)$, then again by Lemma 3

$$d(y,u) \geq d(y,v) + d(w,u) = k + R - 1 > delta,$$

which is a contradiction. □

Claim 2 If *delta* is odd ($delta = 2k - 1$) then $R = k$, i.e. $d(u,C) \leq k$ for any $u \in V$.

Proof of Claim 2 Assume that there exists a vertex for which $d(u,C) \geq k+1$, and let $v \in Pr(u,C)$. Recall that $C = L(y, k-1, z)$, i.e. $d(y,v) = k-1, d(z,v) = k$. Denote by w some vertex of the interval $I(v,u)$ adjacent to v . We distinguish two cases.

Case 1 $w \in I(y,z)$, i.e. $w \in I(y,v) \cup I(v,z)$.

If $w \in I(y,v)$, then $v \in I(w,z)$ and by Lemma 3

$$d(z,u) \geq d(z,v) + d(w,u) \geq 2k,$$

in contradiction with the assumption that y and z are mutually farthest vertices. Now assume that $w \in I(v,z)$. Then from Lemma 3

$$d(y,u) \geq d(y,v) + d(w,u) \geq 2k - 1.$$

Since $z \in D(y)$ the equality holds $d(y, u) = 2k - 1$. From the second part of the same lemma this equality holds iff there is a vertex $x \in I(v, u) \cap I(w, y)$ adjacent to v and w. Since $I(w, y) \subset I(z, y)$ we get $x \in L(y, k - 1, z) = C$ and $d(u, x) < d(u, v)$, contradicting our assumption that $v \in Pr(u, C)$.

Case 2 $w \notin I(y, z)$, i.e. $v \in I(w, y) \cup I(w, z)$.

If $v \in I(w, z)$ then by Lemma 3 $d(u, z) \geq 2k$, contradicting our assumption that $e(z) = delta$. Therefore $v \in I(y, w)$ and $d(w, z) = d(v, z)$. From this equality and Lemma 2 we deduce that there exists a vertex $x \in I(w, z) \cap I(v, z)$ adjacent to w and v. Note that $w \in I(u, x)$, otherwise $x \in I(v, u)$ and we get in conditions of Case 1. Since in addition $v \in I(y, w)$, from Lemma 3 we deduce that

$$d(u, y) \geq d(u, w) + d(v, y) \geq 2k - 1,$$

$$d(z, u) \geq d(z, x) + d(w, u) \geq 2k - 1.$$

Since $e(y) = e(z) = 2k - 1$ we obtain that $d(y, u) = d(z, u) = 2k - 1$. By the second part of Lemma 3 there exist two vertices $s \in I(w, y) \cap I(v, u)$ and $t \in I(x, u) \cap I(w, z)$ adjacent to w, v and w, x correspondently (Fig. 4). Vertices s, t and w are equidistant from u. So by Lemma 2 there exist vertices $u' \in I(s, u) \cap I(w, u)$ and $u'' \in I(w, u) \cap I(t, u)$ adjacent to s, w and w, t respectively. As $u', u'' \in L(w, 1, u)$ then these vertices either are adjacent or coincide. In any case we obtain the cycle (v, s, u', u'', t, x, v). In every triangulation of this cycle the edge (v, x) belongs to at least one triangle. Since $u', u'' \in L(v, 2, u)$ the third vertex of such a triangle is one of the vertices s or t. This means that either $s \in I(v, u) \cap I(z, y)$ or $t \in I(v, u) \cap I(y, z)$ and we get in the conditions of the preceding case. $\square$

Claim 3 If $e(c) > \lfloor delta/2 \rfloor + 1$ then for any vertex $u \in D(c)$ $e(u) > delta$.

Proof of Claim 3 Put $p = \lfloor delta/2 \rfloor + 1$. By Claims 1 and 2 $d(u, C) \leq p$ and therefore $e(c) = p + 1$ and there is a vertex $w \in I(c, u) \cap C$. Since $u \in D^*$ and c is a vertex which belongs to a maximum number of projections of vertices from D^* to the clique C, there is such a vertex $u' \in D^*$ that $c \in I(w, u')$. By Lemma 3 $d(u, u') \geq d(u, w) + d(c, u') = 2p$ and so $e(u) > delta$. $\square$

Our final claim will complete the proof of the theorem.

Claim 4 $c \in C(G)$, except the case when $e(c) > \lfloor delta/2 \rfloor + 1$.

Proof of Claim 4 First we note that $r(G) \geq \lfloor delta + 1 \rfloor / 2$. If $delta$ is odd then either $e(c) = R = \lfloor delta + 1 \rfloor / 2$ and so $c \in C(G)$ or by Claim 3 for any vertex $u \in D(c)$ $e(u) > delta$. Now assume that $delta$ is even. If $e(c) > delta/2 + 1$ then by Claim 3 $e(u) > delta$ for any $u \in D(c)$. Hence suppose that $e(c) \leq delta/2 + 1$. In order to show that $c \in C(G)$ it is enough to consider only the case when $r(G) = delta/2$. Then $d(G) = d(y, z) = 2r(G)$ and by Lemma 5

$$C(G) \subset L(y, r(G), z) = C.$$

On the other hand $R = r(G)$ and all metric projections of vertices from D^* on C have a non–empty intersection. From our choice the vertex c must be in this intersection. Hence $e(c) = r(G)$ and therefore $c \in C(G)$ □

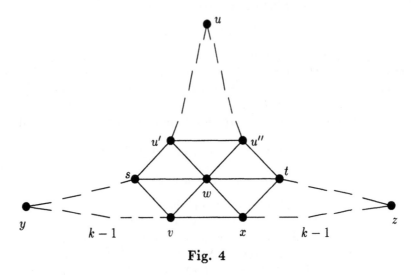

Fig. 4

Remark 1 Using this algorithm we can solve the following more general center problem :

for a given subset $M \subset V$ of vertices of a chordal graph G find a vertex $v \in V$ such that $e_M(v)$ is minimal, where

$$e_M(v) = \max\{d(v, u) : u \in M\}.$$

Remark 2 As a consequence of our algorithm we obtain that the interval $I(u, v)$ between any diametral vertices u and v intersects the center $C(G)$ of a chordal graph G. According to the algorithm either $L(u, \lfloor d(u, v)/2 \rfloor, v) \cap C(G) \neq \emptyset$ or we find a pair of vertices with a larger distance that is impossible in this case.

Open problem Find subquadratic time algorithms for computing the diameter $d(G)$ and the whole center $C(G)$ of a chordal graph G.

References

[1] G.J. Chang and G.L. Nemhauser, The k–domination and k–stability problems on sun-free chordal graphs, *SIAM J. Algebraic Discrete Meth.* 5 (1984) 332–345.

[2] G.J. Chang, Centers of chordal graphs, *Graphs and Combinatorics* 7 (1991) 305–313.

[3] V.D. Chepoi, Some properties of d–convexity in triangulated graphs (in Russian), *Mathematical Researches (Chişinău)* 87 (1986) 164–177.

[4] V.D. Chepoi, Centers of triangulated graphs (in Russian), *Matematiceskie Zametki (English Transl.: Mathematical Notes)* 43 (1988) 143–151.

[5] F.F. Dragan, Centers of graphs and Helly property (in Russian), *PhD Thesis, Moldova State University,* 1989.

[6] P. Duchet, Classical perfect graphs: an introduction with emphasis on triangulated and interval graphs, *Ann. Discrete Math.* 21 (1984) 67–96.

[7] M. Farber and R.E. Jamison, Convexity in graphs and hypergraphs, *SIAM J. Algebraic Discrete Meth.* 7 (1986) 433–444

[8] A.M. Farley and A. Proskurowski, Computation of the center and diameter of outerplanar graphs, *Discrete Appl. Math.* 2 (1980) 85–191.

[9] M.C. Golumbic, Algorithmic Graph Theory and Perfect Graphs, *Academic Press, New York* (1980).

[10] A. Hajnal and J. Suranyi, Über die Auflösung von Graphen in vollständige Teilgraphen, *Ann. Univ. Sci. Budapest Eötvös Sect. Math.* 1 (1958) 113–121.

[11] G. Handler, Minimax location of a facility in an undirected tree graph, *Transportation Sci.* 7 (1973) 287–293.

[12] S.M. Hedetniemi, E.J. Cockayne and S.T. Hedetniemi, Linear algorithms for finding the Jordan center and path center of a tree, *Transportation Sci.* 15 (1981) 98–114.

[13] R. Laskar and D. Shier, On powers and centers of chordal graphs , *Discrete Appl. Math.* 6 (1983) 139–147.

[14] S. Olariu, A simple linear-time algorithm for computing the center of an interval graph *Tech. Rept. TR-89-20 Old Dominion University, Norfolk* (1989).

[15] A. Proskurowski, Centers of 2–trees, *Ann. Discrete Math.* 9 (1980) 1–5.

[16] R.E. Tarjan and M. Yannakakis, Simple linear–time algorithms for test chordality of graphs, test acyclicity of hypergraphs, and selectively reduce acyclic hypergraphs, *SIAM J. Comput.* 13 (1984) 566–579.

The Time Complexity of Updating Snapshot Memories

Amos Israeli[1], and Asaf Shirazi[2]

[1] Dept. of Electrical Engineering, Technion — Israel
[2] Dept. of Computer Science, Technion — Israel

Abstract. An *atomic snapshot memory* enables a set of processes, called *scanners,* to obtain a consistent picture of the shared memory while other processes, called *updaters,* keep updating memory locations concurrently. Implementations of atomic snapshot memory are key tools in designing distributed protocols in shared memory systems. Such an implementation consists of two protocols: An *update protocol* and a *scan protocol,* executed by updaters and scanners, respectively.

It is clear that the time complexity of the scan protocol is at least linear, and there exists an implementation which matches the lower bound. In this paper we show that the time complexity of an optimal update protocol is $\Theta(\min\{u, s\})$, where u is the number of updaters and s is the number of scanners.

1 Introduction

Consider a system of processes communicating through shared memory in which write and read operations are executed instantaneously. At any given time t, each memory cell holds a well defined value which is the value that was most recently written to it (or its initial value if no such write operation occurs before t). A *Snapshot* at time t is the vector of values held by all memory cells at t. An *atomic snapshot memory* (for brevity, a *snapshot memory*) is an object that allows some processes to acquire a snapshot while other processes can update their memory cells *concurrently*. An *implementation* of a snapshot memory consists of two protocols, called the *updater* protocol and the *scanner* protocol. A process that wishes to update one of the memory cells executes the updater protocol; a process that wishes to acquire a snapshot executes the scanner protocol. Implementations of snapshot memory are key tools in designing concurrent protocols and have many applications. Among the applications are randomized consensus, [As90], concurrent timestamping, [DS89], approximate agreement, [ALS90] and a general method for implementing wait-free data structures, [AH90].

Traditionally, a snapshot memory is implemented by means of *locking* — a process that wishes to scan the memory locks it first, so no other process can update any memory cell until the scan operation is completed. This approach is used satisfactorily in many database systems. In contrast to the locking approach, there is an increasing interest in *wait-free* implementations. In these implementations, no process is required to wait for operations of another process while executing its own scan or update protocol. Wait-free implementations have a strong practical motivation: In a multi-processor environment, processors of different speeds frequently need to cooperate. In such cases, it is inefficient to allow a process executing on a fast processor to wait for a process executing on a slow processor. Moreover, in a multi-process environment, processes may be delayed for long periods due to swapping, I/O operations, page faults, etc. Once more, waiting for a delayed process decreases the throughput. In addition, wait-free implementations are resilient to process failures.

The *time complexity* of a wait-free implementation is the maximal number of read and write operations executed during a single execution of an update or a scan operation.

Wait-free snapshot memories were proposed independently by Afek et al in [AAD90] and by Anderson in [An93]. Anderson presented in [An93] an implementation for snapshot memory with exponential-time complexity. An implementation with quadratic time complexity was presented in [AAD90]. Another quadratic-time implementation was presented by Aspnes and Herlihy in [AH90]. A linear-time implementation for a system with *one* scanner was presented by Kirousis, Spirakis and Tsigas in [KST91].

A new approach was proposed by Attiya, Herlihy and Rachman in [AHR92]. In that paper, the authors introduced the notion of *Lattice-Agreement* and showed that one can implement a snapshot memory using an implementation of a lattice agreement object without increasing the time complexity. Then, they presented a lattice agreement implementation whose time-complexity is linear using test and set registers for two processes. This last implementation induces two randomized implementation of a snapshot memory using read write registers. The first implementation uses multi-writer registers and its expected time complexity is $O(n)$, where n is the number of processes in the system. The second implementation uses single-writer registers and its expected time complexity is $O(n \log^2 n)$. Another randomized implementation, with the same expected time complexity as the latter implementation, was presented by Chandra and Dwork in [CD92].

Israeli, Shaham and Shirazi showed, in [ISS93], a method of converting arbitrary snapshot memory implementations to implementations in which the time complexity of one protocol (either update or scan) is linear in the number of updaters, while the time complexity of the second protocol is the sum of the time complexities of the protocols in the original solution. Deterministic implementations of the lattice agreement object (and hence for the snapshot memory object) were proposed independently by Israeli and Shirazi in [IS93] (time complexity $O(n^{1.5} \log n)$) and by Attiya and Rachman in [AR93] (time complexity $O(n \log n)$). The time-complexity of all these implementations except the single scanner implementation of [KST91] is super-linear. A linear-time implementation for a similar object called *Time-Lapse* snapshot memory, was presented by Dwork, Herlihy, Plotkin and Waarts in [DHPW92]. This object, however, is slightly weaker than snapshot memory, since time-lapse snapshots may be inconsistent. Hoepman and Tromp showed in [HT93] that in order to implement a snapshot memory object it suffices to consider a system in which the value fields are single bits. Inoue, Chen, Masuzawa and Tokura showed in [ICMT94] a linear-time implementation of a lattice agreement object that uses multi-writer registers.

Intuitively, it is clear that the time complexity of the scan protocol in any implementation of a snapshot memory is at least linear in the number of updaters, and it is not too difficult to formalize this intuition. In [ISS93], we showed implementations that match the lower bound. Previous implementations usually incorporated a scan protocol as a subroutine of the update protocol, which implied a linear lower bound on the time-complexity of update protocols of that kind. Unlike the scan protocol, there is no clear intuition for the time complexity of an update protocol. It is not inconceivable that there exist implementations of an entirely different nature in which the time complexity of the update protocol is sublinear and perhaps even a constant.

In this paper, we show that for any system, the time complexity of an optimal update protocol is $\Theta(\min\{u, s\})$, where u is the number of updaters and s is the number of scanners. The lower bound is valid even when the updates are *serial*, meaning, no two updates overlap. In addition, the lower bound does not assume that the protocols are uniform or that the processors use bounded registers. To the best of our knowledge, this is the first ever lower bound on time complexity in the wait-free model, and among the few known lower bounds for fault-tolerant distributed systems. The upper bound is obtained by converting an arbitrary implementation of snapshot memory to an implementation in which the time complexity of the update protocol is optimal.

The rest of the paper is organized as follows: In Section 2 we give a formal definition of the model. The lower bound is proven in Section 3. In Section 4 we give a protocol of matching complexity.

2 Model and Requirements

In this section, we define the model of computation and the atomic snapshot memory object. Informally, *processes* are deterministic sequential threads of control that communicate through shared data structures called *objects*. Formally, we model processes and objects using a simplified form of the I/O automata of Lynch and Tuttle, [LT87]. A *system* is a collection of processes and objects. Processes access the objects by executing *operations*. Operations are either *atomic* or *non-atomic*. An atomic operation is executed instantaneously. A non-atomic operation is built from several operations executed one after the other (for convenience, we neglect the internal computation). A process is described by its *protocol* — a non-atomic operation that can be executed several times (possibly an infinite number of times). Each atomic operation corresponds to a single state-transition. Each protocol has a distinguished atomic operation, called the *initial operation*, corresponding to the initial state.

Executions are described under the interleaving model. A global state is described by a *configuration* — a vector containing the state of each process and the state of each object. The system's *initial configuration* contains the processes' and objects' initial states. The execution of an (atomic) operation is called an *(atomic) action*. An execution is a sequence of configurations and atomic actions $E = conf_0, act_1, conf_1, \ldots, conf_{i-1}, act_i, conf_i, \ldots$, where $conf_0$ is the system's initial configuration and for every $i > 0$, $conf_i$ is obtained from $conf_{i-1}$ by executing the atomic action act_i ($conf_i$ is the *occurance configuration* of act_i). We stress that no assumption is made on the relative speeds of the processes. Usually, atomic actions of different processes are interleaved. An execution is *sequential* if atomic actions of different non-atomic operations are not interleaved.

An object is specified by a set of legal sequential executions, [HW90]. The notion of *implementation* is discussed and formally defined in [H88, AAD90]. We omit the formal definitions and give the essence of implementing an object. Intuitively, in order to provide an implementation of an object A from a set of objects, B, one must give a protocol for each operation on the object A. All the operations in the protocols are on objects from B, and executing a protocol is "equivalent" to executing the operation on A. The equivalence notion is made precise by the *linearizability* correctness condition, [HW90]. Informally, we want each non-atomic action to *appear as if* it was executed instantaneously. In addition, the order between non-atomic actions that are not concurrent should be preserved.

Formally, consider any execution, E, of an implementation in which the atomic actions of the processes are interleaved. The execution induces a *partial order* $<_E$ on the non-atomic operations executed in E: Let a, b be two non-atomic actions in E. If a ended before b started, then $a <_E b$. An execution is *linearizable* if $<_E$ can be extended to a *complete* order $<_S$, where $<_S$ is the order induced by some sequential execution of the non-atomic operations in E and S is one of the legal sequential executions. Clearly, since $<_S$ extends $<_E$, the order between non-atomic operations that are not concurrent in E is preserved. An implementation is *linearizable* if all its executions are linearizable.

Let a be a non-atomic operation executed in E. We denote the start and end configurations of a in E by $start_E(a)$ and $end_E(a)$, respectively (if a does not complete in E then $end_E(a) = \infty$). The *execution interval* of a in E includes all the configurations in the interval $[start(a), end(a)]$. In order to prove that E is linearizable, it is sufficient to assign a *linearization point* to each operation in E, a, such that the linearization point of a lies in the execution interval of a in E. The obtained sequential execution must be one of the legal sequential executions.

An implementation is *wait-free* if in all its executions, the number of atomic actions executed in each protocol is bounded, where the bound may depend on the number of processes in the system. We require our implementations to be wait-free.

We consider two types of shared objects: *atomic registers*, henceforth *registers*, and *atomic snapshot memories*. Processes access registers by executing *write* and *read operations*. A write operation stores a new value in the register and a read operation obtains the value stored in the register. In the initial configuration, each register holds its *initial value*. Each operation is executed instantaneously[3] and each read operation returns the value written by the most recent, preceding, write operation, or the initial value if no such write operation exists. The registers are *single writer multi reader* registers. That is, each register is associated with one process, called its *owner*, which is the only process that can write to it. However, any process can read the register.

An atomic snapshot memory, henceforth a *snapshot memory*, is defined with respect to a set of *cells* (this set is called a compound register in [An93]). The processes are divided into two groups: *updaters* and *scanners*. Updater i, denoted U_i, owns cell i and can change the value of its cell in an *update operation*. Scanner j, denoted S_j, obtains the value of *all* the cells in a *scan operation*. Though the two groups of processes are not necessarily distinct, it is convenient to assume that they are. This does not harm the generality of our results since a process which is both an updater and a scanner can be viewed as two processes. We denote the number of updaters and scanners by u and s respectively. A snapshot memory object must be implemented from registers. The cell of each updater is one field in the register of the updater, called the *data field*. An implementation of a snapshot object consists of two protocols, one for updaters and one for scanners.

The time complexity of a snapshot implementation, $\mathcal{I}$, is a pair of functions $(f_{\mathcal{I}}(u, s), g_{\mathcal{I}}(u, s))$, where $f_{\mathcal{I}}(u, s)$ and $g_{\mathcal{I}}(u, s)$ are the maximal number of atomic operations on registers performed in an update and scan operations respectively.

3 Lower Bound for Update Protocols

In this section, we establish a lower bound of $\Omega(\min\{u, s\})$ on the time complexity of an update protocol in any implementation of a snapshot memory, where u is the number of updaters and s is the number of scanners. In order to do so, we prove that for the well studied case of n updaters and n scanners, the time complexity of the update protocol in any snapshot implementation is $\Omega(n)$. This suffices since if $u \neq s$, we can choose $\min\{u, s\}$ updaters and $\min\{u, s\}$ scanners to execute alone. We prove the lower bound for a model in which atomic actions of an update action are not interleaved with atomic actions of other actions (either update or scan actions). The adversary in this model is *weaker* than the one usually assumed, hence, the lower bound is stronger. We stress that the processes do not use randomization.

The proof strategy is as follows: We assume, by way of contradiction, that there exists an implementation whose update protocol has time complexity $o(n)$. Since the updaters do a small amount of work in each update operation, they also read the registers of few processes. After formalizing this observation, we construct an infinite execution in a stage by stage fashion. In each stage, a complete update operation is executed and then one scanner executes one atomic

---

[3] It is possible that an atomic register is implemented from weaker registers, [La86a, La86b, VA86, ILV87, IS92]. However, in such cases the linearizability of the implementation allows us to assume that the operations appear as if they are executed instantaneously, [HW90].

operation. Let E be an infinite execution in the structure described above. The infinite execution E_i, $i \in \mathcal{N}$, is identical to E, except that the $(i+1)$st update action is moved just before the ith update action. The essence of the proof is to choose the updater and scanner of each stage in a way that guarantees that no process can notice the difference between E and E_i. The proof is then complete, since no scanner can ever stop with a valid snapshot.

We now turn to the formal proof. Consider a system of n updaters and n scanners. Let $\mathcal{I}$ be an implementation of snapshot memory, and let $f_{\mathcal{I}}(n)$ be the time complexity of the update protocol in $\mathcal{I}$. We look at any implementation of snapshot memory such that the there exists $n \geq 24$ for which $f_{\mathcal{I}}(n) < \frac{n}{12}$ (the bound can be slightly improved, at the cost of more complicated expressions). Let $conf$ be any configuration and assume that updater i, U_i, performs a complete update operation starting at $conf$. The set of registers read in this update action is denoted by $regs(conf, U_i)$. This set is well defined by the assumption that updates execute in exclusion.

Definition 1. We say that process P (either scanner or updater) *affects* updater U_i at $conf$ if at least one of P's register is in $regs(conf, U_i)$.

We first prove that for any configuration, the majority of processes affect few updaters.

Lemma 2. Given a configuration $conf$, the number of processes such that each of them affects at least a quarter of the updaters at $conf$ is less than $n/3$.

Proof. To simplify calculations, we assume that every updater reads its register during every update it performs. Every updater reads at most $f_{\mathcal{I}}(n)$ registers in a single update operation. Thus, n updaters perform at most $n \cdot f_{\mathcal{I}}(n)$ reads. Hence, the number of processes (either scanners or updaters) that affect at least a quarter of the updaters at $conf$ is bounded by $\frac{n \cdot f_{\mathcal{I}}(n)}{n/4} = 4 \cdot f_{\mathcal{I}}(n) < n/3$

$\square$

Next, we construct the infinite execution, E. The configuration before the ith stage begins is denoted by $conf_i$ ($conf_0$ is the initial configuration). Before the ith stage, $i \in \mathcal{N}$, a single updater (denoted $U_{\alpha(i)}$) and a single scanner (denoted $S_{\beta(i)}$) are chosen. In the ith stage, $U_{\alpha(i)}$ performs a complete update operation and then $S_{\beta(i)}$ performs a single atomic operation. We choose both $U_{\alpha(0)}$ and $S_{\beta(0)}$ such that they affect at most a quarter of the updaters. At least $\frac{2}{3} \cdot n$ scanners and $\frac{2}{3} \cdot n$ updaters are suitable. Once $conf_{i+1}$ is reached, we choose a scanner $S_{\beta(i+1)}$ that affects at most a quarter of the updaters at $conf_{i+1}$, and an updater $U_{\alpha(i+1)}$ that satisfies the following five requirements (the maximal number of updaters that do not comply with each restriction is written in parenthesis):

- $U_{\alpha(i+1)}$ affects at most a quarter of the updaters at $conf_{i+1}$ $(n/3)$.
- $U_{\alpha(i+1)}$ is not affected by $U_{\alpha(i)}$ at $conf_i$, in particular, $U_{\alpha(i+1)} \neq U_{\alpha(i)}$ $(n/4)$.
- $U_{\alpha(i+1)}$ is not affected by $S_{\beta(i)}$ at $conf_i$ $(n/4)$.
- $U_{\alpha(i)}$ is not affected by $U_{\alpha(i+1)}$ at $conf_i$ $(n/12)$.

– $S_{\beta(i)}$ does not read the register of $U_{\alpha(i+1)}$ in its next atomic operation at $conf_i$ (1).

By Lemma 2, and since an updater is affected by at most $\frac{n}{12}$ processes at any configuration, at least $\frac{2}{3} \cdot n$ scanners and $\frac{n}{12} - 1$ updaters comply with the restrictions.

Lemma 3. *Let E be an infinite execution built as described above. The infinite execution E_i, $i \in \mathcal{N}$, is identical to E, except that the $(i+1)$st update operation is executed just before the ith update operation. It holds that no process can distinguish between E and E_i.*

Proof. The construction of the execution guarantees that $U_{\alpha(i)}$, $S_{\beta(i)}$ and $U_{\alpha(i+1)}$ cannot notice the change. This suffices since only these processes execute in the changed interval. Therefore, other processes can notice the change only by looking at their registers.

$\square$

Theorem 4. *For any implementation of snapshot memory with n updaters and n scanners, the time complexity of the update protocol is $\Omega(n)$.*

Proof. Assume, towards a contradiction, that the theorem is not correct. Let $\mathcal{I}$ be an implementation of a snapshot memory that contradicts the theorem, and choose $n \geq 24$ such that $f_{\mathcal{I}}(n) < \frac{n}{12}$. Let E be an infinite execution constructed as described above. Since the implementation is wait-free, it follows that there exists a scanner S that halts in E after a finite number of actions. Assume that the last update operation whose value is included in the returned scan of S is the update operation executed in the ith stage of E. Clearly, $i \geq 0$ since the update operation in stage 0 completes before any scan begins. From the previous lemma, S does not notice the difference between E and E_i. Therefore, S halts in E_i with the same snapshot, which is no longer valid, a contradiction.

$\square$

Using the above theorem, we immediately obtain the following corollary.

Corollary 5. *For any implementation of snapshot memory with u updaters and s scanners, the time complexity of the update protocol is $\Omega(\min\{u, s\})$.*

4 An Optimal Snapshot Implementation for Updaters

In this section, we describe a method to convert an arbitrary implementation of snapshot memory to another implementation with an *optimal time* update protocol. As we present a conversion of snapshot implementations, one of our objects is a snapshot memory. The original implementation and its two protocols are called the *elementary implementation* and *elementary protocols*, respectively. Since the elementary implementation is linearizable, we can assume that operations on this object occur instantaneously, [HW90]. However, since we measure the time complexity in the number of operations on registers, the time complexity incurred by these operations is the time complexity of the elementary protocol used. Throughout the rest of this paper, it is understood that the read, write, elementary scan and elementary update operations are atomic, while the scan and update operations are not. For brevity, we denote the elementary update and elementary scan operations by *eupdate* and *escan*, respectively.

In addition to the elementary implementation, we use the time-lapse snapshot implementation of [DHPW92], where the u updaters execute a time-lapse update operation and the s scanners execute a time-lapse scan operation. The complexity of the resulting update protocol is $O(s)$. Israeli, Shaham and Shirazi showed, in [ISS93], a conversion method that yields an implementation in which the time complexity of the update protocol is u. Therefore, for any system there exists a snapshot implementation such that the time complexity of the update protocol is $O(\min\{u, s\})$. The underlying idea of the conversion method is as follows: A scanner begins by obtaining a time-lapse snapshot of the memory. Time-lapse snapshots should not be returned as snapshots because they are only *partially* ordered. To achieve a *complete* order, each scanner eupdates its time-lapse snapshot and then performs an escan operation to obtain the time-lapse snapshots held by the other scanners. Finally, the scanner chooses the latest value for each updater and obtains a snapshot. The presented construction is unbounded. However, it is easily verified that the construction can be bounded using the method of [DHW93], without increasing the time complexity.

4.1 Time-Lapse Snapshot Memory

In this subsection, we briefly survey the results of [DHPW92] which are used in our implementation of snapshot memory. Recall that updaters and scanners execute time-lapse update operations and time-lapse scan operations, respectively.

Time-lapse snapshot memory is defined in a similar way to snapshot memory. However, two time-lapse scans TLS and TLS' are allowed to disagree on the order of two time-lapse updates TLU and TLU' *if the four operations are concurrent with one another* (the intersection of their execution intervals is not empty).

Let TLU_i^a (TLS_i^a) denote the ath time-lapse update (time-lapse scan) of U_i (S_i), and let val_i^a denote the value written during TLU_i^a. The following properties hold for any execution E of the implementation presented in [DHPW92].

- *Regularity*: For any value val_i^a returned by TLS_j^b, TLU_i^a begins before TLS_j^b terminates, and there is no TLU_i^c such that $TLU_i^a <_E TLU_i^c <_E TLS_j^b$.
- *Monotonicity of time-lapse scans*: Let TLS and TLS' be two time-lapse scans satisfying $TLS <_E TLS'$. For every U_k, the value of U_k returned by TLS' is not older than the value of U_k returned by TLS.
- *Monotonicity of time-lapse updates*: Let TLU_i^a and TLU_j^b be two time-lapse updates satisfying $TLU_i^a <_E TLU_j^b$ (possibly $i = j$), and let TLS be a time-lapse scan operation (possibly concurrent with both time-lapse update

operations) such that val_j^b is included in TLS. It holds that val_i^a (or a later value of U_i) is included in TLS.

- The only time val_i^a is written during TLU_i^a is in its last atomic operation.
- The time complexity of TLU_i is $O(s)$.
- The time complexity of TLS_i is $O(u)$.

4.2 Description

The protocols appear in Figure 1. Updater i keeps an internal variable, $count_i$, which is initialized to zero and incremented by one at the beginning of every update operation. The ath update operation of U_i is denoted by U_i^a, and the value updated in U_i^a is denoted by val_i^a. Operation U_i^a is simply the time-lapse update operation with the input value $(val_i^a, count_i)$. We denote the last write operation in the time-lapse update by w_i^a. The complexity of the update protocol is $O(s)$.

The scanners communicate among themselves using the elementary implementation. The data field of scanner j, called $view_j$, consists of an array of u entries in which each entry is a pair of the form $(value, count)$. The k-th entry of $view_j$ always holds a $(value, count)$ pair of U_k. The new scan protocol consists of two parts: In the first part, the scanner performs a time-lapse scan operation and obtains a *local view*. In the second part, the scanners reach a complete order on the local views using the elementary snapshot protocols: Each scanner executes an eupdate operation of its local view, followed by an escan operation on the views of all scanners. Following the elementary scan operation, the local view is recomputed by choosing the latest value for each updater. This yields a snapshot which is returned. The bth scan action of S_j is denoted by S_j^b. The actions of S_j^b are denoted by TLS_j^b, eu_j^b and es_j^b. We denote the view eupdated in eu_j^b and the snapshot returned by S_j^b by $view_j^b$ and $snap_j^b$ respectively. The complexity of the scan protocol is $O(u)$ plus the sum of the complexities of the elementary protocols.

4.3 Linearization Scheme

For the rest of this section we consider an arbitrary execution, E, of the converted implementation and prove that the execution is linearizable. For brevity, we omit the reference to the execution throughout the reminder of this section. In particular, $<$ and $\leq$ should be read as $<_E$ and $\leq_E$ respectively. For any non-atomic action a, the linearization point of a, denoted $lin(a)$, is related to some specific configuration. This configuration is called the *linearization configuration* of a and is denoted by $lin\_conf(a)$. Sometimes, more then one action is linearized by the same configuration. In some of the arguments below, it is simpler to consider the linearization configuration. The starting and ending configurations of a are denoted by $start(a)$ and $end(a)$ respectively. For an atomic action a, $occ(a)$ denotes the occurance configuration of a. We stress that operations on time-lapse snapshot memory are *not* atomic.

```
Update_i(value)
    count_i ← count_i + 1
    TLU_i(value, count_i)

Scan_j
    lview ← TLS_j
    eupdate_j(lview)
    s[1...u] ← escan_j
    for k ← 1 to s
        for ℓ ← 1 to u
            if s[k][ℓ].count > lview[ℓ].count then lview[ℓ] ← s[k][ℓ]
        Endfor
    Endfor
    Return lview
```

Fig. 1. Protocols for U_i and S_j

We define a domination order on views returned as snapshots in the following natural way: Let *snap* and *snap'* be two local views returned as snapshots. *snap* dominates *snap'* if for every i, $1 \le i \le u$, the count field in the ith entry of *snap* is not less than the count field in the ith entry of *snap'*.

Definition 6. The linearization configuration of scan action S_i^a is defined to be the minimum between $end(S_i^a)$ and the occurrence configuration of the first write action, w_k^c, for which c is larger then the count of $snap_i^a[k]$.

$$lin\_conf(S_i^a) = \min\{end(S_i^a), \min_{j,b}\{occ(w_j^b) \mid b > snap_i^a[j].count\}\}$$

In the first case, the linearization point is the same as the linearization configuration. In the second case, we say that S_i^a is linearized by w_k^c, and the linearization point is *just before* the linearization configuration. If two scan actions are linearized by the same write action then they are further linearized by domination order of the views they return as snapshots. If the snapshots are equal, the scan actions are linearized arbitrarily.

The validity of the above definition is guaranteed by Lemma 8, which shows that all views returned as snapshots are ordered by domination and the fact that two scan actions of the same scanner cannot be linearized by the same write action.

Definition 7. The linearization configuration of update action U_j^b is defined to be the minimum between $end(U_j^b)$ and the linearization configuration of the first scan action S_ℓ^d which returns val_j^b.

$$lin\_conf(U_j^b) = \min\{end(U_j^b), \min_{k,c}\{lin\_conf(S_k^c) \mid val_j^b \in snap_k^c\}\}$$

In the first case, the linearization point is the same as the linearization configuration. In the second case, we say that U_j^b is linearized by S_ℓ^d, and the linearization point is *just before $lin(S_\ell^d)$*, where ties are broken arbitrarily.

4.4 Correctness Proof

The correctness of the converted implementation is proven by the following lemmas. The proofs are omitted for lack of space.

Lemma 8. *All views returned as snapshots are ordered by domination.*

Lemma 9. *Every action is linearized within its execution interval.*

Lemma 10. *Snapshot $snap_i^a$ is the valid snapshot at $lin(S_i^a)$.*

References

[An93]   J. Anderson, Composite Registers, *Distributed Computing*, Vol. 3, No. 3, 1993, pp. 141-154.

[As90]   J. Aspnes, Time- and Space-Efficient Randomized Consensus, *Proceedings of the 9th Annual ACM Symposium on Principles of Distributed Computing*, 1990, pp. 15-29.

[AAD90] Y. Afek, H. Attiya, D. Dolev, E. Gafni, M. Merritt, and N. Shavit, Atomic Snapshots of Shared Memory, *Proceedings of the 9th Annual ACM Symposium on Principles of Distributed Computing*, 1990, pp. 1-13.

[AH90]   J. Aspnes and M. Herlihy, Wait-free Data Structures in the Asynchronous PRAM Model, *Proceedings of the 2nd Annual Symposium on Parallel Algorithms and Architectures*, 1990, pp. 340-349.

[AHR92] H. Attiya, M. Herlihy and O. Rachman, Efficient Atomic Snapshots Using Lattice Agreement, *Proceedings of the 6th International Workshop on Distributed Algorithms and Graphs*, 1992, pp. 35-53, Lecture Notes in Computer Science #647, Springer-Verlag, 1992.

[ALS90] H. Attiya, N.A. Lynch and N. Shavit, Are Wait-Free Algorithms Fast? *Proceedings of the 31st IEEE Symposium on Foundations of Computer Science*, 1990, pp. 55-64.

[AR93]   H. Attiya and O. Rachman, Atomic Snapshots in $O(n \log n)$ Operations, *Proceedings of the 12th Annual ACM Symposium on Principles of Distributed Computing*, 1993, pp. 29-40.

[CD92]   T.D. Chandra and C. Dwork, personal communications.

[DHPW92] C. Dwork, M. Herlihy, S. Plotkin and O. Waarts, Time-Lapse Snapshots, *Proceedings of the 1st Israeli Symposium on Theory of Computing and Systems*, 1992, pp. 154-170, Lecture Notes in Computer Science #601, Springer-Verlag, 1992.

[DHW93] C. Dwork, M. Herlihy and O. Waarts, Bounded Round Numbers, *Proceedings of the 12th Annual ACM Symposium on Principles of Distributed Computing*, 1993, pp. 53-64.

[DS89]   D. Dolev and N. Shavit, Bounded Concurrent Time-Stamp Systems are Constructible!, *Proceedings of the 21st Annual ACM Symposium on Theory of Computing*, 1989, pp. 454-465.

[H88]   M. P. Herlihy. Impossibility and universality results for wait-free synchro-
          nization, *Proceedings of the 7th Annual ACM Symposium on Principles of
          Distributed Computing*, 1988, pp. 276-290.

[HT93]   J.H. Hoepman and J. Tromp, Binary Snapshots, *Proceedings of the 7th Inter-
          national Workshop on Distributed Algorithms and Graphs*, 1993, pp. 18-25,
          Lecture Notes in Computer Science #725, Springer-Verlag, 1993.

[HW90]   M.P. Herlihy and J.M. Wing, Linearizability: A Correctness Condition for
          Concurrent Objects, *ACM Transactions on Programming Languages and
          Systems*, Vol. 12, No. 3, 1990, pp. 463-492.

[ICMT94] M. Inoue, W. Chen, T. Masuzawa and N. Tokura, Linear-Time Snapshot
          Using Multi-writer Multi-reader Registers, to appear in *Proceedings of the
          8th International Workshop on Distributed Algorithms and Graphs*, 1994.

[ILV87]   A. Israeli, M. Li, and P. Vitanyi, Simple Multireader registers using Time-
          Stamp schemes, Report no. CS-R8758, Center for Mathematics and Com-
          puter Science, Amsterdam, Holland, 1987.

[IS92]   A. Israeli and A. Shaham, Optimal Multi-Writer Multi-Reader Atomic Reg-
          isters, *Proceedings of the 11th Annual ACM Symposium on Principles of
          Distributed Computing*, 1992, pp. 71-82.

[IS93]   A. Israeli and A. Shirazi, Efficient Snapshot Protocol Using 2-Lattice Agree-
          ment, preprint.

[ISS93]   A. Israeli, A. Shaham and A. Shirazi, Linear-Time Snapshot Protocols for
          Unbalanced Systems, *Proceedings of the 7th International Workshop on Dis-
          tributed Algorithms and Graphs*, 1993, pp. 18-25, Lecture Notes in Computer
          Science #725, Springer-Verlag, 1993.

[KST91]   L.M. Kirousis, P. Spirakis and P. Tsigas, Reading Many variables in One
          Atomic Operation: Solutions with Linear or Sublinear Complexity, *Pro-
          ceedings of the 5th International Workshop on Distributed Algorithms and
          Graphs*, 1991, pp. 229-241, Lecture Notes in Computer Science #579,
          Springer-Verlag, 1992.

[La86a]   L. Lamport, On Interprocess Communication. Part I: Basic Formalism, Dis-
          tributed Computing, Vol. 1, No. 2, 1986, pp. 77-85.

[La86b]   L. Lamport, On Interprocess Communication. Part II: Algorithms, Dis-
          tributed Computing, Vol. 1, No. 2, 1986, pp. 86-101.

[LT87]   N. Lynch and M. Tuttle, Hierarchical Correctness Proofs for Distributed
          Algorithms, In *Proceedings of the 6th Annual ACM Symposium on Principles
          of Distributed Computing*, 1988, pp 137-151.

[VA86]   P. Vitanyi, and B. Awerbuch, Atomic Shared Register Access by Asyn-
          chronous Hardware, Proceedings of the 27th IEEE Symposium on Foun-
          dations of Computer Science, 1986, pp. 233-243.

Non-Exploratory Self-Stabilization for Constant-Space Symmetry-Breaking

Giuseppe Parlati[1] and Moti Yung[2]

[1] Columbia University and Università di Salerno
[2] IBM Research Division, T. J. Watson Center, Yorktown Heights, NY 10598

Abstract. We introduce the notion of *non-exploratory* self-stabilizing algorithms. The notion minimizes what we call "exploration"– which is the additional (overhead) messages sent in an already stable system in order to assure stabilization maintenance. A non-exploratory algorithm implies significant reduction in overall communication complexity. We demonstrate the applicability of non-exploratory algorithms on the problems of randomized round-robin constant-space token-management, and symmetry breaking (leader election), solved on ring networks for hardware oriented systems (that is, constant space, constant message-size and uniform systems).

1 Introduction

Round-robin token-management was considered in 1974 by Dijkstra [Dij74] where the fundamental notion of *self-stabilization* was introduced. The goal of self stabilization is to achieve self-organization (or self-healing) so that the system reaches "normal" operation starting from an arbitrarily faulty initial state (The importance of this notion is advocated in [Lam84, LL90]).

Naturally, self-stabilizing solutions are most suitable for practical systems when the fault frequency is low, which means that the stabilizing time is smaller than the time between recurring faults. What we observed is that if the stabilization communication complexity is c_1 and the stable phase communication complexity is c_2 then if only an ϵ fraction of the time is devoted to stabilization, the actual overall communication complexity is $\{\epsilon c_1 + (1 - \epsilon)c_2\}$. Thus, c_2 is eventually dominating. In many cases we noticed that the maintenance of stable state involves some overhead effort. Our goal in this work is to introduce the notion of *non-exploratory* self-stabilizing algorithms, where the communication complexity of maintaining stability is minimized, as opposed to certain solutions where maintenance requires extra overhead communication (note that we assume message passing systems).

Following [Dij74], the most popular self stabilization problem has been token management on ring networks. This problem attempts to converge from any given state (with arbitrary number of tokens) to a state with a single token traveling in a round-robin fashion in the ring. Dijkstra's original work showed that given certain conditions (i.e. mutual-exclusion, uniformity, and no-deadlock (namely, token always exists)), there is no deterministic solution, while with a special leader processor (non-uniform systems) a solution does exist. Thus, in fact, leader election (symmetry breaking) is impossible as well via deterministic solutions. This is due to the ring's symmetry which is a fundamental barrier in distributed systems [Ang80, IR81]. On asymmetric rings (i.e., prime number size rings) a solution was given in [BP88] (see also [LS92]). In Brown, Gouda, and Wu [BGW89] a solution was proposed using a leader. They introduced the use of additional control signals besides the token in the algorithm (i.e. exploration, as we call it). The first to propose the use of randomization in solving token problems on self–stabilizing uniform systems was the influential work of Israeli and Jalfon [IJ90a]. They gave up the round-robin property (needed for coordination activities like fairness in access rights) by initiating the use of random-walks on the ring for self stabilization. Other works include Herman [Her90] who gave a solution requiring strong synchrony and limited to odd-size rings, and Afek and Brown [AB89] who gave token-management in the message-passing model (all other works assumed shared memory) and round-robin with the use of randomization, but they still assumed that a leader exists).

Motivation: We are motivated by the recent development of high-speed networks, where processing of control signals at a switch level needs to be done in short delay (of a few bits, e.g. [CS92]). In addition, there are only few bits devoted by communication standards to control bits (ATM's generic flow control allows for four bits only). This implies that future architectures will employ hardware-oriented constant-time and constant-area control algorithm (i.e., finite automata) to support fast on-line processing at a low-layer protocol. Note that many future fast (e.g., gigabit/sec.) LAN designs: FDDI [R86], Meta-Ring [CO90], Cambridge LAN [HN88], Magnet [LTG90], ATM-ring [OMS89], etc., have ring-based topology (this is followed by a significant theoretical interest which is due to this architecture's inherent symmetry [AAHK86, ASW85, Ang80, Dij74, FL84, IR81, It90, PKR82, V84]). Having self-stabilizing mechanism can prevent costly centralized/duplicated monitoring and recovery protocols in rings, see e.g., [BCK+83]. A particular important operation is the mechanism for *fair round-robin network access regulation*, needed for access-control regulation (e.g., token ring) and future LAN's, e.g. IBM's experimental prototyped MetaRing network [CO90]. The need for constant–space solution has motivated [MOOY92] to initiate a constant bit message model, and to suggest a solution to the round-robin token management problem. We assume their model. The importance (from a systems perspective) of reducing the message complexity when handling fast control signals was communicated to us by Ofek [Of93].

Definition and Properties: what constitutes a good solution? Let us review the problems and properties involved in self-stabilizing solutions to round-robin token management and to leader election on a ring with constant space processing. First note that we need a fast solution (i.e. low polynomial time) after which the system is stable. The next problem to handle is the communication deadlock [DIM91], i.e. when no tokens are present in the system (as high-speed networks are message passing systems). As was noticed, this problem is inherent– which means that, in practice, a system needs to employ a fault-detection (i.e. a time-out mechanism on a monitor which checks message passing via the node). Another source of deadlock was found in [IJ90a] who showed that deadlock-freeness implies that $\Omega(\log n)$ bits are needed to represent a token (where n is the ring size) for any deadlock-free solution. Thus the *constant-space property* of our system (as the one of [MOOY92]) implies that the system does not stabilize if initially no tokens are present in the ring. In this case, the *non-deadlocking property* [MOOY92] of the system has a relevant importance. That is, starting from a non-deadlock (live) state the system never deadlocks and converges to a state with a single token. Another important parameter for such a system is the *responsiveness*. This property can be seen as the maximum time (*delay*) consumed by the token between two consecutive visits at the same processor. It relates to the relative fairness of the solution. The best case is when the delay is exactly n and the system is called *fully-responsive*. If the delay is unbounded (or/but assured only with high probability) the system is called *non-responsive* (e.g., a random walk). Systems in between but with linear delay are called *responsive*.

There are two main differences between the system in [MOOY92] and our system. While [MOOY92] is fully-responsive, our algorithm has a delay equal to $2 \cdot n$ namely, it is responsive. However, our algorithm possesses the property of being *non-exploratory* (a property we highlight in this work). The system designed in [MOOY92] is highly exploratory. In fact, after stabilization, it uses extra messages called *probes* that have the task to explore the current state and eventually change the state of the tokens; as mentioned, the algorithm in [BGW89] is exploratory as well. Exploratory nature makes an algorithm's communication complexity high as explained above (in [MOOY92], $O(n)$ probes can be present in the system). In particular, after stabilization, the extra messages keep on being sent. Instead, our system uses extra messages only *prior* to stabilization, and after stabilization only one message is present in the system, namely the much needed "token".

We note that various existing algorithms are non-exploratory such as the non-responsive random walk [IJ90a]. Also, in the model of local-checking self-stabilization initiated in [AKY90] (where the node is allowed to constantly generate local checks with its neighbors rather then being only token-activated), some solutions are non-exploratory (i.e., after stabilization only local checking signals are sent) [AKY90], while other systems are faster but are inherently exploratory as in [AKMPV].

Our results: We achieve a non-exploratory, non-deadlocking, responsive, constant space, self-stabilizing algorithm for token management on uniform systems (sec. 2) and prove its correctness (sec. 3). A crucial part of this work is the analysis (sec. 4), showing $O(n^2 \log n)$ average stabilization time. The solution also elects a leader.

2 The Algorithm

2.1 High-Level Description

To illustrate the algorithm's ideas, we present a game played by t players on a circular n-sized array. The game is called **sentry game** and we can view the players as tokens and the array as the ring:

- The sentry game is played on a circular array composed of n unnamed cells. There are t players and each one plays against all the others. The numbers n and t are unknown to the players and the game is over when only one player is alive. Each player can be seen as a sentry that has its own domain to be protected by patrolling. Each sentry guards its domain in a pendulum-like manner, i.e. it goes from one end of its domain to the other (*swinging property*). Each time one end of the domain is reached, the sentry gains a cell, widening its domain by one unit. Of course, the adjacent domain (if it exists) decreases its length by one (*overrunning property*). When a sentry s realizes the presence of another sentry it defends its domain by transforming into a **warrior**.

So, it starts chasing its enemies (the other players) invading their domains and all the sentries it meets are eliminated. Last crucial game's property is the *single-elimination* rule: if two players of the same kind (sentry/warrior) meet face to face then only one of them can be eliminated. Hence, when two warriors meet they randomly agree on who will be eliminated and who stays alive. Same rule holds when two sentries meet. The single-elimination rule ensures that starting with a non-zero number of players at least one player is always present on the ring.

Some more details are needed to understand the correspondence between the sentry game and the algorithm. The first problem we want to solve is how a player can recognize its own domain (and its bounds) and when it can assume that other players are present. The idea is that each player signs the cells of its domain with a footprint randomly chosen in a set S ($|S| > 2$) of marks. Unfortunately, this idea does not work when there is only one player; in fact, in this case, there is only one domain whose both ends coincide and hence the player is not able to hold the pendulum-like swinging property. The trick is that a sentry has a mark with positive sign when going clockwise and negative one otherwise (wlog the array is oriented, see[IJ90a]). So, each sentry will take care only of the cells marked with the same sign. When such a cell is visited and the mark is the same, the sentry had gone around all the array reaching the other end of the domain. Thus, no other player is present in the ring. Unfortunately, the sentry cannot be sure about this claim since the number of players (t) can be much more than the number of marks ($|S|$) (that is, many players can have the same mark). A solution can be that the sentry, before changing its direction (and hence its sign), randomly chooses a new mark in S. Thus, if another player is present, perhaps after a number of unlucky (identical) choices, the adjacent marks will be different and so at least one of the sentries can realize the presence of the other. Hence, when a visited cell has the same sign and a different value, the sentry suspects the presence of another player and, as already said, it transforms into a warrior that kills all the players it meets. During its travelling, the warrior signs the visited cells with a special mark (the same for each warrior). As for the sentries, a warrior assumes that a total round of the array has been done when the special mark is encountered; so it transforms back into a sentry. Another trick we employ solves the problem that when a sentry transforms into a warrior it leaves a previously marked domain. So, in order to clean the traces of its presence, the warrior sends back on its domain a new kind of "passive" player (a proxy) called *mimetic* whose task is to sign with the warrior's mark all the cells it visits; i.e., a mimetic tries to mimic a warrior but since it is very weak it dies as soon as another message (maybe the same that created it) is met. We have to show that this combative state of affairs actually leads to stabilization starting from *every* state which includes an active participant.

More Details: Before the formal description, we give more details. Each processor i stores a value $V(i) \in \{-3, -2, -1, 0, 1, 2, 3\}$ that represents the mark footprint of the last message that visited it. As explained, the messages are divided into three classes: sentries (s), warriors (w) and mimetics (m). The sentries are the messages carrying a non-zero value ($V(s) \in \{-3, -2, -1, 1, 2, 3\}$); each of them can go clockwise (if $V(s) > 0$) and counter-clockwise (if $V(s) < 0$). The values $1, 2$ and 3 that $|V(s)|$ can assume can be seen as three different kinds of shoes and so $|V(s)|$ determines the kind of shoes actually worn by s. When s passes through a processor p such that $V(s) \cdot V(p) < 0$ then it leaves the footprint of the worn shoes (i.e. $V(s)$); if $V(p) = V(s)$ or $V(p) = 0$ then s wears a different kind of shoes, changes its sign, leaves on p the new footprint and returns back. This behavior defines the sentry's domain, which bounds grow step by step whenever the sentry changes direction. When $V(s) \cdot V(p) > 0$ (and $V(p) \neq V(s)$) the sentry s transforms into a warrior (with the same direction), creates a mimetic and sends it in the opposite direction. The warriors can wear only one kind of shoes corresponding to the zero value ($V(w) = 0$). Their only function is to eliminate any sentry they meet on their path; they are present only in the stabilizing phase. In contrast, a mimetic just leaves a zero on the visited processors; mimetics are also present only till stabilization.

The *"stabilized"* state is the state composed of exactly one token (it will be a sentry) and its domain (which spans the entire ring). The goal is to reach the stabilized state starting from any, possibly *"chaotic"* state. But, first we have to ensure that when the stabilized state is reached the ring never leaves it. Intuitively, the only sentry present in the ring just goes around its domain leaving its footprint. When it meets its own footprints it changes direction as described. The sentry chooses its new value uniformly among colors different from the current one. This strategy achieves a round-robin responsive management of the token.

To complete the algorithm behavior, we have to show that starting from a non-deadlock state (i.e. a state that contains at least one sentry or one warrior) it converges to the stabilized state. The idea is that starting from any chaotic state we first reach a state with a certain level of order where all the domains contain a message (i.e. two adjacent domains are divided by at most a sequence of processors with value 0). Then, with a non-zero probability, the number of sentries and warriors decreases till it gets to one and so the stabilized state is obtained. We do not care about the number of mimetics because when the stabilized state is reached, they will visit at most n (ring's size) processors before being eliminated. Only a slight modification is now needed due to the symmetry that arises when the system has only two sentries and their respective domains. In this case, in fact, since the domains are adjacent the messages are always of the same kind (both sentries or both warriors), and so the system never evolves (sentry turn into a warrior, bumps into the marks of the other warrior and turns into a sentry again, etc. etc.). The solution to this symmetry is that when $V[s] \neq V[i]$ and $V[s] \cdot V[i] > 0$ s chooses to transform into a warrior or to stay a sentry with probability p and $1 - p$ respectively, constant $p = 1/2$ will work. We note that one difficulty lies in showing that the set of local rules based on constant bits suffices for global fast stabilization.

2.2 A Formal Description

The following is the formal description of the algorithm. It respects all the rules we described in the previous sections.

The data structures and the procedures are the following:

$V(i)$: value of node i ($0 \leq |V(i)| \leq 3$);
$V(s)$: value of sentry s ($1 \leq |V(i)| \leq 3$);

$V(w)$: value of warrior w ($V(i) = 0$);

$D(w)$: direction of warrior w ($V(i) \in \{-1, 1\}$);

send-back(m): sends the message m in the opposite direction it was coming from;

send-ahead(m): sends the message m in the same direction it was coming from;

create-mimetic(m): creates the mimetic m;

delete(m): deletes the message m.

The function $rnd(x)$ chooses a random value $y \in \{1, 2, 3\} - \{|x|\}$ and returns $y \cdot sign(x)$. If $x = 0$ then $rnd(0)$ returns a value in $\{1, 2, 3\}$.

Sentry s on Processor i :

```
if( V[i]= 0 )
    V[i]←V[s]← -rnd( V[s] );
    send-back( s );
else
  if( V[s]*V[i]< 0 )
    V[i] ← V[s];
    send-ahead( s );
  else
    if( V[s]=V[i])
       // s stays sentry //
       V[s]←V[i]← -rnd( V[s] );
       send-back( s );
    else
       tmp=rnd( 3 );
       if( tmp=1 )
          // s stays sentry //
          V[s]←V[i]← -rnd( V[s] );
          send-back( s );
       else
          // s stores the direction //
          D[s]←V[s]/|V[s]|
          // s transforms into a warrior //
          V[s]←V[i] ←0;
          send-ahead( s );
          create-mimetic( m );
          send-back( m );
       endif
    endif
  endif
endif
```

Mimetics m1 and m2 collide:

```
delete( m1 );
delete( m2 );
```

Warrior w on Processor i :

```
if( V[i]=0 )
    // the warrior w becomes a sentry //
    V[i]←V[b]← -D[b]·rnd(0);
    send-back( w );
else
    V[i]←V[b];
    send-ahead( w );
endif
```

Mimetic m on Processor i:

```
V[i]←0;
send-ahead( m );
```

Warrior w and Sentry s collide :

```
delete( s );
```

Warrior w and Mimetic m collide :

```
delete( m );
```

Warriors w1 and w2 collide :

```
tmp=rnd( 3 );
if( tmp=1 )
    delete( w1 );
else
    delete( w2 );
endif
```

Sentries s1 and s2 collide:

```
tmp=rnd( 3 );
if( tmp=1 )
    delete( s1 );
else
    delete( s2 );
endif
```

Sentry s and Mimetic m collide :

```
delete( m );
```

3 Correctness

3.1 Definitions

For the formal proof we need definitions based on local view and global view of the system, like "global system sate". In order to classify the state-space we need definitions based on local information of the ring (i.e. local-definitions like segment, activity, etc.). The classification reflects several sub-goals that the algorithm obtains at any execution before stability is reached. Formally:

LOCAL DEFINITIONS:

1. **segment:** is a sequence of nodes $\sigma(i,j) \equiv \prec i, i+1, \cdots, i+j-1 \succ$, $j \geq 0$
   s.t.: $V(i) = V(i+1) = \cdots = V(i+j-1) = V(\sigma) \neq 0$, $V(i-1) \neq V(\sigma)$ and $V(i+j) \neq V(\sigma)$.
   Nodes i and $i+j-1$ are respectively the left-end and the right-end of $\sigma(i,j)$.
   Value j is the length of the segment and $\sigma(i,0)$ is a *null* segment.
2. **wall:** is a sequence of nodes $\delta(i,j) \equiv \prec i, i+1, \cdots, i+j-1 \succ$, $j > 0$ s.t.:
   $V(i) = V(i+1) = \cdots = V(i+j-1) = 0$, $V(i-1) \neq 0$ and $V(i+j) \neq 0$.
   Value j is the length of the wall. δ (i,0) is a *null* wall.
3. **active segment:** is a segment $\sigma(i,j)$ s.t. if $V(\sigma) > 0$ then at least one of the following conditions is true:
   - there is a sentry s on node i+j-1 s.t. $V(s) = V(\sigma)$,
   - there is a negative sentry or a mimetic on node i+j,
   - there is a warrior on node i-1,
   For negative segment the definition is symmetric. If no condition is satisfied $\sigma(i,j)$ is **inactive**.
4. **domain:** let $+$ be the concatenation of sequence; the sequence of nodes $\rho(i,j,k) \equiv \sigma(i,j) + \sigma(i+j, i+j+k) \equiv \prec i, i+1, \cdots, i+j-1, i+j, \cdots, i+j+k-1 \succ$ with $j, k \geq 0$ and $j \cdot k \neq 0$, is a domain if $\sigma(i,j)$ and $\sigma(i+j, i+j+k)$ are active, $V(\sigma(i,j)) > 0$ and $V(\sigma(i+j, i+j+k)) < 0$.
   Nodes i and $i+j+k-1$ are respectively the left-end and the right-end of $\rho(i,j,k)$. We say that ρ is the domain of the message m if m makes ρ active.
5. **walled domain:** ρ is left (right) walled if $V(i-1) = 0$ ($V(i+j+k) = 0$). If ρ is both left and right walled then ρ is a walled domain.

GLOBAL DEFINITIONS:

1. **state:** is defined as:
   - for each message (sentry, warrior or mimetic) m its position on the ring and the values $V(m)$ eventually it carries;
   - for each node n the information $V(n)$ it stores.
   Let S be a state of the system; we indicate with:
   - $s(S)$ the number of sentries in S;
   - $w(S)$ the number of warriors in S;
   - $m(S)$ the number of mimetics in S;

- $t(S)$ the sum $s(S) + w(S)$ (t=total number of tokens);
- $i(S)$ the number of inactive segments in S.

2. **transition:** a state transition from a state S to a state S' is a successor function $\xi : S' = \xi(S)$ where S' results from S by applying one of the following transitions:
   - **go-ahead:** can be performed by any kind of message; the message leaves its value on the current processor and goes (ahead) to the next one;
   - **go-back:** can be performed only by sentries; the sentry changes its value and sign, leaves its new value on the current processor and goes (back) to the previous one;
   - **transform:** can be performed only by sentries and warriors; a sentry transforms into warrior taking value zero, leaves its new value on the current processor, sends back a new mimetic signal m and goes ahead; a warrior transforms back into sentry selecting a new value, leaves it on the current processor and goes ahead;
   - **delete:** can be performed by any kind of message after a collision (when two message with opposite direction meet on the ring; for implementation of collision see [MOOY92]); if two mimetics collide then both are deleted; in all the other cases only one message is deleted;

   Let $\xi^i(S)$ be the ith successor of S.

3. **stabilized state:** S is stabilized if $s(S) = 1$, $w(S) = m(S) = 0$, there is only one domain and it is active, there are no walls;
4. **chaotic state:** the state S is chaotic if at least one segment is inactive ($i(S) > 0$).
5. **ordered state:** the state S is ordered if all the segments are active ($i(S) = 0$).
6. **t-state:** an ordered state S is called t-state if $t(S) = t \geq 2$.
7. **quiet state:** an ordered state S is quiet if $w(S) = 0$.
8. **excited state:** an ordered state S is excited if $w(S) > 0$.
9. **non-deadlock state:** S is a non-deadlock state if $t(S) > 0$.
10. **walled state:** S is walled if all domains are walled.

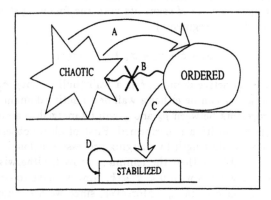

Fig. 1. Correctness Proof-Scheme.

3.2   Correctness Proof

To ease the proof of the algorithm dealing with arbitrary faults, we have broken it into pieces, identifying intermediate steps.   The first easy lemmas prove the non-deadlocking property of the algorithm and the fact that the token number does not increase (and are proved directly from the algorithm's transitions).

Lemma 1 *For every state S for which $t(S) \geq 1$ and $\forall l > 0$: $t(\xi^l(S)) \geq 1$.*

Lemma 2 *For every state S and $\forall l > 0$: $t(S) \geq t(\xi^l(S))$.*

Next we show that

Lemma 3 *For every state S and $\forall l > 0$: $i(S) \geq i(\xi^l(S))$.*

Proof. By definition, we observe that every message m makes active its own domain. Moreover, by the definition of transition, m never creates inactive segments.

Lemma 3 proves that the level of chaos in the system never increases. Moreover we get:

Corollary 1 *For every state S for which $i(S) = 0$: $\forall l > 0$ we have that $i(\xi^l(S)) = 0$.*

This means that if the system reaches an ordered state then it never "goes back" to a chaotic one (edge B of Figure 1).

Lemma 4 *For every state S for which $i(S) > 0$: $\exists l > 0$ s.t. $i(S) + t(S) > i(\xi^l(S)) + t(\xi^l(S))$.*

Proof. Since we are in a non-deadlock state $(t(S) > 0)$ there exists at least an active segment. Moreover, by hypothesis, there exists at least an inactive segment. So we can find at least one sequence of segments with the following structure (recall that + means concatenation of segments):

$$\rho_a + \delta + \sigma_1 + \sigma_2 + \cdots + \sigma_k + \chi$$

where $k > 0$, ρ_a is an active domain, δ is a wall (null or not), σ_j $(1 \leq j \leq k)$ are inactive segments and χ can be either a wall or an active domain. We want to show that either all processors between ρ_a and χ will be merged to an active domain (i.e. ρ_a or χ) or a delete transition is performed. First of all we observe that the active domain χ can increase its length by merging processors belonging to σ_k, σ_{k-1} and so on. If χ includes also σ_1 then the lemma is proved. Otherwise, we just have a similar sequence of segments with $0 < k' < k$. Now we have to consider all possible behaviors of the message m in ρ_a . Moreover, since we are studying the quantity $i(S) + t(S)$ which is not affected by the mimetics' behaviour we do not consider the case in which m is a mimetic.

We start by considering m to be a sentry (s) and all the transitions it can perform:

- go-ahead/go-back: if s only performs (alternatively) these two transitions it means that s remains sentry and that δ and all the inactive segments σ_j are merged to ρ_a till χ is reached $\Rightarrow$ the number of inactive segments decreases by k.
- transform: if s transforms into warrior on the left-end of ρ_a then a mimetic (say m') is sent to the right-end in order to clean ρ_a but m' also cleans all the inactive segments till a message (i.e. an active segment) is met. If s transforms into warrior on the right-end of ρ_a (δ is null) then the warrior cleans all inactive segments till a message or a wall is met. In both cases the number of inactive segments decreases by k.
- delete: obviously the quantity $t(S)$ decreases. (In some cases the inactive segments are also cleaned so also the number $i(S)$ decreases.)

If m is a warrior (w) we have to consider what is its direction. Let l and r be respectively the left-end and the right-end of ρ_a . If w is clockwise then its position is $l - 1$. So, if the wall δ is not null then the warrior w transforms back into a sentry and we have the previous case; otherwise w cleans all the inactive segments σ_j $\Rightarrow$ the number of inactive segments decreases by k. If w is counter-clockwise then its position is $r + 1$. In this case w cleans ρ_a reaching l and then it will goes-ahead till it collides with a message (and so $t(S)$ decreases) or it meets a wall and transforms back into a sentry (previous case).

Lemmas 3 and 4 help us to prove the following lemma which ensures that starting from a chaotic state the system (after all the processors have been visited) goes into an ordered state (edge A of Figure 1):

Lemma 5 *For every state S for which $i(S) > 0 : \exists l > 0 \ s.t. \ i(\xi^l(S)) = 0$.*

Proof. By lemmas 3 and 4 and observing that, by lemma 1, if $t(S) = 1$ then $t(S)$ does not decrease.

It is easy to see that the condition $i(\xi^l(S)) = 0$ is reached when all the processors have been visited at least once and, as we will see in lemma 11, this happens after at most $O(n^2)$ steps. Another aspect of lemma 5 is that after all processors have been visited the system enjoys the consistency property, i.e., the value of each processor is consistent with the value of at least one message in the ring. All this implies that after at most $O(n^2)$ steps, all the tokens are adjacent, i.e., each pair of consecutive domains (tokens) has a contact point or they are separated by a wall. We assume, through the rest of this section, that at least $O(n^2)$ steps have been performed so that the system enjoys the consistency property.

Next step is to prove that stability is maintainable, i.e., edge D of Figure 1. This is proved by following the unique sentry's actions.

Lemma 6 *For every stabilized state S and $\forall l > 0: S' = \xi^l(S)$ is also a stabilized state and only one message is present in the system.*

In order to prove that the system stabilizes starting from any ordered state, i.e., edge C of Figure 1, we need the following two lemmas. They follow the proof-scheme represented in Figure 2.a ($t > 1$).

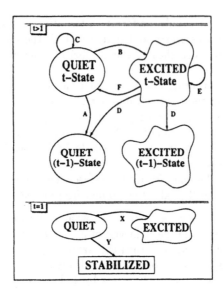

Fig. 2. Transition-State in the Ordered States.

Lemma 7 *For every quiet t-state S, there exists a finite $l > 0$ s.t. $(t(\xi^l(S)) < t(S)) \vee (\Pr\{w(\xi^l(S)) > 0\} > 0)$.*

Lemma 8 *For every excited t-state S, $\exists$ a finite $l > 0$ s.t. $\Pr[t(\xi^l(S)) < t(S)] > 0$.*

Proof. By the hypothesis S has at least one warrior w. Without losing generality we can assume that w is clockwise. Let ρ be its domain. We have to consider all the possible cases for the other message m on ρ :

1. if m is a counter-clockwise sentry then w deletes it $(t(\xi^l(S)) < t(S))$ (edge D in Figure 2.a);
2. if m is a clockwise sentry then m reaches the right-end of ρ and so: it becomes a clockwise warrior (see case 4) with probability $2/3 \cdot 1/2 = 1/3$ or a counter-clockwise sentry (see case 1) by performing a go-back transition with probability $1/3 + 2/3 \cdot 1/2 = 2/3$;
3. if m is a counter-clockwise warrior then w or m is deleted and the surviving warrior transforms back into a sentry (edge D in Figure 2.a). Observe that this case means that some message has been already deleted;
4. if m is a clockwise warrior then w collides with the counter-clockwise mimetic created by m and it transforms back into a sentry. This case can only occur as a subcase of case 2 and hence it has probability $1/3$ (edge E or F in Figure 2.a);

Finally, by lemmas 7 and 8 we claim the following lemma, proving that an ordered state stabilizes, which is edge C in Figure 1:

Lemma 9 *For every ordered t-state S, $\exists$ a finite $l > 0$ s.t. $\Pr[t(\xi^l(S)) < t(S)] > 0$.*

Since a t-state has been defined for $t > 1$, we have to consider the case in which $t = 1$. The following lemma shows that this state stabilizes with probability one. Figure 2.b gives a sketch of the proof.

Lemma 10 *For every ordered state S for which $t(S) = 1$, there exists a finite $l > 0$ s.t. $\Pr[\xi^l(S)$ is stabilized$] > 0$.*

Proof. By hypothesis $t(S) = 1$ and $i(S) = 0$; moreover these quantities never increase, so we have only to show that $\xi^l(S)$ has no walls. First we observe that, since $i(S) = 0$, only one wall can exists. If $w(S) = 1$ then the warrior collides with the mimetic that it created or it meets its own footprint. In both cases the warrior transforms back into a sentry (edge X in Figure 2.b). If $s(S) = 1$ then the sentry cleans the only present wall reaching a stabilized state (edge Y in Figure 2.w).

Theorem 1 *The algorithm presented is a randomized self-stabilizing algorithm for fair round-robin token management on an asynchronous bidirectional ring of uniform (nameless) finite automata processors, using constant size buffers and messages, starting from an arbitrary non-deadlocking state. Stability is reached with prob. 1 (where only one control-signal is present in the ring).*

Proof. By lemma 5 $i(\xi^{l_1}(S)) = 0$ for a finite l_1 and by corollary 1 this condition holds for every state $\xi^l(S)$ ($l \geq l_1$). Since by lemma 2 the number of tokens never increases we have only to show that the system never gets stuck at the same number of tokens for an infinite number of steps. In fact, let k be the actual number of tokens and let us consider (for contradiction) an infinite sequence of states where the system is stuck at k. Since the number of different ordered k–states in the system is finite there exists a k–state S that is visited infinitely often. But by lemma 9 there exists a finite l_2 ($l_2 > l_1$) such that $S' = \xi^{l_2}(S)$ where $t(S') = k' < k$ (i.e., there is non-zero probability that the system in S moves to S'. Thus, with probability 1 the system goes into the state S' (a contradiction).

This argument can be repeated for each $k > 1$ until $k' = 1$. By lemma 10 the system reaches stability eventually, and by lemma 6 the system remains stabilized and only one control-signal will be present on the ring.

Corollary 2 *The algorithm solves the leader election problem.*

Proof. Once the system is stabilized the leader is the only processor that apply a send-back operation.

4 Analysis

Next, we analyze the system and compute its stabilization time, i.e., the time-interval from the last transient fault till the system reaches a legal state. The technique we use comprises of two steps. First we show that the expected stabilization time (considering the local worst-case progress probability) is $T(n, t) = O(tn^2)$, where t is the initial number of tokens in the ring. This requires a careful investigation of the structure of our state-space. This analysis implies that if $t = O(\log n)$ then $T(n) = O(n^2 \log n)$. Then, we show that if t is big (i.e., $\log n \leq t \leq n$), then the number of tokens decreases quickly and reaches $\log n$ in $O(n^2 \log n)$ expected time, (we identify parallel independent events as Bernoulli trials, and employ Chernoff bounds). This will prove the following:

Theorem 2 *The expected worst-case stabilization time of the system is $O(n^2 \log n)$.*

The communication complexity in the entire initial fault overcoming phase is $O(n^3)$, while in the stable phase it is one message per round.

Figure 1 shows that the self-stabilizing time of the system is given by $T(t) = T_{c,o}(t) + T_{o,s}(t)$, where $T_{c,o}(t)$ is the transition time from a chaotic state with t tokens to an ordered t-state, and $T_{o,s}(t)$ is the transition time from an ordered t-state to a stabilized state. The next lemma gives $T_{c,o}$.

Lemma 11 *Starting from any chaotic state with $t \geq 1$ tokens, the system reaches an ordered t-state in at most $O(n^2)$ steps; i.e., $T_{c,o}(t) = O(n^2)$.*

Proof. Our basis is the proof of lemma 4. Any inactive domain σ has at least one active domain as neighbor. Each $O(|\rho|)$ steps at least one processor in σ part of ρ. Thus, after at most $O\left(\sum_{i=1}^{|\sigma|}(|\rho| + i)\right) = O\left(|\sigma| \cdot |\rho| + |\sigma|^2\right)$ all the inactive segments become active. The lemma follows since all domains' size are at most n.

Next, we compute $T_{o,s}$. The scheme in figure 3.a gives all the possible ways in which the system can reach stability starting from an ordered state. First of all we note that, in the worst case, each transition from a t-state to a t'-state ($t \geq t'$) takes at least $O(n^2)$. This means that the worst case for $T_{o,s}$ is obtained when the system goes from a t-state to a $(t-1)$-state (scheme in figure 3.b). By defining $T_{tr}(i)$, ($i \geq 2$), as the time to transit from an i-state to an $(i-1)$-state and T_{st} as the time for stabilizing from a 1-state, we have that $T_{o,s}(t) \leq \sum_{i=2}^{t} T_{tr}(i) + T_{st}$.

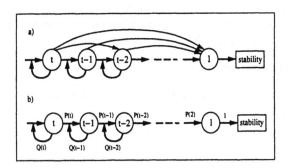

Fig. 3.

First of all, we compute $T_{o,s}(1)$. By the same arguments used in the proof of lemma 11, it is easy to show that:

Lemma 12 *Given any ordered state with one token, i.e. a 1-state, the system reaches stability in at most $O(n^2)$ steps; i.e., $T_{o,s}(1) = O(n^2)$.*

Our next goal is to compute the expectation of $T_{tr}(t)$ and by linearity of expectation, we will get the expectation of $T_{o,s}$.

Our first step is to compute the probability to go from a t-state to a $(t-1)$-state, that is indicated by $P(t)$ in figure 3.b. As usual, we set $Q(t) = 1 - P(t)$ and we determine P by computing Q. Figure 2 magnifies an ordered t-state and shows all the transitions performable by the system. Edges A and D are the only transitions leading to a $(t-1)$-state, but we have to ignore edge A since it only happens when two sentries collide and hence it is a "lucky" event (in the sense that it only depends on the speeds of the two sentries). Instead of computing the probability that edge E is executed, we compute the probability that E is not executed. Recall that edge E represents the delete transition between two warriors or between a sentry and a warrior. First observe that the worst case for the execution of edge E is when the state is quiet and its tokens (all sentries) have walled domains (*walled state*). In fact, the probability that edge E is executed is zero when the state is quiet and walled. After $O(n^2)$ steps, the walls have been destroyed and hence this probability becomes non-zero. By these considerations we can say that the worst cases causing edge E not to be executed are only two:

Event S(t)) while staying in a quiet and walled t-state, no sentry is transformed into warrior (edge C);

Event W(t)) while staying in a walled and quiet t-state, all sentries are transformed into warriors (edges B and E) having the same direction and then all warriors are transformed back into sentries (edge F).

We can write that $P(t) = 1 - P_s(t) - P_w(t)$, where $P_s(t) = \Pr[S(t)]$ and $P_w(t) = \Pr[W(t)]$.

We observe that the execution of edge C takes $O(1)$ and the execution of edges B,E and F takes $O(n)$.

4.1 Computing P_s and P_w

In this section we compute the probabilities of P_s and P_w by enumerating all the possible states. In order to compute the probability space we need the following definition. Given an array $A(1, \ldots, k)$ over the set $\{1, 2, 3\}$, the predicate $\mathcal{P}(k, x)$ is true iff $\bigwedge_{j=1}^{k-1}(A(j) \neq A(j+1)) \wedge (A(k) \neq x)$. Moreover, given $x, y \in \{1, 2, 3\}$ we define the function $\mathcal{F}(k, y, x)$ as the number of vectors $A(1, \ldots, k)$ over $\{1,2,3\}$, such that $(\mathcal{P}(k, x) = \text{True}) \wedge (A(1) = y)$.

Finally, we define the function $\mathcal{N}(k)$ as the number of vectors $A(1, \ldots, k) \in \{1, 2, 3\}^k$ such that $\bigwedge_{j=1}^{k-1}(A(j) \neq A(j+1)) \wedge (A(k) \neq A(1))$. Function $\mathcal{F}$ helps us computing $\mathcal{N}$:

Lemma 13 *For each integer $k \geq 1$, $\mathcal{N}(k) = 2^k + 2(-1)^k$.*

Proof. First observe that, given an integer $x \in \{1, 2, 3\}$, $\mathcal{N}(k) = 3 \cdot \mathcal{F}(k, x, x)$. Let $\overline{x}$ be an integer such that $\overline{x} \in \{1, 2, 3\} - \{x\}$. Observing that

$$\begin{cases} \mathcal{F}(1, x, x) = 1; \\ \mathcal{F}(1, \overline{x}, x) = 0; \end{cases} \quad \text{and} \quad \begin{cases} \mathcal{F}(k, x, x) = 2 \cdot \mathcal{F}(k-1, \overline{x}, x); \\ \mathcal{F}(k, \overline{x}, x) = \mathcal{F}(k-1, x, x) + \mathcal{F}(k-1, \overline{x}, x), \end{cases}$$

by induction it follows that $\mathcal{F}(k, x, x) = \frac{2^k + 2(-1)^k}{3}$ and hence the lemma holds.

Now we can compute the cardinality of the probability space.

Lemma 14 *Let $\Omega(t)$ be the probability space of states conditioned on having t tokens. Then its size is $|\Omega(t)| \geq 5^t + 2(-1)^t$.*

Proof. Since we want an upper bound on the size of this probability space, we will only consider the configurations where all tokens have the same direction. Thus, we can forget about the tokens' directions and we can represent each state by the tokens' value vector $V(1, \ldots, t, t+1)$, where $V(t+1) = V(1)$. Recall that when a sentry has the same value as its neighbor domain, it stays a sentry. Instead, if its neighbor domain has a different value then it tosses a coin and has equal probability to become a warrior or stay a sentry. This means that for each vector V such that there exist k integers $i_1, \ldots, i_k$ such that $V(i_j) \neq V(i_j + 1)$, $1 \leq j \leq k$, we have to add 2^k events in the probability space. By lemma 13 there are $\mathcal{N}(k)$ such vectors (just shrink all consecutive elements of V with the same value to one element with that value) and $\binom{t}{k}$ possible choices for the integers $i_1, \ldots, i_k$. Hence we have that

$$|\Omega(t)| \geq \sum_{k=0}^{t} \binom{t}{k} 2^k (2^k + (-1)^k 2) = \sum_{k=0}^{t} \binom{t}{k} 4^k + 2 \sum_{k=0}^{t} \binom{t}{k} (-2)^k = 5^t + 2(-1)^t.$$

The last two steps of this section are the computations of the probabilities P_w and P_s. We have to count the number of ways in which the events $W(t)$ and $S(t)$ can happen. For $W(t)$, it is very easy since it can happens only if $\bigwedge_{i=1}^{t}(V(i) \neq V(i+1))$ and hence $|W(t)| = \mathcal{N}(t)$. This proves the following lemma

Lemma 15 *For each integer $t \geq 2$, $P_w(t) \leq \frac{2^t + 2(-1)^t}{5^t + 2(-1)^t}$.*

Proof. In fact, $P_w(t) = \Pr[W(t)] = \frac{|W(t)|}{|\Omega(t)|} \overset{\text{(by lemmas 13 and 14)}}{\leq} \frac{2^t + 2(-1)^t}{5^t + 2(-1)^t}$.

Lemma 16 *For each integer $t \geq 2$, $P_s(t) \leq \frac{3^t}{5^t + 2(-1)^t}$.*

Proof. In fact, for each vector V such that there exist k integers $i_1, \ldots, i_k$ such that $V(i_j) \neq V(i_j + 1)$, $1 \leq j \leq k$, there exists exactly one event in the probability space $\Omega(t)$ for which $S(t)$ is true (the one for which for each i such that $V(i) \neq V(i+1)$ the sentry does not transform into warrior), and hence $|S(t)| = 3^t$. Thus we have that $P_s(t) = \Pr[S(t)] = \frac{|S(t)|}{|\Omega(t)|} \leq \frac{3^t}{5^t + 2(-1)^t}$.

Lemma 17 *For each integer $t \geq 2$, $P(t) \geq \frac{5^t - 2^t - 3^t}{5^t + 2(-1)^t}$.*

Proof. By lemmas 15 and 16, and recalling that $P(t) = 1 - P_s(t) - P_w(t)$.

4.2   The Time Complexity

We are now ready to compute the expectation of $T(t)$.

Lemma 18 *Given any initial state with t tokens, the expected stabilizing time of the system is $E[T(t)] = O(tn^2)$.*

Proof. First of all we recall that $T(t) = T_{c,o}(t) + T_{o,s}(t) \leq T_{c,o}(t) + \sum_{i=2}^{t} T_{tr}(i) + T_{st}$. By lemmas 11 and 12 we obtain $T(t) \leq O(n^2) + \sum_{i=2}^{t} T_{tr}(i)$ and hence $E[T(t)] \leq O(n^2) + E[\sum_{i=2}^{t} T_{tr}(i)]$. Thus, it only remains to compute $E\left[\sum_{i=2}^{t} T_{tr}(i)\right]$. As already observed, this can be done by studying the Markov's chain represented in figure 3.b. It can also be seen as a random variable having negative binomial distribution with probabilities $P(t)$ to go from a t-state to a $(t-1)$-state (head) and $Q(t) = 1 - P(t)$ to stay in the t-state (tail). In order to ease the stochastic analysis, we can substitute the probability $P(t)$ with its minimum $P(2) = 12/27 = p$ (P is increasing) and hence $Q(t)$ with its maximum $q = 15/27$. So, the number of tosses $\mathcal{X}_n$ (including the last one) taken to achieve t successes has distribution:

$$\Pr[\mathcal{X}_t = t + k] = \binom{t+k+1}{k} q^k p^t$$

and the mean is $E[\mathcal{X}_t] = tq/p = 15t/12$ (we can view the distribution as the sum of t independent geometric distributions using linearity of expectation). Since each coin toss costs at most $O(n^2)$, we have that $E\left[\sum_{i=2}^{t} T_{tr}(i)\right] = O(tn^2)$.

4.3 Improving $T_{o,s}$'s Analysis

Our hypothesis is that the system is in an ordered t-state. We would like to show that if $t \geq \log n$ then w.h.p. the number of tokens decreases linearly every $O(n^2)$.

In order to reach this goal, we focus our attention on one token of the ring, and we compute the probability that it is involved in a delete transition. We need some new local definitions. Let $\overrightarrow{s}\,(\overrightarrow{w})$ and $\overleftarrow{s}\,(\overleftarrow{w})$ indicate sentries (warriors) going on the ring with clockwise and counter-clockwise direction respectively. Then, let us call *bi-state* a pair $(a,b) \in \{\overrightarrow{s}, \overleftarrow{s}, \overrightarrow{w}, \overleftarrow{w}\}^2$ of consecutive tokens and *bi-transition* any event such that the bi-state (a,b) goes into a new bi-state (a',b') with a certain probability p (indicated by $(a,b) \overset{P}{\hookrightarrow} (a',b')$). Since we are only considering ordered states, such events can only happen when a or b reach a boundary processor of their own domain.

Lemma 19 *Given a token τ, the probability that after $O(n^2)$ steps it is involved in a deletion is at least 4/27.*

Proof. The token τ can be either a warrior or a sentry. In the former case, the worst time is obtained when the warrior does not meet any other token and hence after at most $O(n)$ steps it transforms back into a sentry.

Thus, we only have to study the case when τ is a sentry. The worst case is when τ's domain is bounded. But, as we often said, after $O(n^2)$ steps it has destroyed its left or right wall. Without lost of generality, we can assume that the right wall has been destroyed. This means that $\tau = \overleftarrow{s}$ and t's domain has a contact point with a domain of another sentry τ'. Hence, $(\tau, \tau') \in \{(\overleftarrow{s}, \overleftarrow{s}), (\overleftarrow{s}, \overrightarrow{s})\}$. Next step is to prove that there is a non-zero probability that τ or τ' are deleted. The diagram in Figure 4 gives all the possible bi-transitions and their activation probabilities. It shows that starting from $(\overleftarrow{s}, \overleftarrow{s})$ or $(\overleftarrow{s}, \overrightarrow{s})$ there is a non-zero probability to reach a bi-state leading to a delete transition.

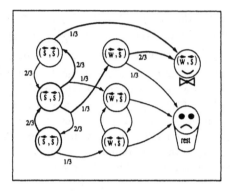

Fig. 4.

Lemma 20 *Let t be the number of tokens present on the ring. After $O(n^2)$ steps, with very high probability $1 - O(1/n)$ the number of tokens is $(79/81)t$.*

Proof. Let $t \geq \log n$ be the number of tokens present on the ring. We consider $m = \lfloor t/3 \rfloor$ independent experiments between separated groups of three contiguous tokens. Each experiment is successful if after $O(n^2)$ steps at least one token of the group is deleted (considering only deletion in the group). By lemma 19 the probability of success is at least $p = 4/27$, and since the experiments are independent, the expected number of successes (unsuccesses) is pm $((1-p)m)$. Each success (unsuccess) leaves on the ring 2 (3) tokens and hence after $O(n^2)$ steps the number of expected tokens is $2pm + 3(1-p)m \leq (26/27)t$. When $t > c \cdot \log n$ (for appropriate constant) we can apply Chernoff bound to show that the number of successes is at least half of the expected number, with high probability (of at least $(1 - d \cdot (1/n))$ for appropriate constant d). This implies a reduction of the number of tokens by a constant factor of $79/81$ with high probability in $O(n^2)$ stages. Let s be the (random variable) number of success and its expectation be E_s. The bound gives:

$$Prob\{E_s < (1-\delta)pm\} < exp\{-\delta^2(pm)/2\}$$

In our case, $(p = 4/27, m > c \log n)$ $(m > t/3 - 2,$ $(t$ number of messages), and we choose $\delta = 1/2)$. Then we multiply the number of successes achieved with high probability by 2 and non-successes by 3 and we get that the new number of tokens is reduced by the $f = 79/81$ factor (with probability at least $1 - O(1/n)$).

Lemma 21 *The expected number of tokens after $O(n^2 \log n)$ time units is $O(\log n)$*

Proof. We look at system states after an interval of $O(n^2)$ time units. Given that we start from $t \leq n$, after the interval of n^2 concurrent steps, we know that the expected number of tokens is reduced by a constant $(79/81)$ factor (or more) with probability at least $1 - d(1/n)$.

Concatenating the stages (interval), can be viewed as a Markov chain on the space of states, each state characterized by the number of tokens, where transitions are always non-increasing in token number (since a state can fix or reduce its number of tokens but not increase them). Furthermore, at any stage the sum of probabilities of transitions (a transition is actually n^2 steps of the algorithm) to states with substantially reduced number of tokens (by at least a factor f) is at least $1 - O(1/n)$ as computed in lemma 20. Thus, it is easy to see that the expected walk length (number of transitions) in this directed graph with self loops is $O(\log n)$.

Proof of Theorem 2. By lemma 21 after at most $O(n^2 \log n)$ the expected number of tokens is $t = O(\log n)$. Once we have at most $e \cdot \log n$ (for appropriate constant e) tokens, the previous analysis in lemma 18 implies another factor of $O(n^2 \log n)$ steps.   □

Acknowledgement. It is a pleasure to thank Yoram Ofek for suggesting the problem and for helpful contributions. G. Parlati would like to thank his parents for having made his studies most comfortable and for their unrepayable love and huge sacrifices that have been most crucial to his career.

References

[AAHK86] K. ABRAHAMSON, L. ADLER, L. HIGHAM, AND D. KIRKPATRICK, Probabilistic solitude verification on a ring. *Proc. 5th ACM Symp. on Principles of Distributed. Computing* (1986).

[AB89] Y. AFEK AND G.M. BROWN, Self-stabilization over unreliable communication media, *Distrib. Comput.*, 7: 27 - 34, 1993. (Previous version appeared as: Self-stabilization of the alternating-bit protocol, *Proc. 8th IEEE Symp. on Reliable Distributed Systems* (1989), 10–12.

[AKY90] Y. AFEK, S. KUTTEN, AND M. YUNG, Memory-efficient self-stabilization on general networks, *Proc. 4th International Workshop on Distributed Algorithms, Lecture Notes in Computer Science*, Vol 486, Springer-Verlag, New York, (1989), 12–28.

[AKMPV] B. AWERBUCH, S. KUTTEN, Y. MANSOUR, B. PATT-SHAMIR, AND G. VARGHESE, Time Optimal Self-stabilizing Synchronization, *Proc. 25th ACM Symp. on Theory of Computing* (1993), 652–661.

[Ang80] D. ANGLUIN, Local and global properties in networks of processors, *Proc. 12th ACM Symp. on Theory of Computing* (1980), 82–93.

[ASW85] C. ATTIYA, M. SNIR AND M. WARMUTH, Computing on an anonymous ring, *Journal of the ACM* 35(4) (1988), 845–875.

[BGW89] G.M. BROWN, M.G GOUDA, AND C.L. WU, Token systems that self-stabilize, *IEEE Transactions on Computers* 38(6) (1989), 845–852.

[BP88] J.E. BURNS AND J. PACHL, Uniform self-stabilizing rings, *ACM TOPLAS* 11(2) (1989), 330–344.

[BCK+83] W. BUX, F.H. CLOSS, K. KÜMMERLE, H.J. KELLER AND H.R. MÜLLER, Architecture and design of a reliable token-ring network, *IEEE J. on Selected Areas in Comm.*, 1(5), (1983), 756–765.

[CO90] I. CIDON AND Y. OFEK, MetaRing - A full-duplex ring with fairness and spatial reuse, *IEEE Transactions on Communications* 41(1):110 - 120 (1993).

[CS92] R. COHEN, AND A. SEGAL, Distributed Priority Algorithms under One-Bit-Delay Constraint *Proc. 11th ACM Symp. on Principles of Distributed. Computing* (1992), 1–12.

[Dij74] E.W. DIJKSTRA, Self-stabilizing systems in spite of distributed control, *Communications of the ACM* 17(11) (1974), 643–644.

[DIM91] S. DOLEV, A. ISRAELI AND S. MORAN, Resource bounds for self stabilizing message driven protocols, *Proc. 10th ACM Symp. on Principles of Distributed. Computing* (1991), 281–293.

[FL84] G.N. FREDERICKSON AND N. LYNCH, The impact of synchronous communication on the problem of electing a leader. *J. of the ACM* **34**(1) (1987), 98–115.

[Her90] T. HERMAN, Probabilistic self-stabilization, *Information Processing Letters* **35** (1990), 63–67.

[HN88] A. HOPPER AND R. M. NEEDHAM, The Cambridge fast ring networking system, *IEEE Transactions on Computers*, **37**(10) (1988), 1214–1223.

[IJ90a] A. ISRAELI AND M. JALFON, Token management schemes and random walks yield self stabilizing mutual exclusion, *Proc. 9th ACM Symp. on Principles of Distributed. Computing* (1990), 119–130.

[IJ90b] A. ISRAELI AND M. JALFON, Self-Stabilizing Ring Orientation, *Proc. 4th International Workshop on Distributed Algorithms, Lecture Notes in Computer Science*, Vol 486, Springer-Verlag, New York, (1990), 1–14.

[IR81] A. ITAI AND M. RODEH, Symmetry breaking in distributive networks, *Information and Computation* **88** (1990) 60–87.

[It90] A. ITAI, On the power needed to elect a leader *Proc. 4th International Workshop on Distributed Algorithms, Lecture Notes in Computer Science*, Vol 486, Springer-Verlag, New York, (1989), 29–40.

[Lam84] L. LAMPORT, Solved problems, unsolved problems and non-problems in concurrency, *Proc. 3rd ACM Symp. on Principles of Distributed Computing*, (1984), 1–11.

[LL90] L. LAMPORT AND N.A. LYNCH, Distributed computing models and methods, *Handbook on Theoretical Computer Science*, MIT Press/Elsevier, Ed. J. van Leeuwen, 1990, 1159–1199.

[LTG90] A. A. LAZAR, A. T. TEMPLE AND R. GIDRON, MAGNET II: A metropolitan area network based on asynchronous time sharing, *IEEE J. on Selected Areas in Comm.*, **8**(8), (1990), 1582–1594.

[LS92] C. LIN, AND J. SIMON, Observing Self-Stabilization *Proc. 11th ACM Symp. on Principles of Distributed. Computing* (1992), 113–124.

[MOOY92] A. MAYER, Y. OFEK, R. OSTROVSKY AND M. YUNG, Self-Stabilizing Symmetry Breaking in Constant-Space, *Proc. 24th ACM Symp. on Theory of Computing* (1992), 667–678.

[Of93] Y. OFEK,, Personal communication.

[OMS89] H. OHNISHI, N. MORITA, AND S. SUZUKI, ATM ring protocol and performance, *Proc. ICC'89, 13.1*, (1989), 394–398.

[PKR82] J. PACHL, E. KORACH AND D. ROTEM, A technique for proving lower bounds for distributed maximum-finding algorithms, *Proc. 14th ACM Symp. on Theory of Computing* (1982), 378–382.

[R86] F. E. ROSS, FDDI - a Tutorial, *IEEE Communication Magazine*, **24**(5), (1986), 10–17.

[V84] P. VITANYI, Distributed elections in an Archimedian ring of processors, *Proc. 16th ACM Symp. on Theory of Computing* (1984), 542–547.

On-Line Distributed Data Management

Carsten Lund[1], Nick Reingold[1], Jeffery Westbrook[2]*, and Dicky Yan[3]**

[1] AT& T Bell Laboratories, Murray Hill, NJ 07974, USA.
[2] Dept. of Computer Science, Yale University, New Haven, CT 06520-2158, USA.
[3] Dept. of Operations Research, Yale University, New Haven, CT 06520-0162, USA.

Abstract. We study competitive on-line algorithms for data management in a network of processors. A data object such as a file or page of
virtual memory is to be read and updated by various processors in the
network. Our goal is to minimize the communication costs incurred in
serving a sequence of such requests. Awerbuch et al. [2] obtain an optimal $O(\log n)$-competitive algorithm for general networks. We study distributed data management on important classes of networks — trees, bus
based networks and small-diameter networks such as the hypercube. We
obtain optimal algorithms with constant competitive ratios and matching lower bounds. Our algorithms are based on work functions [6] and
illustrate the technique of "factoring". For hypercube-derived networks,
we show that the algorithm in [2] is $O(\log \log n)$-competitive.

1 Introduction

The management of data in a distributed network is an important and much
studied problem in management science, engineering, computer systems and
theory [8, 10]. Dowdy and Foster [8] give a comprehensive survey of research
in this area, listing eighteen different models and many papers. A data object,
F, such as a file, or page of virtual memory, is to be read and updated by a
network of processors. Each processor may store a copy of F in its local memory,
so as to reduce the time required to read the data object. All copies must be
kept consistent, however, so having multiple copies increases the time required
to write to the object. As read and write requests occur at the processors, an
on-line algorithm has to decide whether to replicate, move, or discard copies of
F after serving each request, while trying to minimize the total cost incurred
in processing the requests. The server has no knowledge of future requests, and
no assumptions are made about the pattern of requests. We apply competitive
analysis [4] to such an algorithm.

Let σ denote a sequence of read and write requests. A deterministic on-line
algorithm A is said to be c-competitive, if, for all σ, $C_A(\sigma) \leq c \cdot OPT(\sigma) + C$
holds, where $C_A(\sigma)$ and $OPT(\sigma)$ are the costs incurred by A and the optimal
off-line solution respectively, and C is a constant independent of σ. If A is a
randomized algorithm, we replace $C_A(\sigma)$ by its *expected* cost and consider two
types of adversaries: the *oblivious* adversary chooses σ in advance, and the more
powerful *adaptive* on-line adversary builds σ on-line, choosing each request with
knowledge of the random moves made by A on the previous requests. (See Ben-
David et al. [4] for a full discussion of different types of adversaries.) An algorithm
is *strongly* competitive if it achieves the best possible competitive ratio.

* Research partially supported by NSF Grant CCR-9009753.
** Research partially supported by University Fellowships from Yale University.

In this paper, we focus on two important classes of networks: trees and the uniform network. A tree is a connected acyclic graph on n nodes and $(n-1)$ edges; the uniform network is a complete graph on n nodes with unit edge weights. We obtain optimal deterministic and randomized on-line algorithms for these classes.

Our algorithms are based on offset functions and use the "factoring" technique. Competitive on-line algorithms based on offset functions have been found for the 3-server [6] and the migration problems [7]. An advantage of these algorithms is they do not need to record the entire history of requests and the server, since decisions are based on the current offset values which can be updated easily. Factoring is first observed in [5] and used in [7, 14]. The idea is to break down an on-line problem on a tree into single edge problems. Thus optimal strategies for a single edge is generalized to a tree. Our algorithms are optimal for specific applications and networks, and also illustrate these two useful techniques.

1.1 Problem Description

We study three variants of distributed data management: *replication* [1, 5, 14], *migration* [5, 7, 17] and *file allocation* [2, 3]. They can be can be described under the same framework. We are given a weighted undirected graph $G = (V, E)$ on n nodes, where each node represents a processor. Let F represent a data file or a page of memory to be stored in the processors. At any time, let $R \subseteq V$, the residence set, represent the set of nodes that contain a copy of F. We always require $R \neq \emptyset$. As time goes on, read and write requests occur at the processors. A read request requires access to a copy of F; a write request requires updating all copies of F. After a request is served, the on-line server can decide how to reallocate the multiple copies of F. Let $D \in Z^{+4}$ represent the size of F. The costs for serving the requests and redistributing the files are as follows.

> *Service Cost:* Suppose a request occurs at a node v. If it is a read request, it is served at a cost equal to the shortest distance from v to a nearest node in R; if it is a write request, it is served at a cost equal to the size of the minimum Steiner tree that contains all the nodes in $R \cup \{v\}$.
> *Movement Cost:* The algorithm can replicate a copy of F from a node a to a node b at a cost D times the shortest distance between the two nodes; it can discard a copy of F at no cost.

Initially, only a single node v contains a copy of F and $R = \{v\}$. The replication and migration problems are special cases of file allocation. For migration, we require $|R| = 1$. For replication, all the requests are reads; we can assume that all pages are not discarded. The (off-line) optimization problem is to specify R after each new request is served so that the total cost incurred is minimized. (For on-line replication, we only consider competitive algorithms that have $C = 0$ in the inequality above, otherwise a trivial 0-competitive algorithm exists [5].)

1.2 Previous and New Results

Table 1 gives the competitive ratios of the best known deterministic and randomized algorithms against an oblivious adversary for the three problems on a

---

[4] R and Z^+ represent the sets of reals and positive integers, respectively.

tree and a uniform network, where $e_D = (1 + 1/D)^D$. They are all optimal. We summarize other related results below.

	Replication	Migration	File Allocation
Deterministic uniform	2 [5]	3 [5]	3 [3]
tree	2 [5]	3 [5]	3*
Randomized uniform	$e_D/(e_D - 1)^*$	$2 + 1/(2D)^*$	?
tree	$e_D/(e_D - 1)^*$	$2 + 1/(2D)$ [7]	$2 + 1/D^*$
* this paper			

Table 1. The State of the Art: Trees and Uniform Networks

Replication: Koga [14] gives randomized algorithms that are asymptotically $(1 + 1/\sqrt{2})$-competitive against an oblivious adversary on trees, and 2-competitive and 4-competitive against an adaptive on-line adversary on trees and circles, respectively.

Migration: For general networks, Awerbuch *et al.* [2] give a deterministic 7-competitive algorithm against a lower bound of $(85/27)$ [7]. Westbrook [17] finds a strongly 3-competitive randomized algorithm against an adaptive on-line adversary, and an asymptotically $(1+\phi)$-competitive algorithm against an oblivious adversary, where $\phi \approx 1.62$ is the golden ratio.

File Allocation: For general networks, Awerbuch *et al.* [2] and Bartal *et al.* [3] give $O(\log n)$-competitive deterministic and randomized algorithms against an adaptive on-line adversary, respectively. Westbrook and Yan [18] show that Bartal *et al.*'s algorithm is $O(\log d(G))$-competitive on an unweighted graph with diameter $d(G)$, and there exists a $O(\log^2 d(G))$-competitive deterministic algorithm. Bartal *et al.* also find a $(3 + O(1/D))$-competitive deterministic algorithm on a tree, and strongly 3-competitive randomized algorithms against an adaptive on-line adversary on a tree and uniform network. Since replication is a special case of file allocation, these upper bounds are also valid for replication.

- For on-line file allocation, we give a strongly 3-competitive deterministic algorithm and a $(2 + 1/D)$-competitive randomized algorithm against an oblivious adversary for file allocation on a tree, and show that this is optimal even if G is an edge.
- For file allocation on general networks, we show that Awerbuch *et al.*'s [2] deterministic algorithm, **FA**, is $O(\log d(G))$-competitive on an unweighted graph. This is optimal in $d(G)$, up to a constant factor [3, 11]. Thus **FA** achieves a competitive ratio better than $O(\log n)$ on small diameter networks, *e.g.* it is $O(\log \log n)$-competitive on n-node hypercube and butterfly networks.
- For migration, we give a strongly $(2 + 1/(2D))$-competitive randomized algorithm against an oblivious adversary on the uniform network.
- For replication, we show that the off-line problem is NP-complete. We obtain randomized algorithms that are $(e_D/(e_D - 1))$-competitive against an oblivious adversary on a tree and a uniform network; this is optimal even if G is a an edge. (Albers and Koga [1] have independently obtained the same results for on-line replication using a different method.)

- We give a polynomial time algorithm for solving the off-line file allocation problem on a uniform network.

2 Preliminaries

We use the technique of work functions and offset functions introduced by Chrobak and Lamore [6]. Let S be a set of states, one for each possible residence set and $Y = \{v^r, v^w | v \in V\}$ be the set of possible requests where v^r and v^w represent read and write requests at node v. A request sequence $\sigma = (\sigma_1, \ldots, \sigma_p)$ is revealed to the server, each $\sigma_i \in Y$. Suppose the file system is in state s when σ_i arrives. The server will be charged a service cost of $acc(s, \sigma_i)$, where $acc(s, \sigma_i) : S \times Y \longrightarrow \mathcal{R}$ is as described in Sect. 1.1. After serving σ_i, the server can move to a different state t at a cost $tran(s, t)$, where $tran : S \times S \longrightarrow \mathcal{R}$ is the minimum cost of moving between the two residence sets.

The work function $W_i(s)$ is the minimum cost of serving requests 1 to i, terminating in state s. Given σ, a minimum cost solution can be found by the dynamic programming algorithm:
$$\forall\, s \in S, i \in \mathcal{Z}^+, \quad W_i(s) = \min_{t \in S}\{W_{i-1}(t) + acc(t, \sigma_i) + tran(t, s)\}$$
with suitable initializations. Let $opt_i = \min_{s \in S} W_i(s), i \geq 1$, be the optimal cost in serving the first i requests. We call $\omega_i(s) = W_i(s) - opt_i$ the *offset function* value at state s after request i has been revealed. Define $\Delta opt_i = opt_i - opt_{i-1}$.

Our on-line algorithms make decisions based on the current offset values, $\omega_i(s), s \in S$. Note that to compute the $\omega_i(s)$'s and opt_i's, it suffices to know only the $\omega_{i-1}(s)$'s. Since $OPT(\sigma) = \sum_{i=1}^{|\sigma|} \Delta opt_i$, to show that an algorithm A is c-competitive, we need only show that for each reachable combination of offset function, request and the current state of the system, $\Delta C_A + \Delta \Phi \leq c \cdot \Delta opt_i$ holds, where ΔC_A is the cost incurred on A and $\Delta \Phi$ is the change in some defined potential function. If the total change in Φ is always bounded or non-negative, summing up the above inequality over σ, we have $C_A(\sigma) \leq c \cdot OPT(\sigma) + C$ where C is some bounded value.

$tran(t, s)$		s				$acc(t, \sigma_i)$		σ_i		
		a	b	ab			a^r	a^w	b^r	b^w
	a	0	D	D		a	0	0	1	1
t	b	D	0	D	t	b	1	1	0	0
	ab	0	0	0		ab	0	1	0	1

Table 2. Transition and Service Costs

3 Deterministic Algorithms for On-Line File Allocation

3.1 An Optimal Deterministic Edge Algorithm

Let $G = (a, b)$ be an edge and $S = \{a, b, ab\}$ the set of states the system can be in — only node a has a copy, only node b has a copy, and both a and b have a copy, respectively. We can assume G is of unit length, otherwise the offsets and costs functions can be scaled to obtain the same results. We write the offset functions as a triplet $\omega_i(S) = (\omega_i(a), \omega_i(b), \omega_i(ab))$ and similarly for the work functions. Suppose the starting state is a. Then $W_0(S) = (0, D, D)$. The acc and $tran$ functions are given in Table 2. By the definition of the offset functions and since it is free to discard a copy of F, we always have $\omega_i(ab) \geq \omega_i(a), \omega_i(b)$, and at least one of $\omega_i(a)$ and $\omega_i(b)$ is zero. Without loss of generality, we assume a starting offset function vector of $\omega_i(S) = (0, k, l)$, $0 \leq k \leq l \leq D$. In Table 3, we give the changes in offsets for different combinations of requests and offsets. Let s be the current state of the system. Our algorithm specifies the new required residence set, R, where a is assumed to be a zero-offset state.

Case 1: $k \geq 1$:

σ_{i+1}	$\omega_{i+1}(a)$	$\omega_{i+1}(b)$	$\omega_{i+1}(ab)$	Δopt_{i+1}
a^r	0	$\min(k+1, l)$	l	0
a^w	0	$\min(k+1, D)$	$\min(l+1, D)$	0
b^r	0	$k-1$	$l-1$ '	1
b^w	0	$k-1$	l	1

Case 2: $k = 0$:

σ_{i+1}	$\omega_{i+1}(a)$	$\omega_{i+1}(b)$	$\omega_{i+1}(ab)$	Δopt_{i+1}
a^r	0	$\min(1, l)$	l	0
a^w	0	1	$\min(l+1, D)$	0
b^r	$\min(1, l)$	0	l	0
b^w	1	0	$\min(l+1, D)$	0

Table 3. Changes in Offsets

Algorithm Edge:
(1) if $s \neq ab$ and $\omega_{t+1}(s) = \omega_{t+1}(ab)$, include both a and b in R
    and update the state variable for the edge, *i.e.* set $s = ab$.
(2) if $s = ab$ and $\omega_{t+1}(ab) = D$, remove b from R.

Theorem 1. Edge *is strongly 3-competitive.*

Proof. We prove that for each request σ_j, $\Delta C_{Edge} + \Delta \Phi \leq 3 \cdot \Delta opt_j$ (∗) holds. Let a be a zero-offset state and we have $\omega_i(S) = (0, k, l)$. At any time, we define the potential function:

$$\Phi(k, s) = \begin{cases} 2 \cdot D - 2 \cdot k & \text{if s=a} \\ 2 \cdot D - k & \text{if s=b} \\ D - k & \text{if s=ab} \end{cases}$$

Initially, $\Phi = 0$ and we always have $\Phi \geq 0$. It can be shown by a simple case analysis that (∗) holds for all combinations of offsets, requests, and the state of the system.

3.2 An Optimal Deterministic Tree Algorithm

We can limit our attention to algorithms that maintain a connected residence set and only consider on-line algorithms that have this property.

Theorem 2. *There exists an optimal (off-line) algorithm that always maintains a connected residence set, R.*

Proof. Given any deterministic algorithm A, we can simulate A on σ and obtain algorithm A' such that $R(A')$ is the minimum size connected set that satisfies $R(A) \subseteq R(A')$. The set $R(A')$ is defined and unique. It can be shown that $\forall \sigma, C_{A'}(\sigma) \leq C_A(\sigma)$.

A basic tool [5, 7] in handling on-line optimization on trees is the fact that any sequence of requests σ and any tree algorithm can be "factored" into $(n-1)$ individual algorithms, one for each edge of the tree, in such a way that the total cost in the tree algorithm is equal to the sum of the costs in each individual edge game. For edge (a, b) we construct an instance of the two-processor file allocation as follows. The removal of edge (a, b) divides T into two subtrees T_a and T_b, containing a and b respectively. A read or write request in σ from a node in T_a is replaced by the same kind of request from a, and a request from a node in T_b is replaced by the same request from b. Let A be an algorithm with residence set $R(A)$. Algorithm A induces an algorithm on edge (a, b) as follows: if $R(A)$ falls entirely in T_a or T_b then the edge algorithm is in state a or b, respectively, otherwise, the edge algorithm is in state ab. This factoring approach is used in our algorithms for file allocation on a tree. For the rest of this paper, given an edge (a, b), we use T_a and T_b to represent the subtrees described above, s to denote the state the edge is in, and let $\omega_i(S) = (\omega_i(a), \omega_i(b), \omega_i(ab))$.

Lemma 3. *Given algorithm A, request sequence σ, let A_{ab} be the algorithm induced on edge (a, b). Let σ_{ab} be the request sequence induced on (a, b). Then $A(\sigma) = \sum_{(a,b) \in E} A_{ab}(\sigma_{ab})$.*

Lemma 4. *Let $OPT(\sigma_{ab})$ be the cost incurred by an optimal edge algorithm for (a, b) on sequence σ_{ab}. Then $\sum_{(a,b) \in E} OPT(\sigma_{ab}) \leq OPT(\sigma)$.*

By Lemmas 3 and 4, if A is an on-line algorithm such that on any σ and for each edge (a, b), A_{ab} is c-competitive on σ_{ab}, then A is c-competitive on σ.

Algorithm Tree: Apply step (1) of Algorithm **Edge** to each edge (a, b) using the induced request sequence σ_{ab}. For any edge, if **Edge** requires adding a node to the residence set, add it to the residence set for the tree problem. Similarly, apply step (2) of **Edge** to each (a, b), using σ_{ab}, and remove a node from the residence set for the tree problem if **Edge** requires so for (a, b).

Algorithm **Tree** gives a new residence set, say R'. We reach R' from R by replicating to all the nodes in $T(R \cup R') - T(R)$, where $T(s)$ is the unique minimum Steiner tree containing the nodes in s, and then discard all the copies of F not in R'. To show that **Tree** is 3-competitive, we need to show that it induces the algorithm **Edge** on each edge and R remains a connected set. The next lemma characterizes the offset distribution between two adjacent edges; it can be proved by a case analysis and showing an algorithm for locating the root node r.

Lemma 5. *The following properties are maintained throughout:*

(A) *At any time, there exists a root node r, such that $R = \{r\}$ corresponds to the zero offset state for all the edges.*

(B) *For any edge (x, y) on the tree, define $A_i(x, y) = \omega_i(xy) - \omega_i(x)$*
Then for any adjacent edges (x, y) and (y, z), $A_i(x, y) \leq A_i(y, z)$ holds, $\forall i$.

Corollary 6. **(C)** *Let (x, y) be an edge in T such that r is in T_x and $z \neq x$ is adjacent to y. Then (x, y) and (y, z) have offsets of the form $(0, k_{xy}, l_{xy})$ and $(0, k_{yz}, l_{yz})$ where*

  (C.1) $l_{xy} \leq l_{yz}$; **(C.2)** $l_{xy} - k_{xy} \geq l_{yz} - k_{yz}$;
  (C.3) $k_{xy} \leq k_{yz}$, *and* **(C.4)** *if $k_{yz} = 0$, then $k_{xy} = 0$ and $l_{xy} = l_{yz}$ hold.*

Theorem 7. *Algorithm* **Tree** *is strongly 3-competitive.*

Proof. Step (1) of **Edge** − *Replication:* when $\omega_{t+1}(s) = \omega_{t+1}(ab)$ and $s \neq ab$. **Edge** will require both a and b to be included in R'. Using **(A)** and **(C.3)**, it can be shown that under **Edge**, all the nodes along the path from the current set R to (a, b) are also required to be included in R', thus forming a connected set R'.

Step (2) of **Edge** − *Discard a Copy:* when $\omega_{t+1}(s) = D$ and $s = ab$ Under **Edge**, b will have to be removed from R. Using **(A)** and **(C.1)**, it can be shown that all the other nodes in T_b that are in the current R are also required by **Edge** to be removed from R; thus R' remains connected set.

Other Cases: The set R' is such that any edge not satisfying the **if** conditions in steps (1) and (2) of **Edge** will remain in the same state as before, which is required by **Edge**.

4 Randomized Algorithms for On-Line File Allocation

4.1 An Optimal Randomized Edge Algorithm

We consider a randomized algorithm along an edge (a, b), against an oblivious adversary. Our approach to designing randomized algorithms differs from the typical method. A randomized algorithm consists of a set of internal states and a probabilistic mapping from triples of internal state, system state, and request point to a set of possible new system and internal states. After a sequence of requests, there is a probability distribution on the possible states the algorithm can be in. Our algorithm is defined directly in terms of desired probability distributions. Let $p_{ab}[a]$, $p_{ab}[b]$, $p_{ab}[ab]$ be the probabilities of being in states a, b, and ab, respectively. Algorithm **RandEdge** is designed so that whenever the offset function is $(0, k, l)$, the probability distribution is exactly

$$p_{ab}[a] = \min\{\frac{l}{D}, \frac{D+k}{2D}\}; \quad p_{ab}[b] = \min\{\frac{l-k}{D}, \frac{D-k}{2D}\}; \quad p_{ab}[ab] = \max\{1 - \frac{2 \cdot l - k}{D}, 0\}$$

Given the desired mapping from offsets to distributions, it is not hard to construct an algorithm that achieves that mapping. Suppose **RandEdge** requires that before the arrival of σ_i, the current state is $s \in S$ with probability $q(i, s)$. Then the minimum expected movement cost between the two required distributions before and after σ_i has arrived is given by

$$\min \sum_{s, s' \in S} f(s, s') \cdot tran(s, s')$$

s.t. (1) $\forall \ s \in S, \sum_{s' \in S} f(s, s') = q(i, s)$,

(2) $\forall \ s' \in S, \sum_{s \in S} f(s, s') = q(i + 1, s')$, and

(3) $\forall \ s, s' \in S, f(s, s') \geq 0$

Let $f^*(\cdot, \cdot)$ be the optimal $f(\cdot, \cdot)$ values found above[5]. If **RandEdge** is in state s before request σ_i, then it moves to state $s' \in S$ with probability $f^*(s, s')/q(i, s)$, so that the overall probability that **RandEdge** is in state s' after the request is $q(i + 1, s')$.

Theorem 8. *Algorithm* **RandEdge** *is* $(2 + 1/D)$-*competitive against an oblivious adversary.*

---

[5] The function $f^*(\cdot, \cdot)$ is the "minimum transportation cost" in the general model of randomized on-line algorithms in [15].

Proof. Given the offset vector $(0, k, l)$, we define the potential function

$$\Phi(k, l) = \sum_{1 \le j \le (D-l)} \frac{j}{D} + \sum_{1 \le j \le \min\{(l-k), (D-l)\}} (1 + \frac{j}{D}) + \sum_{[2(D-l)+1] \le j \le (D-k)} (\frac{1}{2} + \frac{j}{2D})$$

Clearly, $\forall\, k, l, \Phi(k, l) \ge 0$ and initially, $k = l = D, \Phi(D, D) = 0$. We show that in response to each possible ith request, the sum of the expected access cost $(\mathbf{E}[\Delta Cost_i])$, the expected movement cost due to a change in distribution $(\mathbf{E}[\Delta M_i])$ and the change in potential function $\Delta \Phi$, is no more than $(2 + 1/D)$ times Δopt_i (see Table 3), *i.e.* $\mathbf{E}[\Delta Cost_i] + \mathbf{E}[\Delta M_i] + \Delta \Phi \le (2 + 1/D) \cdot \Delta opt_i$ and the theorem follows. The proof involves a case analysis and considers all possible requests and offset combinations.

4.2 An Optimal Randomized Tree Algorithm

We extend algorithm **RandEdge** to one for file allocation on a tree, T, with n nodes. Let $p_{ab}[a], p_{ab}[b]$ and $p_{ab}[ab]$ be the probability values described in Sect. 4.1 for an edge (a, b). Our algorithm maintains $O(D \cdot n)$ tree configurations which define the state space. Each configuration, t, is a subtree of T of size at least one. After a request has arrived, algorithm **RandTree** redistributes probability masses on the set of configurations so that they induce **RandEdge** on each edge. **RandTree** also decides how the probability masses are transferred so that the induced expected movement cost on each edge e is no more than that incurred on e by **RandEdge** on the induced request sequence. It then follows from Theorem 8 that **RandTree** is $(2 + 1/D)$-competitive. Algorithm **RandTree** has the following properties:

- After request $i \ge 1$, **RandTree** defines S_i, the set of tree configurations and the probability masses placed on them. Initially $S_0 = \{v_0\}$, where v_0 is the node that contains the single copy of F and v_0 has a probability mass of 1.
- The probability mass placed on a single node is always $1/(2D)$ and that placed on a tree configuration that has at least one edge is always $1/D$. The same configuration may appear more than once in an S_i. In that case, the probability mass placed on it is the sum of the masses placed on the same configuration.

RandTree reacts differently to a *read* and a *write* request. We describe how **RandTree** generates S_i after σ_i has arrived, in particular, how the probability masses are transferred from configurations in S_{i-1} to those in S_i. The values of these transferred masses correspond to the variables $f(\cdot, \cdot)$ in the linear program in Sect. 4.1. Suppose that our algorithm requires a transfer of mass of $f^*(t, t')$ from $t \in S_{i-1}$ to $t' \in S_i$ after σ_i has arrived, so that there is a total mass of $q(i+1, t')$ at t'. Similar to the single edge case, we move to subtree configuration t' with probability $f^*(t, t')/q(i, t')$. The algorithm's responses to a write and a read request, σ_i, are given in Fig. 1 and 2, respectively. The final tree configurations obtained after running the algorithm gives S_i. The offset values described here and in the algorithm are the values *before* σ_i arrives. We assume T is rooted at the request node. Let r be the requesting node and $e = (a, b)$ be any edge in T with a nearer to r than b.

Algorithm RandTree (*write request*)

1. Let $\mathcal{F}$ be the forest formed from all the edges in T that have offsets $(0, k, l)$ with $(D + k - 2l) > 0$. Call **DoForest**$(\mathcal{F}, S_{i-1})$.
2. Let $\mathcal{F}$ be the forest of trees formed from all the edges with $(D + 2l - k) \leq 0$ and (i) $k < D$ or (ii) $k = D$ and the request comes form the non-zero offset state.
   for each tree in $\mathcal{F}$, **do**
     Perform, in a bottom-up fashion, for each edge (x, y), where $x \neq r$ and y is the parent of x, a transfer of a probability mass of $1/(2D)$ from node x to node y.

Procedure DoForest$(\mathcal{F}, S)$
while there is a tree $T' \in \mathcal{F}$ with at least one edge, **do**

1. Let x be the root node of T' and $X \in S$ be such that both T' and X contains edge (x, w) where w is some child of x.
2. **if** $(X - T')$ contains no edge, remove X from S and deposit a mass of $1/D$ at node x, **otherwise**, replace X by $(X - T')$ in S.
3. In $\mathcal{F}$, replace T' by the trees in $T' - X$.

Fig. 1. Algorithm RandTree for Write Requests

Write Requests: Given a write request, σ_i, and the induced request sequence on (a, b), algorithm **RandEdge** would require the following changes: (1) if e has offsets satisfying $(D + k - 2l) > 0$, then $p_e[ab]$ decreases by $1/D$, $p_e[a]$ increases by $1/D$, and $p_e[b]$ is unchanged; (2) if e has offsets satisfying $(D + k - 2l) \leq 0$ and (i) $k < D$ or (ii) $k = D$ and the request comes from the non-zero offset state, then $p_e[ab]$ is unchanged, $p_e[a]$ increases by $1/2D$, and $p_e[b]$ decreases by $1/2D$, and (3) there should be no change in the probability values for other edges. It can be checked that our algorithm is feasible and implements the required changes mentioned above when σ_i is a write request.

Lemma 9. *The algorithm satisfies the above requirements for all edges.*

Read Requests: Suppose σ_i is a read request. Edges with offsets satisfying $(D + k - 2l) < 0$ are said to be in *write mode*, otherwise, they are in *read mode*. Edges that are in the read mode but (i) has offset $k = l = 0$ or (ii) $k = l \geq 1$ and do not lie along the path from the offset root to r, are said to be in *idle mode*. Given the induced request sequence on e, algorithm **RandEdge** requires that after σ_i has arrived, (1) if e is in read mode but not in idle mode, $p_e[b]$ decreases by $1/D$, $p_e[ab]$ increases by $1/D$, and $p_e[b]$ is unchanged; (2) if e is in write mode, then after the request, $p_e[ab]$ is unchanged, $p_e[a]$ increases by $1/2D$, and $p_e[b]$ decreases by $1/2D$, and (3) otherwise, there should be no change in the probability values for e.

The following lemmas describe properties maintained by **RandTree**. They can be proved by induction and using the relationships between offsets in Corollary 6.

Let $e = (a, b)$ be an edge with $\omega(a) = 0$, $\omega(b) = k$, and $\omega(ab) = l$. Define the *write bias* of e, B_e, as $\max\{0, 2l - (D + k)\}$. The following lemma holds after every (read or write) request.

Lemma 10. *For vertex b, let $e_i = (b, a_i)$, for $1 \leq i \leq k$ be set of incident edges. Let $p[b]$ denote the sum of the probability masses on configurations consisting of the single node b. Then for each i, $p[b] \geq \frac{1}{2D} \max_i \{B_i - \sum_{j \neq i} B_j\}$*

Algorithm RandTree (*read requests*)

1. Let $\mathcal{F}$ be a forest of trees formed from edges with $(D + k - 2l) < 0$.
   for each tree in $\mathcal{F}$, **do**
   Perform, in a bottom-up fashion, for each edge (x, y), where $x \neq r$ and y is the parent of x, a transfer of a probability mass of $1/(2D)$ from x to y.
2. Break T into different subtrees at the request node. Let $\mathcal{F}$ be the forest formed from edges e in these subtrees that satisfies the following conditions:
   (i) e has offset with $(D + k - 2l) \geq 0$ and (ii) if e has $k = l$, it is along the path from the offset root to the request node and satisfies $l \geq 1$.
   Call **DoForest2**$(\mathcal{F}, S_{i-1})$.

Procedure DoForest2$(\mathcal{F}, S)$
while there is a tree $T' \in \mathcal{F}$ with at least one edge, **do**
1. Let r' be the root node for T', x a leaf node of T', and P the path from x to r'.
2. **if** x has a probability mass $\geq 1/D$, transfer mass of $1/D$ from x to P,
   otherwise there exists a subtree configuration, $Q \in S$, that
          (i)   lies in the subtree rooted in x,
         (ii) includes edge (x, w), where w is a child of x, and
       (iii) the offset on the edge is $(0, k', k')$.
          In this case, replace the configuration Q in S by $Q \cup P$.
3. Replace T' in $\mathcal{F}$ by the trees formed by $T' - P$.

Fig. 2. Algorithm RandTree for Read Requests

Lemma 11. *Let y and $w_1, \ldots, w_q$, $q \geq 1$ be neighbors of x where (y, x) is a read mode but not an idle mode edge, and all (x, w_j)'s are in idle or write mode. Then (1) if all (x, w_j)'s are in idle mode, then either $p[x] \geq 1/D$ or there exists a tree configuration that does not include (y, x) but includes one of the (x, w_j)'s; (2) if one of (x, w_j)'s is in write mode and the other (x, w_j)'s are in the idle mode, then either $p[x] \geq 1/(2D)$ or the same as in (1) holds.*

The next lemma can be proved using Lemmas 10 and 11.

Lemma 12. RandTree *induces* **RandEdge** *on all the edges when σ_i is a read.*

Lemmas 9 and 12 and Theorem 8 imply the next theorem.

Theorem 13. *Algorithm* **RandTree** *is $(2 + 1/D)$-competitive agains an oblivious adversary.*

4.3 The Lower Bound

We show that the competitive ratio, $(2 + 1/D)$, obtained in the randomized algorithms above are the best possible for file allocation against an oblivious adversary, even if G is a single edge.

Theorem 14. *No on-line algorithm for the file allocation problem on two points (a, b) is c-competitive, for any $c < 2 + 1/D$.*

Proof. Let $\mathcal{A}$ be any randomized algorithm for the file allocation problem on two points. We define a potential function Ψ, and give a strategy for generating adversary request sequences such that (1) For any C there is a request sequence with optimum cost $\geq C$; (2) Ψ is bounded; (3) On each request, the amortized cost to **RandEdge** (using the potential given in Sect. 4.1) is exactly $(2 + 1/D)$ times the cost to $\mathcal{OPT}$, the optimal off-line server, on that request; and (4) For each request generated by this adversary, $\Delta C_{\mathcal{A}} + \Delta \Psi \geq \Delta C_{\textbf{RandEdge}}$ (#).

Summing (#) over an adversary sequence gives $C_A(\sigma) \geq (2 + 1/D) \cdot OPT(\sigma) + b$ where b is bounded. Since the adversary can make $OPT(\sigma)$ arbitrarily large, A cannot be c-competitive for $c < 2 + 1/D$.

We assume that both the on-line and off-line algorithms start at a. Our adversary will only generate requests that result in offset functions of the form $(0, i, i)$, where $0 \leq i \leq D$. A *zero-cost self-loop* is a request such that the offset function is unchanged and $\Delta opt = 0$. By a theorem of [15], there is always an optimal on-line algorithm that incurs 0 expected cost on a zero-cost self loop. Hence we assume A has this property.

We now define the adversary's strategy. Suppose that the current offset function is $(0, i, i)$, and let p_i be the mass at a for **RandEdge**. Suppose A has mass q at a. If $q < p_i$ the adversary requests a^w, otherwise the adversary requests b^r. When $i = D$ we will have $q = 1$ (a^w is a zero-cost self-loop if $i = D$) and so the adversary will request b^r. Similarly, when $i = 0$, $q = 0$ (a^r is a zero-cost self-loop) and the adversary requests a^w. Notice that the only cycles that cost OPT nothing are zero-cost self-loops. Since the adversary never uses these requests, it can generate a sequence of arbitrarily large optimum cost.

Define Ψ to be $D \max\{0, q - p_i\}$. Notice that this is bounded. As was shown in the proof of Theorem 8, for request sequences generated by our adversary, the amortized cost to **RandEdge** on each request is exactly $(2 + 1/D)$ times the cost to OPT. It can be verified that (#) holds by a case analysis.

5 FAP on a Uniform Network — Off-line Solution

FAP on a uniform network, U, on n nodes can be converted to a min-cost max-flow problem on an acyclic layered network, with $O(n \cdot |\sigma|)$ nodes and $O(n^2|\sigma|)$ arcs. *Sketch:* There are $(2n - 1)$ nodes in each of the $(|\sigma| + 1)$ layers which represent the file configurations after each request. All node and arc capacities are one. Nodes in each layer are divided into two types, v_i and u_i, of size n and $(n - 1)$, respectively. The v_i nodes represent actual nodes in U. A flow of one unit into a v_i node represents the presence of F at the corresponding node in U; a flow of one unit into a u_i represents a copy of F not present in the network. With suitable choices of edge costs dependent on σ and D, FAP can be solved as a min-cost max-flow problem using the algorithm in [16].

Theorem 15. *FAP on U can be solved in $O(n^3|\sigma| \log_{(n+2)} |\sigma|)$ time.*

6 On-line Algorithms for Replication

Bartal *et al.* [3] find interesting relationships between the on-line Steiner tree problem [11, 18] and on-line FAP. It can be shown that the (off-line) replication problem is NP-hard by using a reduction from the Steiner tree problem [9, 13].

Theorem 16. *Replication is NP-hard, even if G is bipartite or planar.*

Let $e_D = (1 + 1/D)^D$ and $\beta_D = e_D/(e_D - 1)$. So $\beta_D \xrightarrow{D \to \infty} e/(e - 1) \approx 1.58$. We describe randomized algorithms that are β_D-competitive against an oblivious adversary on the uniform network and trees. First, consider a single edge (r, b) of unit length. Initially, only r contains a copy of the page.

Algorithm EDGE: After the ith request at b, replicate to b with probability p_i, where $\forall i, 1 \leq i \leq D$, $p_i = [(D + 1)/D]^{i-1}/[D(e_D - 1)]$, and other $p_i = 0$.

It can be shown that **EDGE** is strongly β_D-competitive using an analysis similar to that used by Karlin *et al.* [12] for their optimal randomized algorithm for the snoopy caching problem on two caches. **EDGE** can be applied to a uniform network by replicating F to each node v after the ith request at v, with a probability of p_i; it can also be extended to a randomized algorithm on a tree as in [5] — another example of factoring.

Theorem 17. *There exists strongly β_D-competitive randomized algorithms, vs. an oblivious adversary, for replication on a tree and a uniform network.*

7 Migration on a Uniform Network

We describe a $(2+1/(2D))$-competitive randomized algorithm against an oblivious adversary for migration on a uniform network. This competitive ratio is optimal even for a single edge [7]. Let G be a complete graph on n nodes labeled 1 to n. Initially, only node 1 has a copy of F. Our algorithm is based on the offsets calculated on-line. Let the $S = \{1,\dots,n\}$ and the algorithm is in state s if the single copy of F is at node s. We have the cost functions

$$acc(t,\sigma_i) = \begin{cases} 0 \text{ if } t = \sigma_i \\ 1 \text{ otherwise} \end{cases} \quad tran(t,s) = \begin{cases} 0 \text{ if } t = s \\ D \text{ otherwise} \end{cases}$$

and initially $W_0(S) = (0, D, \dots, D)$. Suppose the ith request is served and the new offset for each node, s, is calculated. Let $v_s = D - \omega_i(s)$ and $\delta = 2D - \sum_{m=1}^{k} v_m$, where $k \in argmax\{j \in S | \sum_{m=1}^{j} v_m < 2D\}$.

Algorithm Migrate: The probability, $p[s]$, that a node, s, contains F is as follows:

- $\forall\, s \le k,\ p[s] = v_s/(2D)$.
- If $k < n$, $p[k+1] = \delta/(2D)$.
- If $k = n$, add a probability mass of $\delta/(2D)$ to the unique professor j such that $v_j = D$.

The next theorem can be proved using an approach similar to the proof for Theorem 8, with the potential function: $\Phi = \sum_{s=1}^{n} \sum_{j=1}^{v_s}(\frac{1}{2} + \frac{j}{2D})$. It implies that Algorithm **Migrate** is strongly $(2 + 1/(2D))$-competitive against an oblivious adversary.

Theorem 18. *Given any request sequence, the expected cost incurred by **Migrate**, $\mathbf{E}_{mig}(\sigma)$ satisfies,* $\mathbf{E}_{mig}(\sigma) \le [2 + 1/(2D)] \cdot OPT(\sigma)$

8 File Allocation in General Networks

Awerbuch *et al.* [2] design a $O(\log n)$-competitive algorithm, **FA**. Combining the proof in [2] with ideas from [18], we can show the following result.

Theorem 19. *Algorithm **FA** is $O(\log d(G))$-competitive on an unweighted graph with diameter $d(G)$.*

References

1. S. Albers and H. Koga. New On-Line Algorithms for the Page Replication Problem. In *Proceedings of the Fourth Scandinavian Workshop on Algorithmic Theory*, Aarhus, Denmark, July 1994.

2. B. Awerbuch, Y. Bartal, and A. Fiat. Competitive Distributed File Allocation. In *Proceedings of the 25th ACM Symposium on Theory of Computing*, pages 164–173, 1993.

3. Y. Bartal, A. Fiat, and Y. Rabani. Competitive Algorithms for Distributed Data Management. In *Proceedings of the 24th Annual ACM Symposium on the Theory of Computing*, pages 39–50, 1992.

4. S. Ben David, A. Borodin, R. Karp, G. Tardos, and A. Wigderson. On the Power of Randomization in Online Algorithms. In *Proceedings of the 22nd ACM Symposium on Theory of Computing*, pages 379–386, 1990.

5. D. L. Black and D. D. Sleator. Competitive Algorithms for Replication and Migration Problems. Technical Report CMU-CS-89-201, Department of Computer Science, Carnegie Mellon University, 1989.

6. M. Chrobak and L. L. Larmore. The Server Problem and On-line Games. In *Proceedings of the DIMACS Workshop on On-Line Algorithms*. American Mathematical Society, February, 1991.

7. M. Chrobak, L. L. Larmore, N. Reingold, and J. Westbrook. Page Migration Algorithms Using Work Functions. In *Proceedings of the 4th International Symposium on Algorithms and Computation, ISAAC '93*, volume 762 of *Lecture Notes in Computer Science*, pages 406–415, Hong Kong, 1993. Springer-Verlag.

8. D. Dowdy and D. Foster. Comparative Models of The File Assignment Problem. *Computing Surveys*, 14(2), 1982.

9. M. R. Garey and D. S. Johnson. The Rectilinear Steiner Tree problem is NP-complete. *SIAM J. Appl. Math.*, 32:826–834, 1977.

10. B. Gavish and O. R. L. Sheng. Dynamic File Migration in Distributed Computer Systems. *Communications of the ACM*, 33(2):177–189, 1990.

11. M. Imaze and B. M. Waxman. Dynamic Steiner Tree Problem. *SIAM J. Disc. Math.*, 4(3):369–384, 1991.

12. A. R. Karlin, M. S. Manasse, L. A. McGeoch, and S. Owicki. Competitive Randomized Algorithms for Non-Uniform Problems. In *Proceedings of the 1st ACM-SIAM Symposium on Discrete Algorithms*, pages 301–309, 1990.

13. R. M. Karp. Reducibility Among Combinatorial Problems. In R. E. Miller and J. W. Thatcher, editors, *Complexity of Computer Computations*, pages 85–103. Plenum Press, New York, 1972.

14. H. Koga. Randomized On-line Algorithms for the Page Replication Problem. In *Proceedings of the 4th International Symposium on Algorithms and Computation, ISAAC '93*, volume 762 of *Lecture Notes in Computer Science*, pages 436–445, Hong Kong, 1993. Springer-Verlag.

15. C. Lund and N. Reingold. Linear Programs for Randomized On-Line Algorithms. In *Proceedings of the 5th ACM-SIAM Symposium on Discrete Algorithms*, 1994.

16. R. E. Tarjan. *Data Structures and Network Algorithms*. SIAM, Philadelphia, Pennsylvania, 1983.

17. J. Westbrook. Randomized Algorithms for Multiprocessor Page Migration. In *Proceedings of the DIMACS Workshop on On-Line Algorthms*. American Mathematical Society, February 1991.

18. J. Westbrook and D. C. K. Yan. Greedy Algorithms for the On-Line Steiner Tree and Generalized Steiner Problems. In *Proceedings of the 3rd Workshop in Algorithms and Data Structures*, volume 709 of *Lecture Notes in Computer Science*, pages 622–633, Montréal, Canada, 1993. Springer-Verlag.

A Unified Scheme for Detecting Fundamental Curves in Binary Edge Images

Tetsuo Asano,[1] Naoki Katoh[2] and Takeshi Tokuyama[3]

[1] Osaka Electro-Communication University, Hatsu-cho, Neyagawa, 572 Japan
[2] Kobe University of Commerce, Gakuen-Nishimachi, Nishi-ku, Kobe, 651-21 Japan
[3] IBM Tokyo Research Laboratory, Yamato-shi, Kanagawa, 242 Japan

Abstract. We present a unified scheme for detecting digital components of various planar curves in a binary edge image. We introduce a measure d to reflect the complexity of a family of curves. For instance, $d = 2, 3$, and 5 for lines, circles, and ellipses, respectively. Our algorithm outputs all eligible curve components (a component is *eligible* if its size is at least a threshold k) in $O(n^d)$ time and linear space, where n is the number of points. Our only primitive operations are algebraic operations of bounded degrees and comparisons. We also propose an approximate algorithm with α−perfectness, which runs in $O((\frac{n}{(1-\alpha)k})^{d-1}n)$ time and outputs $O((\frac{n}{(1-\alpha)k})^d)$ curves.

1 Introduction

One of the most fundamental problems in pattern recognition is how to detect all the digital line components in a given binary edge image. Although a number of methods such as the Hough Transform [3, 4, 6, 18] have been proposed for this purpose, none of them seems able to guarantee that all line components will be detected. The algorithm based on the L_1-dual transform presented by the first two authors of the present paper [2] is the first one that can guarantee perfect detection of all line components of a given set of n edge points (or black dots) in an $N \times N$ grid in $O(nN^2)$ time and $O(n)$ space.

In this paper, we present a unified scheme for detecting all *eligible* components of various planar curves, and show that it has several properties desirable in algorithms for detecting curve components.

1.1 Definition of curve detection problems

Suppose that we are given a two-dimensional lattice plane G of size $N \times N$, a *binary edge image* consisting of *edge points* (or black dots), which form a subset S of G, and a family $\mathcal{F}$ of planar curves defined over the square region containing G, which can be expressed by linear combination of $d + 1$ linearly independent polynomial expressions for x and y with linear parameters. For a curve $f(x, y) = 0 \in \mathcal{F}$ (we often abbreviate to $f \in \mathcal{F}$), its digital image $Im(f)$ is defined to be a set of all the lattice points in G whose horizontal or vertical distance to the curve is at most 0.5 (see Figure 1). If there exist a subset P of

S and an $f \in \mathcal{F}$ such that $P \subseteq Im(f)$ and $p \notin Im(f)$ for any $p \in S - P$, P is a digital curve component of $\mathcal{F}$. For practical purposes, we are usually interested in a digital component whose size is greater than or equal to a given threshold k. Such a component is called an *eligible component*.

With these settings, the curve component detection problem consists of enumerating all eligible digital curve components of $\mathcal{F}$.

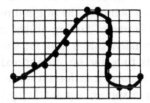

Fig. 1. A digital image of a planar curve, consisting of all lattice points with horizontal or vertical distance to the curve $\leq 1/2$.

1.2 Desirable features of a curve detection scheme

A good scheme for finding digital curve components, should include the following features:

1. **Detection ability:** It should be guaranteed that all eligible digital curve components can be detected.
2. **Linear space complexity:** It is crucial to develop algorithms with small (preferably, linear) space complexity in practical applications.
3. **Small output size:** We need a method to compress outputs, since the total number of possible digital components usually becomes very large. One possible way is to output only *maximal components*, where a digital component is *maximal* if there is no other digital component properly including it. The second is not to output digital components, but to output only equations of curves f for which a digital curve component P with $P \subseteq Im(f)$ exists. The third is to output neither each digital component nor the corresponding curve equation, but to aggregate information on digital components in a compact form by outputting the region of parameter space defining $\mathcal{F}$ for which a digital component exists.
4. **Simple operations:** It is desirable that an algorithm should require only algebraic operations and comparisons. That is, we should avoid having to solve nonlinear equations.
5. **Efficient approximation:** If it is too expensive to output all eligible components, an approximation algorithm must be considered. For each eligible digital curve component P, a curve f such that $Im(f)$ contains a large portion of P should be output. It is desirable that the performance of the approximation algorithm can be theoretically guaranteed. Moreover, the computing cost should be cheap and the output should be small.

1.3 Defects of the standard method

Detecting lines in a digital image is the first step in recognizing some geometric shape. Among the various existing techniques for this purpose, the standard algorithm is based on the Hough Transform [3, 4, 6, 18]. It transforms an edge point (x_i, y_i) into a sinusoidal curve $\rho = x_i \cos\theta + y_i \sin\theta$ in the $\rho - \theta$ plane. Since ρ is the orthogonal distance to the line passing through (x_i, y_i) with a normal vector of angle θ, the intersection of two such curves corresponds to the line passing through the two points in the original plane. Thus, one might easily imagine that the problem is one of finding all the intersections at which many such curves meet. But actually, this is not so, since we are interested not in exact fitting of edge points to a line but in approximate fitting. This is one of the difficulties. A more standard transformation in computational geometry is the Duality Transform for mapping a point (x_i, y_i) to a line $v = x_i u + y_i$. In fact, the original algorithm proposed by Hough[10] was based on the duality transform. One advantage of the sinusoidal transformation is that the range in the $\rho - \theta$ plane can be limited to a small region (note that $|\rho| \leq \sqrt{2}N$ and $0 \leq \theta < \pi$). This fact suggests a bucketing technique. We partition the $\rho - \theta$ plane into equal-sized $M \times M'$ buckets (usually both M and M' are $O(N)$). Then, for each edge point (x_i, y_i) we compute $\rho_j = x_i \cos\theta_j + y_i \sin\theta_j$ for each angle $\theta_j \in (\theta_1, \theta_2 ..., \theta_{M'})$ to put a vote in the bucket specified by (ρ_j, θ_j). After voting is over we choose buckets having more votes than a given threshold and output lines corresponding to those buckets.

Here a question arises. What are we trying to detect? Although the word "line components" seems to be clear enough, it is vague from a mathematical point of view. This may be the reason for the scarcity of theoretical studies of the performance guarantees of line detection algorithms. We start with the above mentioned reasonable definition of a line component. Then, we will find that the Hough Transform outputs less concrete defined digital lines.

Another disadvantage of the Hough Transform is that as long as a bucketing technique is used, it cannot be guaranteed that all possible line components will be detected. To overcome this disadvantage, the first two authors proposed an algorithm that can guarantee perfect detection of all line components in $O(nN^2)$ time and $O(n)$ space [2]. They also introduced a new notion of α−perfectness concerning detection ability, which relaxes the 100% detection ability by a factor of $(1 - \alpha)$ (see Section 3 for its definition), and proposed an approximate algorithm with α−perfectness that requires $O(nN \log n / (1 - \alpha))$ time. However, these algorithms may output non-maximal line components.

There have been several papers [3, 6, 13, 18, 19] on the problem of detecting digital components of planar curves such as circles and ellipses. However, every existing algorithm requires more than linear space and none guarantees that all possible digital curve components will be detected.

1.4 Our results

Generalizing the idea of [2], we treat in a uniform manner the problem of detecting digital curve components by reducing it to one of sweeping a two-dimensional arrangement of lines in a two-parameter space in order to enumerate appropriate cells by applying the topological walk algorithm [1], and we shall propose algorithms for the problem that have most of the desirable properties listed above.

[Lines:] For line detection, we shall propose an algorithm that requires $O(n^2)$ time and $O(n)$ space, and outputs only maximal digital line components. This algorithm is essentially the same as the one proposed in a previous paper [2] but it explicitly uses the standard dual transformation between points and lines to transform the problem into one of sweeping arrangements of $O(n)$ lines in two-parameter space. One major advantage of this algorithm over the previous one [2] is that it outputs only maximal components. This feature is achieved by associating appropriate information with each line in the dual plane, and by applying the topological walk algorithm [1].

We shall also propose an $\alpha-perfect$ approximation algorithm that runs faster than the previous one [2] by using range search techniques.

[Planar curves:] For the detection of planar curve components, we shall propose a unified scheme for detecting all digital curve components, except very singular ones, for a family $\mathcal{F}$ of curves. Let us consider families of curves defined by expressions $f[a_1, .., a_d](x, y) = 0$ that are polynomials for x, y, and d parameters $a_1, .., a_d$, together with parameter constraints which are systems of inequalities and equalities independent of x and y. We can assume that $f[a_1, .., a_d](x, y)$ is linear with respect to the parameters, since we can linearize the expressions by increasing the number of parameters and constraints. For example, the family of conchoids defined by $(x - a)^2(x^2 + y^2) = b^2 x^2$ is expressed as $(x^4 + x^2 y^2) + a_1(x^3 + xy^2) + a_2(x^2 + y^2) + a_3 x^2 = 0$, $a_1^2 = 4a_2, a_3 \leq 0$; hence we need three parameters, although the number of degrees of freedom of the parameters is two. However, for many families of curves in practical use, for example, circles, hyperbolas, ellipses, cubic curves, cissoids, and lemniscates, we have expressions such that the constraints do not decrease the number of degrees of freedom of the parameters.

The algorithm to be presented runs in $O(n^d)$ time and $O(n)$ space for such a family of curves. One nice feature of our algorithm is that it does not need to solve a nonlinear equation; the only primitive operations required are comparisons and algebraic operations of bounded degrees. This is achieved by the introduction of the notion of a *regular digital image*. We can show that for most circles and ellipses that we are concerned with in applications, regular digital images coincide with digital images. Moreover, all the eligible digital images of circles and ellipses can be detected by using the algorithm detecting regular digital images, unless the threshold is extreamly low.

Because of the linearity of parameters, the problem of detecting curves whose regular digital images contain digital components is naturally transformed into one of examining a $d-dimensional$ arrangement to find appropriate cells. Focusing on critical cases of such curves, the problem can be reduced to one of searching an arrangement of lines (not "curves"!) in a two-parameter space, which enables us to apply the topological walk algorithm [1].

When specialized to ellipses (resp. circles), our algorithm computes all digital elliptic (resp. circular) components in $O(n^5)$ (resp. $O(n^3)$) time and $O(n)$ space. A number of algorithms have been proposed for detecting digital circle or ellipse components [13, 14, 17, 19], but to the authors' knowledge, no existing algorithm guarantees to detect every such component. In this sense, therefore, ours is the first algorithm with guaranteed performance and linear space complexity. Moreover, it is conjectured that these computational complexities are optimal within a constant factor.

We also consider α-perfect approximate solutions. We shall propose a Las-Vegas algorithm that runs in $O((\frac{n}{(1-\alpha)k})^{d-1}n)$ time, outputting $O((\frac{n}{(1-\alpha)k})^d)$ curves. Here k is the threshold on the size of components to be output, and $0 < \alpha < 1$ is the approximation factor. In many practical cases, n/k is small; thus, the above algorithm becomes efficient.

[Polygonal chains:] Finally, we shall apply our idea to the case in which $\mathcal{F}$ is a family of polygonal chains, obtained by transforming a given polygonal chain by a set of affine transformations. If translation, rotation and scaling transformations are permissible, the problem is to detect all digital components of a polygonal chain transformed by a combination of these three transformations. This problem can be similarly solved in $O(m^4 n^4)$ time and $O(mn)$ space, where m is the number of edges of a polygonal chain.

The problem is strongly related to geometric matching problems [11, 9, 12]. Because of the limitation of space in this extended abstract, we shall omit all the details.

2 Curve Detection

2.1 Line detection and circle detection

We start with an algorithm for detecting all line components. By the definition above, an edge point (x_i, y_i) is contained in the digital image of a line $y = ux + v$ for $|u| \leq 1$ if and only if $ux_i + v - 0.5 \leq y_i \leq ux_i + v + 0.5$. This is equivalent to the inequality $-x_i u + y_i - 0.5 \leq v \leq -x_i u + y_i + 0.5$, which corresponds to a strip defined by two parallel lines in the $u - v$ plane. Since we have n edge points, we have n such strips, which together define an arrangement of lines. If k such strips intersect at a cell in the arrangement, the cell corresponds to a line component of size k. Thus, the problem is reduced to sweeping the arrangement to detect all the cells corresponding to the maximal subsets of edge points. Although we have $2n$ lines and $O(n^2)$ cells, we can visit all the cells in $O(n^2)$ time by using topological sweep [7] or topological walk [1]. Although it is not a trivial extension, we can implement topological walk to output only maximal line components (or equations of the corresponding lines). Maximal line components corresponding to lines with slope u, $|u| > 1$, can be detected in a similar manner by exchanging x and y coordinates.

Circle detection is harder than line detection, especially if we require an algorithm with perfect detection ability. Using a voting technique again, one can design a naive algorithm for detecting circles. The idea is as follows: Consider a cone $z = (x - x_i)^2 + (y - y_i)^2$, $0 \leq z \leq N$. If three such cones intersect at one point (x, y, z), the circle of radius z with its center at (x, y) passes through the corresponding three points. Thus, if we put votes in a three-dimensional array according to such cones we can detect circle components. The time complexity is $O(nN^2)$, and a large amount of space $O(N^3)$ is required. Because of the large space complexity, this approach seems impractical. Another disadvantage is that it cannot guarantee perfect detection.

A digital image of a circle $f(x, y) = x^2 + y^2 + ax + by + c = 0$ is defined by the four circles obtained by shifting the circle in the four orthogonal directions (East, West, South and North). Since we can define the interior of the circle by the inequality $x^2 + y^2 + ax + by + c \leq 0$, a condition that an edge point (x_i, y_i) is contained in the digital image of the circle $f(x, y) = 0$ is described (if the radius of the circle is not less than 0.5) by
$[f(x_i+0.5, y_i)f(x_i, y_i) \leq 0] \vee [f(x_i-0.5, y_i)f(x_i, y_i) \leq 0] \vee [f(x_i, y_i+0.5)f(x_i, y_i) \leq 0] \vee [f(x_i, y_i - 0.5)f(x_i, y_i) \leq 0]$, which is equivalent to

$$[f(x_i + 0.5, y_i)f(x_i - 0.5, y_i) \leq 0] \vee [f(x_i, y_i + 0.5)f(x_i, y_i - 0.5) \leq 0] \qquad (1)$$

The inequalities (1) define a polygonal region (union of a pair of wedges) in the (abc)-space. Thus, again it suffices to examine an arrangement of planes in the three-dimensional space. Unfortunately, no existing algorithm visits every cell in the arrangement in $O(n^3)$ time and $O(n)$ space. In our approach, we fix one edge point on the boundary of a digital image of a circle so that one of the variables a, b, and c can be deleted. With this constraint it suffices to examine an arrangement in the two-dimensional plane. Thus, we obtain an algorithm that runs in $O(n^3)$ time and $O(n)$ space.

In this approach, for each edge point p_i, we can detect all maximal circle components on condition that the point p_i lies on the boundaries of digital images, which are not always maximal without this condition. This is a disadvantage. However, the perfect detection ability is presented.

2.2 General Curve Detection

We generalize the idea expressed in previous subsections in order to develop algorithms to detect all the digital components of more complex curves such as ellipses. To deal with various planar curves, we consider a family of planar curves defined by linear combination of $d + 1$ linearly independent polynomial expressions for x and y with d parameters. More precisely, the family is specified by an equation of the form

$$f[a_1, ..., a_d](x, y) \equiv t_0(x, y) + \sum_{i=1}^{d} a_i t_i(x, y) = 0, \qquad (2)$$

where each $t_i(x, y)$, $0 \leq i \leq d$ is a polynomial function for x and y, and each a_i, $1 \leq i \leq d$ is a real parameter. We assume that d is a fixed constant. This includes most of the interesting families of 2D curves such as lines, circles, ellipses, hyperbolas, cissoids, and lemniscates.

A family of planar curves defined by (2) is denoted $\mathcal{F}$. When understood from the context, a member of $\mathcal{F}$ is simply written as $f(x, y) = 0$. In most applications, we need to impose some constraints on parameters represented by polynomial inequalities. Let $C(\mathcal{F})$ denote the set of constraints for $\mathcal{F}$.

For example, a family $\mathcal{E}$ of ellipses is specified by the functional form

$$x^2 + a_1 xy + a_2 y^2 + a_3 x + a_4 y + a_5 = 0 \tag{3}$$

with

$$C(\mathcal{E}) = \{a_1^2 - 4a_2 < 0\}. \tag{4}$$

Another example is a family C of circles given by

$$(x^2 + y^2) + a_1 x + a_2 y + a_3 = 0 \tag{5}$$

with $C(\mathcal{C}) = \emptyset$. Notice that with these definitions of $C(\mathcal{E})$ and $C(\mathcal{C})$, families $\mathcal{E}$ and $\mathcal{C}$ contain imaginary ellipses and circles. However, as will be seen in Section 4, our algorithm does not produce any such imaginary curves without explicitly taking care of constraints. Thus, such constraints are neglected in our definition.

We assume throughout this section that the threshold for the size of a digital component that we want to detect is larger than d (i.e., the number of parameters). Notice that this assumption is natural and is satisfied in most applications.

Let D be the square region of size $N \times N$ containing G. Here we can observe the following fact: The curve $f(x, y) = 0$ may partition D into a number of regions. The sign of the value of f is identical at every point in each such region. That is, the domain is partitioned into regions of positive signs and negative signs.

Here we make an assumption concerning the form of a function. That is, whenever two regions are adjacent to each other, their signs must be different. The assumption is reasonable, since it is always satisfied except when the function has an irreducible component that is the sum of even powers of polynomials.

[**Definition 1**] Let f be an instance of a family $\mathcal{F}$ of (2). The planar curves obtained by shifting $f = 0$ right, left, up, and down by 0.5 are denoted by $f_l = 0, f_r = 0, f_t = 0$, and $f_b = 0$, respectively. In other words, $f_r(x, y) \equiv f(x-0.5, y)$, for example. For notational convenience, $f_0(x, y)$ is defined to be $f(x, y)$. Five curves $f_\gamma(x, y) = 0$, $\gamma \in \{0, l, r, t, b\}$, are referred to as the *characteristic curves* of f.

$RI_r(f), RI_l(f), RI_t(f)$, and $RI_b(f)$, defined below, are called the *right, left, top*, and *bottom* **regular images**, respectively.

$$RI_\gamma(f) = \{(x, y) \in D \mid f_\gamma(x, y) \times f(x, y) \leq 0\}, \tag{6}$$

where $\gamma \in \{r, l, t, b\}$. The union

$$RI(f) = RI_r(f) \cup RI_l(f) \cup RI_t(f) \cup RI_b(f) \tag{7}$$

is simply called the *regular image* of the curve $f = 0$ (see Figure 2). Clearly, $G \cap RI(f) \subseteq Im(f)$ holds.

Fig. 2. Regular image of a curve defined by four characteristic curves.

[Definition 2] For a given instance $f = 0$ of $\mathcal{F}$, an edge point $p = (x, y) \in S$ is called a *regular point* of $Im(f)$, if $p \in RI(f)$. A subset P of S is called a *regular digital component* of $\mathcal{F}$ if there exists an instance $f(x, y) = 0$ of $\mathcal{F}$ such that $P = S \cap RI(f)$. A curve $f(x, y) = 0$ is *regular* if $RI(f) \supseteq Im(f)$, and is *irregular* otherwise.

It is easy to see that the digital component of a regular curve is a regular digital component. Also, the set of regular points of the digital component of a irregular curve is a regular digital component.

Lemma 1 *Any line is regular and a circle is regular if its radius is at least* 0.5. *(2) Let E be an ellipse* $ax^2 + by^2 = c$, $b > a > 0$. *If the inequality* $\sqrt{c/b} \cdot \sqrt{2a/(a + b)} \geq 0.5$ *holds, any rotated copy of E is regular.*

It follows from the above lemma that all ellipses except very skinny ones are regular. Indeed, an eligible circle component is regular if the threshold is larger than 4. An eligible component of threshold k corresponding to an irregular ellipse either contains a regular eligible ellipse component of threshold $k/2$ or a (regular) eligible component of a segment (which is the long axis of the ellipse) of threshold $k/2$. In practice, there will be no problem in concentrating on detecting only regular digital components

Let $\mathcal{F}$ be a family of planar curves with a set of constraints $C(\mathcal{F})$. If there is a set of parameters such that for any values of them we can adjust other parameters so as to satisfy $C(\mathcal{F})$, the set of parameters is called a *free parameter set*. We are interested in a maximal free parameter set, whose size is denoted by $free(\mathcal{F})$.

For example, for a family $\mathcal{E}$ of ellipses described by (3) with constraint set $C(\mathcal{E})$ of (4). $\{a_2, a_3, a_4, a_5\}$ is a free parameter set, and hence $free(\mathcal{E}) = 4$. For a family $\mathcal{C}$ of circles described by (5), the maximal free parameter set is clearly $\{a_1, a_2, a_3\}$ and $free(\mathcal{C}) = 3$ by $C(\mathcal{C}) = \emptyset$.

Algorithm: We shall briefly describe an $O(n^d)$ time and $O(n)$ space algorithm for detecting all regular digital components of planar curves.

For an edge point $p = (x, y) \in S$, the condition with resect to parameters that there exists an instance $f \in \mathcal{F}$ such that $p \in T(f)$ is expressed by

$$\bigcup_{\gamma \in \{r, l, t, b\}} \{(a_1, \ldots, a_d) \mid (f_\gamma(x, y) \geq 0 \wedge f(x, y) \leq 0) \vee (f_\gamma(x, y) \leq 0 \wedge f(x, y) \geq 0)\}$$

$$(8)$$

from (7). Thus, the region in d-dimensional parameter space satisfying (8) forms a union of polyhedra. Thus, the following theorem is immediate from this observation.

Theorem 1. *Let $\mathcal{F}$ be a family of planar curves defined by the expression (2), and let P be an arbitrary regular digital component of $\mathcal{F}$, and assume that the size of P is greater than d. Then, there always exists an instance f of $\mathcal{F}$ with $P \subseteq T(f)$ such that for some r with $0 \le r \le free(\mathcal{F})$, (i) r edge points of P lie on one of its five characteristic curves, and (ii) $free(\mathcal{F}) - r$ parameters can be arbitrarily fixed.*

It can happen that $r < free(\mathcal{F})$. For example, if we consider the family $a_1 x(2x-1)(2x+1) + a_2 x^2(2x-1)(2x+1) + 4y + 1 = 0$, the origin $(0,0)$ is never located on a characteristic curve, although it can be contained in every digital component. Hence, if we choose $P = \{(0,0), (-1,1)\}$, $r = 1$ and we can fix one of a_1 and a_2 arbitrary.

On the basis of this theorem, the algorithm to be presented below can be applied to a family of curves with d parameters such that $free(\mathcal{F}) \ge d - 2$. First, the algorithm chooses a set of $d - 2$ free parameters, denoted A, among $free(\mathcal{F})$ parameters. Without loss of generality, let $A = \{a_1, a_2, \ldots, a_{d-2}\}$. In the algorithm we do the following procedure for every subset P of size $d - 2$ of S.

Suppose P consists of points $p_1 = (x_1, y_1), \ldots, p_{d-2} = (x_{d-2}, y_{d-2})$. For each point $p_i \in P$, we choose one of five characteristic curves and associate it with p_i. (We try all 5^{d-2} possible combinations). Let $f_{\gamma_i} = 0$ be the characteristic curve associated with p_i.

We eliminate the parameters $a_1, .., a_{d-2}$ from the equation $f[a_1, .., a_d](x, y) = 0$ by using the equations $f_{\gamma_i}(x_i, y_i) = 0$ and Theorem 1.

Once we have substituted every parameter of A expressed as a linear combination of a_{d-1} and a_d into (8), the condition concerning a parameter set $(a_1, \ldots, a_d)$ that determines the existence of f with $p \in T(f)$ for each edge point p of $S - P$ can be expressed by a (possibly not connected) polygonal region, denoted $R(p)$, in the (a_{d-1}, a_d) plane. (Recall that it is just a generalization of a strip in the $u - v$ plane defined for line detection.) Notice that $R(p)$ is a union of cells of an arrangement of five lines in the (a_{d-1}, a_d) plane, since $f_\gamma(x, y) \ge 0$ (or ≤ 0) for an edge point $p = (x, y)$ and for some $\gamma \in \{0, r, l, t, b\}$ corresponds to a halfplane in the plane.

In this way we have $n - d + 2$ such polygonal regions in the (a_{d-1}, a_d) plane. The set of lines bounding these regions makes an arrangement of $O(n - d + 2)$ lines in the plane, and subdivides the plane into $O(n^2)$ cells. Then, all we need to do is to enumerate all the cells at which k or more such regions intersect. This task can be done in $O(n^2)$ time using linear space by applying the topological walk algorithm [1].

Since we need to test every $d - 2$ points from n edge points and to associate each of the chosen $d-2$ points with one of five characteristic curves, the algorithm requires a total of $O(n^d)$ time and $O(n)$ space.

Theorem 2. *We can detect all regular eligible digital components for a family of curves $\mathcal{F}$ in $O(n^d)$ time and $O(n)$ space.*

3 α-perfect curve detection

Although the algorithm in the previous section outputs all digital components in $O(n^d)$ time, the time complexity is often too expensive. In this section, therefore, we consider much cheaper approximation algorithms. For a digital curve component P containing m points and for $0 < \alpha < 1$, its α-*approximation* is a curve f of $\mathcal{F}$ such that $Im(f)$ contains at least αm points of P.

We consider eligible components of a point set S with respect to a family $\mathcal{F}$ of curves with d parameters. Let k be the threshold. A subfamily $\mathcal{Y}(S)$ of $\mathcal{F}$ is called an α-perfect family of S if for any eligible component P of S, there exists an α-approximation of P in $\mathcal{Y}(S)$.

Theorem 3. *There exists an α-perfect family containing $O((\frac{n}{(1-\alpha)k})^d)$ curves. Such a family can be computed in $O((\frac{n}{(1-\alpha)k})^{d-1}n)$ randomized time.*

The algorithm is efficient if the threshold k is large, as is often the case in practice.

In order to prove Theorem 3, let us consider the space R^d of the parameters $(a_1, .., a_d)$. The condition that an edge point $p = (x_0, y_0)$ lies on $f_\gamma(x, y)$ defines a hyperplane in R^d for each $f \in \mathcal{F}$ and $\gamma \in \{r, l, t, b, 0\}$. Thus, we have an arrangement of $5n$ hyperplanes corresponding to the point set S. A $1/r$-*cutting* of n hyperplanes is a space subdivision, such that at most n/r hyperplanes intersect each cell of the subdivision. The following theorem is known. [5]

Theorem 4. *If d is a constant, there exists an r-cutting of size $O(r^d)$, which can be computed in $O(nr^{d-1})$ randomized time, together with the set of hyperplanes intersecting each cell.*

We set $r = 5n/(1 - \alpha)t$, compute an $1/r$-cutting, select a representative point $v(c)$ (for example, a vertex) in each cell c, and output the set $\mathcal{X}$ of all curves associated with the selected points, whose digital images contain at least αk points (counting points in the digital images can be done in $O(nr^{d-1})$ time in total).

For any digital curve component P, consider the associated point p in the space of parameters. The point p is located in a cell c of the cutting. Let $Q(c)$ be the digital curve component associated with the representative point $v(c)$. Because of the definition of the cutting, $P \backslash Q(c)$ contains at most $5n/r = (1-\alpha)k$ points. Therefore, $\mathcal{X}$ is an α-perfect family of size $O((\frac{n}{(1-\alpha)k})^d)$. This proves Theorem 3. $\square$

We remark that, if we further continue searching in the cells of the cutting in which dual points of the eligible components can exist, we can obtain an exact algorithm that is sensitive to the number of eligible components with respect to the threshold αk. However, the number of eligible components may be large even if the threshold is high, unlike the exact fitting problem [8].

The following approximation method (outputting more curves) is easier, although it may output wrong answer in very unlucky cases.

Theorem 5. *There exists a Monte-Carlo type randomized algorithm that outputs an α-perfect family of size $O((\frac{n \log n}{(1-\alpha)k})^d)$ in $O((\frac{n \log n}{(1-\alpha)k})^d)$ time with high probability. The time complexity is increased to $O((\frac{n \log n}{(1-\alpha)k})^{d-1}n)$ if we want to output only curves whose digital images contain at least αk points.*

Proof. Instead of Chazell's cutting, we apply random sampling. □

The above methods are useful only if the threshold k is large. However, if $n \gg N$, we can efficiently compute an $1/2$-perfect family of digital *line* components even if the threshold k is small.

Theorem 6. *There exists a $1/2$-perfect family consisting of $O(N^2)$ digital line components. Such a family, together with the size of digital components of members of the family, can be computed in $O(N^{4/3}n^{3/2} \log n)$ time.*

Proof. Consider the family of $O(N^2)$ lines penetrating two grid points located on the boundary of the grid frame. It is not difficult to show that the associated family of digital components contains a $1/2$-perfect family. We can compute the size of a digital component $S \cap Im(f)$ by applying slab-range searching, because $Im(f)$ is a slab. If we apply a slab-range search data structure with $O(N^{4/3}n^{2/3})$ space [15], the query time is $O((n/N)^{2/3} \log n)$ and the preprocessing time is $O(N^{4/3}n^{2/3} \log n)$. □

4 Concluding remarks

Besides circle and ellipse, a parabola is also an important curve and a fundamental piece of quadratic splines frequently used in CAD systems. A family of parabolas can be described by the same equations as (3) with a constraint of $4a_1^2 = a_2$. Our scheme gives an $O(n^5)$ time algorithm to detect all regular parabolic components. If we are allowed to solve a quadratic equation in constant time (or less than $O(\log n)$ time), we can obtain an $O(n^4 \log n)$ time and linear space algorithm, and an $O(n^4)$ time randomized algorithm.

We can also detect digital components of polynomial surface in the three-dimensional lattice space by using our framework. For example, since a plane is expressed by using three parameters, we obtain $O(n^3)$ time and $O(n)$space algorithm to detect all digital planes.

As is discussed in the literature [16], there may be various manners of defining a digital image of a planar curve. We have adopted the L_∞ distance, which is natural for the line detection. However, orthogonal distance may be useful in many cases for practical application. When we apply a definition of the digital image suitable to the orthogonal distance measure, the problem becomes a little harder in general. For line detection and circle detection, we need to deal with arrangements of quadratic curves. Moreover, for higher-order curves such as ellipses, it looks much harder to have a practically efficient algorithm, since the defining equation of the trajectory of the points which have distance 0.5 from a curve (say, ellipse) is complicated.

References

1. T. Asano, L. Guibas, and T. Tokuyama, Walking in an Arrangement Topologically, *Int. J. of Comput. Geom. and Appl.* 4-2 (1994).

2. T. Asano and N. Katoh, Number Theory Helps Line Detection in Digital Images, *Proc. ISAAC'93, Springer LNCS 762*, (1993), pp.313-322.

3. D.H. Ballard, Generalizing the Hough Transform to Detect Arbitrary Shapes, *Pattern Recognition*, 13-2 (1981), pp. 111-122.

4. C.M. Brown, Inherent Bias and Noise in the Hough Transform, *IEEE Trans. Pattern Anal. Machine Intell.*, PAMI-5, No.5 (1983), pp. 493-505.

5. B. Chazelle, An Optimal Convex Hull Algorithm and New Results on Cuttings, *Proc. 32nd IEEE Symp. on Foundation of Computer Science*, (1991), pp.29-38.

6. R. O. Duda and P.E. Hart. Use of the Hough Transformation to Detect Lines and Curves in Pictures, *Comm. of the ACM*, 15, January 1972, pp.11-15.

7. H. Edelsbrunner and L.J. Guibas, Topologically Sweeping an Arrangement, *J. Comp. and Sys. Sci.*, 38 (1989), 165-194.

8. L. Guibas, M. Overmars, and J-M. Robert, The Exact Fitting Problem for Points, *Proc. 3rd CCCG*, (1991), pp.171-174.

9. P.J. Heffernan and S. Schirra. Approximate Decision Algorithms for Point Set Congruence, *Proc. 8th Symposium on Computational Geometry*, (1992), pp. 93-101.

10. P.V.C. Hough, Method and Means for Recognizing Complex Patterns, *US Patent 3069654*, December 18, (1962).

11. D.P. Huttenlocher and K. Kedem, Computing the Minimum Housdorff Distance of Point Sets under Translation, *Proc. 6th Symposium on Computational Geometry*, (1990), pp. 340-349.

12. K. Imai, S. Sumino, and H. Imai, Minimax Geometric Fitting of Two Corresponding Sets of Points, *Proc. 5th Symposium on Computational Geometry*, (1989), pp. 276-282.

13. C. Kimme, D.H. Ballard, and J. Sklansky, Finding circles by an array of accumulators, *Comm. ACM*, 18 (1975), pp. 120-122.

14. H. Maitre, Contribution to the Prediction of Performance of the Hough Transform, *IEEE Trans. Pattern Anal. Machine Intell.*, PAMI-8, No.5 (1986), pp. 669-674.

15. J. Matoušek, Range Searching with Efficient Hierarchical Cuttings, *Proc. 8th ACM Computational Geometry*, (1992), pp.276-285.

16. A. Rosenfeld and R.A. Melter, Digital Geometry, Tech. Report, CAR-323, Center for Automatic Research, University of Maryland, 1987.

17. P. Sauer, On the Recognition of Digital Circles in Linear Time, *Computational Geometry: Theory and Applications* , 2 (1993), pp. 287-302.

18. J. Sklansky, On the Hough technique for curve detection, *IEEE Trans. Comp.*, C-27, 10 (1978), pp. 923-926.

19. S. Tsuji and F. Matsumoto, Detection of ellipses by a modified Hough transformation, *IEEE Trans. Comp.*, C-27, No. 8 (1978), pp. 777-781.

How to Compute the
Voronoi Diagram of Line Segments:
Theoretical and Experimental Results

Christoph Burnikel   Kurt Mehlhorn   Stefan Schirra

Max-Planck-Institut für Informatik, 66123 Saarbrücken, Germany

Abstract. Given a set of non-intersecting (except at endpoints) line segments in the plane we want to compute their Voronoi diagram. Although there are several algorithms for this problem in the literature [Yap87, For87, CS89, BDS$^+$92, KMM90] nobody claims to have a correct implementation. This is due to the fact that the algorithms presuppose *exact* arithmetic and that the Voronoi diagram of segments requires to compute with non-rational algebraic numbers. We report about a detailed study of the numerical precision required for evaluating the geometric test exactly and about first experimental experiences. More specifically, we improve the precision bound implied by classical root separation results by more than two orders of magnitude and we compare the implementation strategies suggested by our theoretical results.

1   Introduction

The Voronoi diagram for a set S of sites partitions the plane into a set of regions one for each site. The region of site $s \in S$ contains all points in the plane that are closer to s than to any other site in S. The Voronoi diagram is one of the most useful structures in computational geometry, cf. [Aur91, OBK92] for a survey. In this paper we concentrate on the Voronoi diagram of line segments, i.e., S is a set of points and open line segments. We assume that for each segment s in S its two endpoints also belong to S.

Several algorithms for the Voronoi diagram of line segments are known; they are based on divide-and-conquer [Yap87], plane sweep [For87] and randomized incremental construction [CS89, BDS$^+$90, KMM90]. *In this paper we discuss the implementation of these algorithms.* The descriptions of all algorithms mentioned above presuppose exact real arithmetic for the implementation of the underlying geometric tests. The so-called *incircle test* is a basic tool in all algorithms: Given a Voronoi vertex defined by three sites and a fourth site decide whether the fourth site intersects, touches, or avoids the clearance circle centered at the Voronoi vertex, cf. Figure 1. Note that the coordinates of Voronoi vertices are *non-rational algebraic numbers* even when all segment endpoints have rational coordinates. How can one implement the incircle test or more generally the geometric primitives needed in an algorithm? There are three approaches.

The first approach uses floating point arithmetic. In order to perform an incircle test the (approximate) coordinates of the Voronoi vertex v are computed

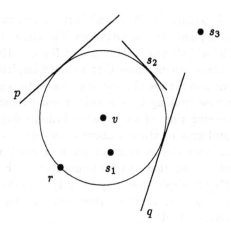

Fig. 1. The Incircle Test: The Voronoi vertex v is defined by sites p, q, and r. The three sites s_1, s_2, and s_3 intersect, touch, and avoid the clearance circle centered at v.

first and then the distances between v and one of the defining sites and between v and the new site are compared. If the distances differ less than some parameter ε then they are considered equal. The choice of the parameter ε is witch-craft. Butz and Rech [But94] and Seel [See94] have implemented the algorithm of Yap [Yap87] and Klein, Mehlhorn, and Meiser [KMM90] using this approach. The implementations work on many examples but are also known to fail on some examples. Similar experiences were reported to us by Drysdale and Dobrindt.

The second approach also uses floating point arithmetic but resolves the outcome of "doubtful" tests in a different way. Whenever floating point arithmetic does not suffice to determine the outcome of a test the outcome is fixed arbitrarily *but in a way that is consistent* with previous tests. In this way the state of the program is always legal, i.e., reachable for some input. It is not clear, however, whether this input is "close" to the real input for an appropriate definition of "close". This approach was used by Li and Milenkovic [LM92] for the convex hull problem, by Fortune [For92] for the Delaunay triangulation of point sets and by Sugihara, Ooishi, and Imai [SOI90] for the Voronoi diagram of line segments. Li and Milenkovic can prove that their algorithm computes an approximate convex hull, Fortune can show that his algorithm computes a planar graph and an embedding of it such that the embedded graph is close to the true Delaunay triangulation (however, the embedding is not necessarily planar), and Sugihara et al compute a planar graph and an embedding of it and verify experimentally that the diagram is close to the true Voronoi diagram (again, the embedding is not planar, in general). Note that only the solution for the convex hull problem can be considered fully satisfactory. However, even in this case the algorithm and in particular its analysis is quite involved so that it seems unlikely that the second approach can be carried out for many geometric problems.

The third approach uses exact arithmetic. This approach was used by Karasick, Lieber, and Nackmann [KLN91] and Fortune and van Wyk [Fv93] for the Delaunay diagram of points and by Ottmann, Thiemt, and Ullrich [OTU87] and Benoumer et al [BJMM93] and Mehlhorn and Näher [MN] for the intersection of line segments. These papers report that a careful implementation of arbitrary precision arithmetic and the use of a floating point filter yields a *provably correct* implementation whose running time is only a small constant factor (less than five) above the running time of an implementation using floating point arithmetic. For both problems mentioned above all objects computed have rational coordinates. In our case the coordinates are in general non-rational algebraic numbers. It is well known that exact computation with algebraic numbers is possible [Loo82, Mig92, Yap94] but only recently, Yap [Yap93], advocated their actual use in geometric computations. However, Yap did not report yet about any actual experiences. We do.

Theoretical Results: Let us assume that all point sites (recall that we assume that the endpoints of all segment sites are point sites) have integral homogeneous coordinates with at most k binary digits ($\equiv$ precision k). Classical root separation results (see for example [Mig92, Sec.4.3]) imply that precision about $9000k$ suffices to compute the incircle test exactly, i.e., if two distances agree in the first $9000k$ binary digits then they are equal. We improve this bound to $48k$ for the incircle test by observing that the radicals involved in the incircle test obey some algebraic identities and that the incircle test can therefore be resolved by repeated squaring.

The proof of the improved root separation bound suggests one way to compute the incircle test: Use repeated squaring and arbitray precision integer arithmetic (BIGNUM). Another possibility is to use arbitrary precision floating point arithmetic (BIGFLOAT) to compute the first $48k$ binary digits of the quantities to be compared. Clearly, if the quantities are different the difference is, on average, not in the last digit and hence much smaller precision should suffice on average. However, if the quantities are equal, the full precision is always required. This will be the case whenever the input is highly degenerate.

Blömer [Bl93] has shown that testing whether a sum of radicals is zero can be done in polynomial time. His result applies to the incircle test. In his algorithm [Bl93], the algebraic numbers involved in the incircle test have to be approximated to precision $100k$. We show that a zero test can be done by a computation with $12k$-bit integers.

This result suggests the following alternative strategy for the incircle test. Decide first whether the fourth site touches the clearance circle of the given three sites (computation with $12k$-bit integers). If not then compute the two distances to be compared (note that we know already that they are different) with precision k, $2k$, $4k$, $8k$, $16k$, $32k$, $48k$ until they are different. Both implementation strategies have their merit: The repeated squaring approach starts out with moderate precsision and needs the full precision only for the final steps of the computation. The BIGFLOAT approach might be able to decide a test by evaluating it with small accuracy. But in order to get the final result with some accuracy a larger accuracy is needed for the intermediate results.

We performed experiments with various types of data sets. The results are shown in Section 3.

2   Theory

We describe our theoretical results in the context of the randomized incremental algorithm by Klein, Mehlhorn, and Meiser [KMM90]. Following Sugihara, Ooishi, and Imai [SOI90] we first insert all point sites in random order and then all open segments in random order. This simplifies some of the geometric tests. Whenever a site s is added it is necessary to determine all features of the current diagram that conflict with s. A vertex v conflicts with s if s touches or intersects the clearance circle centered at v and an edge e conflicts with s if the Voronoi region of the new site s intersects e. The latter intersection can have one of four types: the entire edge, a single segment incident to one of the endpoints, two segments incident to the endpoints, a single segment not incident to any endpoint. In fact, the last case cannot occur since we insert all point sites first. Conflicts between vertices and sites can be decided by the incircle test, conflicts between edges and sites can be determined similarly; we are not going to treat them in this extended abstract. Once all conflicts are known it is easy to update the Voronoi diagram. This is a purely combinatorial operation. Since we insert all point sites first the standard analysis of randomized incremental constructions does not apply. However, we can still show:

Lemma 1. *The expected running time of the algorithm is $O(n \log n)$.*

Proof. (sketch): This is clear for the insertion of the point sites. For the insertion of the segment sites we observe that all conflicts can be found in time proportional to their number (there is no need to search through the history graph) and that the expected number of conflicts of the i-th segment is $O(n/i)$. The total structural change is therefore $O(n \log n)$. ☐

Let us turn to the incircle test next. Let v be a Voronoi vertex; v is defined by three sites. Here we will only discuss the case that two of the defining sites are lines and one is a point. All other cases are no more demanding computationally. We represent a point p by its homogeneous coordinates (x_p, y_p, z_p). For lines we use the equation $ax + by + c = 0$. We assume that x_p, y_p, z_p and a, b, c are k-bit integers. The bisector of a point $p = (x_p, y_p, z_p)$ and a line $\ell : ax + by + c = 0$ is given by the parabola

$$(ax + by + c)^2 = N((x - \tfrac{x_p}{z_p})^2 + (y - \tfrac{y_p}{z_p})^2),$$

and the bisector of intersecting lines $\ell_1 = a_1x + b_1y + c_1$ and $\ell_2 = a_2x + b_2y + c_2$ is given by the lines

$$\frac{a_1x + b_1y + c_1}{D_1} = \pm \frac{a_2x + b_2y + c_2}{D_2}$$

where $D_i = \sqrt{N_i}$ and $N_i = a_i^2 + b_i^2$.

Assume now that v is defined by the lines $\ell_1 = \bar{a}_1 x + \bar{b}_1 y + \bar{c}_1$ and $\ell_2 = \bar{a}_2 x + \bar{b}_2 y + \bar{c}_2$ and the point $p = (x_p, y_p, z_p)$. We transform the coordinate system such that $p = (0, 0, 1)$. This transforms the equation of line ℓ_i into

$$(\bar{a}_i z_p)x + (\bar{b}_i z_p)y + (\bar{a}_i x_p + \bar{b}_1 y_p + \bar{c}_1 z_p) = 0.$$

Hence the coefficients of the transformed lines are $(2k+2)$-bit integers. We use a_i, b_i, c_i to denote the coefficients of the transformed lines.

The Voronoi vertices (there are two) defined by the lines ℓ_1, ℓ_2, and point p are given by the intersection of a parabola and a line and have coordinates

$$\left(J + \alpha \sqrt{G}, I - \alpha \operatorname{sign}(rs)\sqrt{F}, X\right) \tag{1}$$

where $\alpha \in \{-1, 1\}$,

$$
\begin{aligned}
I &= b_1 c_2 + b_2 c_1 & J &= a_1 c_2 + a_2 c_1 \\
F &= 2c_1 c_2(D_1 D_2 + b_1 b_2 - a_1 a_2) & G &= 2c_1 c_2(D_1 D_2 + a_1 a_2 - b_1 b_2) \\
r &= a_1 D_2 - a_2 D_1 & s &= b_1 D_2 - b_2 D_1 \\
X &= D_1 D_2 - a_1 a_2 - b_1 b_2
\end{aligned}
$$

Note that $D_1 D_2 = \sqrt{(a_1^2 + b_1^2)(a_2^2 + b_2^2)}$, i.e., the coordinates involve nested roots. It is useful to observe that $|r|\sqrt{G} = |s|\sqrt{F}$. In particular $\sqrt{FG}$ is an integer. This observation is crucial for what follows. The two Voronoi vertices differ in the cyclic order in which the three defining sites occur on the clearance circle. We do not discuss here how to select α given the cyclic ordering but only mention that our approach also handles that problem.

Assume now that the fourth site is a line ℓ_3 with the equation $a_3 x + b_3 y + c_3 = 0$. a_3, b_3, c_3 are again $2k$-bit integers. We compare the square of the distance of $v = (x_s, y_s, z_s)$ and $p = (0, 0, 1)$ with the square of the distance of v and ℓ_3. So we need to decide

$$\left(\frac{x_s}{z_s}\right)^2 + \left(\frac{y_s}{z_s}\right)^2 < \left(\frac{a_3 \frac{x_s}{z_s} + b_3 \frac{y_s}{z_s} + c_3}{D_3}\right)^2. \tag{2}$$

Inequality (2) is the algebraic formulation of the considered incircle test. It can be reformulated as

$$D_3^2\left((J + \alpha\sqrt{G})^2 + (I - \alpha\operatorname{sign}(rs)\sqrt{F})^2\right) < \left(a_3(J + \alpha\sqrt{G}) + b_3(I + \alpha\operatorname{sign}(rs)\sqrt{F}) + c_3 X\right)^2$$

After rearranging terms we get an expression of the form

$$A_1 + A_2\sqrt{N} < A_3\sqrt{F} + A_4\sqrt{G} + \sqrt{N}(A_5\sqrt{F} + A_6\sqrt{G}) \tag{3}$$

where A_1 is a $12k$, A_2, A_3, A_4 are $8k$, and A_5, A_6 are $4k$-bit integers. $N = N_1 N_2$ is a $8k$-bit integer. After squaring we get an expression of the form

$$A_7 + A_8\sqrt{N} < A_9 + A_{10}\sqrt{N} \tag{4}$$

since $\sqrt{FG}$ is an integer and

$$(A_3\sqrt{F} + A_4\sqrt{G})^2 = A_3^2 F + A_4^2 G + 2A_3 A_4 \sqrt{FG}$$

$$\left(\sqrt{N}(A_5\sqrt{F} + A_6\sqrt{G})\right)^2 = N(A_5^2 F + A_6^2 G + 2A_5 A_6 \sqrt{FG})$$

$$\sqrt{N}(A_3\sqrt{F}+A_4\sqrt{G})(A_5\sqrt{F}+A_6\sqrt{G}) = \sqrt{N}(A_3 A_5 F+(A_3 A_6+A_4 A_5)\sqrt{FG}+A_4 A_6 G)$$

Rearranging (4) and squaring again leads to an integer relation. This integer relation can be decided using roughly $48k$-bit integers. Note that we have to determine the sign of both sides of a relation before we square them, but this can be done analogously and requires less precision.

Our argumentation also implies a bound on the precision required for evaluating the test exactly with floating point arithmetic. It is sufficient to compute the first $48k + O(1)$ binary digits of both sides of relation (3):

Lemma 2. *Let l be the lefthand side and r be the righthand side of relation (3). Then either $l = r$ or there are $u, v \in \mathbf{Z}$ such that $2^u \geq |l|, |r|$ and $|l - r| \geq 2^{u-v}$ where $v = 48k + O(1)$.*

Proof. (sketch): We first observe that squaring affects the number of relevant bits only by an additive constant, i.e., if the difference of a^2 and b^2 shows up in the t-th most significant bit then the difference of a and b shows up in the $(t + 1)$-th significant bit.

Lemma 3. *Let $0 \leq a < b$ with $a^2, b^2 \leq 2^s$ and $b^2 - a^2 \geq 2^{s-t}$. Then $a < b \leq 2^{s/2}$ and $b - a \geq 2^{s/2-t-1}$.*

Proof. Clearly $a, b \leq 2^{s/2}$. Also $2^{s-t} \leq (b^2 - a^2) = (b-a)(b+a) \leq (b-a) \cdot 2^{s/2+1}$. □

In the case of relation (3) we obtained an integer relation after squaring twice. So $t = s = 48k + O(1)$ after the second squaring. Applying Lemma 3 twice proves Lemma 2. □

How good is this bound? Our bound is more than two orders of magnitude better than one implied by classical root separation results. The *measure* of an algebraic number α gives a simple bound on the precision that is sufficient to decide the sign of α, see [Mig92, Sec. 4.3]. For a polynomial P with complex roots $z_1, z_2, \ldots, z_d$ and leading coefficient c_d define the measure of P by

$$M(P) = |c_d| \prod_{i=1}^{d} \max(1, |z_i|).$$

The measure $M(\alpha)$ of an algebraic number α is the measure of its minimal polynomial over $\mathbf{Z}$. We have $M(\alpha) > |\alpha| > M(\alpha)^{-1}$. However, we don't know a minimal polynomial for the algebraic numbers whose sign we are interested in. We only have a formula involving sums (and products) of radicals. The following estimates can be used to compute a bound on the degree and the measure of an algebraic number, where the degree $\deg(\alpha)$ of an algebraic number α is the degree of its minimal polynomial. Let β and β' be algebraic numbers of degrees d and d' respectively.

$$M(\beta \cdot \beta') \leq M(\beta)^{d'} \cdot M(\beta')^d \qquad \deg(\beta \cdot \beta') \leq \deg(\beta)^{d'} \cdot \deg(\beta')^d$$
$$M(\beta + \beta') \leq 2^{dd'} M(\beta)^{d'} \cdot M(\beta')^d \qquad \deg(\beta + \beta') \leq \deg(\beta)^{d'} \cdot \deg(\beta')^d$$
$$M(\sqrt[k]{\beta}) \leq M(\beta) \qquad \deg(\sqrt[k]{\beta}) \leq k \cdot \deg(\beta)$$

For $p/q \in \mathbb{Q}$, $p, q \in \mathbb{Z}$, p, q relative prime, $M(p/q) = pq$ and $\deg(p/q) = 1$. Using the estimates above we can compute a bound on the measure of an algebraic number given as an arithmetic expression involving roots. Direct application of the estimates above to the expression $A_1 + A_2\sqrt{N} + A_3\sqrt{F} + A_4\sqrt{G} + \sqrt{N}(A_5\sqrt{F} + A_6\sqrt{G})$ of relation (3) gives a bound on its measure of $2^{86016k+1024}$. Rearranging the expression to $A_1 + A_2\sqrt{N} + \sqrt{F}(A_3 + A_5\sqrt{N}) + \sqrt{G}(A_4 + A_6\sqrt{N})$ leads to a "better" bound. of $2^{9216k+1024}$, i.e., roughly 9000-fold precision is sufficient to decide relation (3). It is likely that this bound can be improved using arguments of Dube and Yap [DY94]. In contrast our bound is $48k + O(1)$.

Can our bound be improved further? There is some indication that it can. Our argumentation applied to the incircle test for four points yields a precision bound of $20k$. However, this particular test can be done by lifting the points onto a paraboloid of revolution and checking the orientation of the lifted points. This approach requires only precision $8k$.

The derivation of the precision bound directly translates into an exact algorithm for the incircle test (3). We only have to use a BIGNUM package and to go through the steps of squaring and rearranging sums outlined above. We report on our experiences with that method below.

An alternative method to implement the incircle test is to compute the first $48k + O(1)$ binary digits of both sides of relation (3) using a BIGFLOAT package. It is reasonable to expect that if the two sides of (3) are distinct then much smaller precision will suffice to detect that fact. We report on some experimental evidence below.

However, if the sides of (3) are equal the full $48k$-bit precision is required to detect that fact. Blömer has shown that testing whether a sum of radicals is zero can be done in polynomial time [Bl91, Bl93]. A similar result for determining the sign of such a sum is not known. This indicates that testing for equality might be easier then determining the sign and that it could be wise to test an expression involving algebraic numbers for zero before a more expensive computation of its sign. The algorithm of Blömer can be applied in our case. Applying his algorithm directly does not give an improvement. Corollary 6.4 in [Bl93] implies that all roots appearing in relation (3) have to be approximated to precision $112k$. We already know that precision $48k$ suffices to decide relation (3). However, using a distinct and more careful analysis we can show that the equality test in relation (3) can be performed by a computation involving only $12k$-bit integers. Details will be given in the full paper.

3   Experimental Results

We have implemented two algorithms RS and FF for the incircle test of Equation (3) in C++.

Algorithm RS uses exact integer arithmetic provided by a BIGNUM package. In order to reduce the overhead of this BIGNUM package we first try to decide relation (3) by computing with doubles. For each appearing type of algebraic numbers we introduced appropriate C++ classes and overloaded the operations $+, -, *$. In addition, we had routines that compute the sign of these numbers by repeated rearranging, sign testing and squaring of the involved expressions.

Algorithm FF is based on a BIGFLOAT package. An element of the corresponding class bigfloat is a floating point number whose maximal number of bits can be chosen by the user. The operations $+, -, *, \sqrt{}$ are carried out within this chosen precision. In order to decide whether the BIGFLOAT evaluation of inequality (3) with precision p leads to a reliable sign test, we used static error analysis similar to the error analysis used in [FvW93] and [MN]: The absolute error in a BIGFLOAT evaluation of an expression E was estimated by

$$Ind(E) * Max(E) * 2^{-p} \qquad (5)$$

where $Max(E)$ is an a priori bound on the value of E, p the current precision and $Ind(E)$ a (small) constant. Our error analysis implies that precision $2k+3$ is always required. We started algorithm FF with this precision and then increased the precision to $3k, 4k, \ldots$ until the sign was reliable.

Experimental Results: We tested the implementations with three types of input data: easy data, realistic data, and difficult data. The endpoints of the line segments were given by their cartesian coordinates (x, y) where x, y are k-bit integers.

- Easy data: We considered $k = 10$, 20, and 30. Point sites and segment endpoints defining the lines involved were chosen randomly and uniformly.
- Realistic data: We used polygonal environments that have been created as examples for a motion planning algorithm that moves a disc along the Voronoi diagram of the polygonal environment, see Figures 2–13. In all examples $k = 10$.
  We run Seel's implementation of the randomized incremental construction of Klein et al. [KMM90] on these examples and kept a log of all incircle tests performed. The log-file was then used as the input file for algorithms RS and FF.
- Difficult data: We considered again $k = 10, 20, 30$. In all these tests, relation (3) was satisfied with equality so that we always needed full precision.

Let us first consider the easy data (Table 1). The items in the columns $p = c \cdot k$ are the numbers of tests which could be decided with precision $p = c \cdot k$ but not with precision $(c - 1) \cdot k$. Although algorithm FF required very low precision, algorithm RS was faster, since it could make all decisions using doubles and never needed BIGNUMs.

		Precision p				Time	
k	#tests	2k + 3	3k	4k	5k	RS	FF
10	10000	2781	6801	398	20	26	80
20	10000	9962	38			25	38
30	10000	all				26	70

Table 1. Easy data: number of tests decided with precision p (in algorithm FF) and running times of both algorithms in seconds (on a SUN SPARCstation 10)

file	#tests	≤ 3k	4k	5k	6k	7k	8k	9k	RS	FF
B1	387	8	122	136	75	9			3.9	2.5
B2	398	45	204	134	12	3			2.3	1.9
B3	918	10	224	620	64				6.3	7.8
B4	1126	35	449	628	14				6.6	6.8
B5	4149		125	2310	1697	17			47	44
B6	13021		23	1266	10376	1310	42	4	141	138
B7	140	25	80	35					0.73	0.95
B8	276	3	63	197	13				2.1	2.4
B9	376		17	191	143	13	12		4.0	3.8
B10	414		46	235	119	14			4.3	3.8
B11	912	11	21	217	641	18	4		11.5	8.9
B12	385			2	7	341	30	5	4.9	15.2

Table 2. Realistic data: number of tests decided with precision p (in algorithm FF) and running times of both algorithms in seconds (on a SUN SPARCstation 10)

The results for the realistic data are shown in Table 2. The last 6 examples were specifically created to contain near-degeneracies. These examples used higher precision on average, but the required precision is quite low: precision $9k$ was sufficient for all of our tests. If we take into account that the points were not given in homogeneous form (i.e. their z-coordinate was 1), the general precision bound derived in Section 2 is $32k$ instead of $48k$. This is still much more than we actually needed. Furthermore the running times of algorithms FF and RS are comparable.

The difficult data (Table 3) show the worst case behaviour of the algorithms. Algorithm FF makes a lot of lower precision computations before it reaches the maximal precision.

Conclusion. In practice all our computations for non–degenerate inputs could be made with very moderate precision. Both the approach using exact integer arithmetic and the approach using BIGFLOATs have their advantages. The BIGFLOAT solution is more flexible and is supposed to work for other problems

k	#tests	Time RS	Time FF
10	1000	11	40
20	1000	12	60
30	1000	14	97

Table 3. Difficult data: running times of both algorithms in seconds (on a SUN SPARCstation 10)

as well while the effort for the adaptation of the symbolic computation machinery is higher. On the other hand, the integer solution is better suited for degenerate inputs.

References

[Aur91]   F. Aurenhammer. Voronoi diagrams: a survey of a fundamental geometric data structure. *ACM Comput. Surv.*, 23:345–405, 1991.

[BDS+90]   J.D. Boissonnat, O. Devillers, R. Schott, M. Teillaud, and M. Yvinec. Applications of random sampling to on-line algorithms in computational geometry. Technical report, INRIA, 1990.

[BDS+92]   J.-D. Boissonnat, O. Devillers, R. Schott, M. Teillaud, and M. Yvinec. Applications of random sampling to on-line algorithms in computational geometry. *Discrete and Computational Geometry*, 8:51–71, 1992.

[BJMM93]   M.O. Benouamer, P. Jaillon, D. Michelucci, and J-M. Moreau. A "lazy" solution to imprecision in computational geometry. In *5th Canadian Conf. on Computational Geometry*, pages 73–78, 1993.

[Bl91]   J. Blömer. Computing sums of radicals in polynomial time. In *FOCS91*, pages 670–677, 1991.

[Bl93]   J. Blömer. Computing sums of radicals in polynomial time. Technical Report B93-13, Freie Universität Berlin, 1993.

[But94]   B. Butz. Robuste Implementierung eines Algorithmus zur Berechung eines Voronoi-Diagrams für Polygone. Diplomarbeit, 1994.

[CS89]   K. L. Clarkson and P. W. Shor. Applications of random sampling in computational geometry, II. *Discrete and Computational Geometry*, pages 387–421, 1989.

[DY94]   T. Dube and C.K. Yap. A basis for implementing exact computational geometry. extended abstract, 1994.

[For87]   S. J. Fortune. A sweepline algorithm for Voronoi diagrams. *Algorithmica*, 2:153–174, 1987.

[For92]   S. Fortune. Numerical stability of algorithms for 2d Delaunay triangulations and Voronoi diagrams. In *ACM Symposium on Computational Geometry*, volume 8, pages 83–92, 1992.

[Fv93]   S. Fortune and C. van Wyk. Efficient exact arithmetic for computational geometry. *Proc. of the 9th Symp. on Computational Geometry*, pages 163–171, 1993.

[FvW93]   S. Fortune and C. van Wyk. Efficient exact arithmetic for computational geometry. In *Proc. of the 9th ACM Symp. on Computational Geometry*, pages 163–172, 1993.

[KLN91]   M. Karasick, D. Lieber, and L.R. Nackman. Efficient Delaunay triangulation using rational arithmetic. *ACM Transactions on Graphics*, 10(1):71–91, 1991.

[KMM90]   R. Klein, K. Mehlhorn, and S. Meiser. On the construction of abstract Voronoi diagrams ii. In *Proc. SIGAL Symp. on Algorithms*, Tokyo, 1990. Springer Verlag. LNCS 450.

[LM92]   Z. Li and V. Milenkovic. Constructing strongly convex hulls using exact or rounded arithmetic. *Algorithmica*, 8:345–364, 1992.

[Loo82]   R. Loos. Computing in algebraic extensions. In B. Buchberger, G.E. Collins, and R. Loos, editors, *Computer Algebra*, pages 173–187. Springer Verlag, 1982.

[Mig92]   M. Mignotte. *Mathematics for Computer Algebra*. Springer Verlag, 1992.

[MN]   K. Mehlhorn and S. Näher. Implementation of a sweep line algorithm for the segment intersection problem. manuscript.

[OBK92]   A. Okabe, B. Boots, and Sugihara K. *Spatial tessellations: concepts and applications of Voronoi diagrams*. Wiley, New York, 1992.

[OTU87]   T. Ottmann, G. Thiemt, and C. Ullrich. Numerical stability of geometric algorithms. In *Proc. of the 3rd ACM Symp. on Computational Geometry*, pages 119–125, 1987.

[OY85]   C. O'Dúnlaing and C. Yap. A "retraction" method for planning the motion of a disk. *Journal of Algorithms*, 6:104–111, 1985.

[See94]   M. Seel. Ausarbeitung und Implementierung eines Algorithmus zur Konstruktion abstrakter Voronoi-Diagramme. Diplomarbeit, 1994.

[SOI90]   K. Sugihara, Y. Ooishi, and T. Imai. Topology-oriented approach to robustness and its applications to several Voronoi-diagram algorithms. In *Proc. 2nd Canad. Conf. Comput. Geom.*, pages 36–39, 1990.

[Yap87]   C. Yap. An $O(n \log n)$ algorithm for the Voronoi diagram of a set of simple curve segments. *Discrete and Computational Geometry*, 2:365–393, 1987.

[Yap93]   C.K. Yap. Towards exact geometric computation. In *5th Canadian Conf. on Computational Geometry*, pages 405–419, 1993.

[Yap94]   C.K. Yap. *Fundamental Problems in Algorithmic Algebra*. Princeton University Press, 1994.

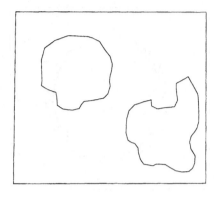

Fig. 2. B1

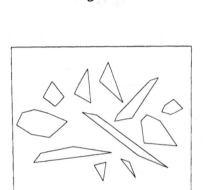

Fig. 3. B2

Fig. 4. B3

Fig. 5. B4

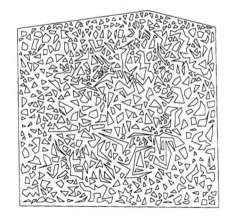

Fig. 6. B5

Fig. 7. B6

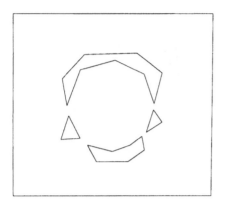

Fig. 8. B7

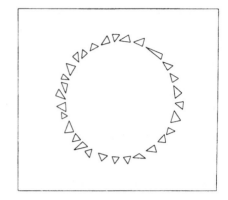

Fig. 11. B10

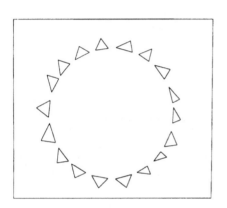

Fig. 9. B8

Fig. 12. B11

Fig. 10. B9

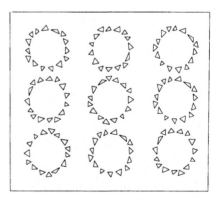

Fig. 13. B12

Range Searching and Point Location among Fat Objects *

Mark H. Overmars   A. Frank van der Stappen

Dept. of Computer Science, Utrecht University,
P.O. Box 80.089, 3508 TB Utrecht, The Netherlands.

Abstract. We present a data structure that can store a set of disjoint fat objects in 2- and 3-space such that point location and bounded-size range searching with arbitrarily shaped regions can be performed efficiently. The structure can deal with either arbitrary (fat) convex objects or non-convex polygonal/polyhedral objects. For dimension $d = 2, 3$, the multi-purpose data structure supports point location and range searching queries in time $O(\log^{d-1} n)$ and requires $O(n \log^{d-1} n)$ storage, after $O(n \log^{d-1} n \log \log n)$ preprocessing. The data structure and query algorithm are rather simple. The results are likely to be extendible in many directions.

1 Introduction

Fatness turns out to be an interesting phenomenon in computational geometry. Several papers present surprising combinatorial complexity reductions [2, 13, 19] and efficiency gains for algorithms [15, 20] if the objects under consideration have a certain fatness. Fat objects are compact to some extent, rather than long and thin. Fatness is a realistic assumption, because in many practical instances of geometric problems the considered objects are fat. The aim of studying fatness is to find new fast and simple algorithms or to demonstrate enhanced efficiency of existing algorithms for such practical instances. In this paper we study two fundamental problems in computational geometry in a context of fat objects: point location and range searching. The point location problem aims at preprocessing a set of disjoint geometric objects for efficiently reporting the specific object containing a query point. The objective of the (general) version of the range searching problem is to preprocess a set of geometric objects for quickly reporting all objects intersecting some query range (e.g. a rectangle, a simplex, or a hypersphere). We show that sets of disjoint fat arbitrary convex and/or polygonal/polyhedral non-convex objects in two- or three-dimensional space ($d = 2, 3$) can be preprocessed in time $O(n \log^{d-1} n \log \log n)$ into a data structure of size $O(n \log^{d-1} n)$ which supports point location queries *and* range searching queries with arbitrarily-shaped but bounded-size regions in time $O(\log^{d-1} n)$. The data structure is based on the structure by Overmars for point location in fat subdivisions [15]. Let us review some relevant results in both point location and range searching to place our results in a broader perspective.

Point location in 2-space has been studied extensively and is solved in a satisfactory way for many types of scenes, as several solutions achieve logarithmic query time and

---

* Research is supported by the Dutch Organization for Scientific Research (N.W.O.) and by the ESPRIT III BRA Project 7141 (ALCOM II).

(near-)linear storage, after (near-)linear preprocessing time. In 3-space, on the contrary, efficient solutions are available only for restricted problem instances. Chazelle [4] obtains $O(\log^2 n)$ query time and an $O(n)$ storage requirement for the case where the stored geometric objects are the 3-cells of a more general spatial subdivision, consisting of a total of n facets and satisfying the restrictive constraint that the vertical dominance relation on its cells is acyclic. Preparata and Tamassia [18] consider point location in a set of disjoint convex polyhedra with total complexity n. Their data structure uses $O(n \log^2 n)$ storage and is capable of answering point location queries in time $O(\log^2 n)$ after $O(n \log^2 n)$ preprocessing time. (The storage requirement can be improved to $O(n \log n)$ without affecting the query time [8].) Apparently, efficient solutions for sets of non-convex or non-polyhedral objects in 3-dimensional space are lacking.

Nearly all papers on range searching discuss how to preprocess a very elementary class of geometric objects, namely points, for efficiently answering range search queries with specific range types. The most extensively studied type of query range is the orthogonal range, and a long-established result says (see e.g. [17]) that a set of n points can be preprocessed in time $O(n \log^{d-1} n)$ into a data structure of size $O(n \log^{d-1} n)$, which is capable of answering an orthogonal range query in time $O(\log^{d-1} n)$. (Actually, some slight further improvements are possible.) More complicated types of ranges, like simplices, are much harder to deal with. For example, the best data structures for triangular range queries in a set of planar points need $O(\sqrt{n})$ query time when only near-linear storage is allowed. To obtain poly-logarithmic query time close to quadratic preprocessing and storage is required. In 3-space the bounds become even worse. Problems become even more complicated once the set of objects no longer consists of points only. (See [1, 6, 7, 10, 11, 12] for some results.) In practice, we normally do not need the generality that these structures provide, as the objects to be stored typically have favorable properties like fatness, and also the query ranges are normally not completely general. For example, in many applications the query ranges are rather small. Hence, it is important to study these problem under such restrictions.

Overmars [15] discusses a data structure for efficient and simple point location in fat subdivisions or sets of disjoint fat objects with total complexity n. The structure supports queries in time $O(\log^{d-1} n)$ and uses $O(n \log^{d-1} n)$ storage. In this paper we show that, at least for $d = 2, 3$, the data structure can be used to answer range queries with arbitrarily-shaped but bounded size ranges as well. To this end we prove that, under the condition of fatness of the stored objects, each bounded-size range query can be solved using a constant number of point location queries with carefully chosen query points, leading to a query time of $O(\log^{d-1} n)$. We also show that, using such range queries, we can efficiently construct the data structure (a problem left open in [15]).

The paper is organized as follows. In Section 2 we give the definition of fatness we use, along with a key result for scenes of fat objects. Section 3 discusses Overmars' data structure for point location among fat objects, while Section 4 shows how the structure can be used for fast range searching among useful classes of objects in 2- and 3-space. The results are subsequently used to support the incremental construction of the multi-purpose data structure, starting from the largest object and repeatedly adding the next largest object. Section 5 summarizes the results and points out potential generalizations to e.g. higher dimensions and sets of slightly penetrating objects.

2 Fatness

Contrary to many other definitions of fatness in literature, the definition given below [19] applies to general shapes in arbitrary dimension. It involves a parameter k which supplies a qualitative measure of fatness: the smaller the value of k, the fatter the object must be.

Definition 1. Let $E \subseteq \mathbf{R}^d$ be an object and let k be a positive constant. The object E is k-fat if for all hyperspheres S centered inside E and not fully containing E:

$$k \cdot volume(E \cap S) \geq volume(S).$$

The definition forbids fat obstacles to be long and thin or to have long or thin parts.

Before we give a crucial result on the local object density in scenes of fat objects, we emphasize that we find the size, or more specifically the radius, of the minimal enclosing hypersphere of an object, the most convenient among the many ways to express or measure the size of an object.

Theorem 2. *Let $\mathcal{E}$ be a set of non-intersecting k-fat objects in $\mathbf{R}^d$ with minimal enclosing hypersphere radii at least ρ, and let $c \in \mathbf{R}^+$ be a constant. Then the number of objects $E \in \mathcal{E}$ intersecting any region R with minimal enclosing hypersphere radius $c \cdot \rho$ is bounded by a constant.*

Informally, the theorem states that the number of k-fat objects intersecting a region, that is not too large compared to the objects, is constant.

3 Point Location among Fat Objects

This section discusses a data structure for point location among disjoint constant-complexity fat objects by Overmars [15]. Overmars presents the data structure as a structure for solving the problem of point location in subdivisions of d-dimensional space into fat constant-complexity cells. The answer to the query is the specific cell containing the query point. The point location problem in fat subdivisions can be seen as an instance of the following, more general, formulation of the point location problem: *Given a set $\mathcal{E}$ of non-intersecting constant-complexity k-fat objects in $\mathbf{R}^d$ and a query point $p \in \mathbf{R}^d$, report the object $E \in \mathcal{E}$ that contains p, or report that no such object exists.* If the objects in $\mathcal{E}$ entirely cover $\mathbf{R}^d$ the we obtain a fat subdivision like in Overmars' paper. In the more general setting, the complement of the objects need not be fat, nor does it have constant complexity.

Overmars' paper only presents a data structure for efficiently answering point location queries; the issue of building the structure remains untouched. We will give a solution for this problem for convex and polygonal/polyhedral objects in $\mathbf{R}^2$ and $\mathbf{R}^3$. The solution relies on the ability to do efficient range searching queries among these objects. Before discussing this problem and its solution and their application to building the point location structure, this subsection simply summarizes the results in [15].

Let us assume that the constant-complexity k-fat objects in $\mathcal{E}$ are ordered by radius of their minimal enclosing hyperspheres: $E_1, \ldots, E_n$. Let furthermore ρ_i be the radius of

E_i's minimal enclosing hypersphere. Let us denote the axis-parallel enclosing hypercube of the minimal enclosing hypersphere of an object E_i by C_i. By construction, the hypercubes are ordered by increasing size. Note that the side length of the hypercube C_i is $2\rho_i$. Notice that the hypercubes C_i may be overlapping, although the objects E_i are disjoint. Furthermore, let V_i be defined as follows:

$$V_i = \{E_j \in \mathcal{E} | E_j \cap C_i \neq \emptyset \wedge j \geq i\}.$$

Hence, V_i is the set of objects that are larger than E_i (i.e., with larger minimal enclosing hypersphere) intersecting the box C_i. Theorem 2 immediately supplies a useful property of the sets V_i, namely that $|V_i| = O(1)$, for all $1 \leq i \leq n$.

Assuming that the hypercubes C_i and the sets V_i for all $1 \leq i \leq n$ are available, we proceed as follows to find the answer to a point location query with a point $p \in \mathbf{R}^d$. Determine the smallest hypercube, if any, containing p. If no hypercube contains p than p lies in no object; if, on the contrary, C_i is the smallest hypercube containing p, then the set V_i provides the answer to the query (see [15] for a proof). To this end, we check the objects in V_i for containment of p. Note that the check for containment of a point in any object $E_j \in V_i$ takes constant time due to the constant complexity of $E_1, \ldots, E_n$ and the constant cardinality of V_i. If no object in V_i contains p then no object in $\mathcal{E}$ contains p; otherwise the unique object $E_j \in V_i$ containing p obviously is the answer to the query. As the inspection of V_i takes constant time, the point location query time is dominated by the time to find the smallest hypercube C_i containing the query point p. The point location problem is now essentially reduced to the following priority point stabbing problem among (intersecting) hypercubes: *Given a set of hypercubes C in $\mathbf{R}^d$ and a query point $p \in \mathbf{R}^d$, report the smallest hypercube $C \in C$ containing p, or report that no hypercube in C contains p.*

To solve a priority point stabbing query among hypercubes, Overmars proposes a d-level data structure in which the upper $d - 1$ levels are segment trees and the lowest level is a list or a balanced binary tree. Searching the multi-level data structure takes $O(\log^d n)$ time, which can be improved by applying fractional cascading [5] to $O(\log^{d-1} n)$. The structure uses $O(n \log^{d-1} n)$ storage.

Theorem 3. *A set $\mathcal{E}$ of non-intersecting constant-complexity k-fat objects in $\mathbf{R}^d$ can be stored in a data structure of size $O(n \log^{d-1} n)$, such that, for a query point $p \in \mathbf{R}^d$, it takes $O(\log^{d-1} n)$ time to report the object $E \in \mathcal{E}$ that contains p, or to conclude that no object contains p.*

Overmars is unable to construct the data structure efficiently. Actually, the construction of the multi-level data structure can easily be performed in time $O(n \log^{d-1} n)$ using standard techniques. The problem lies in the computation of the sets V_i. Finding the objects E_j with $j \geq i$ intersecting the hypercube C_i requires a range search query with C_i. The query is not an ordinary range search query, since we are only interested in objects with a certain minimal size (or index). Performing a range search query among all objects and subsequently filtering out the smaller objects is not a good idea, as the answer to the range query might by orders of magnitude larger than V_i. A better idea would be to perform a range search query with V_i only among objects that are larger than E_i. Surprisingly, we show in the next section that it is possible, in most interesting cases, to use the point location structure itself for solving the range search query efficiently. This suggests an approach where we add the hypercubes from large to small, meanwhile computing the sets V_i in the following (incremental) way: use, before adding the hypercube C_{n-m}, the multi-level data structure storing the hypercubes $C_{n-m+1}, \ldots, C_n$ to perform a range search query among the objects $E_{n-m+1}, \ldots, E_n$ (and, hence, to compute V_{n-m}) with the hypercube C_{n-m}. See Section 4.3 for details.

4   Range Searching by Point Location

In this section we use the point location data structure to tackle the range searching problem: *Given a set $\mathcal{E}$ of disjoint constant-complexity k-fat objects in $\mathbf{R}^d$ with minimal enclosing hyperspheres with radii at least ρ and a constant-complexity query region R with diameter at most $h \cdot \rho$ for some positive constant h, report all objects $E \in \mathcal{E}$ that intersect R.* Using Theorem 2 it is easy to verify that the answer to the query with a region R, satisfying the diameter bound, is a set of objects of constant cardinality. Let us define the set $Q(R)$ of objects intersecting the region R:

$$Q(R) = \{E \in \mathcal{E} | E \cap R \neq \emptyset\}.$$

We will now show how to use the point location structure to solve the stated problem in time $O(\log^{d-1} n)$ for $d = 2, 3$ and $\mathcal{E}$ being a set of convex or polygonal/polyhedral objects. The solution relies on local properties of fat objects. The definition of fatness requires a k-fat object E to have a large 'density' in the vicinity of any point p in the object: $1/k$-th of a hypersphere centered at p is covered by E. One gets the feeling that this property makes it possible to hit any object or object part with a certain minimum size, regardless of its exact location, with at least one point from a sufficiently dense, but not too large, pattern of sample points. To structure the problem and the shape of its solution, we restrict the sample points to be arranged as a regular orthogonal grid.

Definition 4. A regular orthogonal grid $\mathcal{G}_r$ with resolution r is defined by: $\mathcal{G}_r = \{(z_1 r, \ldots, z_d r) | z_1, \ldots, z_d \in \mathbf{Z}\}$.

We show the existence of a grid resolution such that a small subset of the corresponding regular orthogonal grid is guaranteed to hit any object E having non-empty intersection with the query region R. As the ability to find a bound on the resolution is very much influenced by the type of object under consideration, we treat the different types of objects one by one. We examine 2D arbitrary convex objects, 2D polygonal nonconvex objects, 3D arbitrary convex objects, and 3D polyhedral objects. The following general property, however, is helpful in each of the cases.

Property 5. *Any hypersphere with radius at least r contains at least one point of the orthogonal grid $\mathcal{G}_{2r/\sqrt{d}}$.*

The main implication of these results is that the intersection query with the range R can be solved by a bounded number of point location queries, each taking $O(\log^{d-1} n)$ time using the data structure for point location among fat objects.

4.1 Two-dimensional Objects

The aim in this subsection is to find a grid resolution r establishing that each planar object $E \in Q(R)$ is hit by at least one point in the subset $\Pi \subset \mathcal{G}_r$, where, preferably, the size of Π depends on k and h (and the complexity of the individual objects) only. Before we focus on the different types of objects in 2D, we first give a result that eases the task of finding a grid resolution. It investigates the largest possible sphere that will fit in any triangle with some minimum area A. Clearly, we can construct a very long and thin triangle with area A that is not guaranteed to contain any circle with positive radius. So, the restriction on the area alone is not enough. We add the restriction that no triangle edge is longer than δ.

Lemma 6. *Let T be a triangle with $area(T) \geq A$ and let none of its edges be longer than δ. Then T contains a circle with radius at least $\frac{2A}{3\delta}$.*

Below we use Property 5 and the above lemma to find a small subset Π of a sufficiently dense grid, so that the points of Π hit all objects $E \in Q(R)$.

Convex 2D Objects Let E be a two-dimensional convex k-fat object intersecting the query region R, and let $m \in E \cap R$. The circle $S_{m,\rho}$ (centered at m and with radius ρ) does not fully contain E because otherwise it would contradict the assumption that all minimal objects enclosing circles have radii at least ρ. Hence, the k-fatness of E yields:

$$k \cdot area(E \cap S_{m,\rho}) \geq area(S_{m,\rho}) = \pi\rho^2. \tag{1}$$

The shape $E \cap S_{m,\rho}$ is convex as it is the intersection of the convex objects E and $S_{m,\rho}$. The following lemma holds for any convex two-dimensional shape.

Lemma 7. *Any convex object $E \subseteq \mathbf{R}^2$ contains a triangle T with $4 \cdot area(T) \geq area(E)$.*

Application of Lemma 7 to the shape $E \cap S_{m,\rho}$, satisfying (1), implies the containment of a triangle $T \subseteq E \cap S_{m,\rho}$ such that

$$area(T) \geq \frac{\pi\rho^2}{4k}. \tag{2}$$

The length of the edges of T is bounded by 2ρ because T is contained in the circle $S_{m,\rho}$. Lemma 6 implies, in combination with the upper bound on the length of the edges of T and the lower bound on the T's area given by (2), that T contains a circle S with radius at least $\frac{\pi\rho}{12k}$. Subsequent application of Property 5, yields that such a circle is hit by at least one point from the regular orthogonal grid with resolution $\frac{\pi\rho}{6k\sqrt{2}}$, or $\mathcal{G}_{\pi\rho/6k\sqrt{2}}$ for short. Hence, at least one point from $\mathcal{G}_{\pi\rho/6k\sqrt{2}}$ hits $E \cap S_{m,\rho} \supseteq T \supseteq S$.

So far, we have only bothered about finding a sufficiently high resolution for a grid to hit all objects $E \in Q(R)$. Clearly, it is unnecessary and even undesirable to perform point locations with a too large subset of the grid, both because it increases the query time and because it would lead to many accidental hits of objects $E \notin Q(R)$. Fortunately, the size of the sample set (and the number of accidental hits) can be adequately limited by

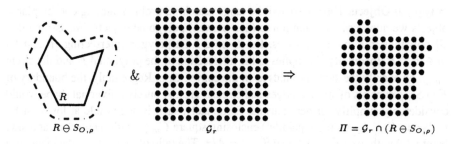

Fig. 1. The construction of the sample set Π from the query region R and the grid $\mathcal{G}_r$.

a quick glance at the deduction of the grid resolution: the resolution is chosen such that any object $E \in Q(R)$ is hit inside a circle $S_{m,\rho}$ with $m \in R$. This circle lies entirely inside the region $R \ominus S_{O,\rho}$, where $\ominus$ denotes the Minkowski difference operator[2] and O is the origin of the Euclidean coordinate frame. Hence, the grid point hitting E lies in $R \ominus S_{O,\rho}$. As a result, it suffices to restrict the set of sample points Π to:

$$\Pi = \mathcal{G}_{\pi\rho/6k\sqrt{2}} \cap (R \ominus S_{O,\rho}).$$

Lemma 8. *Let $\mathcal{E}$ be a set of two-dimensional convex k-fat objects with minimal enclosing circles with radii at least ρ and let R be a region with diameter $h \cdot \rho$. The set $Q(R)$ of objects $E \in \mathcal{E}$ intersecting R can be found by point location queries with the points from $\mathcal{G}_{\pi\rho/6k\sqrt{2}} \cap R \ominus S_{O,\rho}$.*

The data structure presented in Section 3 allows us to perform point location queries among the objects of $\mathcal{E}$ with all points in Π in time $O(|\Pi| \log n)$. The resulting set $\{E \in \mathcal{E} | \Pi \cap E \neq \emptyset\}$ of answers, which clearly has at most $|\Pi|$ elements, is a superset of the answer $Q(R)$ to the range query with R. For each of the objects $E \in \{E \in \mathcal{E} | \Pi \cap E \neq \emptyset\}$, additional constant time suffices to verify the membership $E \in Q(R)$ by a simple test for the non-emptiness of $E \cap R$, provided that R and all $E \in \mathcal{E}$ have constant complexity. Hence, the computation of $Q(R)$ takes $O(|\Pi| \log n)$ time.

To get an idea of the number of elements in Π and the factors that influence the size of Π, realize that $R \ominus S_{O,\rho}$ fits entirely in a square with side length $(h+2)\rho$. As a result, the number of elements in Π is bounded by the number of grid points in the enclosing square, leading to $|\Pi| \leq (6\pi^{-1}k\sqrt{2} \cdot (\lfloor h \rfloor + 2))^2 = O(k^2 h^2)$. In a setting where all objects are k-fat for some constant k and the diameter of the query region R does not exceed a constant multiple h of ρ, the cardinality of Π is constant: $|\Pi| = O(1)$, which implies an $O(\log n)$ time bound for range searching with a bounded-size region R.

Theorem 9. *Let $k > 0$ and $h \geq 0$ be constants, and let $\mathcal{E}$ be a set of disjoint two-dimensional convex k-fat constant-complexity objects with minimal enclosing circles with radii at least ρ. A range query with a region R with diameter $h \cdot \rho$ among $\mathcal{E}$ takes $O(\log n)$ time.*

---

[2] The Minkowski difference of two sets A and B is defined by $A \ominus B = \{a - b | a \in A \wedge b \in B\}$.

Polygonal Objects Having solved the range searching problem among convex planar objects we now turn our attention to (non-convex) polygonal objects. We assume that all polygons in $\mathcal{E}$ have c vertices. Let E be a k-fat polygon intersecting R, and let $m \in E \cap R$. Inequality (1) applies to the intersection of the polygonal E and the circle $S_{m,\rho}$, on exactly the same grounds as in the convex case. Regrettably, the boundary of $E \cap S_{m,\rho}$ consists of straight edges and circular arcs. A purely polygonal shape would considerably simplify further analysis. Fortunately, such is achievable at low cost by replacing $S_{m,\rho}$ by its (axis-parallel) enclosing square $C_{m,\rho}$ (with 'center' m and side length 2ρ) with $area(C_{m,\rho})/area(S_{m,\rho}) = 4/\pi$. The ratio of these areas, the inequality (1), and the obvious inequality $area(E \cap C_{m,\rho}) \geq area(E \cap S_{m,\rho})$, together yield

$$\frac{4k}{\pi} \cdot area(E \cap C_{m,\rho}) \geq area(C_{m,\rho}) = 4\rho^2. \tag{3}$$

The intersection $E \cap C_{m,\rho}$ has a polygonal shape. A vertex of the boundary of $E \cap C_{m,\rho}$ is either one of the four corners of $C_{m,\rho}$ or contributed by one of the edges of E and their endpoints. It is easily verified that each edge e and its two endpoints can contribute at most two vertices to the boundary of $E \cap C_{m,\rho}$, each vertex being either an endpoint of e or an intersection point of $C_{m,\rho}$ and e's interior. In summary, $E \cap C_{m,\rho}$ is a polygon with at most $2c + 4$ vertices. As a result, the largest triangle T in the triangulation of $E \cap C_{m,\rho}$ satisfies, using (3):

$$area(T) \geq \frac{1}{2c + 4} \cdot area(E \cap C_{m,\rho}) \geq \frac{\pi\rho^2}{k(2c + 4)}. \tag{4}$$

The largest triangle T lies entirely inside $C_{m,\rho}$, which implies that the length of its edges cannot exceed $2\rho\sqrt{2}$. The bounded edge lengths and area of T make it suited for application of Lemma 6. This lemma establishes that T contains a circle with radius $\frac{\pi\rho}{6k\sqrt{2}(c+2)}$. Application of Lemma 5 yields that such a circle is hit by the regular orthogonal grid with resolution $\frac{\pi\rho}{6k(c+2)}$. The grid point that hits an object $E \in Q(R)$ lies inside the square $C_{m,\rho}$ which, in turn, due to $m \in E \cap R$, lies completely inside the Minkowski difference $R \ominus C_{O,\rho}$. Hence, we may define Π to be as follows:

$$\Pi = \mathcal{G}_{\pi\rho/6k(c+2)} \cap R \ominus C_{O,\rho}.$$

Lemma 10. *Let $\mathcal{E}$ be a set of two-dimensional k-fat polygons with at most c vertices and minimal enclosing circles with radii at least ρ and let R be a region with diameter $h \cdot \rho$. The set $Q(R)$ of objects $E \in \mathcal{E}$ intersecting R can be found by point location queries with the points from $\mathcal{G}_{\pi\rho/6k(c+2)} \cap R \ominus C_{O,\rho}$.*

Similar to the convex case, the sequence of point locations with all points in Π takes $O(|\Pi| \log n)$ time and results in the set $\{E \in \mathcal{E} | \Pi \cap E \neq \emptyset\}$. The extraction of $Q(R)$ from this set takes $O(|\Pi|)$ time under the additional assumption that R and all $E \in \mathcal{E}$ have constant complexity, so c must be constant. To bound the number of elements in Π, we notice that $R \ominus C_{O,\rho}$ also fits completely in some enclosing square with side length $(h + 2)\rho$. The number of grid points in the square bounds the number of elements in Π by $|\Pi| \leq (6\pi^{-1}k(c + 2) \cdot (\lfloor h \rfloor + 2))^2 = O(k^2h^2c^2)$. In our framework of constant c, k, and h this reduces to $|\Pi| = O(1)$, which implies logarithmic query time for range searching with bounded-size ranges among fat objects.

Theorem 11. *Let $k > 0$ and $h \geq 0$ be constants, and let $\mathcal{E}$ be a set of disjoint two-dimensional k-fat constant-complexity polygons with minimal enclosing circles with radii at least ρ. A range query with a region R with diameter $h \cdot \rho$ among $\mathcal{E}$ takes $O(\log n)$ time.*

4.2 Three-dimensional Objects

Most of the ideas that are used in the previous subsection to solve planar range searching queries among convex and polygonal objects generalize quite smoothly into three-dimensional space where our aim is to find a grid resolution r establishing that each convex or polyhedral object $E \in Q(R)$ is hit by one or more points from $\Pi \subset \mathcal{G}_r$, where, preferably, the size of Π depends on k and h (and the complexity of the individual objects) only. Lemma 12 is the spatial counterpart of Lemma 6: it identifies a sphere in a tetrahedron with bounded edge length and minimum volume with a certain minimal radius, depending on the volume and length bounds.

Lemma 12. *Let T be a tetrahedron with $volume(T) \geq V$ and let none of its edges be longer than δ. Then T contains a sphere with radius at least $\frac{3V}{2\delta^2}$.*

Below we use Property 5 and the above lemma to solve the range searching problem among convex and polyhedral k-fat objects by a sequence of point locations.

Convex 3D Objects The arguments that lead to our results for range searching among 3D convex objects are straightforward generalizations of the arguments that led to the result for the planar convex case. Therefore, we omit the details and confine ourselves to reporting the two basic tools that yield the results; the tools also play a role in handling the polyhedral case. First of all, the fatness of E implies that for all $S_{m,\rho} \in U_E$ with $m \in E \cap R$:

$$k \cdot volume(E \cap S_{m,\rho}) \geq volume(S_{m,\rho}) = \frac{4\pi\rho^3}{3}. \tag{5}$$

Lemma 13 is the spatial counterpart of Lemma 7.

Lemma 13. *Any convex object $E \subseteq \mathbf{R}^3$ contains a tetrahedron T with $24 \cdot volume(T) \geq volume(E)$.*

It follows that at least one point of the orthogonal grid with resolution $\frac{\pi\rho}{24k\sqrt{3}}$ hits any object $E \in Q(R)$ in a sphere $S_{m,\rho}$ with the property $m \in R$. It suffices to restrict the set Π of sample points to be the grid points in this Minkowski difference

$$\Pi = \mathcal{G}_{\pi\rho/24k\sqrt{3}} \cap R \ominus S_{O,\rho}.$$

Lemma 14. *Let $\mathcal{E}$ be a set of three-dimensional convex k-fat constant-complexity objects with minimal enclosing circles with radii at least ρ and let R be a region with diameter $h \cdot \rho$. The set $Q(R)$ of objects $E \in \mathcal{E}$ intersecting R can be found by point location queries with the points from $\mathcal{G}_{\pi\rho/24k\sqrt{3}} \cap R \ominus S_{O,\rho}$.*

The $O(\log^2 n)$ point location query time and the constant cardinality of the set Π lead to Theorem 15.

Theorem 15. *Let $k > 0$ and $h \geq 0$ be constants, and let $\mathcal{E}$ be a set of disjoint three-dimensional convex k-fat objects with minimal enclosing spheres with radii at least ρ. A range query with a region R with diameter $h \cdot \rho$ among $\mathcal{E}$ takes $O(\log^2 n)$ time.*

Polyhedral Objects Finally, we turn our attention to scenes of polyhedral objects and assume that all polyhedra in $\mathcal{E}$ have at most c vertices. We use an alternative approach of subdividing into convex parts instead of tetrahedralizing which is mainly due to the lack of exact bounds on the number of tetrahedra in a tetrahedralization of a polyhedron. Therefore, we cut up the polyhedron into convex parts according to Chazelle [3], where planes through the reflex edges subdivide the polyhedron into convex parts. (An edge is reflex if its two incident faces define an angle that is larger than π w.r.t. to the interior of the polyhedron.) Lemma 16 bounds the number of convex parts [3].

Lemma 16. *A polyhedron with m reflex edges can be decomposed into $m^2/2 + m/2 + 1$ convex parts.*

Let E be a k-fat polyhedron intersecting R, and let $m \in E \cap R$. Inequality (5) applies to the intersection of E and the sphere $S_{m,\rho} \in U_E$, on exactly the same grounds as in the convex case. The enclosing cube $C_{m,\rho}$ of $S_{m,\rho}$, satisfying $volume(C_{m,\rho})/volume(S_{m,\rho}) = 6/\pi$, provides the convenient property that its intersection with E is polyhedral. The ratio of these volumes, the inequality (5), and the obvious inequality $volume(E \cap C_{m,\rho}) \geq volume(E \cap S_{m,\rho})$, jointly result in

$$\frac{6k}{\pi} \cdot volume(E \cap C_{m,\rho}) \geq volume(C_{m,\rho}) = 8\rho^3. \tag{6}$$

To apply Theorem 16 to the polyhedron $E \cap C_{m,\rho}$, a bound on the number of (reflex) edges should be known. The following lemma gives a, probably non-optimal, bound on the number of vertices of the intersection $E \cap C_{m,\rho}$.

Lemma 17. *The intersection of a polyhedron E with c vertices and a cube has at most $18c - 28$ vertices.*

Euler's formula [17] bounds the number of edges of any polyhedron with at most $18c - 28$ vertices by $3(18c - 28) - 6 = 54c - 90$. So, the number of reflex edges of $E \cap C_{m,\rho}$ is definitely bounded by $54c - 90$ as well. By Lemma 16, the intersection $E \cap C_{m,\rho}$ is decomposable into $\gamma(c) = (54c - 90)^2/2 + (54c - 90)/2 + 1 (= O(c^2))$ convex parts. The convex part E' with the largest volume satisfies

$$volume(E') \geq \frac{1}{\gamma(c)} \cdot volume(E \cap C_{m,\rho}). \tag{7}$$

Lemma 13 states that E' contains a tetrahedron T with $24 \cdot volume(T) \geq volume(E')$. Combination with the inequalities (6) and (7) then yields

$$volume(T) \geq \frac{\pi \rho^3}{18k \cdot \gamma(c)}. \tag{8}$$

The length of the edges of T does not exceed $2\rho\sqrt{3}$ because T is contained in the cube $C_{m,\rho}$. Application of Lemma Lemma 12 to T with its bounded-length edges and bounded volume, shows that T contains a sphere with radius at least $\frac{\pi\rho}{144k\cdot\gamma(c)}$. By Property 5, at least one grid point of the regular orthogonal grid with resolution $\frac{\pi\rho}{72k\sqrt{3}\cdot\gamma(c)}$ hits the sphere contained in T. The grid point hitting the object $E \in Q(R)$ in the way described above lies in the square $C_{m,\rho}$ with $m \in R$, and hence in the Minkowski difference $R \ominus C_{O,\rho}$. The set Π of sample points may therefore be chosen

$$\Pi = \mathcal{G}_{\pi\rho/72k\sqrt{3}\gamma(c)} \cap R \ominus C_{O,\rho}.$$

Lemma 18. *Let $\mathcal{E}$ be a set of three-dimensional k-fat polyhedra with at most c vertices each and minimal enclosing spheres with radii at least ρ, and let R be a region with diameter $h \cdot \rho$. The set $Q(R)$ of objects $E \in \mathcal{E}$ intersecting R can be found by point location queries with the points from $\mathcal{G}_{\pi\rho/72k\sqrt{3}\gamma(c)} \cap R \ominus C_{O,\rho}$.*

The data structure of Section 3 supports point location queries with all points of Π in time $O(|\Pi|\log^2 n)$. The sequence of queries results in a set $\{E \in \mathcal{E}|\Pi \cap E \neq \emptyset\} \supseteq Q(R)$ of answers with at most $|\Pi|$ elements. Additional $O(|\Pi|)$ time suffices to eliminate all objects $E \in \{E \in \mathcal{E}|\Pi \cap E \neq \emptyset\}$ satisfying $E \notin Q(R)$, provided that all R and all $E \in \mathcal{E}$ have constant-complexity. The outlined computation of $Q(R)$ takes $O(|\Pi|\log^2 n)$ time.

In analogy to the planar approach, we bound the number of elements in Π by the number of grid points in the enclosing cube of $R \ominus C_{O,\rho}$ with side length $(h + 2)\rho$, giving $|\Pi| \leq (72\pi^{-1}k\sqrt{3} \cdot \gamma(c) \cdot (\lfloor h \rfloor + 2))^3 = O(k^3 h^3 c^6)$. In our framework of constant c, k, and h, this reduces to $|\Pi| = O(1)$, leading to $O(\log^2 n)$ query time for range searching with bounded-size ranges among fat polyhedral objects.

Theorem 19. *Let $k, c > 0$ and $h \geq 0$ be constants, and let $\mathcal{E}$ be a set of disjoint three-dimensional k-fat polyhedra with at most c vertices and minimal enclosing spheres with radii at least ρ. A range query with a region R with diameter $h \cdot \rho$ among $\mathcal{E}$ takes $O(\log^2 n)$ time.*

4.3 Building the Data Structure

The results of the previous section make the incremental construction of the point location structure outlined in Section 3 a feasible approach. Let us assume we are given the d-level data structure for priority point stabbing queries among hypercubes from Section 3, storing the m largest hypercubes $C_{n-m+1}, \ldots, C_n$, and the corresponding constant cardinality sets $V_{n-m+1}, \ldots, V_n$. We refer to the mentioned partial priority point stabbing structure as T_m. Hence, the objective is to eventually compute T_n from some initial structure T_0. The outline of the incremental construction is as follows.

compute T_0;
$m := 0$;
while $m < n$ **do**
    1. compute V_{n-m} by querying T_m and inspecting the sets $V_j, n - m < j \leq n$;
    2. compute T_{m+1} by inserting C_{n-m} into T_m.

We study both steps in the loop, starting with the second step, as the implications of its solution influence the first step as well.

Computation of T_{m+1}: The problem with the insertion of a hypercube into the d-level priority point stabbing structure lies in the use of fractional cascading, which was incorporated to improve the point location and range search query time from $O(\log^d n)$ to $O(\log^{d-1} n)$. Unfortunately, insertions into the multi-level data structure do not benefit from fractional cascading, so an insertion into the structure T_m would require $O(\log^d n)$ time instead of $O(\log^{d-1} n)$. Moreover, a sequence of insertions into the multi-level data structure with the static fractional cascading part is likely to increase the time for a query back to $O(\log^d n)$ as the fractional cascading part no longer 'suits' the updated multi-level data structure. Building the data structure would, even with fractional cascading, require $O(n \log^d n)$ time. Fortunately, Mehlhorn and Näher describe in [14] a dynamic version of fractional cascading. Incorporation of dynamic fractional cascading in the data structure during the construction phase improves the preprocessing time to $O(n \log^{d-1} n \log \log n)$. Details will be given in the full paper.

Computation of V_{n-m}: The computation of $V_{n-m} = \{E_j \in \mathcal{E} | E_j \cap C_{n-m} \neq \emptyset \wedge j \geq n - m\}$ is based on a sequence of point location queries. In the static point location structure, the query time was found to be $O(\log^{d-1} n)$ due to the incorporation of fractional cascading. Throughout the incremental construction of the point location structure, however, we use dynamic fractional cascading instead of fractional cascading to achieve efficient insertions, which leads to a query time of $O(\log^{d-1} n \log \log n)$.

The computation of the set V_{n-m} benefits from the fact that the hypercubes are inserted into the data structure from large to small in the sense that at the time of the set's computation, the intermediate data structure only stores hypercubes and objects from the appropriate index range $[n - m + 1, \ldots, n]$. Therefore, we may restrict ourselves to finding the objects in the data structure that intersect the query hypercube C_{n-m} without having to bother about the sizes of these objects. Moreover, note that future additions of hypercubes and their corresponding objects do not affect the earlier computed sets V_j. To apply the range searching results from Section 4, we must verify the validity of the constant ratio between the diameter of the search region (the hypercube C_{n-m}) and the lower bound on the radii of the minimal enclosing hyperspheres of the stored objects. The radii of the minimal enclosing hyperspheres of the objects in $\{E_{n-m+1}, \ldots, E_n\}$ are bounded from below by ρ_{n-m+1}, and, by the ordering on the radii, also by ρ_{n-m}. The query region C_{n-m} is the axis-parallel enclosing hypercube of the minimal enclosing hypersphere of E_{n-m} with radius ρ_{n-m}. As a result, the diameter of C_{n-m} is $2\sqrt{d} \cdot \rho_{n-m}$. The combination of the diameter of the query region and the lower bound on the hypersphere radii allow us to apply Theorems 9, 11, 15, and 19, which yield, taking into account the modified point location query time due to dynamic fractional cascading, a worst-case time bound of $O(\log^{d-1} n \log \log n)$ for the computation of V_{n-m}.

The time bounds for the computation of both T_{m+1} and V_{n-m} show that each of the $O(n)$ steps in the incremental construction of the d-level data structure for point location and range searching among 2D and 3D convex or polygonal/polyhedral takes $O(\log^{d-1} \log \log n)$ time, resulting in a time bound of $O(n \log^{d-1} \log \log n)$ for the computation of T_n from the skeleton T_0.

Theorem 20. *Let $\mathcal{E}$ be a set of disjoint convex or polygonal/polyhedral constant-complexity k-fat objects in dimension $d = 2, 3$. Then the d-level point stabbing structure can be built in time $O(n \log^{d-1} \log \log n)$.*

After the construction of the data structure, the query time can be improved back to $O(\log^{d-1} n)$. To this end we rebuild the structure using static fractional cascading. Because we now know all sets V_i this can easily be achieved in time $O(n \log^{d-1} n)$.

5 Summary of Results and Extensions

In this paper we have presented a data structure for both point location and range searching with bounded-size ranges in scenes of fat objects in 2-space and 3-space. We summarize the results in 2-space and 3-space in one powerful theorem, combining Theorems 3, 9, 11, 15, 19, and 20.

Theorem 21. *Let $k > 0$ and $h \geq 0$ be constants and let $\mathcal{E}$ be a set of disjoint constant-complexity convex or polygonal/polyhedral k-fat objects $E \subseteq \mathbf{R}^d$ ($d = 2, 3$) with minimal enclosing circle/sphere radii at least ρ. Then the set $\mathcal{E}$ can be stored in time $O(n \log^{d-1} n \log \log n)$ in a data structure of size $O(n \log^{d-1} n)$ which supports point location queries and range searching queries with ranges R with diameter $h \cdot \rho$ among the objects of $\mathcal{E}$ in time $O(\log^{d-1} n)$.*

As a first straightforward extension of the results, we mention that the above theorem also applies to mixed scenes of convex objects and polytopes. Clearly, one should take care to choose the highest of the two required grid resolutions for range searching, in order to be sure to hit both types of objects.

Moreover, the results in Section 4 on hitting fat objects with a finite resolution orthogonal grid seem extendible to other types of objects. The only thing we need is a bound on size of the required grid of points. Intuitively it is clear that such grids do exist for any type of fat objects in any dimension. In the full paper [16], we generalize the results of Theorem 21 to convex objects and polytopes in arbitrary dimension. We are still working on different types of objects, like two-dimensional objects bounded by algebraic polygonal curves of bounded degree.

Throughout the paper, the assumption that the objects in $\mathcal{E}$ are non-intersecting only plays a role in showing that the number of larger objects E' intersecting the enclosing hypercube C of some object E is bounded by a constant. No other lemma or theorem relies on the disjointness of the objects. As a consequence, all results remain valid if we drop the requirement of disjointness and instead impose the weaker restriction upon $\mathcal{E}$ that each enclosing hypercube C of $E \in \mathcal{E}$ is intersected at most a constant number of objects E' larger than E. In the generalized setting of intersecting objects, a query point may be contained in more than one obstacle. The answer to a point location, or point stabbing, query should therefore be the collection of objects containing the query point. Note that the new restriction on the data set $\mathcal{E}$ prevents more than a constant number of simultaneous containments. An interesting example of such a set is obtained by enlarging all objects $\mathcal{E}$ by an amount that is proportional to the size of the smallest object in $\mathcal{E}$. This applies for example to molecular models [9].

References

1. P.K. AGARWAL AND J. MATOUŠEK, On range searching with semialgebraic, *Proc. 17th Symp. Mathematical Foundations of Computer Science*, Lecture Notes in Computer Science **629** (1992), pp. 1-13.

2. H. ALT, R. FLEISCHER, M. KAUFMANN, K. MEHLHORN, S. NÄHER, S. SCHIRRA, AND C. UHRIG, Approximate motion planning and the complexity of the boundary of the union of simple geometric figures, *Algorithmica* **8** (1992), pp. 391-406.

3. B. CHAZELLE, Convex partitions of polyhedra: a lower bound and worst-case optimal algorithm, *SIAM Journal on Computing* **13** (1984), pp. 488-507.

4. B. CHAZELLE, How to search in history, *Information and Control* **64** (1985), pp. 77-99.

5. B. CHAZELLE AND L. GUIBAS, Fractional cascading I: A data structuring technique, *Algorithmica* **1** (1986), pp. 133-162.

6. B. CHAZELLE, M. SHARIR, AND E. WELZL, Quasi-optimal upper bounds for simplex range searching and new zone theorems, *Algorithmica* **8** (1992), pp. 407-429.

7. B. CHAZELLE AND E. WELZL, Quasi-optimal range searching in spaces of finite VC-dimension, *Discrete & Computational Geometry* **4** (1989), pp. 467-489.

8. M.T. GOODRICH AND R. TAMASSIA, Dynamic trees and dynamic point location, *Proc. 23rd ACM Symp. on Theory of Computing* (1991), pp. 523-533.

9. D. HALPERIN AND M. OVERMARS, Spheres, molecules, and hidden surface removal, *Proc. 10th ACM Symp. on Computational Geometry* (1994), pp. 113-122.

10. M. VAN KREVELD, *New results on data structures in computational geometry*, Ph.D. Thesis, Dept. of Computer Science, Utrecht University (1992).

11. J. MATOUŠEK, Efficient partition trees, *Discrete & Computational Geometry* **8** (1992), pp. 315-334.

12. J. MATOUŠEK, Range searching with efficient hierarchical cuttings, *Discrete & Computational Geometry* **10** (1993), pp. 157-182.

13. J. MATOUŠEK, J. PACH, M. SHARIR, S. SIFRONY, AND E. WELZL, Fat triangles determine linearly many holes, *SIAM Journal on Computing* **23** (1994), pp. 154-169.

14. K. MEHLHORN AND S. NÄHER, Dynamic fractional cascading, *Algorithmica* **5** (1990), pp. 215-241.

15. M.H. OVERMARS, Point location in fat subdivisions, *Information Processing Letters* **44** (1992), pp. 261-265.

16. M.H. OVERMARS AND A.F. VAN DER STAPPEN, Range searching and point location among fat objects, in preparation.

17. F.P. PREPARATA AND M.I. SHAMOS, *Computational geometry: an introduction*, Springer Verlag, New York (1985).

18. F.P. PREPARATA AND R. TAMASSIA, Efficient spatial point location, *Proc. 1st Workshop on Algorithms and Data Structures*, Lecture Notes in Computer Science **382** (1989), pp. 3-11.

19. A.F. VAN DER STAPPEN, D. HALPERIN, M.H. OVERMARS, The complexity of the free space for a robot moving amidst fat obstacles, *Computational Geometry, Theory and Applications* **3** (1993), pp. 353-373.

20. A.F. VAN DER STAPPEN AND M.H. OVERMARS, Motion planning amidst fat obstacles, *Proc. 10th ACM Symp. on Computational Geometry* (1994), pp. 31-40.

Convex Tours of Bounded Curvature*

Jean-Daniel Boissonnat**   Jurek Czyzowicz***
Olivier Devillers**   Jean-Marc Robert[†]
Mariette Yvinec[‡]

Abstract. We consider the motion planning problem for a point constrained to move along a smooth closed convex path of bounded curvature. The workspace of the moving point is bounded by a convex polygon with m vertices, containing an obstacle in a form of a simple polygon with n vertices. We present an $O(m+n)$ time algorithm finding the path, going around the obstacle, whose curvature is the smallest possible.

1 Introduction

Consider the problem of moving a point robot in the interior of a convex polygon containing a single obstacle. We are looking for a smooth, closed, convex, curvature-constrained path of the point around the obstacle. No source or target position of the point are specified.

The problem of planning the motion of a robot subject to kinematic constraints has been studied in numerous papers in the last decade (cf. [Lat91], [SS90]). For example, Reif and Sharir [RS85] studied the problem of planning the motion of a robot with a velocity bound amidst moving obstacles in two and three-dimensional space. Ó'Dúnlaing [Ó'D87] presented an exact algorithm solving the one-dimensional kinodynamic motion planning problem whereas Canny, Donald, Reif and Xavier [CDRX88] gave the first approximation algorithm solving the two and three-dimensional kinodynamic motion planning for a point amidst polyhedral obstacles.

Another aspect of the motion planning problem in the plane consists in finding paths under curvature constraints. Dubins [Dub57] characterized shortest curvature constrained paths in the Euclidean plane without any obstacle. More recently, Fortune and Wilfong [FW88] gave a decision procedure to verify if the source and target placement of a point robot may be joined by a curvature constrained path avoiding the polygonal obstacles. Their procedure has time

* *This work has been supported in part by the ESPRIT Basic Research Actions Nr. 7141 (ALCOM II) and Nr. 6546 (PROMotion), NSERC, FCAR and FODAR.*

** INRIA, 2004 Route des Lucioles, B.P.109, 06561 Valbonne cedex, France
   Phone : +33 93 65 77 38, E-mails : firstname.name@sophia.inria.fr

*** Dép. d'informatique, Université du Québec à Hull

[†] Dép. d'informatique et de mathématique, Université du Québec à Chicoutimi

[‡] INRIA and CNRS, URA 1376, Lab. I3S, 250 rue Albert Einstein, Sophia Antipolis, 06560 Valbonne, France

and space complexity $2^{O(poly(n,m))}$, where n is the number of obstacle vertices, and m is the number of bits required to specify the positions of these vertices. Jacobs and Canny [JC89] gave an algorithm computing an approximate curvature constrained path, and Wilfong [Wil88] designed an exact algorithm for the case where the curvature constrained path is limited to some fixed straight "lanes" and circular arc turns between the lanes. Finally, Švestka [Š93] applied the random approach introduced by Overmars [Ove92] to compute curvature constrained paths for car-like robots.

Besides heuristic and approximating approaches, an exact algorithmic solution seems to be difficult to find for the general case. An interesting direction of research is to design exact algorithms for some variants of the problem. In this paper, we give an efficient solution for the problem of computing a smooth, convex path going around a single polygonal obstacle with n vertices inside a convex polygon with m vertices. We design an $O(n + m)$ time and space algorithm finding a path of smallest curvature. The idea of the algorithm is to compute the curvature constraints imposed by each obstacle vertex. The maximal such constraint is then used to compute the curve, which must surround the entire obstacle.

We mention some extensions of this solution for the case of numerous obstacles, and for the case of obstacles coming as queries in a dynamic setting.

2 Preliminaries

Let $E \subset \mathbb{R}^2$ be a convex polygon with m vertices and let $I \subset E$ be a simple polygon with n vertices. The region $E \setminus int(I)$ represents the *workspace* W in which the point robot can move. A function $p : [0, L] \to W$ is a *smooth path* if $p(r) = (x_p(r), y_p(r))$ and the functions $x_p, y_p : [0, L] \to \mathbb{R}$ are continuous with continuous derivative. A smooth path p is *closed* if $p(0) = p(L)$ and its right derivative at point 0 is equal to its left derivative at point L. As any smooth path has finite length, we can assume that p is parameterized by arc length. Let $\phi_p(r)$ be the direction of the tangent to $p(r)$. The *curvature* of p at a point r can be defined as $\lim_{r' \to r} \frac{|\phi_p(r) - \phi_p(r')|}{|r - r'|}$. A path p has its *average curvature* bounded by a constant κ if $|\phi_p(r_2) - \phi_p(r_1)| \le \kappa |r_2 - r_1|$ for all r_1, r_2.

A curvature bounded closed path p is a *tour* of I in E if the bounded region of E, delimited by the Jordan curve p, is convex and contains I. Note that the points of boundaries of E and I are allowed to lie on the tour. Finally, a tour is *optimal* if the bound on its curvature is the smallest possible.

The main problem considered in this paper can be formulated as follows. Find an optimal tour of I in E. We first consider the degenerated case where the internal polygon I is a single point.

Lemma 1. *For a given convex polygon E and a point v inside E, let C denote a circle of radius r contained in E, passing through v, and tangent to the boundary of E in two points p_1 and p_2. If the arc $p = p_1 v p_2$ of C is not greater than a semicircle, then any tour of v in E contains a point with curvature at least equal to $1/r$ (see Fig. 1).*

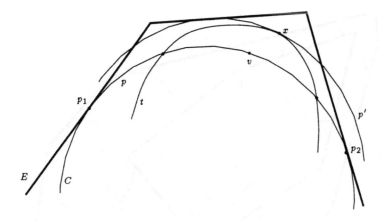

Fig. 1. An optimal tour of a point.

Proof: Any tour of v in E must intersect arc p. Let t denote such a tour. When we translate p along the bisector of the angle defined by the two lines tangent to C at p_1 and p_2, the extremities of p remain outside or on the boundary of E. Let p' be the furthest translated position of p, at which t is tangent to p' at a point x (c.f. Fig. 1). The curvature of t at x is at least equal to $1/r$. $\square$

Following the above lemma, suppose that we have a circle C inscribed in the convex polygon E, being tangent to E in two points p_1 and p_2. We say that C is the *critical circle* of v in E, if arc p_1vp_2 is not greater than a semicircle. Arc p_1vp_2 is called the *critical arc* of v in E. Observe that only points lying outside the largest circle inscribed in E admit critical arcs.

3 Computing tours

3.1 The Case of Given Curvature

Consider the problem of computing, if one exists, a tour of I in E with curvature bounded by some given constant κ. We present in this section an algorithm solving this problem in $O(m + n)$ time. The algorithm proceed by computing a "maximal tour" in E with curvature bounded by κ.

Let S be the set of all circles of radius $1/\kappa$ inscribed in E, and tangent to E in at least two points. Note that S contains at most m circles. The curve ζ, formed by the boundary of the convex hull of S, is a smooth closed path with curvature bounded by κ. Such a path ζ is called a *maximal path* in E. It follows from the proof of Lemma 1 that the convex region bounded by ζ contains any smooth, closed path internal to E with curvature bounded by κ (see Fig. 2). Hence, if ζ is not a tour of I in E, there exists no tour of I in E with curvature bounded by κ.

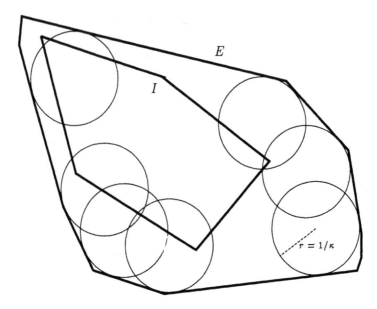

Fig. 2. There is no tour of I in P with curvature bounded by κ

Before we turn our attention to the algorithm verifying the existence of a tour of given curvature, we introduce some useful concepts. Consider the medial axis of E [Pre77]. Since E is a convex polygon, its medial axis corresponds to a tree. Each internal vertex x of this tree is the center of a circle tangent to three edges of E. This circle is called a *Voronoi circle*. We assign to x a weight $w(x)$ corresponding to the radius of the Voronoi circle. Thus, $w(x)$ represents the distance between x and the boundary of E. This weighted tree, rooted at its vertex with the largest weight, is called the *skeleton tree* and is denoted SkT(E). It follows from the definition of the medial axis that each edge of SkT(E) is a straight line segment belonging to the bisector of some two edges of E. It follows also from the definition that each vertex of SkT(E) has at least two descendants. Finally, we can easily prove that the weight of any vertex in SkT(E) is greater than the weights of its descendants.[1] This property will be crucial for our algorithms.

We are now ready to present how to compute the maximal path ζ.

Lemma 2. *Given the skeleton tree SkT(E), the maximal path ζ in E with curvature bounded by κ can be computed in $O(k)$ time, where k is the size complexity of the path.*

Proof: Perform a tree traversal on SkT(E). Each time a vertex x is visited, such that $w(parent(x)) \geq 1/\kappa > w(x)$, output a circle with radius $1/\kappa$, tangent

---

[1] The root may have the same weight as one of its children if E has two parallel edges.

to the boundary of E, and centered on the edge joining x and $parent(x)$. This circle can be computed easily once we know the edges of E defining the edge joining x to $parent(x)$. Then, the subtree of $\mathrm{SkT}(E)$ rooted at x is pruned and the traversal continues from $parent(x)$. In this way, all the k circles with radius $1/\kappa$ inscribed in E are found in order of their appearance on ζ. Hence, joining two consecutive circles by their supporting segment, we get the maximal path ζ, bounding the convex hull of the set circles. The $O(k)$ time complexity of the algorithm follows from the fact that the number of vertices visited during the transversal of $\mathrm{SkT}(E)$ is in $O(k)$. $\square$

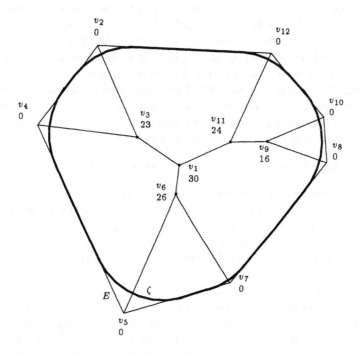

Fig. 3. A skeleton tree and a maximal path ζ of bounded curvature. The arcs of ζ of radius 20 are centered on v_3v_2, v_3v_4, v_6v_5, v_6v_7, $v_{11}v_9$ and $v_{11}v_{12}$.

It should be obvious now how to determine if there exist a tour of I in E with curvature bounded by κ. First, compute the medial axis of E in $O(m)$ time [AGSS89]. Then, compute the maximal path ζ. Finally, determine if I lies completely inside ζ. This latter step can be done easily in $O(n + k)$ time where k is the complexity of ζ. Hence, our algorithm determines if there exists a tour of I in E with curvature bounded by κ in $O(m + n)$ time.

3.2 An Algorithm Computing Optimal Tours

Consider the problem of computing an optimal tour of I in E. An algorithm solving this problem can be sketched as follows. First, find the largest circle inscribed in E. If this circle contains I, it is obviously the optimal tour. Otherwise, for each vertex of I lying outside the largest inscribed circle, compute the radius of its critical circle in E. The minimum of these radii determines the curvature of the optimal tour. Once the curvature of the optimal tour is known, a tour can be computed as we described in the previous section. We present in this section how to implement this algorithm optimally in $O(m + n)$ time.

We first present the data structures used by the algorithm. Let $CH(I)$ be the list of the vertices of the convex hull of I, given in counterclockwise order. Now, let $Arcs(E)$ be the list of arcs defined as follows. Consider the Voronoi circles associated with the internal vertices of $SkT(E)$. The tangent points of these circles with the boundary of E partition each circle into at least three arcs. Each of these arcs is put in $Arcs(E)$ if it is less than a semicircle. We also put in $Arcs(E)$ the leaves of $SkT(E)$. These points represent degenerated arcs. The elements of $Arcs(E)$ are ordered such that the first endpoints of the arcs appear in counterclockwise order on the boundary of E (see Fig. 4). In the next lemma, we show how to build list $Arcs(E)$ efficiently.

Lemma 3. *$Arcs(E)$ can be generated in $O(m)$ time and space.*

Proof: Perform a tree traversal on $SkT(E)$. The traversal can be oriented such that the children of any node are visited in counterclockwise order. An arc is produced each time a vertex x is visited from its parent v. This arc is less than a semicircle, centered at v, and tangent to the two edges of E whose bisector contains the edge vx of $SkT(E)$. Furthermore, a degenerated arc is produced if x is a leaf of $SkT(E)$.

To see that the arcs are produced in the right order, observe that the tree traversal can be performed by moving a point z continuously along the edges of $SkT(E)$. Let $\pi(z)$ be the orthogonal projection of z on the edge of E belonging to the Voronoi cell on the right-hand side of z with respect to the direction of the traversal. Since $SkT(E)$ corresponds to the medial axis of a convex polygon, $\pi(z)$ moves continuously around the boundary of E in the counterclockwise direction. Now, consider the arc computed while z traverses the edge vx of $SkT(E)$. By construction, the first endpoint of this arc corresponds to $\pi(z)$ when z coincides with v. Thus, the arcs are produced during the traversal of $SkT(E)$ such that the first endpoints of the arcs appear in counterclockwise order on the boundary of E.

The $O(m)$ time and space complexities of the algorithm follow from the fact that $SkT(E)$ has at most $2m - 2$ vertices. $\square$

The elements of $CH(I)$ and $Arcs(E)$ will be traversed by the algorithm according to the circular counterclockwise order around a point p inside $CH(I)$. The choice of p do not affect the order within $CH(I)$ or within $Arcs(E)$, but it may affect the position of an element of $CH(I)$ with respect to an element of $Arcs(E)$. Variable V will denote the current element of $CH(I)$ and variable

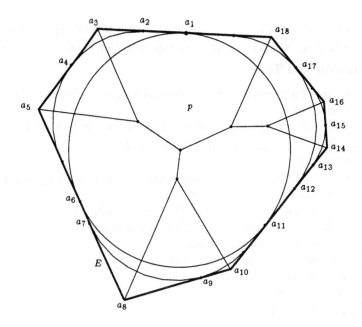

Fig. 4. $Arcs(E)$ is determined according to order of arcs' first endpoints

A will denote the current element of $Arcs(E)$. We say that vertex V is *before* arc A, if it precedes the first endpoint of A in the circular order around p. V is *after* A if it succeeds the second endpoint of A in this order. For V situated neither before nor after A, V is *inside* A if the ray pV reaches V before crossing A, otherwise V is *outside* A. Hence, the first element A of $Arcs(E)$ is set to an arc of the largest circle inscribed in E, and the first element V of $CH(I)$ is set to the first vertex which is *not before* A.

We are now ready to present the algorithm computing an optimal tour of I in E. The aim of the algorithm is to traverse the list of vertices of $CH(I)$ and localize each vertex in the planar map generated by the arcs in $Arcs(E)$ and the boundary of E (cf. Fig. 4). Once the cell containing the current vertex is determined, its critical arc may be computed easily.

Each iteration of the main step of the algorithm performs one among five possible actions. The action depends on the position of V with respect to five regions determined by the current arc A. Let $next(A)$ denote the successor of A in the list $Arcs(E)$ and let α be the smallest arc of the Voronoi circle C extending $next(A)$ and containing all the tangent points between C and E. Notice that α is completely outside A. (see Fig. 5). V falls into $\boxed{1}$, if it is outside A but not outside α, and in Region $\boxed{2}$ if it is outside α. V is in Region $\boxed{3}$ if it is inside A. Finally, V is in Region $\boxed{4}$ if it is after A, and in Region $\boxed{5}$ if it is before A.

Algorithm Optimal Tour

Input: A convex polygon E of m vertices and a simple polygon I of n vertices internal to E.

Output: A tour of I in E with the lowest possible curvature bound κ.

1. Compute $CH(I)$. Choose a point p inside $CH(I)$.
2. Compute $SkT(E)$.
3. Compute $Arcs(E)$ sorted by the arcs' first endpoint around p, starting by an arc of the largest circle inscribed in E.
4. $V \leftarrow first(CH(I))$. $A \leftarrow first(Arcs(E))$. $r \leftarrow radius(A)$.
5. **while** $Arcs(E)$ is not empty and $CH(I)$ is not empty **do**
   case the region containing V **do**
   $\boxed{1}$ $r \leftarrow min(r,$ radius of critical arc of $V)$.
   $\quad\quad V \leftarrow next(V)$.
   $\boxed{2}$ $A \leftarrow next(A)$.
   $\boxed{3}$ $V \leftarrow next(V)$.
   $\boxed{4}$ $A \leftarrow next(A)$.
   $\boxed{5}$ $V \leftarrow next(V)$.
6. Output ζ, the maximal path internal to E, with curvature bounded by $\kappa = 1/r$.

End of the Algorithm

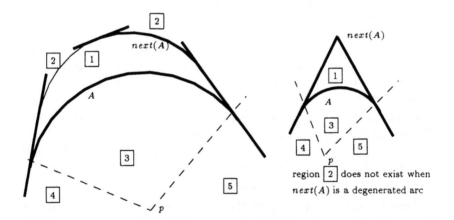

Fig. 5. Illustrating algorithm Optimal Tour

The Correctness of the Algorithm

To prove the correctness of the algorithm, we first have to show that the algorithm finds the minimum among the radii of critical circles of vertices of I. Thus, by Lemma 1, any tour of I in E would have a curvature at least as great as the curvature of that circle.

The aim of the algorithm is to locate the vertices of $C(I)$ in the planar map induced by the arcs of $Arcs(E)$ and the boundary of E. In Case $\boxed{1}$, the extremities of arcs A and α extending $next(A)$ lie on the same two edges of E. This follows from the fact that the Voronoi circles containing A and $next(A)$ are centered on the same edge of $SkT(E)$. These two edges of E define the edge of $SkT(E)$ on which the Voronoi circles are centered. Hence, the cell of the planar map containing V is defined by two edges and two arcs. The critical arc of any point lying in that cell must be tangent to the two edges and can be computed in constant time.

Observe that in case $\boxed{5}$ the predecessor of A in $Arcs(E)$ was an arc of radius smaller than A. Furthermore, at least one of the predecessors of V has been found outside an arc A' of radius smaller than A. It is easy to prove the same in case $\boxed{3}$.

Thus, the algorithm finds the vertices of I whose critical arcs have the smallest radius. Then, the maximal path computed in Step 6 must be a tour of I. Otherwise, there would be a vertex of I lying outside ζ. By Lemma 1, the critical arc of that vertex would have a radius smaller that r which is impossible.

The Complexity of the Algorithm

The first two steps of the algorithm rely on well known optimal algorithms. The convex hull of I can be computed in $O(n)$ time, and the skeleton tree of E can be computed in $O(m)$ time. In Step 3, the list $Arcs(E)$ can be constructed in $O(m)$ time according to Lemma 3. Step 5 represents the core of the algorithm. Each iteration of the loop takes a constant time. However, as each iteration removes one vertex of $CH(I)$ or one arc of $Arcs(E)$, the overall time complexity of this step is in $O(n + m)$. Finally, by Lemma 2, the optimal tour ζ can be computed in $O(k)$ time, where $k \leq m$. Hence, we obtain the following result.

Theorem 4. *An optimal tour of a simple polygon with n vertices in a convex polygon with m vertices can be computed in $O(n + m)$ time and space.*

Note that the algorithm can be adapted to compute a constrained optimal tour of I in E. Suppose that the tour is constrained to be tangent to some given lines when passing through some s given points of $E \setminus I$. Let E' denote the intersection of E with s half-planes delimited by the given lines, and let I' denote the convex hull of I and the given s points. Then, we simply have to compute an optimal tour of I' in E'.

Corollary 5. *An optimal tour of a simple polygon with n vertices in a convex polygon with m vertices, constrained to have given tangents when passing through s given points, can be computed in $O(n + m + s \log s)$ time and $O(n + m + s)$ space.*

4 The Dynamic Setting

The motion planing problem considered in the previous section can be reformulated in a dynamic setting. In this case, we want to preprocess a convex polygon E with m vertices in such a way that for any given query polygon I with n vertices, we can find quickly an optimal tour of I in E.

This dynamic problem can be solved by adapting Algorithm Optimal Tour. In Step 5, if the vertex V lies in Region $\boxed{4}$ with respect to the arc A ,the list $Arcs(E)$ is processed in order but it is clear that V remains in Region $\boxed{4}$ with respect to all other arcs outside A. Those arcs correspond to the subtree of $\text{SkT}(E)$ rooted at a child of the vertex on which A is centered. This subtree can be skipped in the traversal of $Arcs(E)$. Hence, the list $Arcs(E)$ is not produced explicitly in Step 3, but it may be obtained by traversing $\text{SkT}(E)$ in Step 5. The subtree of $\text{SkT}(E)$ effectively traversed is a subset of the subtree of $\text{SkT}(E)$ used to generate an optimal tour in Step 6. Thus, the time complexity of Step 5 of the algorithm can be reduced to $O(n + k)$, where k represents the complexity of the tour. This gives us the following result.

Theorem 6. *It is possible to preprocess a convex polygon E with m vertices in $O(m)$ time and space, so that for any simple polygon I with n vertices, an optimal tour of I in E can be computed in $O(n + k)$ time, where k is the complexity of the tour.*

If the obstacle is given as a set of n points instead of a simple polygon, we simply have to compute the convex hull of these points and to appply the above result.

Corollary 7. *It is possible to preprocess a convex polygon E with m vertices in $O(m)$ time and space, so that for any set S of n points, an optimal tour of S in E can be computed in $O(n \log n + k)$ time, where k is the complexity of the tour.*

If one is just interested in computing the curvature of the optimal tour instead of computing the tour itself, an alternative solution may be used.

The main problem is still to compute the critical circle of each point of I. The discussion from the previous section shows that this problem reduces to a point location problem in the planar map induced by the arcs of $Arcs(E)$ and the boundary of E. For each vertex v of I, we locate v in the map and compute its critical arc in E. Recall that a typical cell of the map is bounded by two arcs and by two portions of edges of E. As a critical arc of any point v falling into the cell must be tangent to both edges, it may be computed in constant time once v has been located in the map.

The planar map has $O(m)$ size and it can be decomposed into trapezoids in $O(m)$ time. Following the idea of [Kir83], this decomposition can be preprocessed in $O(m)$ time and space, so that the point location would be possible in $O(\log m)$ time. The query time would now be $O(n \log m)$ to find the smallest critical circle. Hence, we obtain the following result.

Theorem 8. *It is possible to preprocess a convex polygon E with m vertices in $O(m)$ time and space, so that for any set S of n points, the curvature of an optimal tour of S in E can be computed in $O(n \log m)$ time.*

If m is much smaller than n, this method may be interesting even for computing the tour itself. The following corollary can be used alternatively to Corollary 7.

Corollary 9. *It is possible to preprocess a convex polygon E with m vertices in $O(m)$ time and space, so that for any set S of n points, an optimal tour of S in E can be computed in $O(n \log m + k)$ time, where k is the complexity of the tour.*

Observe that similar generalization may be obtained for the case of many obstacles defined as points or polygons provided they all have to be situated in the interior of the tour. In such cases, we simply have to compute the convex hull of the obstacles and then to compute an optimal tour of that new "obstacle".

5 Conclusions

The paper gives an efficient algorithm computing a smallest curvature motion of a point robot around an obstacle inside a convex polygon. The solution easily generalizes on the case of numerous obstacles. We explore the fact that the resulting path must be convex. In this case, it is sufficient to compute the curvature constraints imposed by obstacles. The maximal constraint κ is used to compute the maximal curve, internal to the workspace, which must surround all the obstacles. The idea works only in the case of convex motion, and it is not clear how it may be generalized on the case of motion admitting left and right turns.

An obvious line of further research is to design algorithms for more general workspace. From the result of [FW88] it is possible to draw a pessimistic inference that a polynomial time algorithm computing curvature-constrained motion of a point in general workspace may not exist. It is natural to ask what are more general settings, that the one studied in this paper, for which the problem of curvature-constrained motion of a point admits an efficient solution, and what are the instances of the problem which are NP-complete.

References

[AGSS89] A. Aggarwal, L. J. Guibas, J. Saxe, and P. W. Shor. A linear-time algorithm for computing the Voronoi diagram of a convex polygon. *Discrete Comput. Geom.*, 4:591–604, 1989.

[CDRX88] J. Canny, B. R. Donald, J. Reif, and P. Xavier. On the complexity of kinodynamic planning. In *Proc. 29th Annu. IEEE Sympos. Found. Comput. Sci.*, pages 306–316, 1988.

[Dub57] L. E. Dubins. On curves of minimal length with a constraint on average curvature and with prescribed initial and terminal positions and tangents. *Amer. J. Math.*, 79:497–516, 1957.

[FW88] S. Fortune and G. Wilfong. Planning constrained motion. In *Proc. 20th Annu. ACM Sympos. Theory Comput.*, pages 445–459, 1988.

[JC89] P. Jacobs and J. Canny. Planning smooth paths for mobile robots. In *Proc. IEEE Internat. Conf. Robot. Autom.*, pages 2–7, 1989.

[Kir83] D. G. Kirkpatrick. Optimal search in planar subdivisions. *SIAM J. Comput.*, 12:28–35, 1983.

[Lat91] J.-C. Latombe. *Robot Motion Planning*. Kluwer Academic Publishers, Boston, 1991.

[Ó'D87] C. Ó'Dúnlaing. Motion-planning with inertial constraints. *Algorithmica*, 2:431–475, 1987.

[Ove92] M. H. Overmars. A random approach to motion planning. Report RUU-CS-92-32, Dept. Comput. Sci., Univ. Utrecht, Utrecht, Netherlands, 1992.

[Pre77] F. P. Preparata. The medial axis of a simple polygon. In *Proc. 6th Internat. Sympos. Math. Found. Comput. Sci.*, volume 53 of *Lecture Notes in Computer Science*, pages 443–450. Springer-Verlag, 1977.

[RS85] J. H. Reif and M. Sharir. Motion planning in the presence of moving obstacles. In *Proc. 26th Annu. IEEE Sympos. Found. Comput. Sci.*, pages 144–154, 1985.

[SS90] J. T. Schwartz and M. Sharir. Algorithmic motion planning in robotics. In J. van Leeuwen, editor, *Algorithms and Complexity*, volume A of *Handbook of Theoretical Computer Science*, pages 391–430. Elsevier, Amsterdam, 1990.

[Š93] P. Švestka. A probabilistic approach to motion planning for car-like robots. Report RUU-CS-93-18, Dept. Comput. Sci., Univ. Utrecht, Utrecht, Netherlands, 1993.

[Wil88] G. Wilfong. Motion planning for an autonomous vehicle. In *Proc. IEEE Internat. Conf. Robot. Autom.*, pages 529–533, 1988.

Optimal Shortest Path and Minimum-Link Path Queries in the Presence of Obstacles*

(Extended Abstract)

Yi-Jen Chiang and Roberto Tamassia

Department of Computer Science
Brown University
Providence, R. I. 02912–1910
{yjc,rt}@cs.brown.edu

Abstract. We present efficient algorithms for shortest-path and minimum-link-path queries between two convex polygons inside a simple polygon, which acts as an obstacle to be avoided. We also extend our results to the dynamic case, and give a unified data structure that supports both queries for convex polygons in the same region of a connected planar subdivision. Performing shortest-path queries is a variation of the well-studied *separation* problem, which has not been efficiently solved before in the presence of obstacles. Also, it was not previously known how to perform minimum-link-path queries in a dynamic environment, even for two-point queries.

1 Introduction

In this paper we present efficient algorithms for shortest-path and minimum-link-path queries between two convex polygons inside a simple polygon, which acts as an obstacle to be avoided. We give efficient techniques for both the static and dynamic versions of the problem.

Let R_1 and R_2 be two convex polygons with a total of h vertices that lie inside a simple polygon P with n vertices. The (*geodesic*) shortest path $\pi_G(R_1, R_2)$ is the polygonal chain with the shortest length among all polygonal chains joining a point of R_1 and a point of R_2 without crossing edges of P. A minimum-link path $\pi_L(R_1, R_2)$ is a polygonal chain with the minimum number of edges (called *links*) among all polygonal chains joining a point of R_1 and a point of R_2 without crossing edges of P. The number of links in $\pi_L(R_1, R_2)$ is called the *link distance* $d_L(R_1, R_2)$.

The related problem of computing the length of the shortest path between two polygons R_1 and R_2 *without obstacle* P has been extensively studied; this problem is also known as finding the *separation* of the two polygons, denoted by $\sigma(R_1, R_2)$. If both R_1 and R_2 are convex their separation can be computed

---

* Research supported in part by the National Science Foundation under grant CCR-9007851, by the U.S. Army Research Office under grants DAAL03-91-G-0035 and DAAH04-93-0134, and by the Office of Naval Research and the Defense Advanced Research Projects Agency under contract N00014-91-J-4052, ARPA order 8225.

in $O(\log h)$ time (see, e.g., [4, 7]); if only one of them is convex an $O(h)$-time algorithm is given in [7]; if neither is convex, an optimal $O(h)$-time algorithm is recently given by Amato [1].

Although there has been a lot of work on the separation problem, the more general shortest-path problem for two objects *in the presence of obstacle P* has been previously studied only for the simple case when the objects are points, for which there exist efficient static [10] and dynamic [5, 9] solutions. The static technique of [10] supports two-point shortest-path queries in optimal $O(\log n)$ time (plus $O(k)$ if the k edges of the path are reported), employing a data structure that uses $O(n)$ space and can be built in linear time. The dynamic technique of [5] performs shortest-path queries between two points in the same region of a connected planar subdivision S with n vertices in $O(\log^3 n)$ time (plus $O(k)$ to report the k edges of the path), using a data structure with $O(n \log n)$ space that can support updates (insertions and deletions of edges and vertices) of S each in $O(\log^3 n)$ time. The recent result of [9] improves the query and update times to $O(\log^2 n)$, with space complexity also improved to $O(n)$.

The minimum-link path problem between two points has been extensively studied. In many applications such as robotics, motion planning, VLSI and computer vision, the link distance often provides a more natural measure of path complexity than the Euclidean distance. For example, in a robot system, a straight-line navigation is often much cheaper than rotation, thus it is desirable to minimize the number of turns in path planning [13, 14]. Also, in graph drawing, it is often desirable to minimize the number of bends [15].

All previously known techniques for the minimum-link path problem are restricted to the static case. The best known results are due to Arkin *et al.* [2]. Their data structure uses $O(n^3)$ space and preprocessing time, and supports minimum-link-path queries between two points and between two segments in optimal $O(\log n)$ time (plus $O(k)$ if the k links are reported). Their technique can also perform queries between two convex polygons, however, in suboptimal $O(\log h \log n)$ time.

Our main results are outlined as follows, where all bounds are worst-case. **(i)** Let P be a simple polygon with n vertices. There exists an optimal data structure that supports shortest-path queries between two convex polygons with a total of h vertices inside P in time $O(\log h + \log n)$ (plus $O(k)$ if the k links of the path are reported), using $O(n)$ space and preprocessing time. **(ii)** Let P be a simple polygon with n vertices. There exists a data structure that supports minimum-link-path queries between two convex polygons with a total of h vertices inside P in optimal time $O(\log h + \log n)$ (plus $O(k)$ if the k links of the path are reported), using $O(n^3)$ space and preprocessing time. **(iii)** Let S be a connected planar subdivision whose current number of vertices is n. Shortest-path and minimum-link-path queries between two convex polygons with a total of h vertices that lie in the same region of S can be performed in time $O(\log h + \log^2 n)$ (plus $O(k)$ to report the k links of the path) and $O(\log h + k \log^2 n)$, respectively, using a fully dynamic data structure that uses $O(n)$ space and supports insertions and deletions of vertices and edges each in $O(\log^2 n)$ time.

The contributions of this work can be summarized as follows. (i) We provide the first optimal data structure for shortest-path queries between two convex polygons inside a simple polygon P that acts as an obstacle. No efficient data structure was known before to support such queries. All previous techniques either consider the case where P is not present or the case where the query objects are points. (ii) We provide the first data structure for minimum-link-path queries between two convex polygons inside a simple polygon P in optimal $O(\log h + \log n)$ time. The previous best result [2] has query time $O(\log h \log n)$ (and the same space and preprocessing time as ours). (iii) We provide the first fully dynamic data structure for shortest-path queries between two convex polygons in the same region of a connected planar subdivision S. No such data structure was known before even for the static version. (iv) We provide the first fully dynamic data structure for minimum-link-path queries between two convex polygons in the same region of a connected planar subdivision S. No such data structure was known before even for two-point queries.

We briefly outline our techniques. Given the available static techniques with optimal query time for shortest paths and minimum-link paths between *two points*, our main task in performing the *two-polygon* queries is to find two points $p \in R_1$ and $q \in R_2$ such that their shortest path or minimum-link path gives the desired path. The notion of *geodesic hourglass* is central to our method. The geodesic hourglass is *open* if R_1 and R_2 are mutually visible, and *closed* otherwise. As for shortest-path queries, the case where R_1 and R_2 are mutually visible is a basic case that, surprisingly, turns out to be nontrivial, and our solution makes use of interesting geometric properties. If R_1 and R_2 are not visible, then the geodesic hourglass gives two points p_1 and p_2 that are respectively visible from R_1 and R_2 such that the shortest path between any point of R_1 and any point of R_2 must go through p_1 and p_2. Then the shortest path $\pi_G(R_1, R_2)$ is the union of the shortest paths between R_1 and p_1, between p_2 and R_2 (both are basic cases), and between two points p_1 and p_2. The geodesic hourglass also gives useful information for minimum-link-path queries. When it is open, a minimum-link path is just a single segment; if it is closed, then it gives two edges such that extending them to intersect R_1 and R_2 gives the desired points p and q whose minimum-link path is a minimum-link path $\pi_L(R_1, R_2)$. However, it seems difficult to compute the geodesic hourglass in optimal time. Interestingly, we can get around this difficulty by computing a *pseudo hourglass* that gives all the information we need about the geodesic hourglass. We also extend these results to the dynamic case, by giving the first dynamic method for minimum-link-path queries between two points.

For the geometric terminology used in this paper, see [12]. A *cusp* of a polygonal chain is a vertex v whose adjacent vertices are both strictly above or below v. Given two points p and q in a simple polygon P, it is well known that their shortest path $\pi_G(p, q)$ is unique while a minimum-link path is not. Adopting the terminology of [14], we define the (unique) *greedy minimum-link path* $\pi_L(p, q)$ to be the minimum-link path whose first and last links are respectively the extensions of the first and last links of $\pi_G(p, q)$, and whose other links are the extensions of the *windows* of the *window partition* of p [14]. In the following we use the term "window" to refer to both a window and its extension.

Given a shortest path $\pi_G(p, q)$, an edge $e \in \pi_G(p, q)$ is an *inflection edge* if its predecessor and its successor lie on opposite sides of e. It is easily seen that an edge $e \in \pi_G(p, q)$ is an inflection edge if and only if it is an internal common tangent of the boundaries of P.

In this extended abstract proofs and technical details are omitted. They can be found in the full version [6].

2   Shortest Path Queries

Now we discuss how to compute the shortest path $\pi_G(R_1, R_2)$. The data structure of [10] computes the length and an implicit representation of the shortest path between any two points inside P in $O(\log n)$ time, using $O(n)$ space and preprocessing time after triangulating P (again in $O(n)$ time by [3]). We modify this data structure so that associated with the implicit representation of a shortest path π_G, there are two balanced binary trees respectively maintaining the inflection edges and the cusps on π_G in their path order. With this data structure, our task is to find points $p \in R_1$ and $q \in R_2$ such that $\pi_G(p, q) = \pi_G(R_1, R_2)$. We say that p and q *realize* $\pi_G(R_1, R_2)$.

To get a better intuition, let us imagine surrounding R_1 and R_2 with a rubber band inside P. The resulting shape is called the *relative convex hull* of R_1 and R_2. It is formed by four pieces: shortest paths $\pi_1 = \pi_G(a_1, a_2)$, $\pi_2 = \pi_G(b_1, b_2)$ ($a_1, b_1 \in R_1$ and $a_2, b_2 \in R_2$), and the boundaries of R_1 and R_2 farther away from each other. We call a_1, b_1, a_2, and b_2 the *geodesic tangent points*. Let $s_1 = (a_1, b_1)$ and $s_2 = (a_2, b_2)$. If we replace R_1 and R_2 with s_1 and s_2, then the relative convex hull of s_1 and s_2 is an *hourglass* $H(s_1, s_2)$ bounded by s_1, s_2, π_1, and π_2. We call $H(s_1, s_2)$ the *geodesic hourglass*. We say that $H(s_1, s_2)$ is *open* if π_1 and π_2 do not intersect, and *closed* otherwise. When $H(s_1, s_2)$ is closed, there is a vertex p_1 at which π_1 and π_2 join together, and a vertex p_2 at which the two paths separate (possibly $p_1 = p_2$); we call p_1 and p_2 the *apices* of $H(s_1, s_2)$ (see Fig. 1(b)). Also, we say that $\pi_G(a_1, p_1)$ and $\pi_G(b_1, p_1)$ form a *funnel* $F(s_1)$. The only internal common tangent ρ_1 of P among all edges of $F(s_1)$ is called the *penetration* of $F(s_1)$, and similarly for ρ_2 in funnel $F(s_2)$ (see Fig. 1(b)). Hereafter we use H_G to denote the geodesic hourglass, and a_1, b_1 ($\in R_1$), a_2, b_2 ($\in R_2$) to denote the geodesic tangent points.

We say that R_1 and R_1 are *mutually visible* if there exists a line l connecting R_1 and R_2 without crossing any edge of P; we call l a *visibility link*. Observe that H_G is open if and only if R_1 and R_2 are mutually visible (see Fig. 1(a)). If H_G is closed, then $\pi_G(R_1, R_2)$ must go through p_1 and p_2 (see Fig. 1(b)), i.e., $\pi_G(R_1, R_2) = \pi_G(R_1, p_1) \cup \pi_G(p_1, p_2) \cup \pi_G(p_2, R_2)$. Note that p_1 and p_2 are respectively visible from R_1 and R_2. In summary, there are two main tasks: (*i*) deciding whether H_G is open or closed, and finding apices p_1 and p_2 when H_G is closed, and (*ii*) computing $\pi_G(R_1, R_2)$ when R_1 and R_2 are mutually visible.

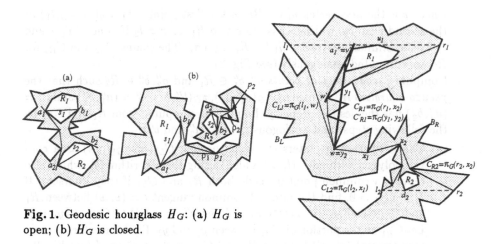

Fig. 1. Geodesic hourglass H_G: (a) H_G is open; (b) H_G is closed.

Fig. 2. A running example for Algorithm *Pseudo-Hourglass* in the case where π has no cusps and $C_{L1} \cap R_1 = \emptyset$.

The Pseudo Geodesic Hourglass We first discuss how to compute the information we need about the geodesic hourglass H_G. In addition to the above information, the data below are also needed for minimum-link paths (see Section 3): a visibility link between R_1 and R_2 when H_G is open, and the penetrations of H_G when it is closed. As shown in [2], computing H_G takes $O(\log h \log n)$ time, which is suboptimal. To achieve optimal time, we do not compute H_G exactly, but rather obtain all the needed information by computing a *pseudo hourglass* H'' such that if H'' is open then H_G is open, and if H'' is closed then H_G is closed with the same penetrations and apices. The following algorithm performs the described function in time $O(\log h + \log n)$.

Algorithm Pseudo-Hourglass

1. Ignore P and compute the ordinary external common tangents (a'_1, a'_2) and (b'_1, b'_2) between R_1 and R_2 using algorithm [11], where $a'_1, b'_1 \in R_1$ and $a'_2, b'_2 \in R_2$. Let $s'_1 = (a'_1, b'_1)$ and $s'_2 = (a'_2, b'_2)$. Compute shortest paths $\pi_1 = \pi_G(a'_1, a'_2)$ and $\pi_2 = \pi_G(b'_1, b'_2)$. If they are disjoint then s'_1 and s'_2 (and also R_1 and R_2) are mutually visible. Compute an internal common tangent l between π_1 and π_2, report {open with visibility link l} and stop.

2. Else (π_1 and π_2 are not disjoint), let u_1 and d_1 be the highest and lowest vertices of R_1, respectively, and similarly define u_2 and d_2 for R_2. Compute shortest paths $\pi_G(u_1, u_2)$, $\pi_G(u_1, d_2)$, $\pi_G(d_1, u_2)$ and $\pi_G(d_1, d_2)$. Take π as the one with the largest number of cusps (break ties arbitrarily). From $R_i, i = 1, 2$, compute horizontal projection points l_i and r_i respectively on the left and right boundaries B_L and B_R of P, depending on the information about the cusps on π (see Fig. 2 and 3(b)).

3. Compute shortest paths $\pi_l = \pi_G(l_1, l_2)$ and $\pi_r = \pi_G(r_1, r_2)$. Extract the "left bounding convex chain" C_{L1} for R_1 as the portion of π_l from l_1 to x,

where x is the first vertex v_1 on B_R or the first point c with $y(c) = y(l_1)$ or the second cusp c_2, whichever is closest to R_1, or $x = l_2$ if none of v_1, c and c_2 exists. Similarly extract C_{R1} for R_1 from π_r. The chains C_{L2} and C_{R2} for R_2 are computed analogously (see Fig. 2).

4. Compute *pseudo tangent points* $a_1'', b_1'' \in R_1$ and $a_2'', b_2'' \in R_2$ such that the *pseudo hourglass* H'' formed by $\pi_G(a_1'', a_2''), \pi_G(b_1'', b_2''), s_1'' = (a_1'', b_1'')$ and $s_2'' = (a_2'', b_2'')$ has the desired property. Point a_1'' is computed from R_1 and C_{L1} by the following steps (and analogously for b_1'', a_2'' and b_2'').

   (a) Check whether R_1 intersects C_{L1} (viewing $C_{L1} = \pi_G(l_1, x)$ as a convex polygon with edge (l_1, x) added) using the algorithm [4], which also reports a common point g inside both R_1 and C_{L1} if they intersect. If $R_1 \cap C_{L1} = \emptyset$, find the internal common tangent $t = (v, w)$ between R_1 and C_{L1} (see Fig. 2). (i) $t \cap C_{R1} = \emptyset$. Set $a_1'' := v$. (ii) $t \cap C_{R1} = \{y_1, y_2\}$. Let C_{R1}' be the portion of C_{R1} between y_1 and y_2. Find the external common tangent (v', w') between R_1 and C_{R1}', and set $a_1'' := v'$ (see Fig. 2).

   (b) Else ($R_1 \cap C_{L1} \neq \emptyset$, with a common point g inside both R_1 and C_{L1}), then there is only one edge of C_{L1} intersecting R_1. Compute this edge (u, b) using the fact that segment (g, u_1) intersects only (u, b) among the edges of C_{L1}. Suppose b is closer to R_2 than u. (i) b is on the left boundary B_L. Compute a_1'' as the tangent point from b to R_1 (see Fig. 3(a)). (ii) b is on the right boundary B_R. Take C as the convex portion of π_l (oriented from R_1 to R_2) from b to z, where z is the first vertex v_1' on B_L again or the first point c' with $y(c') = y(b)$ or the second cusp c_2' after b, whichever is closest to R_1. Find the external common tangent (v'', w'') between R_1 and C, and set $a_1'' := v''$ (see Fig. 3(b)).

5. Compute shortest paths $\pi_1 = \pi_G(a_1'', a_2'')$ and $\pi_2 = \pi_G(b_1'', b_2'')$ to form pseudo hourglass H''. If H'' is open, then compute an internal common tangent l between π_1 and π_2, report {open with visibility link l} and stop. Else (H'' is closed), extract the penetrations ρ_1 and ρ_2 and apices p_1 and p_2 of H'', report {closed with penetrations ρ_1 and ρ_2 and apices p_1 and p_2}, and stop.

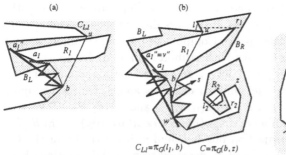

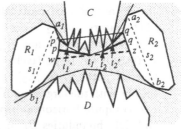

Fig. 3. Steps 4b(i)–(ii) of Algorithm *Pseudo-Hourglass*: (a) $b \in B_L$; (b) $b \in B_R$.

Fig. 4. Lemma 1

The Case of Mutually Visible Query Polygons We now discuss how to compute $\pi_G(R_1, R_2)$ when R_1 and R_2 are mutually visible. Ignoring P and computing the separation of R_1 and R_2, we can find $p' \in R_1$ and $q' \in R_2$ with $length(p', q') = \sigma(R_1, R_2)$ in $O(\log h)$ time. Now we compute $\pi_G(p', q')$. If $\pi_G(p', q')$ has only one link, then (p', q') is the shortest path $\pi_G(R_1, R_2)$. Otherwise $\pi_G(p', q')$ must touch the boundary of P, and there are two cases: (1) $\pi_G(p', q')$ touches only one of $\pi_G(a_1, a_2)$ and $\pi_G(b_1, b_2)$; or (2) $\pi_G(p', q')$ touches both $\pi_G(a_1, a_2)$ and $\pi_G(b_1, b_2)$.

Lemma 1. *Let the geodesic hourglass H_G be open and (p', q') with $p' \in R_1$ and $q' \in R_2$ be the shortest path between R_1 and R_2 without obstacle P. If $\pi_G(p', q')$ touches only one of $\pi_G(a_1, a_2)$ and $\pi_G(b_1, b_2)$, say $\pi_G(a_1, a_2)$, then $\pi_G(R_1, R_2)$ touches $\pi_G(a_1, a_2)$ but does not touch $\pi_G(b_1, b_2)$. (See Fig. 4.)*

Therefore in the above situation (see Fig. 4), if t_1' and t_2' are the points of obstacle C where $\pi_G(p', q')$ first touches C and finally leaves C, respectively, and t_1 and t_2 are the points of C where $\pi_G(p, q)$ first touches C and finally leaves C (recall that $\pi_G(p, q) = \pi_G(R_1, R_2)$), then t_2 is the point where the shortest path $\pi_G(t_1', R_2)$ from t_1' to R_2 finally leaves C, and similarly for t_1. We say that $t_2 \in C$ and $q \in R_2$ realize $\pi_G(t_1', R_2)$, and similarly for the other side. It is clear that $\pi_G(R_1, R_2)$ consists of (p, t_1), $\pi_G(t_1, t_2)$ and (t_2, q). So we only need to independently compute $t_2 \in C$ and $q \in R_2$ that realize $\pi_G(t_1', R_2)$, and t_1 and p that realize $\pi_G(t_2', R_1)$.

The other case where $\pi_G(p', q')$ touches both $\pi_G(a_1, a_2)$ and $\pi_G(b_1, b_2)$ can be handled in the same way:

Lemma 2. *Let the geodesic hourglass H_G be open and (p', q') with $p' \in R_1$ and $q' \in R_2$ be the shortest path between R_1 and R_2 without obstacle P. If $\pi_G(p', q')$ touches both $\pi_G(a_1, a_2)$ and $\pi_G(b_1, b_2)$, say first $\pi_G(a_1, a_2)$ (entering at point t_1 and leaving at point t_3) and then $\pi_G(b_1, b_2)$ (entering at t_4 and leaving at t_2), then $\pi_G(R_1, R_2) = \pi_G(R_1, t_3) \cup (t_3, t_4) \cup \pi_G(t_4, R_2)$. (See Fig. 5.)*

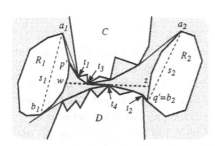

Fig. 5. Lemma 2

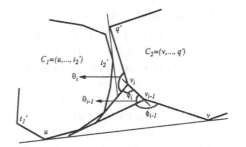

Fig. 6. Lemma 3

We now discuss how to compute $t_2 \in C$ and $q \in R_2$ that realize $\pi_G(t_1', R_2)$ in the situation of Fig. 4. We only need to consider the two convex chains $\pi_G(u, t_2')$

(denoted by C_1) and the clockwise boundary $(v, ..., q')$ of R_2 (denoted by C_2), where (u, v) is the external common tangent between R_2 and the convex hull of C with $u \in C$ and $v \in R_2$. Our algorithm uses the following properties.

Lemma 3. *Let $v_1, v_2, ..., v_k$ be a sequence of points on C_2 in clockwise order. If we draw a tangent to C_1 from each v_i and make angles θ_i and ϕ_i (see Fig. 6), then (1) $\theta_1 < \theta_2 < ... < \theta_k$ and $\phi_1 > \phi_2 > ... > \phi_k$; and (2) if $\phi_i \geq \frac{\pi}{2}$ then $\pi_G(v_i, t'_1) < \pi_G(v_{i-1}, t'_1) < \cdots$, and if $\theta_i \geq \frac{\pi}{2}$ then $\pi_G(v_i, t'_1) < \pi_G(v_{i+1}, t'_1) < \cdots$; and (3) $\min_{w \in C_2} \pi_G(w, t'_1)$ occurs at $w = v_i$ with $\phi_i \geq \frac{\pi}{2}$ and $\theta_i \geq \frac{\pi}{2}$.*

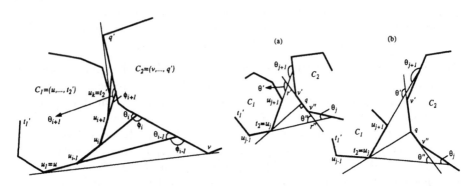

Fig. 7. Lemma 4   **Fig. 8.** Lemma 4 part (2)

Lemma 4. *Let $u_1 = u, u_2, ..., u_k = t'_2$ be the vertices of C_1 in counterclockwise order. If we extend each edge (u_{i-1}, u_i) to intersect C_2 and make angles θ_i and ϕ_i (see Fig. 7), then (1) $\theta_2 < \theta_3 < ... < \theta_k$ and $\phi_2 > \phi_3 > ... > \phi_k$; and (2) if $t_2 \in C_1$ and $q \in C_2$ realize $\pi_G(t'_1, C_2)$, where t_2 is some vertex u_j, then $\theta_j < \frac{\pi}{2}$ and $\theta_{j+1} > \frac{\pi}{2}$ (see Fig. 8).*

The following algorithm computes $t_2 \in C_1$ and $q \in C_2$ that realize $\pi_G(t'_1, C_2)$ in time $O(\log h + \log n)$.

Algorithm Double-Search

1. If either $| C_1 | = 1$ or $| C_2 | \leq 2$ then go to step 3.
2. Else, let v and w be the median vertices of current C_1 and C_2. Prune the wiggly portion(s) as shown in Fig. 9, and go to step 1.
3. Now $| C_1 | = 1$ or $| C_2 | \leq 2$. **(a)** $| C_2 | = 1$. The only vertex of C_2 is q. Compute the tangent from q to C_1 and take t_2 as the tangent point. Report q and t_2, and stop. **(b)** $| C_2 | = 2$. Let $C_2 = (w_1, w_2)$ with the interior of R_2 to the right of ray (w_1, w_2). From w_1 and w_2 compute tangents (w_1, v_1) and (w_2, v_2) of C_1, where $v_1, v_2 \in C_1$. Let $\theta_1 = \angle v_1 w_1 w_2$ and $\phi_2 = \angle v_2 w_2 w_1$. **(i)** $\theta_1 \geq \frac{\pi}{2}$. Report $q = w_1$, $t_2 = v_1$ and stop. **(ii)** $\phi_2 \geq \frac{\pi}{2}$. Report $q = w_2$, $t_2 = v_2$, and stop. **(iii)** $\theta_1 < \frac{\pi}{2}$ and $\phi_2 < \frac{\pi}{2}$. Perform a binary search on C_1 to find vertex t_2 with (t_2, q) tangent to C_1, where q is the projection point of t_2 on (w_1, w_2). Report t_2 and q, and stop. **(c)** $| C_1 | = 1$. The only vertex of C_1 is t_2. Perform a binary search on C_2 to find the point $q \in C_2$ such

that $length(t_2, q)$ is the shortest distance from t_2 to C_2. Report t_2 and q, and stop.

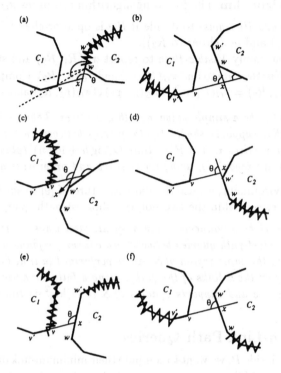

Fig. 9. The cases (a)–(f) in step 2 of Algorithm *Double-Search*, depending on whether x is below, above, or on (w, w') and $\theta \geq \frac{\pi}{2}$ or not.

The algorithm below computes $\pi_G(R_1, R_2)$ for R_1 and R_2 mutually visible in time $O(\log h + \log n)$ (plus $O(k)$ to report the k links of the path).

Algorithm Visible-Path

1. Ignore P and compute $\sigma(R_1, R_2)$ by any method (e.g., [7]), giving also $p' \in R_1$ and $q' \in R_2$ with $length(p', q') = \sigma(R_1, R_2)$.
2. Compute $\pi_G(p', q')$. If $\pi_G(p', q')$ has only one link, then report $\pi_G(R_1, R_2) = (p', q')$ and stop.
3. Otherwise, let (p', t_1') and (t_2', q') be the first and last links of $\pi_G(p', q')$. **(a)** There is no inflection edge in $\pi_G(p', q')$. Let $C = \pi_G(t_1', t_2')$. Find the external common tangent (u, v) between C and R_2 with $u \in C$ and $v \in R_2$; let C_1 be $\pi_G(u, t_2')$ and C_2 be the clockwise boundary $(v, ..., q')$ of R_2. Compute $t_2 \in C$ and $q \in R_2$ that realize $\pi_G(t_1', R_2)$ by applying *Double-Search* on C_1 and C_2, and similarly $t_1 \in C$ and $p \in R_1$ that realize $\pi_G(t_2', R_1)$. Report $\pi_G(R_1, R_2) = (p, t_1) \cup \pi_G(t_1, t_2) \cup (t_2, q)$ and stop. **(b)** There is an inflection edge (t_3, t_4) in $\pi_G(p', q')$. Use *Double-Search* to compute two pairs of points

that realize $\pi_G(R_1, t_3)$ and $\pi_G(t_4, R_2)$. Report $\pi_G(R_1, R_2) = \pi_G(R_1, t_3) \cup (t_3, t_4) \cup \pi_G(t_4, R_2)$ and stop.

The Overall Algorithm The following algorithm computes $\pi_G(R_1, R_2)$.

1. Perform *Pseudo-Hourglass* to decide if H_G is open or closed (with apices p_1 (closer to R_1) and p_2 (closer to R_2)).
2. If H_G is open, apply *Visible-Path* to report $\pi_G(R_1, R_2)$ and stop.
3. Else, apply *Visible-Path* to find $\pi_G(R_1, p_1)$ and $\pi_G(p_2, R_2)$. Compute $\pi_G(p_1, p_2)$, report $\pi_G(R_1, R_2) = \pi_G(R_1, p_1) \cup \pi_G(p_1, p_2) \cup \pi_G(p_2, R_2)$ and stop.

Theorem 5. *Let P be a simple polygon with n vertices. There exists an optimal data structure that supports shortest-path queries between two convex polygons with a total of h vertices inside P in time $O(\log h + \log n)$ (plus $O(k)$ if the k links of the path are reported), using $O(n)$ space and preprocessing time.*

In a dynamic environment, we use the data structure of [9] to support two-point shortest-path queries within the two-polygon shortest-path query algorithm.

Theorem 6. *Let S be a connected planar subdivision whose current number of vertices is n. Shortest-path queries between two convex polygons with a total of h vertices that lie in the same region of S can be performed in time $O(\log h + \log^2 n)$ (plus $O(k)$ to report the k links of the path), using a fully dynamic data structure that uses $O(n)$ space and supports updates of S in $O(\log^2 n)$ time.*

3 Minimum-Link Path Queries

Given R_1 and R_2 inside P, we want to compute their minimum-link path $\pi_L(R_1, R_2)$. The data structure of [2] supports minimum-link-path queries between two points inside P in $O(\log n)$ time (plus $O(k)$ if the k links are reported), using $O(n^3)$ space and preprocessing time. We now show how to perform two-polygon queries in optimal time, using the same data structure.

Lemma 7. *Let H_G be the geodesic hourglass between R_1 and R_2. If H_G is closed, then there exists a minimum-link path $\pi_L(R_1, R_2)$ that uses the penetrations of H_G as the first and last links.*

We now give the algorithm for computing a minimum-link path $\pi_L(R_1, R_2)$.

1. Use *Pseudo-Hourglass* to decide if H_G is open (with a visibility link l) or closed (with penetrations ρ_1 (closer to R_1) and ρ_2 (closer to R_2)).
2. If H_G is open, then report $\pi_L(R_1, R_2) = l$, $d_L(R_1, R_2) = 1$ and stop.
3. Otherwise, extend ρ_1 and ρ_2 to intersect R_1 and R_2 respectively at p and q, and compute $\pi_L(p, q)$ (and $d_L(p, q)$). Report $\pi_L(R_1, R_2) = \pi_L(p, q)$, $d_L(R_1, R_2) = d_L(p, q)$ and stop.

Theorem 8. *Let P be a simple polygon with n vertices. There exists a data structure that supports minimum-link-path queries between two convex polygons with a total of h vertices inside P in optimal time $O(\log h + \log n)$ (plus $O(k)$ if the k links of the path are reported), using $O(n^3)$ space and preprocessing time.*

Now we show that the dynamic data structure of Theorem 6 can also support minimum-link-path queries between two convex polygons.

Dynamic Two Point Queries Let p and q be two points in the same region P of the planar subdivision. Let $e_1, \cdots, e_j$ be the inflection edges of the shortest path $\pi_G(p, q)$. Then $e_1, \cdots, e_j$ partition $\pi_G(p, q)$ into *inward convex* subchains. It is shown that every inflection edge $e \in \pi_G(p, q)$ must be contained in $\pi_L(p, q)$ [2, 8]. Hence, extending each inflection edge of $\pi_G(p, q)$ by ray shooting on both sides, together with the extensions of the first and last links of $\pi_G(p, q)$, we have fixed windows $W_1, \cdots, W_{j+2}$. Now the task in computing $\pi_L(p, q)$ is to connect consecutive fixed windows. In particular, each W_i has a portion $(u, v) \in \pi_G(p, q)$, with u closer to p than v in $\pi_G(p, q)$. Let the endpoints of W_i be u' and v' such that $W_i = (u', u, v, v')$. We call (u', u) the *front* of W_i and (v, v') the *rear* of W_i. We want to connect the rear of W_i with the front of W_{i+1} for each $i = 1, \cdots, j+1$.

Lemma 9. *Let W_i and W_{i+1} be consecutive fixed windows, W the front of W_{i+1}, and w the rear of W_i (or a window between the rear of W_i and W). If the hourglass $H(w, W)$ is closed, then the window w' following w is the penetration of funnel $F(w)$.*

The following algorithm for computing $\pi_L(p, q)$ runs in time $O(k \log^2 n)$, where k is the number of links in $\pi_L(p, q)$.

Algorithm Point-Query

1. Compute $\pi_G(p, q)$. If $\pi_G(p, q)$ has only one link, report $\pi_L(p, q) = (p, q)$, $d_L(p, q) = 1$ and stop.
2. Else, perform ray-shooting queries to extend the inflection edges and the first and last links of $\pi_G(p, q)$ to obtain fixed windows $W_1, \cdots, W_j$.
3. For each pair of non-intersecting W_i and W_{i+1}, repeat step 4 to compute the windows connecting W_i and W_{i+1}.
4. Initially, let w be the rear of W_i. Let $W = (a_2, b_2)$ be the front of W_{i+1} with $a_2 \in \pi_G(p, q)$.
   (a) Assume that $w = (a_1, b_1)$ with a_1 on $\pi_G(p, q)$. Compute $\pi_G(b_1, b_2)$.
   (b) If $\pi_G(b_1, b_2)$ has no inflection edges, then $H(w, W)$ is open. Compute an internal common tangent t between $\pi_G(a_1, a_2)$ and $\pi_G(b_1, b_2)$, set t to be the window following w and exit step 4.
   (c) Else ($\pi_G(b_1, b_2)$ has inflection edges), extract the penetration ρ and apex p_1 of funnel $F(w)$. Extend ρ in both directions (hitting w at v and P at u) to obtain the window (v, u) following w. Set $w := (p_1, u)$ and go to step 4(a).
5. Now there are windows $w_1, \cdots, w_k$ connecting p and q. Let $v_i = w_i \cap w_{i+1}$, $i = 1, \cdots, k-1$. Report $\pi_L(p, q) = (p, v_1, \cdots, v_{k-1}, q)$, $d_L(p, q) = k$ and stop.

Now we can use Algorithm *Point-Query* to support two-point queries within the two-polygon minimum-link-path query algorithm.

Theorem 10. *Let S be a connected planar subdivision whose current number of vertices is n. Minimum-link-path queries between two convex polygons with a total of h vertices that lie in the same region of S can be performed in time*

$O(\log h + k \log^2 n)$ *(where k is the number of links in the reported path), using a fully dynamic data structure that uses $O(n)$ space and supports updates of S in $O(\log^2 n)$ time.*

References

1. Nancy M. Amato. An optimal algorithm for finding the separation of simple polygons. In *Proc. 3rd Workshop Algorithms Data Struct.*, volume 709 of *Lecture Notes in Computer Science*, pages 48–59. Springer-Verlag, 1993.

2. E. M. Arkin, J. S. B. Mitchell, and S. Suri. Optimal link path queries in a simple polygon. In *Proc. 3rd ACM-SIAM Sympos. Discrete Algorithms*, pages 269–279, 1992.

3. B. Chazelle. Triangulating a simple polygon in linear time. *Discrete Comput. Geom.*, 6:485–524, 1991.

4. B. Chazelle and D. P. Dobkin. Intersection of convex objects in two and three dimensions. *J. ACM*, 34:1–27, 1987.

5. Y.-J. Chiang, F. P. Preparata, and R. Tamassia. A unified approach to dynamic point location, ray shooting, and shortest paths in planar maps. In *Proc. 4th ACM-SIAM Sympos. Discrete Algorithms*, pages 44–53, 1993.

6. Y.-J. Chiang and R. Tamassia. Optimal shortest path and minimum-link path queries between two convex polygons in the presence of obstacles. Report CS-94-03, Comput. Sci. Dept., Brown Univ., Providence, RI, 1994.

7. F. Chin and C. A. Wang. Optimal algorithms for the intersection and the minimum distance problems between planar polygons. *IEEE Trans. Comput.*, C-32(12):1203–1207, 1983.

8. S. Ghosh. Computing visibility polygon from a convex set and related problems. *J. Algorithms*, 12:75–95, 1991.

9. M. T. Goodrich and R. Tamassia. Dynamic ray shooting and shortest paths via balanced geodesic triangulations. In *Proc. 9th Annu. ACM Sympos. Comput. Geom.*, pages 318–327, 1993.

10. L. J. Guibas and J. Hershberger. Optimal shortest path queries in a simple polygon. *J. Comput. Syst. Sci.*, 39:126–152, 1989.

11. M. H. Overmars and J. van Leeuwen. Maintenance of configurations in the plane. *J. Comput. Syst. Sci.*, 23:166–204, 1981.

12. F. P. Preparata and M. I. Shamos. *Computational Geometry: an Introduction.* Springer-Verlag, New York, NY, 1985.

13. J. H. Reif and J. A. Storer. Minimizing turns for discrete movement in the interior of a polygon. *IEEE J. Robot. Autom.*, pages 182–193, 1987.

14. S. Suri. On some link distance problems in a simple polygon. *IEEE Trans. Robot. Autom.*, 6:108–113, 1990.

15. R. Tamassia. On embedding a graph in the grid with the minimum number of bends. *SIAM J. Comput.*, 16(3):421–444, 1987.

Fast algorithms for collision and proximity problems involving moving geometric objects

Prosenjit Gupta*   Ravi Janardan*   Michiel Smid†

Abstract

Consider a set of geometric objects, such as points or axes-parallel hyper-rectangles in $\mathbb{R}^d$, that move with constant but possibly different velocities along linear trajectories. Efficient algorithms are presented for several problems defined on such objects, such as determining whether any two objects ever collide and computing the minimum inter-point separation or minimum diameter that ever occurs. In particular, two open problems from the literature are solved: Deciding in $o(n^2)$ time if there is a collision in a set of n moving points in $\mathbb{R}^2$, where the points move at constant but possibly different velocities, and the analogous problem for detecting a red-blue collision between sets of red and blue moving points. The strategy used involves reducing the given problem on moving objects to a different problem on a set of static objects, and then solving the latter problem using techniques based on sweeping, orthogonal range searching, simplex composition, and parametric search.

1 Introduction

Problems involving geometric objects that are in time-dependent motion arise in diverse applications, such as, for instance, traffic control, robotics, manufacturing, and animation, to name just a few. In such problems, we are given a collection of geometric objects, such as points, line segments, or polyhedra, along with a description of their motion, which is usually specified by a low-degree polynomial in the time parameter t. The objective is to answer questions concerning (i) properties of the objects (e.g., the closest pair) at a given time instant t or in the so-called "steady-state", i.e., at $t = \infty$; or (ii) the combinatorics of the entire motion i.e., from $t = 0$ to $t = \infty$ (e.g., the number of topologically different Euclidean minimum spanning trees (EMSTs) determined by a set of moving points); or (iii) the existence of certain properties (e.g., collision) or computing the optimal value of some property (e.g., the smallest diameter) over the entire motion.

The systematic study of such dynamic problems was initiated by Atallah [3]. Examples of problems considered by him include computing the time intervals during

---

*Department of Computer Science, University of Minnesota, Minneapolis, MN 55455, U.S.A. E-mail: {pgupta,janardan}@cs.umn.edu. The research of these authors was supported in part by NSF grant CCR-92-00270.

†Max-Planck-Institut für Informatik, Im Stadtwald, D-66123 Saarbrücken, Germany. E-mail: michiel@mpi-sb.mpg.de. This author was supported by the ESPRIT Basic Research Actions Program, under contract No. 7141 (project ALCOM II).

objects	dimension	# directions	velocities	problem	time
points	d	c	same	$\exists$? collision	$cn\log n$
points	2	2 orthogonal	same	closest distance	$n\log n$
points	2	2 orthogonal	different	$\exists$? collision	$n^{3/2}(\log n)^{4/3}$
points	2	arbitrary	different	$\exists$? collision	$n^{5/3}(\log n)^{6/5}$
points	2	arbitrary	different	L_2-diameter	$n\log^3 n$
boxes	d	d orthogonal	same	$\exists$? collision	$n\log^{2d-3} n$
boxes	d	d orthogonal	different	$\exists$? collision	$n^{3/2+\epsilon}$
segments	2	arbitrary	different	$\exists$? collision	$n^{5/3+\epsilon}$

Table 1: Summary of results for problems on n moving objects in $\mathbb{R}^d$, $d \geq 2$. Each object moves with constant velocity from $t = 0$ to $t = \infty$. The velocities are either the same for all objects, or each object has a possibly different velocity. The trajectories of the objects may come from c different directions, two (or d) orthogonal directions, or may be arbitrary. All bounds are "big-oh" and worst-case. The problems that are indicated are collision detection, computing the closest distance, and minimum L_2-diameter over all times $t \geq 0$. The line segments have arbitrary directions, but each one moves along its supporting line.

---

which a given point appears on the convex hull of a set of moving points and determining the steady-state closest/farthest pair, EMST, and smallest enclosing circle for moving points. Fu and Lee [5, 6] show how to maintain the Voronoi diagram and the EMST of moving points in the plane. In [14], Monma and Suri investigate combinatorial and algorithmic questions concerning the EMSTs that arise from the motion of one or more of the input points. In [10], Huttenlocher et al. consider the problem of computing the minimum Hausdorff distance between two rigid planar point-sets under Euclidean motion. In [8] Golin et al. solve a number of query problems on moving points, such as reporting the closest pair and the maximal points at a given query time instant t.

1.1 Summary of results and contributions

In this paper, we address problems of type (iii) above. Specifically, we consider sets of moving objects such as points or axes-parallel hyper-rectangles in $\mathbb{R}^d$ and are interested in questions such as: "Do two objects ever collide?" and "What is the smallest inter-point distance or smallest diameter ever attained?" Note that since our problems do not involve a query time instant, the answer is determined solely by the input, namely the initial positions of the objects, their velocities, and their trajectories. Throughout, we assume that the objects all start moving at $t = 0$, have constant, but possibly different velocities, and move along straight-line trajectories. Of course, the problems that we consider can be solved easily in quadratic time, by brute-force. The challenge is to do significantly better, which makes the solutions interesting and non-trivial. Table 1 summarizes our results.

Collision detection problems arise very often in robotics. For an overview of these problems, we refer to the book by Fujimura [7] and the references listed there. Ottmann and Wood [15] gave efficient solutions for collision detection for points moving on the real line. They also raised the open question of deciding in $o(n^2)$ time whether there is any collision in a set of n moving points in $\mathbb{R}^2$, where the points move along straight lines at constant but possibly different velocities. A related open problem raised by Atallah [3] is to decide in $o(mn)$ time whether there is a red-blue collision between a set of m red points and n blue points, all moving in the plane with different constant velocities. We answer both these questions affirmatively. We are not aware of any previous work on the closest/farthest pair questions that we address.

Our strategy for solving these dynamic problems is to reduce the problem at hand to a different problem on a set of static objects. We then solve the latter problem using techniques such as sweeping, orthogonal range searching, halfspace range searching, simplex compositions, and parametric search. Due to space constraints we will discuss only a subset of our results here; the full paper is available as [9].

2 Collision detection for points moving with the same velocity

Let S be a set of n points in $\mathbb{R}^d$. At time $t = 0$, all points start moving. We want to decide if any two points of S ever collide, and if so, compute the first time instant at which a collision occurs. In this section, we assume that all points move with the same constant velocity $v > 0$. In Section 2.1, we consider the case where the trajectories are in one of c different directions. In Section 2.2, we treat the special case where the points are in the plane and move in two orthogonal directions. In this case, we can even compute the closest distance among the points over all times $t \geq 0$ efficiently.

2.1 Points moving in c different directions

We consider the case where the points move in the plane and the trajectory of each point is oriented in one of c different directions. The d-dimensional case is solved in the full paper. Note that each trajectory is a ray.

Consider one of the directions. Let $C = (x, y)$ be a coordinate system whose y-axis is parallel to this direction and whose x-axis is orthogonal to it. We call the points of S that move in the positive y-direction *blue points*. Consider one of the other directions, and let ϕ, $-\pi/2 < \phi < \pi/2$, be the angle it makes with the positive x-axis. We call the points of S that move along this direction in such a way that their x-coordinates increase *red points*. We will solve our collision detection problem for these red and blue points. For any blue point b, let (b_x, b_y) be its position at time $t = 0$. Then, at time t, this point is at position $(b_x, b_y + vt)$. Similarly, the position of any red point r at time t can be written as $(r_x + vt \cos \phi, r_y + vt \sin \phi)$.

Lemma 2.1 *Let r and b be red and blue points, respectively, such that $r_x \leq b_x$. Let $\alpha_\phi = (1 - \sin \phi)/\cos \phi$. Then, r and b collide iff $\alpha_\phi b_x + b_y = \alpha_\phi r_x + r_y$. Moreover, if there is a collision between these points, then it takes place at time $t = (b_x - r_x)/(v \cos \phi)$.*

This lemma leads to the following algorithm. We sort the red and blue points according to their x-coordinates. Then we sweep from left to right. During this sweep, we maintain a balanced binary search tree T that stores all red points that have been visited already in its leaves. These points are sorted according to their $(\alpha_\phi r_x + r_y)$-values. Red points for which these values are equal are sorted according to their r_x-value. If the sweep line visits a red point r, then we insert $\alpha_\phi r_x + r_y$ into T. If a blue point b is visited, then we search in T for the rightmost leaf storing the value $\alpha_\phi b_x + b_y$. If this leaf does not exist, then b does not collide with any red point. Otherwise, if r is the red point that corresponds to the value that is stored in this leaf, then r and b collide. Moreover, this point r has maximal r_x-coordinate among all red points that ever collide with b. Therefore, by Lemma 2.1, the first time at which b collides with any red point is $t_b = (b_x - r_x)/(v \cos \phi)$.

Having visited all red and blue points, we know that the minimal t_b-value computed is the first time at which there is a collision between a red and a blue point. If no t_b-value has been computed, no collision will ever take place.

Theorem 2.1 *Consider a set S of n points in the plane, where each point is moving with the same constant velocity along a ray that is oriented in one of c different directions. We can determine in $O(cn \log n)$ time and $O(n)$ space if any two points of S ever collide, and if so, the first time a collision takes place. If c is a constant, then this is optimal.*

Remark 2.1 If all the red points have the same constant velocity v_r, and all the blue points have the same constant velocity v_b, then the same approach works.

Our approach also works if each point p moves along a line segment, i.e., from $t = 0$ to $t = T_p$. This is because the given algorithm finds for each point p the first time instant at which a collision involving p occurs.

2.2 The closest pair over time for points moving in two orthogonal directions

In this section, we consider the case where each point moves in the plane, in the direction of the positive x-axis (red point) or positive y-axis (blue point). We show how to compute the minimum separation over time, i.e., the smallest inter-point distance taken over all times $t \geq 0$.

First we consider the red points. It is clear that the distances among these points do not change over time. Therefore, the minimum separation over time among the red points is determined by the closest pair of red points at $t = 0$. Similarly for the minimum separation among the blue points.

It remains to consider the minimum red-blue separation. Consider any red point r and any blue point b. Their positions at time t are $r(t) = (r_x + vt, r_y)$ and $b(t) = (b_x, b_y + vt)$, respectively. The square of their Euclidean distance at time t is given by

$$Z_{rb}(t) = 2v^2 t^2 + 2v((r_x - b_x) - (r_y - b_y))t + (r_x - b_x)^2 + (r_y - b_y)^2,$$

which is a polynomial in t of degree two. Let t_{rb}^* be the time at which Z_{rb} is minimum, where $-\infty < t_{rb}^* < \infty$. We have $t_{rb}^* = \frac{(r_y - r_x) - (b_y - b_x)}{2v}$. It is straightforward to verify that if $t_{rb}^* > 0$ then $\min_{t \geq 0} Z_{rb}(t) = \frac{1}{2}((r_x + r_y) - (b_x + b_y))^2$. Otherwise, $\min_{t \geq 0} Z_{rb}(t) = (r_x - b_x)^2 + (r_y - b_y)^2$, namely the distance between r and b at $t = 0$. This suggests the following solution. First we compute the minimum of the red-red, blue-blue, and red-blue distance at time $t = 0$, using a standard closest pair algorithm on all the points [17]. Then we compute the minimum red-blue separation over time among all red-blue pairs r, b for which $t_{rb}^* > 0$.

The minimum red-blue separation among all red-blue pairs r, b for which $t_{rb}^* > 0$ is computed as follows. For each red point r, we have to find among all blue points b such that $b_y - b_x \leq r_y - r_x$ the one for which $\frac{1}{2}((r_x + r_y) - (b_x + b_y))^2$ is minimal. Clearly, we can as well minimize $|(r_x + r_y) - (b_x + b_y)|$.

We sort all points p according to their $(p_y - p_x)$-values. Ties are broken such that blue points come first in the ordering. Then we sweep over the points in this ordering. During the sweep, we maintain all blue points that have been visited in the leaves of a balanced binary search tree T, in increasing order of their $(b_x + b_y)$-values.

If the sweep line visits a blue point b, then we insert $b_x + b_y$ into T. If a red point r is visited, then we search in T for the smallest (resp. largest) value $b_x' + b_y'$ (resp. $b_x'' + b_y''$) that is at least (resp. at most) $r_x + r_y$. If $|(r_x + r_y) - (b_x' + b_y')| < |(r_x + r_y) - (b_x'' + b_y'')|$, then we know that $\min_{b \in T} \min_{t \geq 0} Z_{rb}(t) = \frac{1}{2}((r_x + r_y) - (b_x' + b_y'))^2$. Otherwise, this minimum is equal to $\frac{1}{2}((r_x + r_y) - (b_x'' + b_y''))^2$.

Having visited all points, we compute the smaller of the minimum overall distance at $t = 0$ and the smallest computed value $\min_b \min_t \sqrt{Z_{rb}(t)}$, taken over all points r. This gives the minimum separation over time among all points.

Theorem 2.2 *Consider a set S of n points in the plane, where each point is moving with the same constant velocity in the positive x-direction or the positive y-direction. In $O(n \log n)$ time using $O(n)$ space we can compute the closest distance between any two points over all times $t \geq 0$.*

Remark 2.2 Our algorithm can easily be extended such that for each point p, the closest distance to any other point over all times $t \geq 0$, is computed.

3 Collision detection for points moving with different velocities

Let S be a set of n points in the plane that are moving with constant velocities. In contrast to the previous section, the velocities may be different for each point.

3.1 Points moving in two orthogonal directions

We assume that the points only move in the positive x-direction (red points) or positive y-direction (blue points). It is easy to detect a collision among the red points (or among the blue points) in $O(n \log n)$ time. Hence, it remains to consider red-blue collisions. Let $r(t) = (r_x + v_r t, r_y)$ be the position of the red point r at time t. Similarly, let $b(t) = (b_x, b_y + v_b t)$ be the position of the blue point b at time t.

Lemma 3.1 *Let r and b be red and blue points, respectively, such that $r_x \leq b_x$. Then, r and b collide iff $r_x + (1/v_b)v_r r_y - (b_y/v_b)v_r = b_x$.*

Let us represent each red point r of S by the point $r' = (r_x, v_r r_y, -v_r)$ in $\mathbb{R}^3$, and each blue point b of S by the plane b' in $\mathbb{R}^3$ with equation $X + (1/v_b)Y + (b_y/v_b)Z = b_x$.

If $r_x \leq b_x$, then Lemma 3.1 implies that r and b collide iff point r' lies on the plane b'. Note that this plane can be written as the intersection of two halfspaces in $\mathbb{R}^3$. To solve our collision detection problem, we use the following result.

Theorem 3.1 (Matoušek [12]) *Let V be a set of n points in $\mathbb{R}^d$, let m be a parameter such that $n \leq m \leq n^d$, let h be an integer such that $1 \leq h \leq d+1$, and let $\delta > 0$ be any real number. In $O(n^{1+\delta} + m(\log n)^\delta)$ time, we can preprocess the points of V into a data structure of size $O(m)$ such that the points of V lying in the intersection of any h halfspaces can be counted in time $O((n/m^{1/d})(\log \frac{m}{n})^{h-(d-h+1)/d})$.*

We store the red points of S in the leaves of a balanced binary search tree, sorted by their r_x-values. At each internal node w of this tree, we store a data structure $D(w)$, storing the red points that are contained in the subtree of w. This structure supports the following query: Given any blue point b, decide if a collision takes place between b and any red point stored in $D(w)$. By the discussion above, we can represent these red points r by the points r' in $\mathbb{R}^3$, and the query point b by the intersection of two halfspaces in $\mathbb{R}^3$. We take for $D(w)$ the data structure of Theorem 3.1 with $d = 3$, $h = 2$ and $m = n_w^{3/2}$, where n_w is the number of red points in the subtree of w.

Given this augmented binary tree, we can solve our red-blue collision detection problem, as follows: For each blue point b, we determine $O(\log n)$ canonical nodes such that the red points stored in the subtrees of these nodes partition the set of all red points r such that $r_x \leq b_x$. For each of these nodes w, we query the data structure $D(w)$ and count the number of red points that collide with b.

Theorem 3.2 *Consider a set S of n points in the plane, all moving with constant but possibly different velocities along trajectories that are parallel to the x- or y-axis. In $O(n^{3/2}(\log n)^{4/3})$ time and $O(n^{3/2})$ space, we can determine if any two points of S ever collide.*

3.2 Points moving in arbitrary directions

We now consider the case where the points move in arbitrary directions. Our approach is similar to that of the previous section. The position of any point p of S at time $t \geq 0$ is given by $p(t) = (p_x + v_{px}t, p_y + v_{py}t)$. Here, v_{px} and v_{py} are the x- and y-components of p's velocity vector v_p, respectively.

Lemma 3.2 *Let p and q be points of S. These points collide iff (i) $(p_x - q_x)(v_{qy} - v_{py}) = (p_y - q_y)(v_{qx} - v_{px})$, and (ii) $p_x - q_x$ and $v_{qx} - v_{px}$ are both less than, greater than, or equal to zero, and (iii) $p_y - q_y$ and $v_{qy} - v_{py}$ are both less than, greater than, or equal to zero.*

For any point p of S, we define the point p' in $\mathbb{R}^5$ by $p' = (p_x, p_y, v_{px}, v_{py}, p_x v_{py} - p_y v_{px})$ and for any point q in S we define q'' in $\mathbb{R}^5$ by $q'' : -v_{qy}X_1 + v_{qx}X_2 + q_yX_3 - q_xX_4 + X_5 + (q_x v_{qy} - q_y v_{qx}) = 0$. Then, condition (i) of Lemma 3.2 holds iff the point p' is contained in the hyperplane q''.

We store the points of S in a five-level data structure, where the first four levels are balanced binary search trees. The first level stores the points sorted by their p_x-values, the second level by their v_{px}-values, the third level by their p_y-values and the fourth level by their v_{py}-values. Let w be any node of the fourth level and let S_w be the subset of S that is stored in w's subtree. Then w stores the data structure $D(w)$ of Theorem 3.1 with $d = 5$, $h = 2$ and $m_w = n_w^{5/3}$, where $n_w = |S_w|$, for the points $\{p' : p \in S_w\}$. Given this data structure, we can solve our collision detection problem, as follows: For each point q of S, we determine $O(\log^4 n)$ canonical nodes of the fourth level, for each case resulting out of combining conditions 2. and 3. of Lemma 3.2. For each of these nodes w, we query the data structure $D(w)$ and count the number of points not equal to q that collide with q.

Theorem 3.3 *Consider a set S of n points in the plane, all moving with constant but possibly different velocities. In $O(n^{5/3}(\log n)^{6/5})$ time and $O(n^{5/3})$ space, we can determine if any two points of S ever collide.*

Remark 3.1 The subquadratic bound provided by Theorem 3.3 solves an open problem raised by Ottmann and Wood [15]. Theorem 3.3 also holds for determining a red-blue collision between a set of n red points and a set of n blue points that move at constant but possibly different velocities. In this case, we build the data structure for the red points and query it with the blue points. This solves an open problem mentioned by Atallah [3].

We leave open the problem of finding the closest distance taken over all times $t \geq 0$ in a set of n points moving in the plane with constant but possibly different velocities, in $o(n^2)$ time. Viewing the trajectory of each point as a ray in xyt-space, we have to find the shortest segment parallel to the xy-plane that connects two rays.

In [16], Pellegrini gives a randomized algorithm for finding the shortest vertical segment that connects two lines in a set of n lines in 3-space. This algorithm has expected running time $O(n^{8/5+\epsilon})$. It is not clear if his technique can be extended to solve our problem since (i) we are looking for a shortest segment parallel to the xy-plane, (ii) we want $t \geq 0$, and (iii) we want a deterministic solution.

4 Collision detection for orthogonally moving boxes in $d \geq 2$ dimensions

Let $\mathcal{B} = B_1 \cup B_2 \cup \cdots \cup B_d$ be a collection of axes-parallel rectangles in $\mathbb{R}^d$, called d-boxes for short. (For simplicity, we assume that $d \geq 2$ is a small constant.) Each B_i contains n_i elements that move in the positive x_i-direction. The B_i's are disjoint and $\sum_{i=1}^d n_i = n$. The problem is to decide if any two d-boxes of $\mathcal{B}$ ever collide.

4.1 The equal-velocities case

Assume that all boxes move with the same constant velocity v. We consider collisions between boxes in B_1 with boxes in B_2. (Collisions between boxes in B_i with boxes in B_j, $j \neq i$, can be detected in the same way.) The boxes of B_1 and B_2 are colored red and blue, respectively. The position of any red box r is given by $r(t) = [r_1 + vt : r_1 + vt + L_{r1}] \times \prod_{i=2}^d [r_i : r_i + L_{ri}]$. Here, $(r_1, \ldots, r_d)$ is the position of the "lower-left" corner of r at $t = 0$ and L_{ri} is the length of r along the i-th dimension. Similarly, the position of any blue box b is given by $b(t) = [b_1 : b_1 + L_{b1}] \times [b_2 + vt : b_2 + vt + L_{b2}] \times \prod_{i=3}^d [b_i : b_i + L_{bi}]$.

Lemma 4.1 *The boxes r and b collide iff the static $(d+1)$-boxes r' and b' intersect, where $r' = [r_1 + r_2 : r_1 + r_2 + L_{r1} + L_{r2}] \times [r_1 : \infty) \times (-\infty : r_2 + L_{r2}] \times \prod_{i=3}^{d}[r_i : r_i + L_{ri}]$ and $b' = [b_1 + b_2 : b_1 + b_2 + L_{b1} + L_{b2}] \times (-\infty : b_1 + L_{b1}] \times [b_2 : \infty) \times \prod_{i=3}^{d}[b_i : b_i + L_{bi}]$.*

We map the red d-boxes r to the $(d+1)$-boxes r' and the blue d-boxes b to the $(d+1)$-boxes b'. Then our problem of detecting collisions between boxes in B_1 with boxes in B_2 is equivalent to detecting intersections between the red $(d+1)$-boxes and the blue $(d+1)$-boxes.

We first solve this problem for the planar case. So, let $d = 2$. Note that $[r_1 : \infty) \times (-\infty : r_2 + L_{r2}]$ and $(-\infty : b_1 + L_{b1}] \times [b_2 : \infty)$ intersect iff the planar point $(r_1, r_2 + L_{r2})$ lies in the north-west quadrant of the point $(b_1 + L_{b1}, b_2)$.

This observation leads to the following solution. We sort the left and right end-points of the first intervals of all boxes r' and b'. Then we sweep over them in this order. During this sweep, we maintain a priority search tree T.

If the sweep line visits a red left endpoint $r_1 + r_2$, then we insert the point $(r_1, r_2 + L_{r2})$ into T. If a red right endpoint $r_1 + r_2 + L_{r1} + L_{r2}$ is visited, then the point $(r_1, r_2 + L_{r2})$ is deleted from T. If the sweep line visits a blue left endpoint $b_1 + b_2$ or a blue right endpoint $b_1 + b_2 + L_{b1} + L_{b2}$, then we search in T for the red points that are in the north-west quadrant of the point $(b_1 + L_{b1}, b_2)$. Of course, the entire algorithm stops as soon as a collision is detected.

However, in this way, we might miss intersections between boxes r' and b' such that the first interval of r' is completely contained in the first interval of b'. Therefore, we repeat the above sweep algorithm with the roles of r and b interchanged.

It is clear that the entire algorithm correctly solves our problem and that its running time is bounded by $O(n \log n)$ and the space used is $O(n)$ since T has query and update time $O(\log n)$ and uses $O(n)$ space.

We now extend this solution to the d-dimensional case. For $3 \leq i \leq d$, the intervals $[r_i : r_i + L_{ri}]$ and $[b_i : b_i + L_{bi}]$ intersect iff $b_i \leq r_i + L_{ri}$ and $b_i + L_{bi} \geq r_i$. Therefore, we obtain a solution for the d-dimensional problem by adding $2(d-2)$ orthogonal range restrictions (see [18]) to the solution for the planar case.

Theorem 4.1 *Let $\mathcal{B}$ be a collection of n boxes in $\mathbb{R}^d$, where each box moves parallel to one of the coordinate axes. If all the boxes move with the same constant velocity, then we can decide in $O(n \log^{2d-3} n)$ time and $O(n \log^{2d-4} n)$ space whether any two boxes ever collide.*

4.2 The different-velocities case

In this section, we consider the case where the boxes move with constant but possibly different velocities. Our solution uses the method of simplex composition, which was introduced by van Kreveld [11]. We first review this method briefly.

Let S be a set of n geometric objects in $\mathbb{R}^d$ and let T be a data structure for some query problem on S. Suppose that we wish now to solve our query problem not w.r.t. S but w.r.t. a subset S' of S that satisfies some condition. Moreover, suppose that S' can be specified by putting S in 1-1 correspondence with a set $\mathcal{P}$ of n points in $\mathbb{R}^d$ and letting S' correspond to the subset $\mathcal{P}'$ of $\mathcal{P}$ that is contained in some query simplex. Van Kreveld gives an efficient data structure to solve the query problem on S', based on combining cutting trees and partition trees. He calls his technique *a simplex composition on* $\mathcal{P}$ *to* T. His result is as follows. (We only state the part of his result that is relevant to us. Also, the building time stated below is not given in [11] but can be derived easily.)

Theorem 4.2 (van Kreveld [11]) *Let* $\mathcal{P}$ *be a set of* n *points in* $\mathbb{R}^d$, *and let* S *be a set of* n *objects in* $\mathbb{R}^d$ *in correspondence with* $\mathcal{P}$. *Let* T *be a data structure on* S *having building time* $p(n)$, *size* $f(n)$ *and query time* $g(n)$. *Let* ϵ *be a positive constant and let* m *be a parameter such that* $n \leq m \leq n^d$. *The application of simplex composition on* $\mathcal{P}$ *to* T *results in a data structure having building time* $O(m^\epsilon(m + p(n)))$, *size* $O(m^\epsilon(m + f(n)))$ *and query time* $O(n^\epsilon(g(n) + n/m^{1/d}))$.

We now return to the problem at hand. Again, as in Section 4.1, we consider collisions between boxes in B_1 with boxes in B_2. These boxes are colored red and blue, respectively. The position of any red box r is given by $r(t) = [r_1 + v_r t : r_1 + v_r t + L_{r1}] \times \prod_{i=2}^d [r_i : r_i + L_{ri}]$, and the position of any blue box b is given by $b(t) = [b_1 : b_1 + L_{b1}] \times [b_2 + v_b t : b_2 + v_b t + L_{b2}] \times \prod_{i=3}^d [b_i : b_i + L_{bi}]$.

Lemma 4.2 *The boxes* r *and* b *collide iff (i)* $v_r r_2 - v_r(b_2 + L_{b2}) - v_b(b_1 + L_{b1}) + v_b r_1 \leq 0$, *and (ii)* $v_r(r_2 + L_{r2}) - v_r b_2 - v_b b_1 + (r_1 + L_{r1})v_b \geq 0$, *and (iii)* $b_1 + L_{b1} \geq r_1$, *and (iv)* $r_2 + L_{r2} \geq b_2$, *and (v)* $\prod_{i=3}^d [r_i : r_i + L_{ri}]$ *intersects* $\prod_{i=3}^d [b_i : b_i + L_{bi}]$.

The first two conditions of Lemma 4.2 are equivalent to saying that the point $b' = (-(b_2 + L_{b2}), -v_b(b_1 + L_{b1}), v_b)$ is in the lower halfspace $r' : v_r X_1 + X_2 + r_1 X_3 + v_r r_2 \leq 0$, and the point $b'' = (-b_2, -v_b b_1, v_b)$ is in the upper halfspace $r'' : v_r X_1 + X_2 + (r_1 + L_{r1})X_3 + v_r(r_2 + L_{r2}) \geq 0$.

Thus, these two conditions can be handled by doing two halfspace compositions in $\mathbb{R}^3$. Specifically, we apply Theorem 4.2 as follows: We take B_2, compute b' for each $b \in B_2$, and build the partition–tree–cutting–tree structure underlying Theorem 4.2 on these points b'. At each node v of this structure we store a secondary structure $D(v)$, which is the 3-dimensional halfspace range reporting structure given in [1]. $D(v)$ is built on a set b'' of points such that b' is in v's subtree. $D(v)$ can be trivially modified so that for any query halfspace, it returns "false" iff the halfspace contains some point of $D(v)$. Denote this structure for B_2 by T_2. We take B_1 and for each $r \in B_1$ we compute r' and r''. We query the outer structure of T_2 with r' and identify a set of canonical nodes. At each canonical node v, we query $D(v)$ with r''. We report a collision (modulo the remaining conditions (iii), (iv) and (v) of Lemma 4.2) iff the query on some $D(v)$ returns "false". To handle the remaining conditions of Lemma 4.2, we apply to T_2 the $2d - 2$ (orthogonal) range restrictions that these conditions specify.

Let us now analyze the complexity of this solution. Consider the structure T_2. Suppose that $D(v)$ is built on s_v points and let $p(s_v)$, $f(s_v)$, and $g(s_v)$ be the preprocessing time, space, and query time of $D(v)$, respectively. From [1], we have $p(s_v) = f(s_v) = O(s_v \log s_v)$ and $g(s_v) = O(\log s_v)$. Let m_2 be a parameter, where $n_2 \leq m_2 \leq n_2^3$ and let $\epsilon > 0$ be an arbitrarily small constant. Let $P(n_2)$, $F(n_2)$, and $G(n_2)$ be the preprocessing time, space, and query time of T_2, respectively. From Theorem 4.2, we get $P(n_2) = F(n_2) = O(m_2^{1+\epsilon} + m_2^\epsilon n_2 \log n_2) = O(m_2^{1+\epsilon})$. Also, $G(n_2) = O(n_2^\epsilon \log n_2 + n_2^{1+\epsilon}/m_2^{1/3}) = O(n_2^{1+\epsilon}/m_2^{1/3})$. The addition of the range restrictions to T_2 contributes a $\log^{\Theta(d)} n_2$ factor to each of $P(n_2)$, $F(n_2)$, and $G(n_2)$, so that the bounds for the range-restricted T_2 are asymptotically the same as above.

Let $R(n_1, n_2)$ be the time to build the range-restricted T_2 and to query it with each of the n_1 boxes of B_1. Then $R(n_1, n_2) = P(n_2) + n_1 \cdot G(n_2) = O(m_2^{1+\epsilon} + n_1 \cdot n_2^{1+\epsilon}/m_2^{1/3})$. Choosing $m_2 = (n_1 \cdot n_2^{1+\epsilon})^{1/(4/3+\epsilon)}$ gives $R(n_1, n_2) = O((n_1 \cdot n_2^{1+\epsilon})^{(1+\epsilon)/(4/3+\epsilon)}) = O((n_1 \cdot n_2)^{3/4+\epsilon}) = O(n^{3/2+\epsilon})$. It can be verified that the total space is also $O(n^{3/2+\epsilon})$.

Theorem 4.3 *Let B be a collection of n boxes in $\mathbb{R}^d$, where each box moves parallel to one of the coordinate axes with a constant but possibly different velocity. We can decide in $O(n^{3/2+\epsilon})$ time and space whether any two boxes ever collide.*

5 Computing the minimum L_2-diameter over all times $t \geq 0$

Let S be a set of n points in the plane that are moving at constant but possibly different velocities. The diameter of the points at time t is the largest distance among all pairs of points at time t. In this section, we consider the problem of computing the minimum L_2-diameter over all times $t \geq 0$. We will solve this problem using Megiddo's parametric search technique. (See [13].) First we review this paradigm.

5.1 Parametric search

Suppose we have a decision problem $\mathcal{P}(t)$ that receives as input n data items and a real parameter t. Assume that $\mathcal{P}$ is monotone, meaning that if $\mathcal{P}(t_0)$ is true for some t_0, then $\mathcal{P}(t)$ is also true for all $t < t_0$. Our aim is to find the maximal value of t for which $\mathcal{P}(t)$ is true. We denote this value by t^*.

Assume we have a sequential algorithm A_s that, given the n data items and t, decides if $\mathcal{P}(t)$ is true or not. The control flow of this algorithm is governed by comparisons, each of which involves testing the sign of some low-degree polynomial in t. Let T_s and C_s denote the running time and the number of comparisons made by algorithm A_s, respectively. Note that by running A_s on input t, we can decide if $t \leq t^*$ or $t > t^*$: we have $t \leq t^*$ iff $\mathcal{P}(t)$ is true. The parametric search technique simulates A_s generically on the unknown critical value t^*. Whenever A_s reaches a branching point that depends on a comparison operation, the comparison can be reduced to testing the sign of a suitable low-degree polynomial $f(t)$ at $t = t^*$. The algorithm computes the roots of this polynomial and checks each root a to see if it is less than or equal to t^*. In this way, the algorithm identifies two successive roots between which t^* must lie and thus determines the sign of $f(t^*)$. In this way we get an interval I in which t^* can possibly lie. Also the comparison now being resolved, the generic execution can proceed. As we proceed through the execution, each comparison that we resolve results in constraining I further and we get a sequence of progressively smaller intervals each known to contain t^*. The generic simulation will run to completion and we are left with an interval I that contains t^*. It can be shown that for any real number $r \in I$, $\mathcal{P}(r)$ is true. Therefore, t^* must be the right endpoint of I.

Since A_s makes at most C_s comparisons during its execution, the entire simulation and, hence, the computation of t^* take $O(C_s T_s)$ time. To speed up this algorithm, Megiddo replaces A_s by a parallel algorithm A_p that uses P processors and runs in T_p parallel time. At each parallel step, A_p makes at most P independent comparisons. Then our algorithm simulates A_p sequentially, again at the unknown value t^*. At each parallel step, we get at most P low-degree polynomials in t. We compute the roots of all of them and do a binary search among them using repeated median finding to make the probes for t^*. For each probe, we run the sequential algorithm A_s. In this way, we get the correct sign of each polynomial in t^*, and our algorithm can simulate the next parallel step of A_p. The entire algorithm computes t^* in time $O(PT_p + T_sT_p \log P)$.

5.2 Applying parametric search

Let t^* be the time at which the Euclidean diameter of S is minimal. If t^* is not unique, then we take the largest possible value. The position of any point point p of S is given by $p(t) = (p_x + v_{px}t, p_y + v_{py}t)$. Let $Z_{pq}(t)$ denote the square of the Euclidean distance between the points p and q. Note that $Z_{pq}(t)$ is a polynomial of degree two. Let t_{pq}^* be the time at which p and q are closest. Again, if t_{pq}^* is not unique, then we take the largest possible value. Finally, let $D(t)$ denote the diameter of S at time t.

Lemma 5.1 *Let $t \geq 0$. Then $t > t^*$ iff $t > t_{pq}^*$ for all points $p, q \in S$ such that $Z_{pq}(t) = D^2(t)$.*

The decision problem $\mathcal{P}(t)$ we need to solve is as follows: Given n points in the plane, all moving with constant but possibly different velocities and a real number $t \geq 0$, decide if $t > t_{pq}^*$ for all points $p, q \in S$ such that $Z_{pq}(t) = D^2(t)$. Clearly, $\mathcal{P}(t)$ is monotone (with "true" and "false" interchanged) and by Lemma 5.1 t^* is the maximal t for which $\mathcal{P}(t)$ is false. Thus, in order to apply parametric search, we need an algorithm that finds all pairs $p, q \in S$ that achieve the diameter at time t, and checks for each such pair if $t > t_{pq}^*$. The sequential algorithm A_s does the following: It first computes the position of the points at time t. Then it computes all pairs $p, q \in S$ such that $Z_{pq}(t) = D^2(t)$. (See [17].) Finally, for each such pair, it checks if $t > t_{pq}^*$. All this can be done in $T_s = O(n \log n)$ time. (Note that there are only $O(n)$ diametral pairs.) For the parallel algorithm A_p, we use the parallel version of the above algorithm as given in Akl and Lyons [2]. This version uses $P = n$ processors and runs in $T_p = O(\log n)$ parallel time on a CREW PRAM. We run the parametric search with these algorithms. Having found t^*, we run A_s once more with this value to get the diameter at time t^*.

Theorem 5.1 *Consider a set S of n points in the plane, all moving with constant but possibly different velocities. In $O(n \log^3 n)$ time, we can compute the minimum L_2-diameter of S taken over all times $t \geq 0$.*

6 Concluding remarks and open problems

We have given several techniques to solve collision detection and distance problems on moving objects. Our main strategy was to reduce the problem to a problem for other objects that do not move and then to solve the latter by known techniques.

A number of interesting open problems remain. First, can some of our bounds be improved? E.g., can we compute the minimum L_2-diameter over time of n moving planar points in $O(n \log n)$ time? Can we extend our solution to three dimensions? Note that in our approach, we need a parallel algorithm for finding all diametral pairs at a given time t. Fortunately, in 3-space, there are only $O(n)$ such pairs and, moreover, there is an efficient algorithm to compute them [4]. What is lacking is an efficient parallel algorithm. In dimensions higher than three, our approach will not be efficient because the number of diametral pairs is $\Theta(n^2)$. (See [17, pages 182–183].)

Another open problem is to extend our solution of Section 2.2 for computing the minimum inter-point distance over time to the case where the points move with constant but possibly different velocities and directions. Finally, can we generalize our approach to detect collisions in a set of moving simplices?

References

[1] A. Aggarwal, M. Hansen and T. Leighton. *Solving query-retrieval problems by compacting Voronoi diagrams.* Proc. 22nd STOC, 1990, 331–340.

[2] S.G. Akl and K.A. Lyons. *Parallel computational geometry.* Prentice–Hall, Englewood Cliffs, 1993.

[3] M.J. Atallah. *Some dynamic computational geometry problems.* Computers and Mathematics with Applications 11 (1985), pp. 1171–1181.

[4] B. Chazelle, H. Edelsbrunner, L. Guibas and M. Sharir. *Diameter, width, closest line pair, and parametric searching.* Proc. 8th ACM Symp. Computational Geometry, 1992, 120–129.

[5] J.-J. Fu and R.C.T. Lee. *Voronoi diagrams of moving points in the plane.* Intern. J. of Computational Geometry & Applications 1 (1991), pp. 23–32.

[6] J.-J. Fu and R.C.T. Lee. *Minimum spanning trees of moving points in the plane.* IEEE Transactions on Computers 40 (1991), pp. 113–118.

[7] K. Fujimura. *Motion planning in dynamic environments.* Springer-Verlag, 1991.

[8] M.J. Golin, C. Schwarz and M. Smid. *Further dynamic computational geometry.* Proc. 4th Canadian Conf. on Computational Geometry, 1992, 154–159.

[9] P. Gupta, R. Janardan and M. Smid. *Fast algorithms for collision and proximity problems involving moving geometric objects.* Report MPI-I-94-113, Max-Planck-Institut für Informatik, Saarbrücken, 1994.

[10] D.P. Huttenlocher, K. Kedem and J.M. Kleinberg. *On dynamic Voronoi diagrams and the minimum Hausdorff distance for point sets under Euclidean motion in the plane.* Proc. 8th ACM Symp. on Computational Geometry, 1992, 110–119.

[11] M. van Kreveld. *New results on data structures in computational geometry.* Ph.D. Thesis, University of Utrecht, The Netherlands, 1992.

[12] J. Matoušek. *Range searching with efficient hierarchical cuttings.* Discrete & Computational Geometry 10 (1993), pp. 157–182.

[13] N. Megiddo. *Applying parallel computation algorithms in the design of serial algorithms.* Journal of the ACM 30 (1983), pp. 852–865.

[14] C.L. Monma and S. Suri. *Transitions in geometric minimum spanning trees.* Proc. 7th ACM Symp. on Computational Geometry, 1991, 239–249.

[15] T. Ottmann and D. Wood. *Dynamic sets of points.* Computer Vision, Graphics and Image Processing 27 (1984), pp 157–166.

[16] M. Pellegrini. *Incidence and nearest-neighbor problems for lines in 3-space.* Proc. 8th ACM Symp. on Computational Geometry, 1992, 130–137.

[17] F.P. Preparata and M.I. Shamos. *Computational geometry, an introduction.* Springer–Verlag, 1988.

[18] D.E. Willard and G.S. Lueker. *Adding range restriction capability to dynamic data structures.* Journal of the ACM 32 (1985), pp. 597–617.

REVERSE-FIT: A 2-OPTIMAL ALGORITHM FOR PACKING RECTANGLES

Ingo Schiermeyer
Lehrstuhl C für Mathematik
Technische Hochschule Aachen
D-52056 Aachen, Germany

Abstract

We describe and analyze a "level-oriented" algorithm, called "Reverse-Fit", for packing rectangles into a unit-width, infinite-height bin so as to minimize the total height of the packing. For L an arbitrary list of rectangles, all assumed to have width no more than 1, let h_{OPT} denote the minimum possible bin height within the rectangles in L can be packed, and let $RF(L)$ denote the height actually used by Reverse-Fit. We will show that $RF(L) \leq 2 \cdot h_{OPT}$ for an arbitrary list L of rectangles.

Key words: level-oriented packing algorithm, bin-packing, two-dimensional packing, k-rectangle packing problem, rectangle packing conjecture

1 Introduction

We consider the following two-dimensional packing problem, first proposed in [1]: Given a collection of rectangles, and a bin with fixed width and unbounded height, pack the rectangles into the bin so that no two rectangles overlap and so that the height to which the bin is filled is as small as possible. We shall assume that the given rectangles are oriented, each having a specified side that must be parallel to the bottom of the bin. We also assume, with no loss of generality, that the bin width has been normalized to 1.

This problem is a natural generalization of the one-dimensional bin-packing problem. Indeed, if all rectangles are required to have the same height, then the two problems coincide. On the other hand, the case in which all rectangles have the same width corresponds to the well-known makespan minimization problem of combinatorial scheduling theory. Both these restricted problems are known to be NP-complete (cf. [4]), from which it follows that the two-dimensional packing problem is also NP-complete.

However, several approximation algorithms have been established which run in polynomial time, among them the so-called *level-oriented packing algorithms*, which can be described as follows: Assume that the rectangles of an arbitrary list L are ordered by decreasing (actually, nonincreasing) height, then the rectangles are packed in the order given by L so as to form a sequence of *levels*. All rectangles will be placed with their bottoms resting on one of these levels. The first level is simply the bottom of the bin. Each subsequent level is defined by a horizontal line drawn through the top of the first (and hence maximum height) rectangle placed on the previous level. Notice how this corresponds with one-dimensional bin-packing; the horizontal slice determined by two adjacent levels can be regarded as a bin (lying on its side) whose width is determined by the maximum height rectangle placed in that bin. The following two level-oriented algorithms are suggested by analogous algorithms studied for one-dimensional bin-packing:

1. **Next-Fit (NF)**. With this algorithm, rectangles are packed left-justified on a level until there is insufficient space at the right to accomodate the next rectangle. At that point, the next level is defined, packing on the current level is discontinued, and packing proceeds on the new level.

2. **First-Fit (FF)**. At any point in the packing sequence, the next rectangle to be packed is placed left-justified on the first (i.e., lowest) level on which it will fit. If none of the current levels will accomodate this rectangle, a new level is started as in the NF algorithm.

Now for L an arbitrary list of rectangles, all assumed to have width no more than 1, let h_{OPT} denote the minimum possible height within the rectangles in L can be packed. Let $NF(L)$ and $FF(L)$ denote the height actually used by these algorithms when applied to L. The following upper bounds for NF and FF have been shown in [3].

Theorem 1.1 *For any list L ordered by nonincreasing height,*

$$NF(L) \leq 3 \cdot h_{OPT},$$

$$FF(L) \leq 2,7 \cdot h_{OPT}.$$

Note that it is easy to verify that $FF(L) \leq NF(L)$ for all lists L.

In [5] Sleator presents an algorithm that packs in $2,5 \cdot h_{OPT}$ which has been the best *absolute performance bound* known so far. This algorithm can be considered as a modification of First-Fit. In the first step all rectangles of width greater than $1/2$ are stacked on top of one another in the bottom of the bin. If the height to which they reach is called H_0, then all subsequent packing will occur above H_0. Actually, this first step has turned out to be of major importance in order to prove the performance bound.

In this paper we present a packing algorithm, called **Reverse-Fit (RF)**, which will be described in section 2. The basic idea of this algorithm is that, instead of packing every block in a left-to-right manner, it alternately packs blocks from left-to-right and then from right-to-left. In section 3 we will show that Reverse-Fit packs in $2 \cdot h_{OPT}$ followed by a discussion and two rectangle packing conjectures in section 4.

2 The Packing Algorithm

In order to describe the algorithm, the following additional notations will be useful. We shall associate with the bin all points (x, y) of the plane having coordinates $0 \leq x \leq 1$ and $y \geq 0$. Thus the left and the right corner of the bin will be associated with $(0, 0)$ and $(1, 0)$, respectively. Let the list L be given as $r_1, r_2, \ldots, r_n$. For a particular rectangle r_i of width w_i and height h_i, its coordinates of the lower left, lower right, upper right and upper left corner will be given by $(a_i, c_i), (b_i, c_i), (b_i, d_i)$ and (a_i, d_i). Analogously, for a particular rectangle r we shall speak of $w(r), h(r), a(r), b(r), c(r)$ and $d(r)$.

Reverse-Fit

1. Stack all the rectangles of width greater than $1/2$ on top of one another in the bottom of the bin. Call the height to which they reach H_0, and the total area of these rectangles A_0. All subsequent packing will occur above H_0.

2. Sort the remaining rectangles in order of decreasing height. The rectangles will be packed into the bin in this order. Let h_{max} be the height of the tallest of these rectangles.

3. Now pack rectangles from left to right with their bottoms along the line of height H_0 with the first rectangle adjacent to the left wall of the bin, and each subsequent rectangle adjacent to the one just packed. Continue until there is no more room or there are no rectangles left to pack (Figure 1).

4. Let d_1 be the height of the tallest of the remaining rectangles. Now pack rectangles from right to left with their tops along the line of height $H_0 + h_{max} + d_1$ with the first rectangle adjacent to the right wall of the bin. Let this be the second *reverse-level* and now pack each subsequent rectangle according to First-Fit. Hence, each subsequent rectangle is packed left-justified on the first level, if it fits there, or right-justified on the second reverse-level, respectively. Continue until the total width of the rectangles packed on the second reverse-level is at least $1/2$ or there are no rectangles left. In the latter case the algorithm stops. Next all rectangles from the second reverse-level are dropped (each of them by the same amount, say e_1), until (at least) one of them touches some rectangle below. Set $H_1 := h_{max} + d_1 - e_1$. Let r_p and r_q be the right most pair of touching rectangles with r_p placed on the first level and r_q placed on the second level, respectively. Let $m_1 := max(a_p, a_q), m_2 := min(b_p, b_q)$. Then the 'touching-line' $T(r_p, r_q)$ of r_p and r_q is given by $T(r_p, r_q) = \{(x, d_p) \mid m_1 \le x \le m_2\}$.

   If $m_2 \ge 1/2$, then set $v_p := max(1/2, m_1)$ and goto step 5.

   If $m_2 < 1/2$, then note that there are at least two rectangles placed on the second reverse-level because of step 1. Let r_k and r_j be the last and the last but one rectangle packed on the second reverse-level, respectively. Now all rectangles from the second reverse-level, except for r_k, are dropped again (each of them by the same amount, say H_2) until (at least) one of them touches some rectangle below (Figure 6). Again we determine the right most pair of touching rectangles r_p and r_q. Since $H_2 > 0$ we have $r_q \neq r_k$ and thus $m_2 \geq m_1 = max(a_p, a_q) \geq a_j > 1/2$. Now, if $H_2 = h_k$, then the third level is defined by a horizontal line drawn through the top of r_j. Next r_k is moved leftwards along this line as far as possible. Hence, it either touches the left wall or a rectangle placed on the first level. Continue with step 5.

If $H_2 > h_k$ (Figure 7), then (again) the third level is defined by a horizontal line drawn through the top of r_j. This time r_k is packed left-justified on the third level, touching a rectangle placed on the first level, since $H_2 > h_k$.

5. We now continue packing rectangles by a modified First-Fit algorithm. Each subsequent level (starting with the third one) is defined by a horizontal line drawn through the top of the first (and hence maximum height) rectangle packed on the previous level. Note that on such a new level, the first rectangle, which is packed left-justified, need not touch the left wall but a rectangle from the first level.

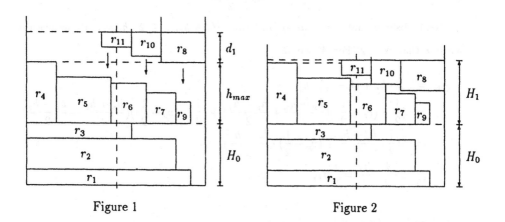

Figure 1   Figure 2

3   The Performance Bound

Some additional notation will be useful for showing the desired performance bound of Reverse-Fit. Packings will be regarded as a sequence of blocks $B_0, B_1, \ldots, B_N$, where the index increases from the bottom to the top of the packing. Let R_i denote the total area of the rectangles in block B_i, and let H_i denote the height of block $B_i, i = 0$ or $3 \leq i \leq N$. The (partial) area of R_i between two vertical lines drawn through $(x, 0)$ and $(y, 0)$ for a pair x, y such that $0 \leq x \leq y \leq 1$ will be denoted by $R_i(x, y)$. Let s_i be the first rectangle placed in block B_i, and let r_{ij} be the last rectangle which has been packed in block B_i at the time r_j is packed.

Theorem 3.1 *Let $RF(L)$ be the height of a packing given by Reverse-Fit. Then*

$$RF(L) \leq 2 \cdot h_{OPT}$$

for any list L ordered by nonincreasing height.

Proof: Since over half of the area of the bin below height H_0 is filled with rectangles we have

$$A_0 \geq \frac{1}{2} H_0.$$

Next observe that $H_0 \leq h_{OPT}$ and thus $H_0 + h_{max} \leq 2 \cdot h_{OPT}$. Hence we may assume that $N \geq 2$. For A_1 we have

$$
\begin{aligned}
A_1 &= v_p \cdot h_p + (1 - v_p) \cdot h_q \\
&= (\frac{1}{2} + (v_p - \frac{1}{2}))(\frac{h_p + h_q}{2} + \frac{h_p - h_q}{2}) + (\frac{1}{2} - (v_p - \frac{1}{2}))(\frac{h_p + h_q}{2} - \frac{h_p - h_q}{2}) \\
&= 2(\frac{1}{2} \cdot \frac{h_p + h_q}{2} + (v_p - \frac{1}{2}) \cdot \frac{h_p - h_q}{2}) \\
&\geq 2 \cdot \frac{1}{2} \cdot \frac{H_1}{2} = \frac{1}{2} H_1,
\end{aligned}
$$

since $h_p \geq h_q, v_p \geq \frac{1}{2}$ and $H_1 = h_p + h_q$.

For A_2 we obtain

$$A_2 \geq (\frac{1}{2} - a_k)H_2 + (a_k - 0)(h_q + H_2 - h_k) \geq \frac{1}{2}H_2,$$

since $h_k \leq h_q$ (Figure 6). Note that $H_2 = 0$ if $v_p = \frac{1}{2}$.

Let r_m be the first rectangle placed on the third level (i.e., $r_m = s_3$). Now suppose that $v_p = \frac{1}{2}$. If $b_m \leq \frac{1}{2}$ (Figure 3 and 6), then

$$> \frac{1}{2}h_m = \frac{1}{2}H_3,$$

since $h_m \leq h_{1m}, h_m \leq h_k, w_m \leq b_m$ and r_m does not fit on the first level. If $b_m > \frac{1}{2}$ (Figure 4 and 7), then

$$A_3 \geq (a_m - 0)h_k + (\frac{1}{2} - a_m)h_m + (b_{1m} - \frac{1}{2})h_{1m}$$
$$\geq b_{1m}h_m > \frac{1}{2}h_m = \frac{1}{2}H_3,$$

since $b_{1m} > \frac{1}{2}$.

Next suppose that $v_p > \frac{1}{2}$. If $0 \leq H_2 \leq h_k$ (Figure 6), then (as above)

$$A_3 \geq (a_m - 0)h_k + (min(\frac{1}{2}, b_m) - a_m)h_m + R_2(\frac{1}{2}, v_p) + (b_{1m} - v_p)h_{1m}$$
$$\geq min(\frac{1}{2}, b_m)h_m + (b_{1m} - \frac{1}{2})h_m$$
$$\geq (w_m + b_{1m} - \frac{1}{2})h_m > \frac{1}{2}h_m = \frac{1}{2}H_3.$$

If $H_2 > h_k$ (Figure 7), then

$$A_3 \geq (a_k - 0)h_j + (\frac{1}{2} - a_k)h_k + (b_{1k} - v_p)h_{1k}$$
$$\geq (\frac{1}{2} + b_{1k} - v_p)h_k > \frac{1}{2}h_k = \frac{1}{2}H_3,$$

since $b_{1k} - v_p > 0$.

If $a(s_4) = 0$, then

$$A_3 \geq (a_m - 0)h_k + w_mh_m + (b_{1m} - \frac{1}{2})h_{1m}$$
$$\geq (b_m + b_{1m} - \frac{1}{2})h_m$$

$$A_4 \geq w_m h_m + (b_{3m} - \frac{1}{2})h_{3m}$$

$$\geq (b_{3m} - \frac{1}{2} + w_m)h_m > \frac{1}{2}h_m = \frac{1}{2}H_4,$$

since w_m does not fit on the third level.

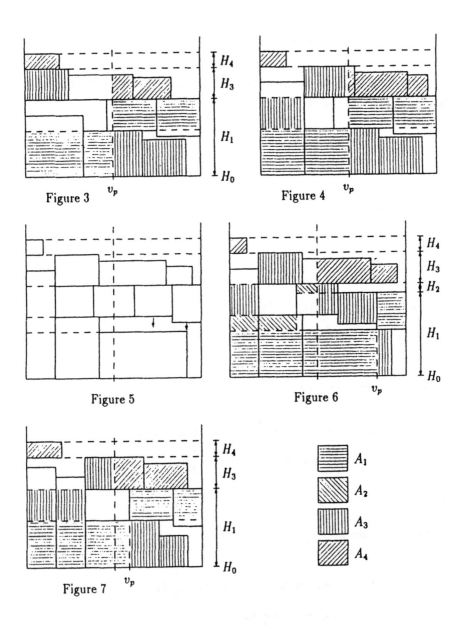

Figure 3

Figure 4

Figure 5

Figure 6

Figure 7

If $a(s_4) > 0$, then (as above)

$$
\begin{aligned}
A_4 &\geq (a_m - 0)h_k + (min(\frac{1}{2}, b_m) - a_m)h_m + (b_{3m} - \frac{1}{2})h_{3m} \\
&\geq (min(\frac{1}{2}, b_m) + b_{3m} - \frac{1}{2})h_m \\
&\geq (b_{3m} - \frac{1}{2} + w_m)h_m > \frac{1}{2}h_m = \frac{1}{2}H_4.
\end{aligned}
$$

In the same way, for each pair of blocks B_i and B_{i+1} with $2 \leq i \leq N - 1$, there is an area A_{i+1} such that

$$
A_{i+1} \geq \frac{1}{2}H_{i+1}.
$$

Altogether we obtain

$$
\sum_{i=0}^{N} R_i \geq \sum_{i=0}^{N} A_i \geq \frac{1}{2} \sum_{i=0}^{N} H_i,
$$

where $RF(L) = H_0 + H_1 + \ldots + H_N$. This completes the proof. $\qquad \square$

4   Discussion and Rectangle Packing Conjectures

First note that by Reverse-Fit at most once rectangles are packed from right-to-left. Hence, with a repeat of this approach, one might expect that the performance bound could be reduced below 2. However, the proof of the performance bound in the previous section already indicates, that it might not be that easy. Especially, step 1 of Reverse-Fit could not be applied any longer in its present form.

Next we consider the following restricted rectangle packing problem which has been suggested by Brucker [2]. We shall call it the **k-rectangle packing problem**.

k-RECTANGLE PACKING PROBLEM

INSTANCE: A set of positive integers w, k, n, w_i, h_i, H such that

$n = \sum_{i=1}^{k} n_i, w_i \leq w$ for $1 \leq i \leq k$, a bin of width w

and infinite height, k pairwise different rectangles r_i

of width w_i and height h_i $((w_i, h_i) \neq (w_j, h_j)$ for

$1 \leq i < j \leq k)$, each of them occuring n_i times.

QUESTION: Is there a packing of the n rectangles into the bin

of height H or less?

For $k = 1$, Next-Fit, First-Fit and Reverse-Fit always find an optimal packing of the n rectangles which can be easily verified. Hence, the **1-rectangle packing problem** can be determined in polynomial time. However, for $k \geq 2$, the complexity of the k-rectangle packing problem seems not to be known so far, as has been mentioned by Brucker and Grötschel [2].

Analyzing several instances for the k-rectangle packing problem we have recognized that the minimum height for a particular instance often depends on the number theoretic properties of the set of integers w, w_i, h_i and n_i. Especially, the difference $h_{OPT} - \frac{1}{w}\sum_{i=1}^{k} n_i w_i h_i$ of h_{OPT} and the (theoretical) lower bound for the height of a packing can be arbitrarily large. On the other hand, because of the (often) 'complicated structures' of optimal packings for particular instances, it seems not to be that easy to reduce another NP-complete or NP-hard problem to the k-rectangle packing problem in order to show that it is NP-complete. All this gives reason for the following two conjectures.

Weak rectangle packing conjecture. The 2-rectangle packing problem can be solved in polynomial time depending on n.

We even believe that this is true for any fixed $k \geq 2$.

Strong rectangle packing conjecture. For any fixed $k \geq 2$ the k-rectangle packing problem can be solved in polynomial time depending on n and k.

References

[1] B. S. Baker, E. G. Coffman, Jr. and R. L. Rivest, Orthogonal packings in two dimensions, SIAM J. Comput., Vol. 9, No. 4 (1980) 846 - 855.

[2] P. Brucker and M. Grötschel, personal communication, January 1994.

[3] E. G. Coffman, Jr., M. R. Garey, D. S. Johnson and R. E. Tarjan, Performance bounds for level-oriented two-dimensional packing algorithms, SIAM J. Comput., Vol. 9, No. 4 (1980) 808 - 826.

[4] M. R. Garey and D. S. Johnson, *Computers and Intractability, A Guide to the Theory of NP-Completeness*, W. H. Freeman and Company, San Francisco, 1979.

[5] D. D. K. D. B. Sleator, A 2.5 times optimal algorithm for packing in two dimensions, Inform. Process. Lett. 10, no. 1 (1980) 37 - 40.

An Optimal Algorithm for Preemptive On-line Scheduling

Bo Chen[1], André van Vliet[1] and Gerhard J. Woeginger[2]

[1] Econometric Institute, Erasmus University Rotterdam, P.O. Box 1738, 3000 DR Rotterdam, The Netherlands
[2] TU Graz, Institut für Mathematik, Kopernikusgasse 24, A-8010 Graz, Austria

Abstract. We investigate the problem of on-line scheduling jobs on m identical parallel machines where preemption is allowed. The goal is to minimize the makespan. We derive an approximation algorithm with worst case guarantee $m^m/(m^m - (m-1)^m)$ for every $m \geq 2$, which increasingly tends to $e/(e-1) \approx 1.58$ as $m \to \infty$. Moreover, we prove that for no $m \geq 2$ there does exist any approximation algorithm with a better worst case guarantee.

1 Introduction

Problem statement. In this note, we discuss the problem of preemptively scheduling a set $J = \{J_1, J_2, \ldots\}$ of jobs on-line on m identical parallel machines. Associated with each job J_i is a *processing time* (or *length*) $p_i = p(J_i)$. At any time, each machine can handle at most one job and each job can be processed by at most one machine. However, preemption is permitted which allows to split a job and spread its processing over several machines. The jobs are not known *a priori*: job J_i becomes only known when J_{i-1} has already been scheduled. As soon as job J_i appears, it must irrevocably be assigned to one or more time slots of one or more machines. The objective is to find a schedule which minimizes the maximum completion time.

Related work. The off-line version of this problem (where all the jobs are fully known in advance) is easily solved to optimality in polynomial time, see McNaughton [3]. On the other hand, the off-line version *without* preemption is known to be NP-complete [2].

Our results. We usually measure the quality of an approximation algorithm H, often called *heuristic*, by its *worst case (performance) ratio*

$$R^H(m) = \sup\{H(L)/\text{OPT}(L) : L \text{ is a list of jobs}\} \qquad (1)$$

where $H(L)$ denotes the makespan of a schedule produced by the heuristic H for scheduling the list L of jobs on m machines, and $\text{OPT}(L)$ denotes the corresponding makespan of some (off-line) optimum schedule.

In this paper we will develop an approximation algorithm with worst case ratio $m^m/(m^m - (m-1)^m)$ for the problem of on-line scheduling with preemption on m machines. As m tends to infinity, this ratio tends to $e/(e-1) \approx 1.58$ where e denotes the Eulerian constant. The main idea of our algorithm is to always preempt a new job in such a way that certain combinatorial invariants of the schedule are maintained. As another result, we will derive lower bounds on the worst-case performance of any on-line algorithm. We will prove that our heuristic is essentially *best possible* since there does not exist any heuristic with smaller worst case ratio.

Organization of the paper. Section 2 gives some definitions and recalls the result of McNaughton for the corresponding off-line problem. Section 3 presents our algorithm and performs its worst case analysis. Section 4 shows that no on-line approximation algorithm with smaller worst case ratio exists. Section 5 finishes with discussion.

2 Preliminaries

For a set of jobs with processing times $p_1, \ldots, p_n$ there are two straightforward lower bounds for the makespan of any preemptive schedule on m machines. First, the makespan must be at least $\max_{1 \le i \le n}\{p_i\}$ since no job can be processed on two machines at the same time. Secondly, the makespan must be at least $(\sum_{i=1}^{n} p_i)/m$, the average machine load. McNaughton [3] proved that there always exists a schedule with makespan at most the maximum of these two bounds.

Theorem 1. *(McNaughton, 1959) For every set of n jobs with processing times $p_1, \ldots, p_n$ and for $m \ge 1$ machines, the length OPT of the optimum preemptive schedule equals $\max\{(\sum_{i=1}^{n} p_i)/m, \max_{1 \le i \le n} p_i\}$.*

We will use the following notation. For $m \ge 2$, we introduce the numbers $\alpha(m) = m/(m-1)$ and

$$r(m) = \frac{\alpha^m}{\alpha^m - 1} = \frac{m^m}{m^m - (m-1)^m}.$$

To simplify the presentation, we will drop the dependence on m and always write α and r instead of $\alpha(m)$ and $r(m)$.

3 The Approximation Algorithm

Throughout this section, the load of machine M_j at step $t \ge 0$ (i.e., immediately after the t-th job has been scheduled) will be denoted by L_j^t. The current optimum makespan in step t (for the job set $\{J_1, \ldots, J_t\}$) is denoted by OPT^t, and the sum $\sum_{i=1}^{t} p_i$ of the processing times of all known jobs is denoted by S^t.

The algorithm will maintain the following three invariants (I1) through (I3):

(I1) At any step t, $L_1^t \le L_2^t \le \ldots \le L_m^t$ holds.

(I2) At any step t, $L_m^t \leq r\text{OPT}^t$.

(I3) At any step t and for any $1 \leq k \leq m$

$$\sum_{i=1}^{k} L_i^t \leq \frac{\alpha^k - 1}{\alpha^m - 1} S^t.$$

Next, we sketch how a new job J_{t+1} with processing time p_{t+1} is scheduled to the m machines with loads $L_1^t, \ldots, L_m^t$. First, we compute the new optimum makespan OPT^{t+1} according to Theorem 1. Then we reserve on every machine M_j a certain time interval $\mathcal{I}_j$ for the potential processing of J_{t+1}:

- On machine M_j, $1 \leq j \leq m-1$, this reserved time interval $\mathcal{I}_j$ is $[L_j^t, L_{j+1}^t]$.
- On machine M_m, the reserved time interval $\mathcal{I}_m$ is $[L_m^t, r\text{OPT}^{t+1}]$.

Properties (I1) and (I2) ensure that all these intervals are well-defined. Now, going from $\mathcal{I}_m$ down to $\mathcal{I}_1$, we put into each reserved interval a part of job J_{t+1} as large as possible until we run out of p_{t+1}. This will lead to completely unused intervals $\mathcal{I}_1, \ldots, \mathcal{I}_{z-1}$, a partially occupied interval $\mathcal{I}_z$ and some fully occupied intervals $\mathcal{I}_{z+1}, \ldots, \mathcal{I}_m$. This uniquely defines the index z (in case no partially occupied interval exists, we choose z to be the index of the first fully occupied interval).

Lemma 2. *If invariant (I3) is fulfilled at step t, then the overall size of all reserved intervals $\{\mathcal{I}_j\}$ is sufficiently large to process the complete job J_{t+1} during step $t+1$.*

Proof. Observe that all $\{\mathcal{I}_j\}$ together cover the contiguous time interval $[L_1^t, r\text{OPT}^{t+1}]$. Hence, it is sufficient to prove $L_1^t + p_{t+1} \leq r\text{OPT}^{t+1}$. Invariant (I3) with $k = 1$ yields $L_1^t \leq (\alpha - 1)S^t/(\alpha^m - 1)$. Let $S^t = \lambda p_{t+1}$ for some $\lambda \geq 0$. Then it is easy to check that

$$\frac{(\alpha - 1)\lambda}{\alpha^m - 1} + 1 \leq r\max\left\{\frac{\lambda + 1}{m}, 1\right\},$$

which, together with the fact that

$$\text{OPT}^{t+1} \geq \max\left\{\frac{S^t + p_{t+1}}{m}, p_{t+1}\right\} = p_{t+1}\max\left\{\frac{\lambda + 1}{m}, 1\right\},$$

leads to

$$L_1^t + p_{t+1} \leq \frac{\alpha - 1}{\alpha^m - 1} S^t + p_{t+1} = \left(\frac{(\alpha - 1)\lambda}{\alpha^m - 1} + 1\right) p_{t+1}$$

$$\leq \left(\frac{(\alpha - 1)\lambda}{\alpha^m - 1} + 1\right)\left(\max\left\{\frac{\lambda + 1}{m}, 1\right\}\right)^{-1} \text{OPT}^{t+1} \leq r\text{OPT}^{t+1}.$$

This completes the argument.

Lemma 3. *If invariant (I3) is fulfilled at step t and if job J_{t+1} is scheduled according to the above rule, then invariant (I3) is still fulfilled at step $t+1$.*

Proof. We must proof that for each $1 \leq k \leq m$ the corresponding inequality in (I3) holds. We will use the fact that all these inequalities were true at step t and we will distinguish two cases depending on the index z of the partially occupied interval.

If $k < z$, all the loads in the left-hand side of the inequality in (I3) did not increase when going from step t to step $t+1$, whereas the sum of processing times in the right-hand side of (I3) did increase by p_{t+1}. This settles the easy first case.

In the remaining case $k \geq z$, we will prove that

$$\sum_{i=k+1}^{m} L_i^{t+1} \geq \frac{\alpha^m - \alpha^k}{\alpha^m - 1} S^{t+1} \tag{2}$$

holds. If $k = m$ then (2) is trivial. Suppose $k < m$. Since $\sum L_i^{t+1} = S^{t+1}$, the inequality (2) is equivalent to the desired inequality in (I3). Observe that by the way we assign job J_{t+1} to the intervals $\{\mathcal{I}_j\}$, $L_j^{t+1} = L_{j+1}^t$ for $z < k+1 \leq j \leq m-1$ and that $L_m^{t+1} = r\mathrm{OPT}^{t+1}$. This transforms the expression in the left-hand side of (2) into

$$\sum_{i=k+2}^{m} L_i^t + r\mathrm{OPT}^{t+1} \geq \frac{\alpha^m - \alpha^{k+1}}{\alpha^m - 1}(S^{t+1} - p_{t+1}) + r\mathrm{OPT}^{t+1}, \tag{3}$$

where the inequality follows from invariant (I3) at step t for $k+1$ and from $S^t = S^{t+1} - p_{t+1}$. Let $p_{t+1} = \mu S^{t+1}$ for some $0 \leq \mu \leq 1$. Then, noticing that $\mathrm{OPT}^{t+1} \geq \max\{S^{t+1}/m, p_{t+1}\}$, we conclude that the right-hand side of (3) is at least

$$\frac{(\alpha^m - \alpha^{k+1})(1 - \mu)}{\alpha^m - 1} S^{t+1} + r \max\left\{\frac{1}{m}, \mu\right\} S^{t+1}. \tag{4}$$

By direct calculation we see that the right-hand side of (4) is greater than or equal to that of (2).

Theorem 4. *The approximation algorithm we described in this section has a worst case ratio at most $r(m) = m^m/(m^m - (m-1)^m) < e/(e-1)$. Furthermore, with a small modification, this algorithm preempts every job at most once.*

Proof. Lemma 2 shows that in case the invariants are fulfilled, every job indeed can be scheduled according to our rules. Invariant (I2) trivially implies the claimed bound on the worst case ratio of the above algorithm.

We argue that all three invariants are fulfilled at all steps t. Clearly, all three invariants are fulfilled at step 0, when all machine loads are zero. Invariant (I1) remains valid throughout the algorithm according to the definition of the algorithm. Invariant (I2) is always fulfilled, since the intervals $\mathcal{I}_j$ do not exceed time point $r\mathrm{OPT}$. Finally, Lemma 3 shows that invariant (I3) remains valid all the time.

It remains to show that we can modify the algorithm so that every job is preempted at most once. At each step t of our algorithm we do the following modification: Find the machine index z as previously described. If $\mathcal{I}_z$ is fully occupied then $z := z - 1$. Assign $r\text{OPT}^{t+1} - L_{z+1}^t$ portion of job J_{t+1} onto machine M_{z+1} and the remaining portion, if any, of the job onto machine M_z. Then renumber the machines by letting $M_{z+1} \to M_m$ and $M_j \to M_{j-1}$ for $j = z + 2, \ldots, m$. It is easy to check that the effect of the modifications is the same as the old algorithm except that job J_{t+1} is preempted at most once.

4 A Matching Lower Bound

In this section, we prove that there cannot exist any on-line algorithm for m machines with worst case ratio smaller than $r(m)$. From the problem statement we see that one has the freedom to reserve idle time for coming new jobs. However, intuition suggests that this sort of idle time is not necessary. The following observation supports this intuition.

Lemma 5. *If there exists an approximation algorithm H_1 for on-line preemptive scheduling with worst case ratio r^*, then there exists another approximation algorithm H_2 with at most the same worst case ratio r^* that never introduces machine idle time between two jobs on the same machine.*

Proof. Given any schedule S_1 (with or without idle time). For any $0 \le j \le m$, we denote by I_j a maximal time interval during which j machines are busy. Let $T_j = \{I_j\}$ and $t_j = \sum\{|I_j| : I_j \in T_j\}$ for $j = 0, \ldots, m$, where $|I_j|$ denotes the length of interval I_j. Then we see that the makespan of S_1 is at least $\sum_{j=1}^{m} t_j$. Consider the following schedule S_2: During time interval $\left[\sum_{j=m-i+2}^{m} t_j, \sum_{j=m-i+1}^{m} t_j \right]$ only machine M_i up to machine M_m are busy, $i = 1, \ldots, m$. Clearly, this simple schedule does not have machine idle time between any two jobs, and its makespan is $\sum_{j=1}^{m} t_j$. Moreover, both schedules S_1 and S_2 have the same $\{t_j\}_{j=0}^{m}$.

Now suppose both S_1 and S_2 are schedules of the same set of jobs, and one more job is added to S_1. Suppose the job is processed within time interval(s) of T_i for q_i time units, $i = 0, \ldots, m-1$. Then in S_2 we can simulate this assignment by letting machine M_{m-i} start processing q_i portion of the job at time $\sum_{j=i+1}^{m} t_j$, $i = 0, \ldots, m-1$. It is easy to see that after adding the job to S_1 and S_2 according to the above way, the makespan of the new S_2 is still less than or equal to that of the new S_1, and both new schedules have the same new $\{t_j\}_{j=1}^{m}$.

It should be clear now that the required approximation algorithm H_2 can be obtained by making H_2 simulate H_1 for assigning every job according to the way we just described. Both H_1 and H_2 start with the same empty schedule.

Next, we introduce the following sequence of jobs, and consider any algorithm's behavior in assigning these jobs.

- At step 0, there arrive m jobs, all with processing times $1/m$.

– At step t, $t \geq 1$, there arrives a new job J_t with processing time $p_t = m^{t-1}/(m-1)^t$.

Lemma 6. *At step t, immediately after the arrival of job J_t, the overall sum of processing times equals α^t and the optimum makespan equals $p_t = \alpha^t/m$.*

Proof. This follows from Theorem 1.

Now suppose that there exists an approximation algorithm H with worst case ratio $r(m) - \varepsilon$ for some real $\varepsilon > 0$. By Lemma 5 we may assume that H never introduces machine idle time. Analogously to the preceding section, we denote by $L_1^t \leq L_2^t \leq \ldots \leq L_m^t$ the machine loads at the time immediately after algorithm H scheduled job p_t. (Note that the machine indices can be adjusted at each step.)

Lemma 7. *For any $t \geq 0$, $L_m^t \leq (r - \varepsilon)\alpha^t/m$ must hold.*

Proof. Since $L_m^t \leq (r - \varepsilon)\text{OPT}^t = (r - \varepsilon)\alpha^t/m$.

Lemma 8. *For all $1 \leq j \leq m-1$ and all $t \geq 0$, $L_{m-j}^{t+j} \leq L_m^t$ holds.*

Proof. We make induction on j. For $j = 1$ we must prove that $L_{m-1}^{t+1} \leq L_m^t$ holds. Suppose the contrary. Then at step $t+1$, two machines have loads larger than L_m^t. Since H does not introduce machine idle time, at the time point L_m^t, the job p_{t+1} was processed on at least two machines with largest and second largest load, which is a contradiction.

Now assume that we have proved the claim up to some fixed value j ($1 \leq j \leq m-2$), i.e., $L_{m-j}^{t+j} \leq L_m^t$. In case $L_{m-j-1}^{t+j+1} > L_m^t \geq L_{m-j}^{t+j}$ would be true, we could argue as above that job p_{t+j+1} is processed on at least two machines at the same time.

Theorem 9. *For all $m \geq 2$, there does not exist any on-line algorithm for pre-emptive scheduling on m machines with worst case ratio strictly smaller than $r(m)$.*

Proof. We are ready to derive a contradiction to the assumption that H has worst case ratio $r - \varepsilon$. We take a closer look at step $(m-1)$.

By Lemma 8, we know that $L_j^{m-1} \leq L_m^{j-1}$ for all $1 \leq j \leq m-1$, which, together with Lemma 7 and Lemma 6, yields that $(r - \varepsilon)$ times the current optimum makespan

$$(r - \varepsilon)\alpha^{m-1}/m \geq L_m^{m-1} = \alpha^{m-1} - \sum_{j=1}^{m-1} L_j^{m-1}$$

$$\geq \alpha^{m-1} - \sum_{j=1}^{m-1} L_m^{j-1} \geq \alpha^{m-1} - \sum_{j=1}^{m-1} (r - \varepsilon)\alpha^{j-1}/m$$

Comparing the last expression in the right-hand side to the left-hand side and doing some easy calculations, we see that $\varepsilon \leq 0$ must hold.

5 Discussion

We derived a (best possible) approximation algorithm for on-line preemptive scheduling on m parallel machines together with a matching lower bound on the worst case performance of any such algorithm.

For $m = 2$ the derived worst case ratio is $4/3$. For $m = 3$ this ratio is $27/19 \approx 1.421$. Bartal, Fiat, Karloff and Vohra [1] investigated *randomized* algorithms for non-preemptive scheduling on m parallel machines. For $m = 2$ they derived a best possible randomized algorithm with worst case ratio $4/3$. For $m = 3$ they proved a 1.4 lower bound on the performance of any randomized on-line algorithm. In fact, we can even generally prove that every randomized algorithm for on-line scheduling a set of jobs on $m \geq 2$ identical and parallel machines has a worst case ratio at least $r(m)$. The question is whether there is some connection between this problem and our problem (e.g., one could try to assign the jobs to machines with probabilities proportional to the lengths of the preempted parts as assigned by our algorithm).

Acknowledgement. This work was supported by the Tinbergen Institute at Erasmus University Rotterdam and by the Christian Doppler Laboratorium für Diskrete Optimierung.

References

1. Bartal, Y., Fiat, A., Karloff, H., Vohra, R.: New algorithms for an ancient scheduling problem. Proc. 24th ACM Symposium on Theory of Computing (1992) 51–58
2. Garey, M.R., Johnson, D.S.: *Computers and Intractability: A Guide to the Theory of NP-Completeness*, Freeman, San Francisco (1979)
3. McNaughton, R.: Scheduling with deadlines and loss functions, Management Sci. **6** (1959) 1–12

Tight Approximations for Resource Constrained Scheduling Problems

Anand Srivastav[1] and Peter Stangier[2]

[1] Institut für Theoretische Informatik, Freie Universität Berlin, Takustr. 9, 1000 Berlin 33, Germany **e-mail:** srivasta@inf.fu-berlin.de
[2] Institut für Informatik, Universität zu Köln, Pohligstr. 1, 50969 Köln, Germany **e-mail:** stangier@informatik.uni-koeln.de

Abstract. We investigate the following resource constrained scheduling problem. Given m identical processors, s renouvable, but *limited* resources, n independent tasks each of *unit length*, where each task needs one processor *and* a task dependent amount of every resource in order to be processed. In addition each task possesses an integer ready-time, which means it cannot be processed before its ready-time. The optimization problem is to assign all tasks to discrete times in $\mathbb{N}$, minimizing the latest completion time C_{max} of any task subject to the processor, resource and ready-time constraints.

The problem is NP-hard even under much simpler assumptions. The best previously known polynomial-time approximation is due to RÖCK and G. SCHMIDT, who gave in 1983 in the case of zero ready-times an $O(m)$-factor algorithm. For the analysis of their algorithm, the assumption of identical (zero-) ready-times is essential.

The main contribution of this paper is to break the $O(m)$ barrier substantially (even for non zero ready-times) and to show the first *constant* factor approximations, which are best possible, unless $P = NP$.

Furtheron we can prove for the apparently simpler problem with zero ready-times the NP-completeness of the problem of deciding whether there exists an *integral* schedule of size C, even if a fractional schedule of size C and an integral schedule of size $C + 1$ are known. The proof is based on an interesting reduction to the chromatic index problem.

1 Introduction

1.1 Complexity of the Problem

An instance of the resource constrained scheduling problem with ready-times consists of:

• Let $T = \{T_1, \ldots, T_n\}$ be a set of *independent* tasks. Each task T_j needs a time of *one time unit* for its completion and cannot be scheduled *before* its ready-time r_j, $r_j \in \mathbb{N}$.

• Let $\mathcal{P} = \{P_1, \ldots, P_m\}$ be a set of *identical* processors. Each task needs one processor.

• Let $\mathcal{R} = \{R_1, \ldots, R_s\}$ be a set of *renouvable*, but *limited* resources. This means that at any time all resources are available, but the available amount of each resource R_i is bounded by $b_i \in \mathbb{N}$. For $1 \leq i \leq s$, $1 \leq j \leq n$ let $R_i(j) \in [0, 1]$ be rational resource requirements, indicating that every task T_j needs $R_i(j)$ amount of resource R_i in order to be processed.

The combinatorial optimization problem is:

• Find a schedule (or assignment) $\sigma : T \mapsto \mathbb{N}$ of *minimal* time length subject to the ready-times, processor and resource constraints.

Since the processor requirements can be decribed by introducing an additional resource R_{s+1} with upper bound $b_{s+1} = m$ and defining $R_{s+1}(j) = 1$, the resource constraints are briefly formalized as

$$\forall z \in \mathbb{N}, i \in \{1, \ldots, s+1\} : \sum_{\{j : \sigma(j) = z\}} R_i(j) \leq b_i,$$

which indicates the *packing* character of the problem.

The problem is NP-hard in the strong sense, even if $r_j = 0$ for all $j = 1, \ldots, n$, $s = 1$ and $m = 3$. In this paper we consider the rational weighted case, when $R_i(j) \in [0, 1] \cap \mathbb{Q}$. According to the standard notation of scheduling problems the unweighted version of our problem $(R_i(j) = 0, 1)$ can be formalized as

$$P | \text{res} \cdots 1, r_j, p_j = 1 | C_{\max}$$

1.2 Previous Work

The known approximation algorithms for the problem class $P | \text{res} \cdots, r_j = 0, p_j = 1 | C_{\max}$ are due to M. R. GAREY et al. and H. RÖCK, G. SCHMIDT [RS83]. Garey et al. constructed with the First-Fit-Decreasing heuristik a schedule of length C_{FFD}, which *asymptotically* is an $O(s)$-factor approximation, i.e.:

$$\lim_{C_{opt} \to \infty} \frac{C_{FFD}}{C_{opt}} \leq s + \frac{7}{10}.$$

Röck and Schmidt showed, employing the polynomial-time solvability of the simpler problem $P2 | \text{res} \cdots, r_j = 0, p_j = 1 | C_{\max}$ with 2 processors, a $\lceil \frac{m}{2} \rceil$-factor approximation algorithm.

Hence for problems with small optimal schedules or many resource constraints or many processors these algorithms can be arbitrary bad.

Note that both approaches are based on the assumption that no ready-times are given, i.e. $r_j = 0$ for all tasks $T_j \in T$. For example, Röck and Schmidt's algorithm cannot be used, when ready-times are given as the problem $P2 | \text{res} \cdots 1, r_j, p_j = 1 | C_{\max}$ is also NP-complete, so their basis solution cannot be constructed in polynomial-time in the presence of ready times.

1.3 The Results

Let C_{opt} be the integer minimum of our scheduling problem and let the integer C denote the size of the minimal schedule, if we consider the LP relaxation, where fractional assignments of the tasks to scheduling times are allowed. We briefly call solutions to the LP relaxation "fractional schedules" and solutions to the original integer problem "integral schedules". This should not cause any confusion: C is always an integer, only the assignments corresponding to C are fractional.

As the one main result we will present the first strongly polynomial approximation algorithm for the problem class $P|\text{res} \cdot \cdot 1, r_j, p_j = 1|C_{\max}$, including its rational weighted version $(0 \le R_i(j) \le 1)$:

• *For every* $\epsilon = \frac{1}{k}$, $k \in \mathbb{N}$, *we can find an integral schedule of size at most* $\lceil (1 + \epsilon)C_{opt} \rceil$, *provided that all resource bounds* b_i *are* $\Omega(\frac{1}{\epsilon^2} \log(Cs))$.

As a surprising consequence we can construct a schedule of length $C_{opt} + 1$, whenever $b_i \in \Omega(C^2 \log(Cs))$ for all resource bounds b_i.

Note that C is at most the sum of n and the maximal ready-time, hence the factor $\log(Cs))$ is within the size of the problem input.

In contrast to Garey et al. and Röck/Schmidt our approximation guarantee is independent of the number of processors or resources and can be used also for small schedules, for example $C_{\max} = 2, 3$ or 4.

Furthermore, we can extend this results to the case of arbitrary integer resource requirements, i.e. $R_i(j) \in \mathbb{N}$, exploiting the fact that we can handle rational weights, hence may rescale. This gives the first polynomial-time approximation algorithm for problems of the form

$$P|\text{res} \cdot \cdot \cdot, r_j, p_j = 1|C_{\max}.$$

Our approximation is most probably best possible as we will show for the simpler problem with zero ready-times that removing the additive time unit is NP-complete, *even if*

- a fractional schedule of size C is known,
- an integral schedule of size $C + 1$ is given.
- $b_i \in \Omega(C^2 \log(Cs))$ for all resource bounds b_i.

1.4 Two Methodical Innovations

Scaled Dice and Random Assignment First we must generate a fractional solution, then we have to define randomized rounding. Let D be the maximum of n and the maximal ready-time. While a fractional optimal schedule is easily constructed by standard methods solving at most $\log D$, linear programs the second problem is far from being trivial: Suppose we have found via linear programming the size C of the optimal fractional schedule along with fractional assignments $\tilde{x}_{jz}$ of the tasks T_j to scheduling times $z \in \{1, \ldots, C\}$. A possible (and suggestive) randomized rounding procedure would be to cast for each task

T_j independently a C-faced dice with face probabilities $\tilde{x}_{jz}$, where the z-th face represents the choice of the time z for task T_j.

Unfortunately, since we have a packing problem it may happen that simple dice casting produces a schedule in which too many tasks are scheduled at the same time requiring more resources than available. In conclusion, the probability distribution in packing problems like resource constraint scheduling should be *asymmetric* rather than *symmetric*. Unfortunately, the Binomial distribution induced by the fractional assignment might not be *asymmetric* enough. A way out could be to decrease the probabilities that too many tasks are scheduled at the same time.

Indeed, such a technique, called scaling, was introduced by P. RAGHAVAN and C. D. THOMPSON [RT87] in a coin-flipping model, i.e. 0-1 random variables:

If we flip a biased coin in order to fix the value of a $0 - 1$ random variable, then such a random experiment can violate packing constraints, if too many ones are generated. Here scaling simply means to decrease the probabilitiy of ones and to increase the probability of zeros by the same amount. The two scaled probabilities sum up to one and we have a nice and more asymetric distribution.

But in a dice model the situation is completely different due to the reason that each dice *necessarily* picks a scheduling time, or in other words, we ever have to generate a 1. Therefore it is not clear at all what a "scaled" dice should be. We propose the following scaling technique, which induces in dice models a more asymmetric probability distribution and in particular for our scheduling problem has an intuitively very clear meaning.

Given $\epsilon = \frac{1}{k}$, $k \in \mathbb{N}$, we enlarge the time interval to $\{1, \ldots, \lceil(1 + \epsilon)C\rceil\}$ and randomly assign the tasks to times in this enlarged interval. Let us call $\{C + 1, \ldots, \lceil(1 + \epsilon)C\rceil\}$ the *compressed image* of $\{1, \ldots, C\}$. For $z \in \{1, \ldots, C\}$ we scale each face probability (or fractional assignment) $\tilde{x}_{jz}$ with a factor δ where $\delta := \frac{1}{1+\epsilon}$. Since casting a so scaled dice would with probability $1 - \delta$ *not* assign the task T_j to any time from $\{1, \ldots, C\}$, we balance the loos defining random assignments to times in the compressed part as follows:

Set $I = \{1, \ldots, \lceil(1+\epsilon)C\rceil\}$, $I_0 = \{1, \ldots, C\}$, $I_1^\epsilon = \{C+1, \ldots, C+\lfloor\epsilon C\rfloor\}$ and $I_2^\epsilon = \{\lceil(1 + \epsilon)C\rceil\}$. Then $I = I_0 \cup I_1^\epsilon \cup I_2^\epsilon$. For integer $l \geq 1$ let g, f be functions

$$f(l) = \frac{l - C}{\epsilon} \text{ and } g(l) = \frac{l - C}{\epsilon} + 1 - \frac{1}{\epsilon}.$$

We define fractional assignments $\hat{x}_{jl}$

$$\hat{x}_{jz} := \begin{cases} \delta\tilde{x}_{j,l} & \text{for } l \in I_0 \\ \sum_{t=g(l)}^{f(l)} (1 - \delta)\tilde{x}_{jt} & \text{for } l \in I_1^\epsilon \\ \sum_{t=g(l)}^{C} (1 - \delta)\tilde{x}_{jt} & \text{for } l \in I_2^\epsilon \end{cases} \tag{1}$$

In section 2 we will see that this is a *feasible* fractional schedule of size $\lceil(1+\epsilon)C\rceil$.

Derandomization Derandomization of our algorithm in usual models of computation, like the RAM-model, is a numerical difficult task, because rational weights are involved, and therefore the upper bounds for the conditional probabilities under consideration are transzendental functions. In [SS94] we showed how this can be done efficiently, proving an algorithmic version of the large deviation inequality due to Angluin and Valiant. There we considered the simpler case of 0-1 random variables. In this paper we extend our result to multivalued random variables, i.e. dice casting experiments.

2 Approximation Algorithms

2.1 Large Deviations

We need the large deviation inequality of Angluin-Valiant.

Let $X_1, \ldots, X_n$ be mutually independent random variables, where X_j is equal to 0 with probability p_j and 1 with probability $1 - p_j$. For $1 \leq j \leq n$ let w_j denote rational weights with $0 \leq w_j \leq 1$ and denote by ψ the random variable

$$\psi = \sum_{j=1}^{n} w_j X_j.$$

Theorem 1 Angluin and Valiant. *[AV79] For $\beta > 0$ we have:*

$$\Pr(\psi > (1 + \beta)\mathrm{E}(\psi)) < \exp\left(-\frac{\beta^2 \mathrm{E}(\psi)}{3}\right)$$

□

We will use Theorem 1 for the estimation of the probability that at any time a resource constraint is violated. For the derandomization of such probabilistic results we cite our result in [SS94], which covers the case of mutually independent dice. The basic theorem from which the following Theorem 2 is derived can also be found in our paper in the proceedings of ESA'93 [SS93]. Due to space limitation for a proof of Theorem 2 we refer to [SS94] or the full paper.

Let n, N be nonnegative integers. We are given n mutually independent random variables X_j with values in $\{1, \ldots, N\}$ and probability distribution $Prob(X_j = k) = \tilde{x}_{jk}$ for all $j = 1, \ldots, n$ and $k = 1, \ldots, N$ and $\sum_{k=1}^{N} \tilde{x}_{jk} = 1$, where the $\tilde{x}_{jk}$ are rational numbers with $0 \leq \tilde{x}_{jk} \leq 1$. Let X_{jk} denote the random variable which is 1 if $X_j = k$ and 0 else. The probability space is

$$\Omega = \left\{ (\omega_1, \ldots, \omega_n) \in \prod_{j=1}^{n} \{0,1\}^N; \omega_j \in \{0,1\}^N, \sum_{k=1}^{N} \omega_{jk} = 1 \right\}.$$

For $1 \leq i \leq m$, $1 \leq j \leq n$ let w_{ij} be rational weights with $0 \leq w_{ij} \leq 1$. For $i = 1, \ldots, m$ and $k = 1, \ldots, N$ define the sums ψ_{ik} by

$$\psi_{ik} = \sum_{j=1}^{n} w_{ij} X_{jk}.$$

Let $0 < \beta_{ik} \leq 1$ be rational numbers. Denote by E_{ik} the event

$$``\psi_{ik} \geq E(\psi_{ik})(1 + \beta_{ik})"$$

Let (E_{ik}) be a collection of mN such events. Since the E_{ik} are sums of Bernoulli trials, we can invoke Theorem 1. Let $f(\beta_{ik}) = \exp(-\frac{\beta_{ik}^2 E(\psi_{ik})}{3})$ be the upper bounds for $\Pr(E_{ik}^c)$ in Theorem 1. We suppose that

$$\sum_{i=1}^{m} \sum_{k=1}^{N} f(\lambda_{ik}) < 1 - \gamma$$

for some $\gamma > 0$. Then $\Pr(\bigcap_{i=1}^{m} \bigcap_{k=1}^{N} E_{ik}) \geq \gamma$, hence $\bigcap_{i=1}^{m} \bigcap_{k=1}^{N} E_{ik}$ is not empty and the derandomization problem is to construct a vector from this intersection in deterministic polynomial-time. The following theorem solves this problem.

Theorem 2. *[SS94] Let E_{ik} and $\gamma > 0$ be as defined above. Then a vector $x \in \bigcap_{i=1}^{m} \bigcap_{k=1}^{N} E_{ik}$ can be constructed in $O\left(Nmn^2 \log \frac{Nmn}{\gamma}\right)$-time.* $\qquad \Box$

2.2 Fractional Assignments

Let C be the size (or length) of an optimal fractional schedule and let C_{opt} be the size of the integral optimal schedule. Let $r_{max} := \max_{j=1,\ldots,n} r_j$ and $D = r_{max} + n$. W.l.o.g we can assume that $D = O(n)$. Then obviously

$$C \leq C_{opt} \leq D.$$

C can be found as follows: Start with an overall integer deadline $\tilde{C} \leq D$ and according to [LST90] check, whether the LP

$$
\begin{aligned}
\sum_j R_i(j)x_{jz} \leq b_i \quad & \forall R_i \in \mathcal{R}, \\
& z \in \{1, \ldots, D\} \\
\sum_z x_{jz} = 1 \quad & \forall T_j \in \mathcal{T} \\
x_{jz} = 0 \quad & \forall T_j \in \mathcal{T}, z < r_j \text{ and} \\
& \forall T_j \in \mathcal{T}, z > \tilde{C} \\
x_{jz} \in [0,1] &
\end{aligned}
$$

has a solution. Using binary search we will find C after having solved at most $\log D$ such LPs. Given the optimal fractional assignments $(\tilde{x}_{jz})$ corresponding to a schedule of size C we enlarge the interval I_0 to $\{1, \ldots, \lceil(1 + \epsilon)C\rceil\}$ and define new fractional assignments $\hat{x}_{jz}$ as in (1). The following lemma shows that this new fractional assignment defines a feasible schedule.

Lemma 3. *The fractional assignments $(\hat{x}_{jz})$ induce a feasible fractional schedule of size $\lceil(1 + \epsilon)C\rceil$.* $\qquad \Box$

The proof is straight forward.

2.3 Rounding Algorithms and Analysis

Our construction has lead us to a feasible fractional schedule of size $\lceil(1+\epsilon)C\rceil$. This fractional solution will be rounded to a feasible *integral* solution of the *same* size. Remember that $C \leq C_{opt}$, so all the following approximations of size $\lceil(1+\epsilon)C\rceil$ are also approximations within $\lceil(1+\epsilon)C_{opt}\rceil$. The rounding algorithm is:

Algorithm DICE

For every task T_j we independently cast a dice with $\lceil(1+\epsilon)C\rceil$-faces, where for $z \in \{1, \ldots, \lceil(1+\epsilon)C\rceil\}$ the z-th face appears with probability $\widehat{x}_{jz}$.

For each pair (j, z), $j = 1, \ldots, n$ and $z \in I$ let X_{jz} be the 0-1 random variable, which is 1, if the j-th dice assigns the task T_j to the time z and is 0 else. Then by definition $Pr(X_{jz} = 1) = \widehat{x}_{jz}$. We have

Theorem 4. *For every $\epsilon = \frac{1}{k}$, $k \in \mathbb{N}$ the algorithm DICE generates an integral schedule of size at most $\lceil(1 + \epsilon)C\rceil$ with probability at least $\frac{1}{2}$ provided that $b_i \geq \frac{3(1+\epsilon)}{\epsilon^2}\lceil\log(4C(s+1))\rceil$.* $\qquad\square$

The proof is based on the fact that the usage of the resources at any time is the sum of independent Bernoulli-trials. So Theorem 1 can be used.

Using our Theorem 2 we get a deterministic polynomial-time algorithm.

Theorem 5. *For every $\epsilon = \frac{1}{k}$, $k \in \mathbb{N}$ we can find with the method of conditional probabilities (Theorem 2.2) an integral schedule of size at most $\lceil(1+\epsilon)C\rceil$, provided that $b_i \geq \frac{3(1+\epsilon)}{\epsilon^2}\lceil\log(4C(s+1))\rceil$.* $\qquad\square$

Refer to our full paper to see how Theorem 2 is used to proof the existence of a deterministic polynomial time algorithm. With $\epsilon = 1$ we have

Corollary 6. *[SS94] If $b_i \geq 6\lceil\log(4C(s+1))\rceil$ then a solution of size at most $2C$ can be found in deterministic polynomial-time.* $\qquad\square$

And with $\epsilon = \frac{1}{C}$

Corollary 7. *If $b_i \geq 3C(C+1)\lceil\log(4C(s+1))\rceil$, then a solution of size at most $C+1$ can be found in deterministic polynomial-time.* $\qquad\square$

Note that we have a *strongly* polynomial-time approxiamtion algorithm in the case of 0-1 weights, i.e. $R_i(j) \in \{0, 1\}$, because then the strongly polynomial LP algorithm of É. TARDOS [Tar86] can be used to solve our linear programming relaxations. In any case, the running time of the LP algorithm dominates clearly the running time of derandomization.

Finally, suppose that the resource requirements $R_i(j)$ are arbitrary integers, i.e. $R_i(j) \in \mathbb{N}$, whereas in the problems above we assumed $R_i(j) \in \{0, 1\}$

Utilizing the fact that we allowed *rational weigthed* resource requirements, we can scale the resource bounds. Just compute for every resource R_i the number $R_{\max}(i) = \max_{J_j \in \mathcal{J}} R_i(j)$, set $R_i(j)' = \frac{R_i(j)}{R_{\max}}$ and $b_i' = \frac{b_i}{R_{\max}}$. Then with Theorem 2.5 we conclude:

Corollary 8. *Let C be the overall completion time of a fractional schedule under resource constraints for n independent tasks using arbitrary amounts of the $s + 1$ resources, including the processors, i.e. $R_i(j) \in \mathbb{N}$ for all $i = 1, \ldots, n$ and $i = 1, \ldots, s + 1$. Let $\epsilon = \frac{1}{k}$ for $k \in \mathbb{N}$. Then an integral schedule of size at most $\lceil (1 + \epsilon)C_{opt} \rceil$ can be found in deterministic polynomial-time, provided that $b'_i \geq \frac{3(1+\epsilon)}{\epsilon^2} \lceil \log (4C(s + 1)) \rceil$.* □

The proof is based on the observation that if the schedule for the scaled problem using b'_i and $R_i(j)'$ is feasible — which can be constructed by our algorithm — the rescaled schedule is feasible for the original problem.

Setting $\epsilon = 1$ gives for problems of the general form $P|\text{res} \cdots, p_j = 1, r_j|C_{\max}$ a 2-factor approximation, setting $\epsilon = \frac{1}{C}$ we can approximate the optimum of an arbitrary resource contrained scheduling problem up to *one* additional time-unit, provided the b'_i are as large as required above.

Note that for any *constant* R_{max} this gives a tight approximation, since then b_i still grows logarithmically in the input size.

3 Negative Result

Under the assumption $b_i = \Omega(C^2 \log(Cs))$, we were able to construct an integral schedule of size at most $C + 1$. This is very close to the truth, because now C_{opt} is either C or $C + 1$.

But that is the end of the story! We will show under the assumption $b_i = \Omega(C^2 \log(Cs))$ that the complexity dramatically jumps, when we try to pass from $C + 1$ to C.

In the rest of the paper we consider the "simpler" problem with zero ready-times. Since then $C \leq n$, the NP-completeness of scheduling problems with $b_i = \Omega(\log(ns))$ implies the NP-completeness of those with $b_i = \Omega(\log(Cs))$.

In particular, we will prove that resource constrained scheduling problems are NP-complete, even for small schedules, i.e. the questions "$C_{\max} = 2$?" or "$C_{\max} = 3$?".

Theorem 9. *Under the assumptions that there exist a fractional schedule of size C and an integral schedule of size $C + 1$, and $b_i = 1$ for all resource bounds, it is NP-complete to decide whether or not there exists an integral schedule of size C, for $C \geq 2$.*

Proof. For $C = 2$ we refer to our result in [SS94], where we reduced the problem to the problem of partitioning a graph into 2 perfect matchings.

For $C > 2$ we give now a reduction to the chromatic index problem, which is known to be NP-complete [Hol81]. The following is known about the chromatic index of a graph G. When $\Delta(G)$ is the maximal node degree of G, then there always exists an edge-colouring that uses $\Delta(G) + 1$ and there certainly never exists a colouring that uses striclty less than $\Delta(G)$ colours. But it is NP-complete to decide whether there exists a colouring that uses $\Delta(G)$ colours, even for cubic graphs, i.e. $\Delta(G) = 3$. Therefore this edge colouring problem is NP-complete for *fix* Δ. This is an important observation, which we will need later in our reductions.

Now to the reduction. Let $G = (V, E)$ be a graph with $|V| = \nu$, $|E| = \mu$ and $\deg(v) \leq \Delta$ for all $v \in V$. We construct a resource constrained scheduling problem asssociated to G as follows.

Introduce for every edge $e \in E$ exactly one task T_e and consider $\mu = |E|$ identical processors. We will freely call the edges tasks and vice versa. For every node $v \in V$ define a resource R_v with bound 1 and resource/task requirements

$$R_v(e) = \begin{cases} 1 \text{ if } & v \in e \\ 0 \text{ if } & v \notin e. \end{cases}$$

It's easy to show that there exists a colouring that uses Δ colours if and only if there is a feasible integral schedule of size Δ.

Note: According to an idea of A. KRÄMER [Krä93] there always exists a fractional schedule of size Δ: Simply set $x_{ez} = \frac{1}{\Delta}\mathbf{r}$. $\square$

In our approximation algorithm we assumed that $b_i > 6\lceil\log(4C(s+1))\rceil$ which we did not respect in the reduction above. In the next two theorems we invoke this assumption.

Theorem 10. *Under the assumption that there exist a fractional schedule of size C and an integral schedule of size $C + 1$, $b_i = \Omega(\log(ns))$ for all resource bounds, and C is fix, it is NP-complete to decide whether or not there exists an integral schedule of size C.* $\square$

The proof is similiar to the proof of the $C = 2$?- result given in [SS94].

Finally, if we allow that some resource requirements are fractional numbers — which can be handeld by our approximation algorithms — the problem remains NP-complete, even if the bounds are as big as required in our algorithms:

Theorem 11. *Under the assumptions that there exist a fractional schedule of size C and an integral schedule of size $C + 1$, $b_i > 6\lceil\log(4n(s + 1))\rceil$ for all resource bounds, C is fix and $R_i(j) \in \{0, \frac{1}{2}, 1\}$, it is NP-complete to decide whether or not there exists an integral schedule of size C.*

Proof. Let $G = (V, E)$ be a graph. For every edge $e \in E$ we define $2\Delta K$ tasks, where $K = 100\log(\mu)$ and $\mu = |E|$. There are $2K$ red and $(\Delta - 1)2K$ blue tasks per edge:

$$T_1^r(e), \ldots, T_{2K}^r(e) \quad \text{and} \quad T_1^b(e), \ldots, T_{2(\Delta-1)K}^b(e)$$

We consider $2\mu K\Delta$ identical processors.

The red tasks of an edge will be use to determine the colour of that edge in an schedule of size Δ. Hence in such a schedule all the red tasks have to be scheduled at the same time. To ensure this we introduce new resources forcing the red tasks of an edge being scheduled at the same time. Having shown this, we can follow the argumenation of the proof of Theorem 9.

For every node $v \in V$ let R_v be a resource with bound $2K$. The resource requirement is

$$R_v(j) = \begin{cases} 1 \text{ if } T_j = T_j^r(e) \text{ and } e \text{ covers } v \\ 0 \text{ else.} \end{cases}$$

These resources correspond to the original resources described in Theorem 9.

First we forbid that more than $2K$ tasks of the same edge are scheduled at the same time: we introduce a resource R_e with bound $2K$ and resource requirement

$$R_e(j) = \begin{cases} 1 \text{ if } T_j = T_i^r(e) \text{ or } T_j = T_i^b(e) \\ 0 \text{ else.} \end{cases}$$

The resource with fractional requirements is:

For every red task $T_i^r(e)$, we choose exactly one other red task $g(T_i^r(e))$ corresponding to e as follows:

$$g(T_i^r(e)) = T_{i+1}^r(e) \text{ for } i < 2K \quad \text{and} \quad g(T_{2K}^r(e)) = T_1^r(e).$$

Let us call $g(T_i^r(e))$ the *buddy* of $T_i^r(e)$. For every red task T_i let R_i be a resource of bound K. The requirements are

$$R_i(j) = \begin{cases} 1 \text{ if } T_i = T_j \\ \frac{1}{2} \text{ if } T_j \text{ corresponds to the same edge as } T_i, \text{ but is not its buddy} \\ 0 \text{ if } T_j \text{ is the buddy of } T_i \text{ or a task corresponding to another edge} \end{cases}$$

It's easy to show that there is a way to schedule the red and blue tasks of the same edge in time Δ, when we do not consider the other edges:

Simply schedule all red tasks of the same edge at the same time and all the blue tasks of the edge in packings of $2K$ tasks in the remaining time, as indicated in figure 1.

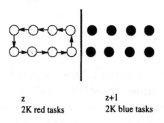

z
2K red tasks

z+1
2K blue tasks

Fig. 1. A way to schedule the red and blue tasks

Now we show that it is impossible to schedule the red tasks corresponding to the same edge at different times. Assume for a moment that at a time z in a schedule of size Δ, $I < K$ red tasks corresponding to the same edge are scheduled at time z as indicated in figure 2.

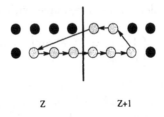

z   Z+1

Fig. 2. An impossible schedule

Then there must be exactly $2K - I$ blue tasks being scheduled at time z in order *not* to violate the constraint of resource R_e at any time. Since there are strictly less than K red tasks scheduled at time z, there is a task $T_i^r(e)$, whose buddy is *not* scheduled at time z. Then we have for resource R_i at time z:

$$\sum_j R_i(j)x_{jz} = 1 + \frac{1}{2}\left((2K - I) + (I - 1)\right) = K + \frac{1}{2}$$

which is bigger than K. Hence the red tasks corresponding to the same edge must be scheduled at the same time.

Setting $K = 100\log(\mu)$ an easy calculation shows that $b_i > 6\lceil\log(4n(s+1)i)\rceil$ Note again that setting $x_{jz} = \frac{1}{\Delta}$ gives a fractional schedule of size Δ. $\quad\square$

With $K = \Delta^c$ for a constant c satifying $\Delta^c > 3\Delta(\Delta+1)\lceil\log(4C(s+1))$ the proof of Theorem 11 implies:

Corollary 12. *Under the assumptions that there exist a fractional schedule of size C and an integral schedule of size $C+1$, $b_i > 3C(C+1)\lceil\log(4C(s+1))\rceil$ for all resource bounds, C is fix and $R_i(j) \in \{0, \frac{1}{2}, 1\}$, it is NP-complete to decide whether or not there exists an integral schedule of size C.* $\quad\square$

4 Conclusion

- We proved that there are polynomial-time algorithms approximating the optimum of a resource constrained scheduling problem, where the makespan is to be minimized. On the other hand we showed that it is NP-complete to guarantee the same approximation quality, when the optimum is slightly smaller. We showed that the problem of finding the chromatic index of a graph is a special case of a resource constrained scheduling problem. An interesting problem is whether our method can be parallelized.

- We hope that our new rounding technique, which we called "scaled dice casting", can be applied to other problems of similar flavour, for example to global routing, where the goal is to find *exactly* one Steiner tree for each net among many choices.

318

5 Acknowledgement

The first author would like to thank Professor LÁSZLÓ LOVÁSZ for helpful discussions during a research stay at Yale University, where a part of this paper has been written. We would also like to thank ANDREAS KRÄMER for helpful discussions.

References

[AV79] D. Angluin and L. G. Valiant. Fast Probabilistic Algorithms for Hamiltonian Circuits and Matchings. *Journal of Computer and System Sciences*, 18:155–193, 1979.

[Hol81] I. Holyer. The NP-completeness of Edge Colouring. *SIAM Journal of Computing*, 10(4):718–720, November 1981.

[Krä93] A. Krämer. Personal Communication, Köln, 1993.

[LST90] J. K. Lenstra, D. B. Shmoys, and É. Tardos. Approximation Algorithms for Scheduling Unrelated Parallel Machines. *Mathematical Programming*, 46:259–271, 1990.

[RS83] H. Röck and G. Schmidt. Machine Aggregation Heuristics in Shop Scheduling. *Mathematics of Operations Research*, 45:303–314, 1983.

[RT87] P. Raghavan and C. D. Thompson. Randomized Rounding: A Technique for Provably Good Algorithms and Algorithmic Proofs. *Combinatorica*, 7(4):365–374, 1987.

[SS93] A. Srivastav and P. Stangier. Integer Multicommodity Flows with Reduced Demands. In Thomas Lengauer, editor, *Proceedings of the First European Symposium on Algorithms*, number 726 in Lecture Notes in Computer Science, pages 360–372. Springer-Verlag, 1993.

[SS94] A. Srivastav and P. Stangier. Algorithmic Chernoff-Hoeffding Inequalities in Integer Programming. In *Proceedings of the Fifth Annual Iternational Symposium on Algorithms and Computation (ISAAC94)*, number to appear in Lecture Notes in Computer Science. Springer-Verlag, 1994.

[Tar86] É. Tardos. A Strongly Polynomial Minimum Algorithm to Solve Combinatorial Linear Programs. *Operations Research*, 34:250–256, 1986.

An algorithm for 0-1 programming with application to airline crew scheduling

Dag Wedelin

Dept. of Computing Science, Chalmers Univ. of Tech.
S-412 96 Gothenburg, Sweden
email dag@cs.chalmers.se

Abstract. We present an approximation algorithm for solving 0-1 integer programming problems where A is 0-1 and where b is integer. The method is based on a simple dual coordinate ascent method for solving the LP relaxation, reformulated as an unconstrained nonlinear problem, and an approximation scheme working together with this method. We report results on solving set covering problems in the CARMEN airline crew scheduling system, used by SAS and Lufthansa.

1 Introduction

In this article we present an approximation algorithm for a particular class of 0-1 integer programming problems. The algorithm is a development of an earlier probabilistic method for combinatorial optimization in [11].

We consider the 0-1 integer programming problem (ILP)

$$\min cx$$

$$Ax = b$$

$$x_j \in \{0, 1\}$$

where A is 0-1, and where b is integer. Inequality constraints are handled by adding binary slack variables. Also with these restrictions, the problem is NP-hard, and several well known NP-hard problems, such as the set partitioning, covering and packing problems, are conveniently stated in this way. The linear programming relaxation (LP) of (ILP) is

$$\min cx$$

$$Ax = b$$

$$0 \le x \le 1$$

which can be efficiently solved in polynomial time. If (LP) happens to have a 0-1 solution, this is also an optimal solution to (ILP), and for certain common and nontrivial subclasses of 0-1 problems, (LP) always has an integer solution.

For more difficult problems particular instances may also be easy in this sense. If however (ILP) is combinatorially complex, the (LP) solution has many non-integer values, and very little information about the solution to the 0-1 problem is obtained in this way.

The algorithm can be viewed as consisting of two interacting components: a simple search algorithm for finding a 0-1 solution to (LP) if such a solution exists, and an approximation scheme that manipulates the cost array c as little as possible, to ensure convergence and to make the modified LP problem have a unique 0-1 solution.

2 Linear programming as nonlinear optimization

If we relax only the equality constraints $Ax = b$ of (LP) the Lagrangean becomes

$$L(x, y) = cx - y(Ax - b) \tag{1}$$

with no sign restriction on y. The solution of (LP) corresponds to a saddle point of $L(x, y)$, and this point can be found by solving the problem

$$\max_y \min_{0 \leq x \leq 1} L(x, y) \tag{2}$$

If we define

$$f(y) = \min_{0 \leq x \leq 1} L(x, y) = yb + \min_{0 \leq x \leq 1} (c - yA)x \tag{3}$$

this problem turns into the unconstrained nonlinear problem

$$\max_y f(y) \tag{4}$$

which is the dual of (LP) with respect to the constraints $Ax = b$ only. Note that this dual problem is different from the usually considered dual problem where all constraints except $x \geq 0$ are relaxed. For any given y we may determine $f(y)$ by setting $x_j = 1$ whenever $c_j - ya_j < 0$ and $x_j = 0$ otherwise. We further observe that $f(y)$ can be viewed as the minimum of a set of linear functions in y, one for each instantiation of x, and that $f(y)$ is therefore concave. In principle, this makes it possible to find the global maximum of $f(y)$ by some ascent method.

Rather than solving the linear constrained optimization problem LP directly, we may instead solve the nonlinear unconstrained optimization problem (4). This is the starting point for the so called subgradient methods for solving linear programs, see [9], as well as for our method.

3 The algorithm for solving (LP)

The algorithm is based on a procedure that considers the reduced cost vector

$$\bar{c} = c - yA \tag{5}$$

Since $\bar{c}x$ for any feasible x differs from cx only by a constant, we may substitute $\bar{c}$ for c in both (LP) and (ILP) without changing the solution. If we could find a y such that the solution to the relaxed unconstrained problem (R)

$$\min_{0 \le x \le 1} \bar{c}x$$

is unique and satisfies $Ax = b$, we would clearly have an optimal 0-1 solution to (LP) as well as to (ILP). A unique solution to (R) requires that all elements of $\bar{c}$ are nonzero, and the solution is then given simply as

$$x_j = \begin{cases} 1 \text{ if } \bar{c}_j < 0 \\ 0 \text{ if } \bar{c}_j > 0 \end{cases} \tag{6}$$

This y would also be a solution to $\max_y f(y)$ since all first derivatives of $f(y)$ exist and are 0 for this point, and the maximum of $f(y)$ would occur on a plateau, or in n dimensions within a polytope with a positive volume, allowing a slack in all of the y_i.

The existence of such a y is determined by the following theorem:

Theorem 1 *A y such that (R) gives a unique solution to (LP) exists if and only if (LP) has a unique solution that is 0-1.*

See [12] for the proof. Note that the theorem holds for arbitrary A and b, so the approach described so far can in principle be applied to any (ILP).

3.1 The dual ascent algorithm

The algorithm works by taking every element of y in turn and iteratively choosing y_i so that the solution x to (3) satisfies constraint i of $Ax = b$. Assuming that A is 0-1 and that b is integer, this can be done by setting y_i so that b_i of the costs for the variables in constraint i become negative. This procedure corresponds exactly to the unconstrained maximization of $f(y)$ through a repeated coordinate search. If we keep all variables except y_i constant the function $f(y)$ may be written

$$f(y_i) = b_i y_i + \min_{0 \le x \le 1} (c - y_i a_i)x + C \tag{7}$$

where C is some constant depending on the other elements of y. In the interval $-r^+ \ge y_i \ge -r^-$ it holds that $a_i x = b_i$, so a small change in the first term of (7) caused by an increase of y_i is cancelled by the same decrease in the second term. Within this interval $f(y)$ is therefore constant, and since $f(y)$ is concave this is also its maximum along y_i.

We now describe how y_i is computed. Since this calulation is repeated many times for all possible i it is useful to maintain and update also the current value of $\bar{c} = c - yA$ during the computation. Let $\bar{c}^i$ be those elements $\bar{c}_j$ of $\bar{c}$ for which it holds that $a_{ij} = 1$ and let

$$r^i = \bar{c}^i + y_i \tag{8}$$

The purpose of (8) is to cancel from $\bar{c}$ the effect of any previous value for y_i. Let r^- be the b_i'th lowest value of r^i, and r^+ the $b_i + 1$'th lowest. The new value of y_i is then chosen as

$$y_i := \frac{r^- + r^+}{2} \tag{9}$$

If r^- and r^+ are nonzero the result after updating of $\bar{c}$ is that exactly b_i of the elements of $\bar{c}^i$ are negative. We note that these computations are very simple and well suited for fast implementation. We also note that it in this case is possible to work with $\bar{c}^i$ directly, rather than r^i, and update y_i differentially.

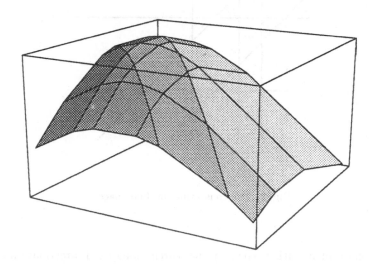

Fig. 1. Plot of $f(y)$.

In figure 1 we show $f(y)$ for the problem

$$\min 6x_1 + 9x_2 + 2x_3 + 3x_4 + 5x_5 + 2x_6 + 5x_7$$

subject to the constraints

$$x_1 + x_2 + x_3 + x_4 + x_5 = 2$$
$$x_1 + x_2 + x_6 + x_7 = 3$$

Since we have only two constraints, $f(y)$ becomes a function of two variables. Note the plateau at the top, this is where the solution to (3) satisfies the constraints. Figure 2 represents an aerial view of $f(y)$, in which the lines $c - yA = 0$ are plotted, and in each area the value of Ax is written. Note that the position of these lines are independent of b, while the value of $f(y)$ itself is not. From any starting point iteration interchangeably proceeds vertically and horizontally until the area 2,3 is reached. The horizontal iteration moves to an area where the first index is 2, and the vertical moves to an area where the second index is 3.

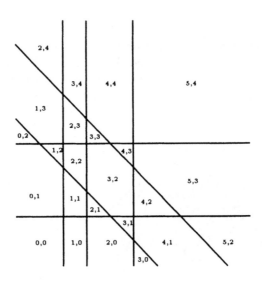

Fig. 2. The constraints in dual space.

A common problem with a repeated line search along fixed directions on a function whose first derivatives are not continuous, is that the procedure may get stuck over a ridge where none of the coordinate directions give an increase in $f(y)$, but where some other direction will. The effect on $\bar{c}$ is that many of the elements converge to 0, and no integer solution is found. The remedy is the approximation scheme of the next section, which provides fast convergence and also makes it possible to find approximate 0-1 solutions.

An example of a situation where convergence is a problem is shown in figure 3. The dark area indicates the area where $f(y)$ reaches its maximum. We see that the speed of convergence depends on the distance between the three parallel lines, and if these were to be placed on top of each other the procedure would be stuck. Note that the dark area would then be reduced to a line, so that no unique 0-1 solution would exist.

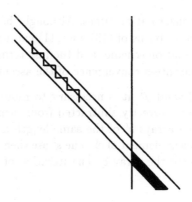

Fig. 3. Slow convergence.

4 The approximation scheme

The idea of the approximation scheme is to manipulate the cost array c as little as possible, so that the modified LP problem has a unique 0-1 solution. The magnitude of this offset is determined by the parameter κ. When $\kappa = 0$ the algorithm is identical to the algorithm of section 3 and there is no approximation at all, and when $\kappa = 1$ the algorithm behaves as a greedy algorithm.

The offset is applied by modifying the ascent algorithm in the following way. Rather than having only one value y_i for each constraint, each constraint is associated with *two different* values y_i^- and y_i^+. The idea is to add a small negative value to elements of $\overline{c}^i$ that are negative after the iteration of y_i, and similarly, a positive value to the elements that are positive. If $\overline{c}_j = 0$ only the costs for variables active in the constraint are affected. The expressions for y_i^- and y_i^+ are

$$y_i^- = y_i + \frac{\kappa}{1-\kappa}(r^+ - r^-) + \delta \tag{10}$$

$$y_i^+ = y_i - \frac{\kappa}{1-\kappa}(r^+ - r^-) - \delta \tag{11}$$

which is to be compared with equation (9). The first term is the same as before, but there is also a difference term, the magnitude of which is governed by κ. The expression $\frac{\kappa}{1-\kappa}$ is just a mapping of the interval 0..1 on 0..∞ and could very well be chosen in some other way. The constant δ is a small constant to ensure that the elements of $\bar{c}$ are nonzero at all times. Although these terms partially have a similar effect, the second terms of (10) and (11) can be interpreted as providing the integer approximation scheme, and the last terms as being a controlled approximation that guarantees convergence of the ascent method.

When we have both y_i^- and y_i^+ it is necessary to modify the calculation of r^i so that the old values are correctly subtracted from each element. An easy way of doing this is to define arrays s^i of the same length as r^i, where the offset is stored individually for each element of $\bar{c}^i$. The s^i are then remembered during the calculation, in addition to the array $\bar{c}$. The iteration of one constraint consists of the following steps:

1.
$$r^i := \bar{c}^i - s^i \tag{12}$$

2.
$$r^- := \text{ the } b_i\text{'th lowest element of } r^i \tag{13}$$
$$r^+ := \text{ the } b_i + 1\text{'th lowest element of } r^i \tag{14}$$
$$y_i^- := y_i + \frac{\kappa}{1-\kappa}(r^+ - r^-) + \delta \tag{15}$$
$$y_i^+ := y_i - \frac{\kappa}{1-\kappa}(r^+ - r^-) - \delta \tag{16}$$
$$s_j^i := \begin{cases} -y_i^- & \text{if } r_j^i \le r^- \\ -y_i^+ & \text{if } r_j^i \ge r^+ \end{cases} \tag{17}$$

3.
$$\bar{c}^i := r^i + s^i \tag{18}$$

The iteration proceeds by repeatedly cycling through the constraints, and for each constraint these steps are performed. We observe that the computation can be organized in a network with one node for each variable and one hyperarc for each constraint. In principle, many of the constraints could then be iterated in parallel, provided that they have no common variables.

In degenerate cases, it may happen that $r^- = r^+$, and some positions of s^i can be chosen in different ways. The array s^i should then be chosen indeterministically among those possible offsets that satisfy constraint i, i.e. so that exactly b_i of the elements of $\bar{c}^i$ are negative.

Since values of $\bar{c}$ are effectively moved away from 0, the slow convergence of the coordinate ascent method when r^- is close to r^+ is effectively eliminated. In terms of $f(y)$ the offsets in $\bar{c}$ eliminate a rigde by splitting it and opening a plateau along the current coordinate. Note that the restrictions on A and b ensure that along each coordinate, the maximum of $f(x)$ always lies between two dual planes, and never on a single plane, so that this is always possible. We therefore obtain improved convergence, at the price of approximation. In practice, convergence in itself does not seem to be a problem, even for very large problems. The difficulty is to approximate as little as possible, to obtain good integer solutions.

4.1 How to choose κ and δ

The value of δ is not very critical and should generally be chosen as large as possible without deteriorating the obtained solutions, to within the right order of magnitude. The most critical factor of the approximation scheme is how to choose κ. Two empirical observations can be made. First, it is difficult to know in advance what is an appropriate value of κ, since different problems converge for different values. Secondly, the lower the value of κ at convergence, the better the result. Therefore, iteration begins with $\kappa = 0$, and we then let κ increase slowly as the iteration proceeds. When a solution is found the process starts over again in a new trial, now with a slower increase around the last convergence point, which often results in a better solution.

4.2 Interpretation of the method for $\kappa = 1/3$ and $\kappa = 1$

For $\kappa = 0$ we have no approximation and the algorithm can only find an optimal solution, although it will frequently fail.

For $\kappa = 1/3$ the expressions for y_i^+ and y_i^- become (assuming $\delta = 0$)

$$y_i^- = r^+ \tag{19}$$
$$y_i^+ = r^- \tag{20}$$

from which it follows that the change in a cost $\bar{c}_j$ is independent of the value itself. If the constraint structure viewed as a hypergraph is acyclic, the algorithm can be interpreted as an algorithm for nonserial dynamic programming (see [2]), and the following theorem holds:

Theorem 2 *If $\kappa = 1/3$ and the constraint structure is acyclic, the algorithm converges to the optimal solution in d steps, where d is the diameter of the hypergraph.*

See [12] for proof and details.

When $\kappa = 1$ the second terms of (10) and (11) will dominate and set each value of $\bar{c}^i$ to $-\infty$ or ∞. Satisfaction of the constraints then completely overrides the cost structure of the problem. The algorithm behaves as a simple greedy algorithm, and in the frequent special case when the equalities are replaced systematically by either $\geq$ or $\leq$, a 0-1 solution is found after exactly one pass over the constraints.

5 Computational results

The algorithm has been extensively tested on large set covering problems arising from airline crew scheduling applications. It is included in the CARMEN system for airline crew scheduling, which is used by SAS and by Lufthansa.

5.1 The airline crew scheduling problem

We desribe the problem in its classical and basic form. Assume that a timetable with all flight legs is given (a flight leg is given flight at a given time). To all these flight legs we wish to assign crews so that every crew receives a reasonable schedule, and so that the cost of the solution is as small as possible.

Every crew has a home base, and the schedule for a crew is described as a sequence of flight legs starting and ending at the home base. Such a sequence is called a pairing. Since pairings must satisfy a large number of legal and contractual rules, the problem is commonly solved in two steps. First a large number of legal pairings are generated, and then a subset of the generated pairings are selected so that every flight in the schedule is included at least once. The second step can be expressed as a set covering problem (SC)

$$\min cx$$

$$Ax \geq 1$$

$$x_j \in \{0, 1\}$$

where A is 0-1. Each row of A corresponds to a flight leg and each column to a pairing. A typical problem deals with about 50 to 2000 flights and 10000 to 1000000 pairings. The normal number of nonzero elements of A is between 5 to 10 per column.

In some approaches to the airline crew scheduling problem there is feedback of dual variables from the set covering LP-relaxation to the pairing generator (see[5, 8]), to allow for a very specific generation of pairings that can improve the present solution. We note that our algorithm can be used to provide similar output.

5.2 Test results

The algorithm has been tested against the CPLEX-system for linear and integer optimization (version 2.1beta), see [4]. CPLEX is an optimization system that is used by several major airlines specifically for the type of problems that are considered here. The size of the problems before and after preprocessing is shown in table 1. In the preprocessing stage as much as possible of the problem is eliminated by simple criteria. We see that for some problems we have a very significant reduction, but for problems with many rows the reduction is more moderate. The remaining "difficult" part of the problem is then solved by an optimizing algorithm and the results are summarized in table 2. The times indicated are total running times for an HP720 workstation, including the time for I/O operations.

In addition to the integer solution, CPLEX provides information about the objective value for the LP relaxation and for the first four problems it can also

	size before preprocessing			size after preprocessing		
problem	rows	columns	nonzero	rows	columns	nonzero
B727scratch	254	282	2675	29	157	372
ALITALIA	1255	3438	41093	118	1165	4155
A320	1255	9678	125156	199	6931	31801
A320coc	1224	43492	546474	235	18753	82229
SASjump	3482	15485	121902	742	10370	42695
SASD9imp2	4116	58850	301614	1366	25032	110881
SASD9imp1	4089	155660	982751	1585	105804	566905

Table 1. Size of problems before and after preprocessing.

	CPLEX				our method	
problem	obj	time	LP	time LP	obj	time
B727scratch	92800	0.2s	92450	0.07s	92800	1.4s
ALITALIA	5017500	1s	5017500	0.7s	5017500	11s
A320	529250	23s	529250	17s	529250	1m4s
A320coc	565000	2m12s	564550	1m40s	565000	5m2s
SASjump	4737768	23m	4734156	7m40s	4737892	3m58s
SASD9imp2	4335780	5h53m	4332348	16m	4333450	7m46s
SASD9imp1	-	-	4329260	3h30m	4329750	36m10s

Table 2. Test results for real set covering problems.

prove that the solution is optimal. We see that for the tested problems the difference between the 0-1 and LP objective is very small. The parameter settings have been recommended by CPLEX specifically for use with these problems.

The times for our algorithm is for 5 trials. The total execution time is essentially proportional to the number of trials, so it is possible to receive solutions considerably faster if fewer trials are specified. All results have been generated with standard parameter settings.

For practical purposes both algorithms give solutions of about the same quality for these problems. By tuning of the parameters individually for each problem both CPLEX and our method can receive marginally better results on SASjump and SASd9imp2, but the here presented results reflect the true behaviour in a production environment. We note however that the running time for CPLEX increases dramatically with the number of constraints in the problem, and that it fails to solve the largest problem in the test set. For our algorithm, the running times are consistently proportional to the size of the problem, and we can without difficulty solve considerably larger problems than those used in this comparison. We also note that for the largest problem, the total running time is considerably lower than the time required by CPLEX to find an LP solution.

We have also tested the algorithms on randomly generated problems, for which the difference between the LP and ILP objective is larger. On these problems preprocessing hardly makes any difference, and virtually no variables in the LP solution have integer values. With our algorithm, and with the same standard parameter settings, we quickly obtain 0-1 solutions that are considerably better than those of any greedy algorithm. No interesting comparison was possible however, since the CPLEX system is not at all suited for such problems, and was only able to find extremely bad solutions.

Considering memory, our implementation has the same memory requirements during the whole run, while CPLEX allocates more memory during the solution process. On a given computer, it is therefore possible to solve problems about 5 times as large with our algorithm than with the version of CPLEX that we had available. The approximate practical limit for a machine with 128 MB of memory is about 1000000 columns for our algorithm and 200000 for CPLEX.

6   Conclusion

We have presented an approximation algorithm for solving 0-1 integer programming problems. The computations are based on very simple operations that can be organized in a network having the same topology as the constraint structure. A parameter determines the compromise between the running time and the quality of the solution. When $\kappa = 0$ we have shown that the algorithm has a close relationship to the corresponding LP-problem. At $\kappa = 1/3$ the algorithm can be interpreted in terms of dynamic programming, when the constraint structure is acyclic. When $\kappa = 1$ it turns into a greedy algorithm.

We have shown that the algorithm is able to obtain high quality solutions to large set covering problems. The tests indicate that it compares well with established methods based on linear programming, especially when the problems are combinatorially difficult. We have also showed that the method works well in a production system for airline crew scheduling.

Acknowledgements

I wish to thank Erik Andersson, who has inspired the development of this algorithm. I also wish to thank Tony Elmroth for detailed comments on the performance of the algorithm.

References

1. Balas E. and Ho A. (1979). Set Covering Algorithms using Cutting Planes, Heuristics and Subgradient optimization: a Computational Study. Man. Sc. Res. Rep. 438, Carnegie-Mellon Univ.

2. Bertele U. and Brioschi F. (1972). Nonserial Dynamic Programming. Academic Press.

3. Chvatal. V (1979). A greedy-heuristic for the Set-Covering problem. Math. Oper. Res. 4, 233-235.

4. CPLEX Reference Manual (1992). CPLEX Optimization Inc.

5. Derosiers J., Dumas. Y, Solomon M.M. and Soumis F. (1993) Time Constrained Routing and Scheduling. To be published in Handbooks in Operations Research and Management Science, volume on networks. North-Holland.

6. Garey M.R and Johnson D.S. (1979) Computers and Intractability: A Guide to the Theory of NP-Completeness. W.H. Freeman & Co.

7. Hu T.C. (1969) Integer Programming and Network Flows. Addison-Wesley.

8. Lavoie S., Minoux M. and Odier E. (1988) A new approach for crew pairing problems with an application to air transportation. European Journal of Operations Research 35, 45-58.

9. Nemhauser G.L. and Wolsey L.A. (1988). Integer and Combinatorial Optimization. Wiley.

10. Syslo M., Deo N. and Kowalik J.S. (1983). Discrete Optimization Algorithms. Prentice-Hall.

11. Wedelin D. (1989). Probabilistic Networks And Combinatorial Optimization. Technical report 49, Dept. of Computer Science, Chalmers Univ. of Tech.

12. Wedelin D. (1993). Probabilistic Inference, Combinatorial Optimization and the Discovery of Causal Structure from Data. PhD Thesis. Dept. of Computing Science, Chalmers Univ. of Tech.

An $o(n)$ Work EREW Parallel Algorithm for Updating MST *

Sajal K. Das

Dept. of Computer Science
University of North Texas
P.O. Box 13886
Denton, TX 76203, USA
E-mail: das@ponder.csci.unt.edu

Paolo Ferragina

Dipartimento di Informatica
Università di Pisa
Corso Italia 40
56125 Pisa, Italy
E-mail: ferragin@di.unipi.it

Abstract. We provide an $o(m)$-work EREW PRAM algorithm to maintain the minimum spanning forest of an undirected graph under edge insertions and deletions. Then, using the sparsification data structure, we improve this result obtaining the first $o(n)$-work EREW parallel algorithm requiring $O(\log n \log \frac{m}{n})$ time for each update. The problem of treating multiple edge updates is also addressed.

1 Introduction

Dynamic graph algorithms are designed to handle graph changes, that is they maintain some property of a graph, after an edge change, more efficiently than a recomputation of the entire graph. Such an algorithm is called *partially dynamic* if only edge insertions are allowed, while it is *fully-dynamic* if both operations – namely, insertions and deletions – can be performed. It is called a *multiple-update* algorithm if it performs simultaneous insertions and/or deletions of a batch of edges. The problem of updating a minimum spanning tree (MST) in a graph involves reconstructing a new MST from the current one when: (1) an edge is deleted or inserted, or (2) a batch of edge updates is performed. Note that weight changes of an edge can be treated as a combination of deletion and insertion operations.

Sequential algorithms for updating an MST have received considerable attention in the past [2, 4, 5, 8, 18]. In particular, Frederickson [8] described an $O(\sqrt{m})$ time sequential algorithm for the single edge update, where m is the number of edges in a graph. Recently, Eppstein *et al.* [4, 5] provided a new technique, called *sparsification*, which can be used to significantly improve Frederickson's algorithm, thus obtaining an $O(\sqrt{n})$ sequential algorithm, where n is the number of nodes in the graph. These algorithms are efficient compared with the known start-over ("from scratch") algorithms for the MST construction [9].

* The first author was supported in part by Texas Advanced Technology Grant TATP-003594031. The research of P. Ferragina was carried out in part while visiting the Dept. of Computer Science, University of North Texas, Denton, with support from M.U.R.S.T. of Italy, and grant TATP-003594031.

In this paper, we are concerned with the design of parallel algorithms for maintaining an MST under edge updates. Although the edge update problem has received considerable attention [13, 11, 14, 15, 17], work–optimal or even "low" work parallel algorithms are deficient. Johnson and Metaxas [11] presented an efficient parallel algorithm for the single update problem on the exclusive-read and exclusive-write (EREW) parallel random access machine (PRAM) model, requiring $O(\log n)$ time and $O(m)$ work. The preceding parallel algorithms are designed to handle single update in the underlying graph but not multiple updates. The multiple edge update problem is particularly important in the context of parallel computing in which the idea is to exploit parallelism while treating all changes simultaneously using all the available processors. There is a source of potential saving, namely the effects of edge insertions and deletions may cancel each other, hence the corresponding updates may be avoided in the batch algorithm.

Pawagi and Kaser [14] proposed a parallel algorithm on the concurrent-read and exclusive-write (CREW) PRAM model, to maintain the MST under batch insertions or deletions of edges. For k-edge insertions, their algorithm requires $O(\log n \log k)$ time and $O(nk)$ work, while for k-edge deletions $O(\log n \log^2 k)$ time and $\Omega(n^2)$ work. Shen and Liang [17] provided an improved multiple update algorithm, running on the CREW PRAM and requiring $O(\log n \log k)$ time and $O(nk)$ work. Clearly, independently of k, the work involved is $\Omega(n)$ according to these approaches.

Recently, Ferragina and Luccio [6] have proposed a new paradigm for sequential multiple update, that uses the sparsification technique. In the domain of parallel computing, they have extended the sparsification data structure for the EREW model showing that, more efficient parallel algorithms can be designed by properly combining the parallel techniques with a "good parallel" data structure. However, their algorithm does not take advantage of the fully-dynamic nature of the MST problem in the sense that, if a "few" number of updates affects a single node in the sparsification tree, it may not be necessary to recompute the certificate (defined in Section 8) from scratch.

In this paper, we provide an $o(m)$-work parallel algorithm on the EREW PRAM model, to maintain the minimum spanning forest of an undirected graph under edge insertions and deletions. The algorithm uses a partition of the minimum spanning tree similar to the approach due to Frederickson [8]. Then we use the sparsification data structure to improve this result, thus obtaining the first $o(n)$-work EREW PRAM parallel algorithm, requiring $O(\log n \log \frac{m}{n})$ time for each update. (By *work* of parallel algorithms, we mean the total number of operations performed during the computation [10].) Finally, the problem of treating multiple edge updates is also addressed, designing an *adaptive* version of our parallel algorithm, which maintains the sparsification data structure in parallel and allows us to significantly improve the existing bounds on the amount of work [6, 14, 17].

2   Preliminaries

There are three cases to be handled in the single edge update problem of a minimum spanning tree (MST). Obviously, if a non–tree edge is deleted, no operations have to be performed because there will be no change in the MST. In the other cases (e.g. insertion or deletion of a tree edge), the MST can be forced to change. However, at most one edge will leave the tree and one edge will enter it (we say that it is a "stable" property). If an edge (v, w) is inserted, then this edge may enter the tree, forcing out some other edge. This case may be detected by determining if the maximum cost of an edge on the fundamental cycle that (v, w) induces in the tree, is greater than that of (v, w). An obvious implementation of this test requires $O(\log n)$ time employing $O(\frac{n}{\log n})$ number of processors on the EREW PRAM. It consists of the application of the Euler tour technique to identify the cycle and the parallel prefix technique to select the maximum cost edge on this cycle [10].

The most interesting case occurs when a tree edge (x, y) is deleted. Here, we have to determine the minimum cost non–tree edge (v, w) that connects the two subtrees created by the removal of (x, y). In the worst case, there can be $O(m)$ edges that are candidates for the replacement. In fact, the algorithm due to Johnson and Metaxas [11] uses $O(m)$ work since it requires the scanning of all the non–tree edges.

We use a data structure similar to the one provided in [8] that allows us to perform the edge–update operation in $o(m)$ work. Our parallel data structures are designed to handle bounded–degree graphs in which no vertex has degree greater than three. Given a graph $G_0 = (V_0, \bar{E}_0)$, a well-known transformation in graph theory can be used to produce a graph $G = (V, \bar{E})$ in which each vertex satisfies the desired degree constraint. Furthermore, the MST of G can be easily transformed into an MST of G_0. In this paper, we deal with graphs of $O(m)$ vertices, each of degree no greater than 3. The bounded (constant) degree of each vertex in the graph will be very useful throughout the paper to perform operations in constant sequential time on the set of outgoing edges of a vertex.

Section 3 presents a parallel graph decomposition technique which is partially derived from [8].

3   Clustering stage

Let G be a graph and G' a connected subgraph of G. We define the *size* of G' as its number of vertices, and denote it by $\mid G' \mid$. Given a rooted tree T and a vertex v, the subtree of T rooted at v will be called a *descending tree* of v, denoted by $T[v]$.

Given a minimum spanning tree T of G, let us partition T into *clusters* C_i, such that $\mid C_i \mid \in [\alpha, 4\alpha]$ and C_i is a connected component. The parameter α (defined later) is chosen to minimize the total work of the parallel algorithm. The clustering stage consists of $O(\log m)$ phases, each requiring $O(\log m)$ time and $O(\frac{m}{\log m})$ processors, where $\mid T \mid = m$. For brevity, the proof of the following two lemmas will be omitted. For details, refer to [3].

Lemma 1. *Given a binary tree T', where $|T'| = m'$, there exists a node v such that the descending tree $T'[v]$ has size $\frac{m'}{4} \leq |T[v]| \leq \frac{m'}{2}$.*

Lemma 2. *Given a tree T' whose size is at least 4α, the vertex v of T' satisfying the property claimed in Lemma 1, partitions T' into two connected components whose size is at least α.*

Let us consider the i-th phase of the clustering stage, in which all the existing components (in particular, they are trees forming a forest, denoted by T_i) are connected to a dummy node. In this way, we obtain a new augmented tree whose root has an outgoing degree equal to the number of connected components in T_i.

The size of the descending tree of each vertex can be computed applying the Euler tour technique. Since the augmented tree has $|T| + 1$ nodes and at most $2|T|$ edges, the Euler tour technique requires $O(\log|T|) = O(\log m)$ time and $O(\frac{|T|}{\log|T|}) = O(\frac{m}{\log m})$ number of processors. The i-th phase consists of finding a set of nodes v_k, one for each connected component C, such that $\frac{|C|}{4} \leq |T[v_k]| \leq \frac{|C|}{2}$. More than one node can have this property in each component C, so that we apply a parallel prefix to select only one of them for each C. The forest T_{i+1} is obtained by removing the edges connecting each v_k to its parent.

Furthermore, the connected components having a size larger than 4α are partitioned into two parts, whose size is a constant fraction of the size of the original tree. We need a logarithmic number of phases so that all the existing components have a size in the interval $[\alpha, 4\alpha]$. Hence this procedure will require $O(\log^2 m)$ time employing $O(\frac{m}{\log m})$ processors. Since each cluster consists of $O(\alpha)$ vertices, we have a total of $O(\frac{m}{\alpha})$ clusters.

4 Data structures

This section describes the data structures that will be used to efficiently maintain the minimum spanning tree subject to an insertion or deletion of an edge. Also it presents a sketch how to compute the maximum cost edge.

An edge is said to be *internal*, if it is a tree-edge connecting two vertices belonging to the same cluster, otherwise it is *external*. A *boundary* vertex is one having an external edge incident on it. Let us introduce a few definitions.

D_1 : Adjacency list of the minimum spanning tree, where each node has a pointer to its cluster.

D_2 : Adjacency list of the original graph. (Recall that each node has degree at most 3.)

D_3 : Each cluster is maintained as an array containing the list of the vertices belonging to it. Thus D_3 is the array containing the pointers to the clusters.

D_4 : The set of external edges of each cluster. Note that D_4 combined with D_3 represents the tree formed by the clusters.

E_{ij} : The list of non–tree edges connecting pair of clusters C_i and C_j. The minimum cost edge within each E_{ij} is also available. We assume that each edge in E_{ij} has a pointer to the head of E_{ij}, and a pointer to its reversal that belongs to E_{ji}.

F_z : It is defined for each cluster C_z and consists of a two-dimensional array indexed by a pair of boundary vertices of C_z. It maintains the maximum cost edge in the tree path connecting each pair of boundary vertices.

All of these data structures can be constructed in $O(m \log \alpha + \frac{m^2}{\alpha^2})$ work (for details see [3]). Recall that the most costly operation will be the clustering stage either in terms of the parallel time complexity or in terms of the number of operations. Given a minimum spanning tree of the graph, the preprocessing stage consists of the following steps:

1. Partition the tree into clusters of size in the range $[\alpha, 4\alpha]$.
2. Build all the data structures listed above.

Due to space limitation, we sketch only the construction of F_z here.

Construction of F_z: For each cluster C_z, a two-dimensional array F_z indexed by its boundary vertices is defined. This construction is done in parallel on all the clusters. With the help of parallel prefix operation, the boundary vertices are enumerated, thus each vertex knows its position inside F_z. The subtree T_z, denoting the cluster C_z, is duplicated as many times as its number of boundary vertices. Then we root the i-th copy of T_z on the i-th boundary vertex $v_i \in T_z$ (obtaining T_z^i) and compute the maximum cost edge on the tree path connecting v_i to each leaf (Lemma 3). The boundary vertices other than $v_i \in C_z$ are leaves of the tree T_z^i. Hence using the information previously computed, we can construct in parallel the elements $F_z[v_i, v_j]$, for each boundary vertex v_j in T_z^i.

The duplication of T_z allows us to avoid read/write conflicts in the construction of F_z. To compute the memory space occupied by F_z, we observe that there are at most $O(\frac{m^2}{\alpha^2})$ pairs of boundary vertices for all the clusters and it is also an upper bound on the number of pairs restricted to boundary vertices within the same cluster. The duplication process can be performed in work-optimal way for each cluster. Since $O(\frac{m}{\alpha})$ is the total number of boundary vertices, we need $O(\frac{m}{\alpha})$ duplications, and thus an overall $O(m)$ work and $O(\log \frac{m}{\alpha})$ time. The computation of maximum cost edge for each tree T_z^i requires $O(\log \alpha)$ time using $O(\frac{\alpha}{\log \alpha})$ processors, as shown in the following. The data structures F for all the clusters can be constructed in $O(\log \frac{m}{\alpha} + \log \alpha)$ time, performing an $O(m + \frac{m^2}{\alpha^2})$ work.

Computing the maximum cost edge: To maintain the F data structure that efficiently allows the single edge insertion, we need to compute for each *boundary* vertex r of the cluster C_r, the maximum cost edge on the tree path connecting r to all the other boundary vertices in C_r. Let us assume that the subtree formed by the vertices of C_r is rooted at r. Hence, the problem can be formulated as follows: For each leaf l of the tree T_r, compute the maximum cost edge on the path connecting l to r.

Our two-stage approach uses the tree contraction algorithm [1] and a different RAKE operation. For brevity we sketch only the first stage (for details see [3]). Given the $(i-1)$-th contraction step, let e be the edge connecting u to v in the contracted tree T_{i-1}. The invariant preserved by the first stage is that, $f(e)$ represents the maximum cost edge on the path connecting u to v in the original tree. (Initially, $f(e)$ is the cost of the edge $e \in T_r$.)

Let u be a node in T_{i-1} with parent t and children w (leaf) and v. By induction on the number of tree contraction steps, we consider the RAKE operation which deletes the node u to contract the leaf w [1]. Using the inductive hypothesis, it is simple to see that the path between t and v in T_{i-1}, consisting of two edges, represents the contracted path between these two nodes in the original tree. Furthermore, let f_{tu} (resp. f_{uv}) be the maximum cost edge on the contracted path between t (resp. u) and u (resp. v). Computing $f = \max\{f_{tu}, f_{uv}\}$, we obtain the maximal cost edge on the contracted path between nodes t and v.

At the end of the contraction process, the final tree consists of the root r and of the rightmost and the leftmost leaves (in the original tree) directly connected to r. Now, using the scheduling provided by the first stage, we can apply back the RAKE operation (called B-RAKE) taking advantage of the f values, and thereby computing the maximum cost edge on the path connecting each leaf to the root of the tree.

Lemma 3. *[3] Given a tree T, the maximum cost edge on the path connecting each leaf l to the root of T can be computed in $O(\log|T|)$ time using $O(\frac{|T|}{\log|T|})$ number of processors.*

In Sections 5 and 6 we design two operations, namely split and merge, which are useful to maintain consistency in the cluster decomposition of the minimum spanning tree under edge updates.

5 Split operation

A split operation is performed when an internal edge (x, y) is deleted, otherwise the partitioning in not affected. First of all, we consider the update process restricted to the data structures $D_i, 1 \leq i \leq 4$. The edge (x, y) and its reversal (y, x) are canceled from the adjacency list of D_1 and D_2 respectively. Then, to identify the new partition, we root in x and y the two subtrees formed into the cluster C_x by the deletion of (x, y). This can be done by using D_3, that specifies the vertices in the cluster, and D_1 that specifies the tree edges. The two subtrees are visited, marking their vertices in order to split the cluster C_x into $C_{x'}$ and $C_{x''}$. The new names of the two clusters are broadcast in D_1 and D_2 to all the edges of C_x. All these operations can be performed in work-optimal way, requiring $O(\log \alpha)$ time and $O(\alpha)$ work.

Now, the data structures E and F have to be updated to reflect the splitting of C_x. Let us mark all the edges in $E_{x-} = \bigcup_z E_{xz}$ according to the two new clusters, and split it by a parallel prefix in $E_{x'-}$ and $E_{x''-}$. Furthermore, using the marks given to the edges in E_{x-} and the cross-pointers, the entries E_{zx} can also be marked and split into $E_{zx'}$ and $E_{zx''}$, for each cluster C_z where $z \neq x$. Recalling that the number of edges outgoing from a cluster is at most $O(\alpha)$, this splitting process requires $O(\log \alpha)$ time and $O(\alpha)$ work in total. During the splitting of the data structure E, we can also recompute the minimum cost edge in the newly formed entries. The split of C_x requires an update of F_x in order to cancel the entries $F_z[v, w]$, where $v \in C_{x'}$ and $w \in C_{x''}$ or vice versa. Since each boundary vertex in C_x has been marked depending on which half it belongs to, this mark is broadcast along the rows and the columns of F_x. Then by scanning F_x, we cancel the entries corresponding to pairs of boundary vertices belonging to different halves of C_x. Since $| F_z | = O(\frac{m^2}{\alpha^2})$i and all these operations can be performed in work-optimal way, we attain $O(\log \frac{m}{\alpha})$ time complexity and $O(\frac{m^2}{\alpha^2})$ work.

Note that the information relative to the pairs of vertices that are inside the same component derived from C_x, are not affected by this update process. We can again enumerate the boundary vertices of $C_{x'}$ and $C_{x''}$ and hence construct the data structures $F_{x'}$ and $F_{x''}$. Thus the split operation requires $O(\frac{m^2}{\alpha^2} + \alpha)$ work and $O(\log \frac{m}{\alpha} + \log \alpha)$ time. Since the splitting may produce clusters of size less than α, we possibly have to merge this small cluster to a "consistent" and adjacent one.

6 Merge operation

Given an external edge (u, v), such that the size of the cluster C_u is less than α, we join C_u to an adjacent cluster and then possibly split it, in order to attain two clusters of "proper" sizes. Note that if a cluster has a size greater than 4α, it can be partitioned in a consistent way, applying the partitioning algorithm in Section 3. Clusters C_u and C_v are merged to form a larger cluster C_{u+v}, scanning the array containing the vertices in C_v to define their mark as C_u, and merging their arrays in D_3. This operation requires $O(\alpha)$ work and $O(\log \alpha)$ time. Then, the external edge (u, v) connecting these two clusters must be canceled from D_4, because it is no more external. Clearly, the vertices u and v are no longer boundary vertices, necessitating an update of the data structure F. The entries $F_u[w, u]$ and $F_v[z, v]$, for each boundary vertex-pair $w \in C_u$ and $z \in C_v$, are deleted. All the other (not deleted) entries of F_v and F_u are copied into the new array F_{u+v} created for the merged cluster C_{u+v}, since they already correspond to pairs of boundary vertices in C_{u+v}. Also the entry $F_u[w, z]$ is inserted in F_{u+v}, with the value $\max\{c_{(u,v)}, F_v[v, z], F_u[w, u]\}$, where $c_{(u,v)}$ denotes the cost of edge (u, v). Observe that the number of entries to be inserted are at most $O(\frac{m^2}{\alpha^2})$.

The merge operation refers also to data structures E_{uz} and E_{vz}, for each cluster C_z. To update E, the two lists E_{uz} and E_{vz} are merged and the minimum cost edge contained in the new list is recomputed. At the same time, all the entries E_{zu} and E_{zv} are merged. This update requires $O(\frac{m}{\alpha})$ work and $O(\log \frac{m}{\alpha})$ time. The new cluster formed by merging C_u and C_v may have a size greater than 4α. If so, it is partitioned into two parts satisfying the size bound within

$[\alpha, 4\alpha]$, applying the partitioning algorithm in Section 3 for a single iteration. Obviously, as for the split operation, this process partitions the data structures E and F according to the newly formed clusters.

7 Single edge update

Single edge insertion: Since the insertion of an edge (u, v) induces a cycle in the minimum spanning tree T, we delete the maximum cost edge in this cycle using the data structure F. Let T_C denote the contracted minimum spanning tree, where the clusters are nodes and the external edges are the edges of the tree.

If the edge to be inserted has end-points belonging to the same cluster C, we can use the simple algorithm in [11] to find the cycle formed by the edge (u, v) in C. Otherwise, the computation is more complex and involves the use of F, as detailed below.

1. Let C_u and C_v be the two clusters for nodes u and v, respectively. Assume that $C_u \neq C_v$.
2. Root the tree of the cluster C_u (resp. C_v) at u (resp. v), obtaining the tree T_{C_u} (resp. T_{C_v}). It requires $O(\log \alpha)$ time and $O(\frac{\alpha}{\log \alpha})$ processors.
3. Find the path P_{uv}^C between C_u and C_v in the contracted tree T_C, using the D_4 data structure, and compute the maximum cost edge M_C on it. This step requires $O(\log \frac{m}{\alpha})$ time and $O(\frac{m}{\log \frac{m}{\alpha}})$ processors.
4. Let V_v and V_u be the boundary vertices where the path P_{uv}^C starts in C_v and C_u, respectively. We can compute the maximum cost edge M_u (resp. M_v) in T_{C_u} (resp. T_{C_v}) on the path connecting u (resp. v) to V_u (resp. V_v).
5. The path in the MST, T, connecting u to v is formed by the external edges (k_j, h_j) of P_{uv}^C and by the paths internal to the clusters $C_{h_j} = C_{k_{j+1}}$, which are traversed by P_{uv}^C and connecting the boundary vertex h_j to k_{j+1}. The maximum cost edge on the internal paths is determined with the help of $F_{h_j}[h_j, k_{j+1}]$. The total computation requires $O(\frac{m}{\alpha})$ accesses to the data structure F, that can be performed in parallel, such that a single processor accesses each entry $F_{h_j}[h_j, k_{j+1}]$ in constant sequential time.
6. Compute the maximum cost edge M_E among all the previously found values M_{h_j}, M_C, M_u, M_v. This requires $O(\frac{m}{\alpha})$ work and $O(\log \frac{m}{\alpha})$ time.
7. If M_E is (u, v),i the minimum spanning tree is not affected by this insertion and so no other operation is needed. Otherwise, delete the edge M_E from the minimum spanning tree, T, possibly splitting the cluster to which it belongs, and add the edge to the data structures E and D_4. As a result of splitting, we have two connected components that are made consistent in order to satisfy the constraint on the size of the clusters.
8. Since the insertion of (u, v) in the tree T modifies the set of boundary vertices of C_u and C_v, the data structure F is modified accordingly (see Section 6).

The deletion of the maximum cost edge, M_E, and possibly the insertion of (u, v) determines the trivial updating of the other data structures. Hence, the insertion operation requires $W_{split} + O(\frac{m}{\alpha}) = O(\frac{m^2}{\alpha^2})$ work and $O(\log\alpha + \log\frac{m}{\alpha})$ time.

Single edge deletion: Let us delete the edge (u, v) belonging to the minimum spanning tree T, otherwise this operation does not affect significantly the data structures. If (u, v) is inside a cluster, we split it and maintain the consistency of the partition. When T is partitioned into two connected components, to recompute the minimum spanning tree, the data structure E is used to find the minimum cost edge connecting them. The detailed procedure follows.

1. Let C_u and C_v be the two different clusters in which u and v are contained after the deletion of edge (u, v). Visit the two halves of the tree T formed by the deletion, and mark properly the clusters (i.e. u's and v's names).
2. Broadcast the marks, denoting the two formed halves, to each entry E_{zw}. This requires $O(\frac{m^2}{\alpha^2})$ work and $O(\log\frac{m}{\alpha})$ time.
3. Find the entries E_{tw} corresponding to clusters C_t and C_w belonging to different halves, avoiding read/write conflicts. That is, each E_{tw} has the mark of the cluster C_t. The processor mapped to this entry takes the first edge of the list E_{tw}, and using its cross-pointer determines its reversal entry E_{wt} containing the mark C_w. In this way we are able to understand if E_{tw} refers to a pair of clusters belonging to different halves.
4. Compute the minimum cost edge e_{min} crossing the cut, by a parallel prefix on all the marked entries of E in $O(\log\frac{m}{\alpha})$ time and $O(\frac{m^2}{\alpha^2})$ work.
5. Since e_{min} is inserted to form the new minimum spanning tree, it is deleted from E requiring the update of the corresponding entry. It also determines two new boundary vertices that must be inserted in F (see Section 6).

The deletion process requires $O(\frac{m^2}{\alpha^2} + W_{split})$ work and $O(\log\alpha + \log\frac{m}{\alpha})$ time.

To minimize the total work of the fully-dynamic algorithm, we fix the parameter $\alpha = m^{\frac{1}{3}}$, attaining $O(m^{2/3})$ work and $O(\log m)$ time to perform a single edge update. Whereas the construction of the entire data structure from scratch requires $O(m\log m)$ work and $O(\log^2 m)$ time.

8 Improved fully-dynamic algorithm

The fully-dynamic algorithm described so far handles a single edge update operation, requiring $o(m)$ work and $O(\log m)$ time complexity.

Our goal is to reduce the total work required to maintain the MST when a single edge insertion or deletion is performed, even if this increases the overall parallel time by a logarithmic factor. In this context, we use the *sparsification* data structure proposed by Eppstein *et al.* (for details see [3, 5]).

A graph $G = (V, \bar{E})$ is partitioned into a collection of *sparse* subgraphs $G_1 = (V, \bar{E}_1), \ldots, G_k = (V, \bar{E}_k)$, $k = \lceil\frac{m}{n}\rceil$, where $|\bar{E}_1| = \ldots = |\bar{E}_{k-1}| = n$, $|\bar{E}_k| \le n$, $\bar{E}_i \cap \bar{E}_j = \emptyset$ for $1 \le i < j \le k$, $\bar{E}_1 \cup \bar{E}_2 \cup \ldots \cup \bar{E}_k = \bar{E}$. The information relevant for each subgraph G_i can be summarized in a even sparser subgraph, called the *sparse certificate* of G_i. Note that, the minimum spanning forest of G_i is chosen as sparse certificate of G_i for the MST problem.

The certificates are merged in pairs in a common parent, producing larger subgraphs which are again made sparse by applying the certificate reduction computed by a *sparse certificate algorithm*. The result is a balanced binary tree of depth $O(\log \frac{m}{n})$, in which each node contains a sparse subgraph (sparse certificate). The original collection $G_1, \ldots, G_k$ and their certificates are allocated in the tree leaves $L_1, \ldots, L_k$, where L_k is the rightmost leaf. An edge update affects the leaf L_i where the edge is inserted or already contained, together with the path from L_i to the root of the tree.

The improved fully-dynamic parallel algorithm uses the algorithm for single update operation on each affected node of the sparsification tree. Let $T(m, n)$ and $P(m, n)$ be the time and the number of processors required by our previous fully-dynamic algorithm, in a graph of n vertices and m edges. Because each node of the sparsification tree is sparse, the algorithm requiring $T(n, n)$ time and $P(n, n)$ processors can be used at each node. When an edge update is performed, the certificates of the nodes on the path $\Pi(l)$, connecting the affected leaf l to the root of the sparsification tree, are recomputed.

A group of $P(n, n)$ processors is mapped to the leaf node l. These processors in parallel maintain the MST certificate of this node requiring $T(n, n)$ time. After this update, a single edge is inserted or deleted from the certificate in the leaf l to produce the new certificate. This update affects the certificate of the parent of l, say $p(l) \in \Pi(l)$. The group of processors used in l is now mapped to $p(l)$ and the update process is repeated walking up along the path $\Pi(l)$. Since the length of this path is $O(\log \frac{m}{n})$ and the update process is executed *sequentially* on the nodes of $\Pi(l)$, the total required time is $O(\log n \, \log \frac{m}{n})$ and the work involved is $O(n^{\frac{2}{3}} \log \frac{m}{n})$. This simple algorithm significantly improves the existing fully-dynamic algorithms [6, 11]. More recently, Ferragina [7] has provided a new technique which reduces the time complexity of this fully-dynamic parallel algorithm to $O(\log n)$, even maintaining the same work.

Multiple edge update algorithm: Let us now present an algorithm for maintaining a minimum spanning tree when a *batch* of edge updates is performed (for details see [3]). Existing algorithms [14, 16, 17] maintain the MST under a batch of edge insertions and/or deletions by using only the existing parallel techniques. They do not take advantage of any data structures that could reduce the number of operations needed to recompute the MST. In this context, it is worth mentioning that the simple approach adopted in [6], where the sparsification data structure is managed by the application of efficient parallel techniques, attains $O(n \operatorname{polylog}(n))$ work. However, the drawback of this approach is the use of a parallel algorithm from scratch for maintaining the certificate of each affected node.

We propose an improved algorithm here. Suppose a batch of $b = O(n)$ edge insertions is performed, so that only a constant number of leaves is to be constructed in the sparsification tree. A *pipelining* computation is used on the affected rightmost path Π to perform b edge insertions. That is, we map a group of $B = \frac{n^{\frac{2}{3}}}{\log n}$ processors to each node Π_i of Π to maintain its certificate C_i, where $i = 1, 2, \ldots, \log \frac{m}{n}$. Clearly, the set of updates for the rightmost leaf is given by the batch of insertions. Each edge update requires $O(\log n)$ time and since the path Π has length $O(\log \frac{m}{n})$, the update process requires $O((b + \log \frac{m}{n}) \log n)$ time and a total of $O(n^{\frac{2}{3}} \log \frac{m}{n} (b + \log \frac{m}{n}))$ work.

In comparison, the approaches proposed in [14, 17] require more work even running on the CREW PRAM model. Furthermore, our proposed algorithm reduces the time complexity of Ferragina and Luccio's algorithm [6] if $b < \log^{\frac{3}{2}} n$, and reduces the work if $b < \sqrt[3]{n} \log^{\frac{3}{2}} n$.

If a mixed batch of edge insertions and deletions is performed on the underlying graph, then the concept of an *adaptive* algorithm can be introduced. This algorithm decides to recompute the certificate from scratch or to apply the fully-dynamic algorithm repeatedly, depending on the number of edge updates affecting a given node. The algorithm used to recompute the certificate from scratch requires $O(\log^{\frac{3}{2}} n)$ time and $O(n)$ processors since each node is sparse [12].

Let b be the size of the batch affecting a single node n_s. In order to reduce the time for each certificate recomputation, the adaptive algorithm applies the fully-dynamic algorithm if $b < \log n$, otherwise it recomputes the certificate from scratch. On the other hand, if the goal is to minimize the total work, it applies repeatedly the fully-dynamic algorithm if $b < \sqrt[3]{n} \log^{\frac{3}{2}} n$, otherwise it performs a recomputation from scratch. Note that the recomputation of the certificate from scratch also requires to rebuild the data structures used by the dynamic algorithm. In the worst-case, for each affected node, the adaptive algorithm requires a work that is given by the algorithm from scratch, as in [6]. In the case of a "small" batch of updates, that is $b < \log n$, our algorithm never recomputes the certificate from scratch, and the worst case is when all the edge updates are performed on the same leaf l (i.e., the time is maximized).

Executing the updates in a pipeline fashion, we attain $O((b + \log \frac{m}{n}) \log n)$ time complexity employing $O(\frac{n^{\frac{2}{3}}}{\log n} \log \frac{m}{n})$ processors. With respect to either the time or the total work, our algorithm performs much better than the algorithm proposed in [6]. Again, the algorithm due to Shen and Liang [17] requires $O(\log n \log b)$ time and $O(\frac{nb}{\log n \log b})$ processors. Although its time complexity is lower than that required by our adaptive algorithm, yet for the same choice of b, the total work is higher than ours by a factor $O(\frac{\sqrt[3]{n} b}{\log \frac{m}{n} (b + \log \frac{m}{n})})$.

Acknowledgments

We are grateful to G. F. Italiano for several fruitful discussions.

References

1. M. Atallah and U. Vishkin. Finding Euler tours in parallel. *Journal of computer and system sciences*, 29:330–337, 1984.

2. F. Chin and D. Houck. Algorithms for updating minimum spanning trees. *Journal of Computer and System Science*, 16:333–344, 1978.
3. S. K. Das and P. Ferragina. A fully-dynamic EREW parallel algorithm for updating MST. *Technical Report CRPDC-94-8*, Dept. of Computer Science, University of North Texas, Denton, May 1994.
4. D. Eppstein, Z. Galil, and G. F. Italiano. Improved sparsification. In *TR 93-20*. Dept. of Information and Computer Science, University of California, Irvine, 1993.
5. D. Eppstein, Z. Galil, G. F. Italiano, and A. Nissenzweig. Sparsification - a technique for speeding up Dynamic Graph Algorithms. *Proc. of IEEE Symposium on Foundations of Computer Science*, 60–69, 1992.
6. P. Ferragina and F. Luccio. Batch dynamic algorithms for two graph problems. In *Proc. PARLE '94, Lecture Notes in Computer Science*, to appear, 1994.
7. P. Ferragina. Parallel dynamic edge deletion for MST. *Technical Report 8/94*, Dipartimento di Informatica, Università di Pisa, Italy, June 1994.
8. G. N. Frederickson. Data structures for on-line updating of minimum spanning trees, with applications. *SIAM J. Computing*, 14(4):781–798, 1985.
9. H. N. Gabow, Z. Galil, T. H. Spencer, and R. E. Tarjan. Efficient algorithms for minimum spanning trees on directed and undirected graphs. *Combinatorica*, 6:109–122, 1986.
10. J. Já Já. *An Introduction to Parallel Algorithms*. Addison-Wesley, 1992.
11. D. B. Johnson and P. Metaxas. Optimal parallel and sequential algorithms for the vertex updating problem of a minimum spanning tree. Technical Report PCS-TR91-159, Dartmouth College, Dept. of Mathematics and Computer Science, 1991.
12. D. B. Johnson and P. Metaxas. A parallel algorithm for computing minimum spanning trees. In *Proc. ACM Symposium on Parallel Algorithm and Architectures*, pp. 363-372, 1992.
13. H. Jung and K. Mehlhorn. Parallel algorithms for computing maximal independent sets in trees and for updating minimum spanning trees. *Information Processing Letters*, 27(5):227–236, 1988.
14. S. Pawagi and O. Kaser. Optimal parallel algorithms for multiple updates of minimum spanning trees. *Algorithmica*, 9:357–382, 1993.
15. S. Pawagi and I. V. Ramakrishnan. An $O(\log n)$ algorithm for parallel update of minimum spanning trees. *Information Processing Letters*, 22(5):223–229, 1986.
16. A. Scháffer and P. Varman. Parallel batch update of minimum spanning trees. *Technical Report COMP-TR90-140*, Rice University, Houston, Texas, 1990.
17. X. Shen and W. Liang. A parallel algorithm for multiple edge updates of minimum spanning trees. In *Proc. International Parallel Processing Symposium*, 310–317, 1993.
18. P. Spira and A. Pan. On finding and updating spanning trees and shortest paths. *SIAM Journal of Computing*, 4:375–380, 1975.

On the structure of *DFS*-forests on directed graphs and the dynamic maintenance of *DFS* on DAG's *

Paolo G. Franciosa[1] and Giorgio Gambosi[2] and Umberto Nanni[3]

[1] Dipartimento di Informatica e Sistemistica, Università di Roma "La Sapienza", via Salaria 113, I–00198 Roma, Italy. pgf@athena.dis.uniroma1.it
[2] Dipartimento di Matematica, Università di Roma "Tor Vergata", via della Ricerca Scientifica, I–00133 Roma, Italy. gambosi@mat.utovrm.it
[3] Dipartimento di Matematica Pura ed Applicata, Università di L'Aquila, via Vetoio, I–67010 Coppito (AQ), Italy. nanni@vxscaq.aquila.infn.it

Abstract. In this paper we provide a characterization of a *DFS*-forest on directed graphs in terms of a relaxed planar embedding of its structure.

We propose an incremental algorithm, based on that characterization, to maintain a *DFS*-forest in a directed acyclic graph with n nodes and m edges, achieving $O(nm)$ total time in the worst case for any sequence of arc insertions, that is $O(n)$ amortized time per arc insertion in a sequence of $\Theta(m)$ such operations. This favorably compares with the time required to recompute DFS from scratch by using Tarjan's $\Theta(n+m)$ algorithm [19].

The graph is represented by means of adjacency lists and the dfs tree is maintained by using for each node a pointer to its parent, and a rank according to a suitable ordering: the whole data structure requires $O(n + m)$ space.

This is the first algorithm for dynamic *DFS* for nontrivial classes of graphs.

1 Introduction

Depth-First Search (*DFS*) is a very basic technique to explore graphs which has been used in several applications and represents a building block for the solution of many graph problems, such as connectivity testing, acyclicity testing, topological sorting, isomorphism of planar graphs, and others (see, e.g. [19, 8, 1, 3]). A *dfs* of a graph can be efficiently performed in linear time [19], that is in time $O(n + m)$ for a graph $G = (V, A)$, where $n = |V|$, and $m = |A|$.

An efficient *dynamic* solution for DFS would provide a valuable tool to deal with several graph problems. A dynamic solution for a given problem P consists in maintaining a data structure so that it is possible to alternatively modify input

---

* Partially supported by the ESPRIT Basic Research Action no.7141 (Alcom II) and by MURST national project "Algoritmi e Strutture di Calcolo".

data and providing a new solution for P, without recomputing it from scratch. A number of graph problems have been given an efficient dynamic solution, including, for example, transitive closure [9, 10, 13, 18], shortest paths [5, 17, 2, 11], planarity test [7, 12], and many others.

Some of the considered problems seem to be much harder to be dealt in the dynamic version. As observed by Reif [16], DFS is one of the problems that, in the worst case, seem to require a new computation from scratch if a single element of the input is modified. Furthermore Reif puts DFS in a class of problems which are reducible each other with constant time updates. In other words a dynamic solution for any of these problems provides a dynamic solution for any other problem in the class, which includes [16]: depth-first numbering, the acceptance of a linear time Turing Machine, path systems, the boolean circuit evaluation problem, unit resolution. Though, no solution is currently known for dynamic DFS in nontrivial cases.

In this paper we provide a characterization for *DFS*-forests for directed graphs in terms of existence of a relaxed version of planar embedding of the graph itself.

Furthermore we consider the *incremental* DFS problem for the class of *directed acyclic graphs* (*dags*). Namely we propose algorithms and data structures for maintaining explicitly a *DFS*-forest under a sequence of arc insertions that leave the graph acyclic in $O(mn)$ total time in the worst case, for a graph with n nodes and m arcs. This means $O(n)$ amortized time per arc insertion in a sequence of $\Theta(m)$ insertions. The achieved bounds favorably compare with the repeated application of Tarjan's $O(n + m)$ algorithm required to compute DFS from scratch: for long sequences of arc insertions on dense graphs the improvement is from $O(n^2)$ worst case to $O(n)$ amortized time per operation. Our space requirements, beside the standard $O(n + m)$ representation of the graph by using adjacency lists, consist in the representation of a suitable numbering of the nodes.

The proposed algorithm exploits the approach introduced in [14], where the authors obtain the same amortized time bound for the incremental maintenance of a topological sort of a *dag*.

The paper is organized as follows: in section 2 some basic definitions are provided, in section 3 an original characterization of *DFS*-forests for general directed graphs is introduced which is exploited in section 4 to provide the incremental algorithm for *dag*'s; finally, some conclusive remarks are given in section 5.

2 Basic definitions

In the following we assume the standard graph theoretical terminology as in [1]. A *directed graph* $G = (N, A)$ (also referred to as a *digraph*) consists of a finite set N of *nodes* and a finite set of ordered pairs $A \subseteq N \times N$ of *arcs*. A *non empty chain* from v_1 to v_k is a sequence $\langle v_1, v_2 \rangle, \langle v_2, v_3 \rangle, \ldots, \langle v_{k-1}, v_k \rangle$ of arcs such that each $\langle v_h, v_{h+1} \rangle$ in the sequence is an arc in A. An *empty chain* connects any node v to itself. A *cycle* is a nonempty chain from a node to itself. A digraph with no cycles is called a *directed acyclic graph* (in short *dag*).

Any depth-first search of a digraph $G = (N, A)$ induces a *dfs* forest $F = (N, A_F)$ such that $A_F \subseteq A$, and an arc $\langle u, v \rangle$ is in A_F if and only if during the search node v is visited the first time by traversing arc $\langle u, v \rangle$ (see [3, 4]).

In the following, for the sake of simplicity, we will consider for any digraph $G = (N, A)$ an extended digraph $G' = (N', A')$ with $N' = N \cup \{r\}$ ($r \notin N$), and $A' = A \cup \bigcup_{v \in N} \{\langle r, v \rangle\}$, i.e. G is extended by adding a new *dummy* node r with an arc to each node in N. Moreover we will only consider *dfs* visits on G' starting from r.

Since any *dfs* forest on G' is actually a *dfs* tree, and there is a one-to-one correspondence between *dfs* trees on G' and *dfs* forests on G, in the rest of the paper we only refer to *dfs* trees, assuming that the visit is performed starting from root r, that may be a dummy node we have inserted.

In the following we also use the notion of topological ordering of the nodes in a *dag*. A *topological ordering* of the nodes of a dag $G = (N, A)$ is any total ordering **ord** of N such that if there exists an arc $\langle x, y \rangle$, then $x \prec_{\mathbf{ord}} y$. In other words a topological ordering is any total order complying the partial order of nodes in a *dag* defined by its edges.

3   A characterization of *DFS*-trees

We characterize *DFS*-trees by means of the existence of a suitable embedding of the graph in the plane.

Definition 1. Let $G = (N, A)$ be a directed graph, and $S \subseteq A$ be any spanning tree of G. S is a *one-sided planar spanning tree* of G if there exists one ordering **ord** of N such that it is possible to draw G with all nodes on a vertical line according to **ord** (the first node at the bottom), arcs in S on the left and all other arcs on the right with the following properties:

1. arcs in S are all directed downward;
2. arcs in S do not intersect each other;
3. for each arc $\langle u, v \rangle$, $\langle u, v \rangle$ is upward if and only if there exists a nonempty chain:

$$\langle x_0, x_1 \rangle, \langle x_1, x_2 \rangle, \ldots, \langle x_k, x_{k+1} \rangle$$

with $v \equiv x_0$, $u \equiv x_{k+1}$, and $\langle x_i, x_{i+1} \rangle \in S$ for $i = 0, \ldots, k$.

An example of one-sided planar spanning tree is shown in figure 1.

The following theorems prove the equivalence of *DFS*-trees and one-sided planar spanning trees. First of all, it is easy to see that the following holds (see, e.g., [19]):

Fact 2 *For any dfs S of G there exists an ordering $\prec_S$ on N such that for any pair of nodes u, v, with $v \prec_S u$, if v, u are children of the same node in S then there exists no arc $\langle p, q \rangle \in A$ with p descendant of v and q descendant of u in S.*

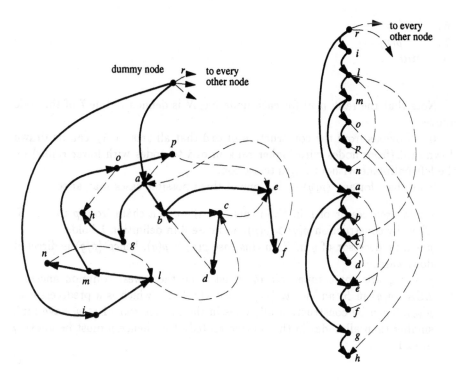

Fig. 1. A digraph and an associated one-sided planar spanning tree. DFS starts from node r.

Lemma 3. *Let $G = (N, A)$ be a digraph, and $S = (N, A_T)$ be any DFS-tree of G. Then S is a one-sided planar spanning tree.*

Proof. S induces a partition of the arcs of G in *tree arcs* (A_T), and *non-tree arcs* $(A_N = A - A_T)$. Furthermore, for any node $x \in V$, it is possible to consider an ordered list of children of x in tree S which is consistent with the ordering in which the edges leaving x are scanned to perform the depth-first search.

An ordering **ord** (required to apply definition 1) can be obtained by the following procedure **Assign_Rank**. We assume that the procedure is invoked the first time by providing root r as the input parameter, and that the global variable i is given the initial value 0.

```
        procedure Assign_Rank(x: node)
1.      begin
2.          mark node x;
3.          let ⟨v₁, v₂, ···, vₖ⟩ be the ordered list of children of x in S
4.          for j = 1 to k
5.              Assign_Rank(vⱼ);
```

```
6.        i := i + 1;
7.        ρ(x) = i
8.    end
```

Note that the rank $\rho(x)$ for each node $x \in N$ is defined in line 7 of the code above.

It is obvious from the construction of **ord** that all arcs in A_T can be drawn downward (from nodes with higher rank in **ord** to nodes with lower ranks) on the left halfplane without any intersection.

Note that, for any $\langle p, q \rangle \in A_N$, one of three possible cases may arise:

- q is a predecessor of p in S: in this case there is a chain from p to q in S. Then by construction $\rho(q) > \rho(p)$, and case 3 in definition 1 holds;
- p is a predecessor of q in S: in this case $\rho(p) > \rho(q)$, hence $\langle p, q \rangle$ is directed downward;
- p and q are not comparable in S: let v be the lowest common ancestor $LCA_S(p, q)$ in S, and let v_i (v_j) be the child of v which is a predecessor of p (q). Then by construction all nodes in the subtree rooted at v_i have rank smaller than all nodes in the subtree rooted at v_j, hence p must be above q in **ord**.

□

Lemma 4. *Let $G = (N, A)$ be a digraph, and S be any one-sided planar spanning tree of G. Then S is a depth-first search tree of G.*

Proof. We show that it is possible to perform a *dfs* visit of G without traversing any arc in the right halfplane.

The *DFS* that we consider is performed by starting from the node with maximum rank (that is, root r), and, for each node, by first traversing outgoing arcs in the left halfplane, starting from those leading to nodes with lower rank.

This implies that tree arcs are traversed first and, by the non-intersection property of tree arcs, tree arcs leading to lower nodes are traversed first.

Let us consider any non tree arc $\langle u, v \rangle$. Two cases are possible:

- $\langle u, v \rangle$ is directed upward: by definition of one-sided planar spanning tree a chain from v to u exists in the left halfplane. This implies that node v has been visited before arc $\langle u, v \rangle$ is considered: thus $\langle u, v \rangle$ is not traversed during the *dfs* visit;
- $\langle u, v \rangle$ is directed downward. Two different subcases may occur:
  - there is a chain from u to v in the left halfplane: the same arguments as in the previous case apply;
  - let $w = LCA_S(u, v)$: then by the visiting order considered above, and since u lies above v, the subtree attached to w containing v is visited before the subtree containing u. This implies that arc $\langle u, v \rangle$ is not traversed.

□

Lemmas 4 and 3 thus give the following characterization of *DFS*-tree:

Theorem 5. *Given a digraph G, a tree T is a DFS-tree of G if and only if T is a one-sided planar spanning tree of G.*

The above characterization also shows that the mapping ρ introduced in the proof of lemma 3 corresponds to the order in which nodes are exited during a *dfs* visit as defined, for example, in [3].

4 Incremental maintenance of a *DFS*-tree in a *dag*

In this section we consider the problem of maintaining a one-sided planar spanning tree of a *dag* G under a sequence of arc insertion. By the characterization provided in Theorem 5, this is equivalent to maintaining a *DFS*-tree for G. Furthermore note that, in the case of a *dag*, it is possible to characterize a *DFS*-tree only by properties 1 and 2 of definition 1.

Corollary 6. *Given a dag $G = (V, E)$, a tree T is a DFS-tree of G if and only if there exists a topological order* ord *of V such that T is a one-sided planar spanning tree of G with ordering* ord.

Hence, in the case of dags, without loss of generality we may restrict our attention only to one-sided planar spanning trees built on the top of a topological ordering.

In the following, for any node $x \in N$, $p(x)$ denotes the parent of node x in the *DFS*-tree S, and, as usual, $\rho(x)$ is the position of node x in some order associated to S, according to definition 1.

In the procedure shown in figure 2, after the insertion of a new arc $\langle u, v \rangle$, a depth-first search is performed on G, starting from arc $\langle u, v \rangle$. Such a search is pruned whenever a node z is entered such that one of the following cases arises:

1. $z = u$:   in this case the insertion of arc $\langle u, v \rangle$ introduces a cycle, and a message is reported;
2. $\rho(z) > \rho(u)$:   then node z is not visited.

In the body of the following procedure, a tree arc will be denoted as *solid*, while a *dotted* arc will be a non-tree arc. At the beginning, it is assumed that all arcs in S are solid, while all other arcs are dotted. We also assume that the current ordering ord with rank $\rho()$ is available. Procedure **Insert** shown in figure 2 carries out the insertion of a new arc $\langle u, v \rangle$, while updating the structure of S and the consistent ordering ord. Figure 3 shows a *dag* and its one-sided planar spanning tree. The effect of our algorithm after the insertion of arc $\langle d, m \rangle$ is shown in figure 4.

Theorem 7. *Let $G = (N, A)$ be a dag, and DFS be any depth-first search tree of G. Then procedure* **Insert** *correctly updates a DFS-tree after the insertion of an arc $\langle u, v \rangle$ in G that does not introduce a cycle.*

```
    procedure Insert(⟨u, v⟩ : arc);
1.  begin
    {all nodes are supposed to be unmarked}
2.      Update(G, ⟨u, v⟩); {the new arc is inserted in the graph}
3.      Perform a dfs starting from arc ⟨u, v⟩
4.          while scanning an arc ⟨w, z⟩
5.              do begin
6.                  if z = u then HALT(''a cycle has been detected'');
7.                  if z is marked or ρ(z) < ρ(u)
8.                  then prune the search at z
9.                  else begin
10.                         mark node z;
11.                         makedotted⟨p(z), z⟩;
12.                         makesolid⟨w, z⟩;
13.                     end;
14.             end
15.     Sort marked nodes according to the postorder dfs visit, and
            move them between node u and the node t with ρ(t) = ρ(u) + 1;
16.     for each marked node z do unmark(z);
17.     makedotted⟨p(v), v⟩;
18.     makesolid⟨u, v⟩; {the new arc is made solid}
19. end.
```

Fig. 2. Maintaining a DFS while inserting an arc $\langle u, v \rangle$.

Proof. Let $A' = A \cup \{\langle u, v \rangle\}$: we will now show that, assuming the partition of A in solid (A_s) and dotted (A_d) arcs plus **ord** induce a one-sided planar spanning tree (N, A_s) on $G = (N, A)$, then the partition of A' in A'_s and A'_d, plus the ordering **ord'** (with rank ρ') obtained by applying procedure **Insert** induce a one-sided planar spanning tree (N, A'_s) on $G' = (N, A')$.

More precisely, we show that:

1. (N, A'_s) is a spanning tree of G';
2. all arcs in A' are directed downward, with respect to **ord'**;
3. arcs in A'_s do not intersect each other, with respect to **ord'**.

Concerning property 1, for each incoming arc of a node z made solid by **Insert** there is an incoming arc that is made dotted and vice-versa. Thus, if A_s is a spanning tree for G then A'_s is a spanning tree for G'.

In order to prove property 2, suppose first $\langle x, y \rangle \in A$: if x and y are both marked by **Insert**, or none of them is, and $\rho(x) > \rho(y)$ then $\rho'(x) > \rho'(y)$. The same holds if only y is marked.

Thus we must only show that if $\langle x, y \rangle \in A$, x has been marked and y has not, then $\rho'(x) > \rho'(y)$. Note that in this case $\rho(v) \geq \rho(x) > \rho(u) > \rho(y)$ must hold (if $\rho(u) = \rho(y)$ a cycle would exist). Let us consider node t such that

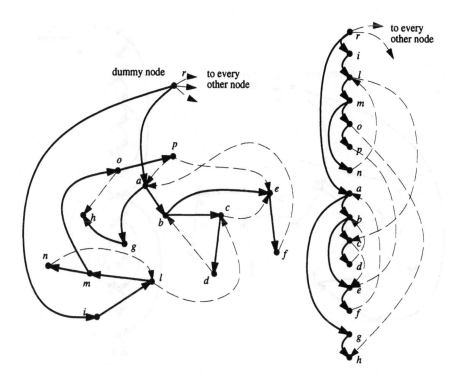

Fig. 3. A *dag* and an associated one-sided planar spanning tree.

$\rho(t) = \rho(u) - 1$ in **ord**. Any marked node x will be moved to a new position $\rho'(x)$ which is bounded by the following inequalities: $\rho'(u) > \rho'(x) > \rho'(t) \geq \rho'(y)$.

Finally, if $\langle x, y \rangle = \langle u, v \rangle$ then $\rho'(u) > \rho'(v)$ by construction.

Let us now sketch the proof of property 3. Let M be the set of nodes marked by procedure **Insert**. It is possible to prove that if x is marked, then each arc in A'_s incident on x comes from a marked node.

Since M, together with arcs made solid by **Insert**, is a *DFS*-tree itself, then by theorem 3 it is a one-sided planar spanning tree for that subgraph. Hence it can be drawn between u and t (the node having $\rho(t) = \rho(u) - 1$) without intersecting arcs incident on nodes in $A - M$. □

Theorem 8. *Let $G = (N, A)$ be a dag, and DFS be any depth-first search tree of G. Let us consider a sequence $\sigma = \langle u_1, v_1 \rangle, \langle u_2, v_2 \rangle, \ldots, \langle u_k, v_k \rangle$ of arcs to be inserted in G such that the final graph $G' = (N, A')$ is still acyclic. Updating the DFS-tree after the insertion of each arc by using algorithm **Insert** requires $O(nm')$ total time for the whole sequence σ, where $n = |N|$, and $m' = |A'|$. Hence, any sequence of $\Theta(m')$ arc insertions in a dag requires $O(n)$ amortized time per update.*

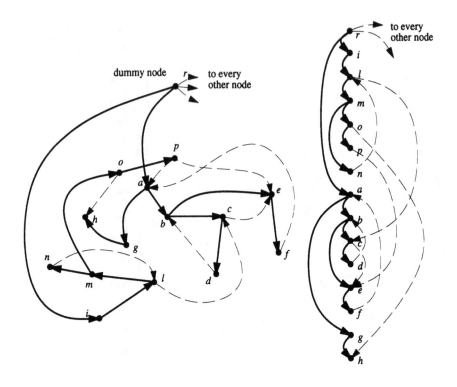

Fig. 4. The *dag* of figure 3 after the insertion of arc $\langle d, m \rangle$, and the associated one-sided planar spanning tree.

Proof. If a node w is marked in stage 1 of algorithm **Insert** due to the insertion of an arc $\langle u, v \rangle$ (with v possibly coincident with w) then a chain from u to w has been introduced. Hence, from now on, $\rho(u) > \rho(w)$, and no further arc insertion can reverse the relative position of u and w: this means that node w will never be marked again due to the insertion of an arc $\langle u, v' \rangle$, for any v'. This lets us conclude that any node is marked at most n times in any sequence of arc insertions.

Since an arc $\langle w, y \rangle$ is scanned by *dfs* in procedure **Insert** only when the node w is marked (otherwise the search is pruned at w), it turns out that any arc is scanned at most n times for any sequence of arc insertions. □

5 Conclusions

In this paper, beside a characterization of depth-first search for general directed graphs, we provide an algorithm for the dynamic maintenance of a *DFS*-forest for directed acyclic graphs under a sequence of arc insertions. Dynamic DFS, due to its applicative and theoretical interest, deserves further attention: the problem for undirected and general directed graphs is still open, as well as to devise a fully dynamic solution, consisting in dealing with alternating arc insertions and deletions.

A different approach to dynamic graph problems might lead to a trade-off between query and update time. In this perspective, the goal is to reduce the worst case time per operation of interleaved queries and input modifications, instead of performing explicit updates in order to answer queries in constant time. In the case of DFS, this might be one way to get efficient fully dynamic solutions: this seems to be the case of other fundamental graph problems, such as transitive closure [18], or shortest paths [11].

References

1. A. V. Aho, J. E. Hopcroft, and J. D. Ullman. *The design and analysis of computer algorithms.* Addison-Wesley, Reading, MA, 1974.

2. G. Ausiello, G. F. Italiano, A. Marchetti-Spaccamela, and U. Nanni, Incremental Algorithms for Minimal Length Paths, *Journal of Algorithms*, **12**, 4 (1991), 615–638.

3. T. H. Cormen, C. E. Leiserson, and R. L. Rivest. *Introduction to algorithms.* MIT Press, Cambridge, MA & McGraw-Hill, New York, NY, 1990.

4. S. Even. *Graph algorithms.* Pittman, London, UK, 1976.

5. S. Even, and H. Gazit, Updating distances in dynamic graphs, *Methods of Operations Research* **49** (1985), 371–387.

6. D. Eppstein, Z. Galil, G. F. Italiano, A. Nissenzweig. Sparsification-A technique for speeding up dynamic graph algorithms. *Proceedings 33th IEEE Symposium on Foundations of Computer Science* (1992), 60–69.

7. Z. Galil, G. F. Italiano, and N. Sarnak. Fully Dynamic Planarity Test. *Proceedings 24th ACM Symposium on Theory of Computing*, (1992), 495–506.

8. J. Hopcroft, R. E. Tarjan. Efficient algorithms for graph manipulation. *Communications of the ACM*, **16** (1973), 372–378.

9. G. F. Italiano. Amortized efficiency of a path retrieval data structure. *Theoretical Computer Science*, **48** (1986), 273–281.

10. G. F. Italiano. Finding paths and deleting edges in directed acyclic graphs. *Information Processing Letters*, **28** (1988), 5–11.

11. P. N. Klein, S. Rao, M. Rauch and S. Subramanian. Faster shortest-path algorithms for planar graphs. *Proceedings 26th ACM Symposium on Theory of Computing*, 1994, to appear.

12. J. A. La Poutré. Alpha-algorithms for incremental planarity testing. *Proceedings 26th ACM Symposium on Theory of Computing*, 1994, to appear.

13. J. A. La Poutré and J. van Leeuwen. Maintenance of transitive closure and transitive reduction of graphs. *Proceedings International Workshop on Graph-Theoretic Concepts in Computer Science (WG 88), Lecture Notes in Computer Science*, **314**, Springer-Verlag (1988), 106–120.

14. A. Marchetti-Spaccamela, U. Nanni, H. Rohnert. On-line graph algorithms for incremental compilation. *Proceedings International Workshop on Graph-Theoretic Concepts in Computer Science (WG 93)*, Utrecht (NL), June 16-18, 1993, *Lecture Notes in Computer Science*, Springer-Verlag.

15. J. H. Reif. Depth-First Search is Inherently Sequential. *Information Processing Letters*, **20**, 1985.
16. J. H. Reif. A Topological Approach to Dynamic Graph Connectivity. *Information Processing Letters*, **25**, 1987.
17. H. Rohnert, A dynamization of the all-pairs least cost path problem, *Proceedings 2nd Annual Symposium on Theoretical Aspects of Computer Science, Lecture Notes in Computer Science*, **182**, Springer-Verlag (1990), 279–286.
18. S. Subramanian. A fully dynamic data structure for reachability in planar graphs. *Proceedings 1st Annual European Symposium on Algorithms* (1993).
19. R. E. Tarjan. Depth-first search and linear graph algorithms. *SIAM Journal on Computing*, **1** (1972), 146–160.

Finding and counting given length cycles [*]

(Extended Abstract)

Noga Alon [**]   *Raphael Yuster* [**]   *Uri Zwick* [**]

School of Mathematical Sciences, Tel Aviv University, Tel Aviv 69978, ISRAEL.

Abstract. We present an assortment of methods for finding and counting simple cycles of a given length in directed and undirected graphs. Most of the bounds obtained depend solely on the number of edges in the graph in question, and not on the number of vertices. The bounds obtained improve upon various previously known results.

1 Introduction

The problem of deciding whether a given graph $G = (V, E)$ contains a simple cycle of length k is among the most natural and easily stated algorithmic graph problems. If the cycle length k is part of the input then the problem is clearly NP-complete as it includes in particular the Hamiltonian cycle problem. For every fixed k however, the problem can be solved in either $O(VE)$ time (Monien [Mon85]) or $O(V^\omega \log V)$ [AYZ94], where $\omega < 2.376$ is the exponent of matrix multiplication.

The main contribution of this paper is a collection of new bounds on the complexity of finding simple cycles of length exactly k, where $k \geq 3$ is a fixed integer, in a directed or an undirected graph $G = (V, E)$. These bounds are of the form $O(E^{\alpha_k})$ or of the form $O(E^{\beta_k} \cdot d(G)^{\gamma_k})$, where $d(G)$ is the *degeneracy* of the graph (see below). The bounds improve upon previously known bounds when the graph in question is relatively sparse or relatively degenerate.

We let C_k stand for a simple cycle of length k. When considering directed graphs, a C_k is assumed to be directed. We show that a C_k in a directed or undirected graph $G = (V, E)$, if one exists, can be found in $O(E^{2 - \frac{2}{k}})$ time, if k is even, and in $O(E^{2 - \frac{2}{k+1}})$ time, if k is odd. For finding triangles (C_3's), we get the slightly better bound of $O(E^{\frac{2\omega}{\omega+1}}) = O(E^{1.41})$, where $\omega < 2.376$ is the exponent of matrix multiplication.

Even cycles in undirected graphs can be found even faster. A C_{4k-2} in an undirected graph $G = (V, E)$, if one exists, can be found in $O(E^{2 - \frac{1}{2k}(1 + \frac{1}{k})})$ time. A C_{4k}, if one exists, can be found in $O(E^{2 - (\frac{1}{k} - \frac{1}{2k+1})})$ time. In particular, we can find an undirected C_4 in $O(E^{4/3})$ time and an undirected C_6 in $O(E^{13/8})$ time.

[*] Work supported in part by THE BASIC RESEARCH FOUNDATION administrated by THE ISRAEL ACADEMY OF SCIENCES AND HUMANITIES.

[**] E-mail addresses of authors: {noga,raphy,zwick}@math.tau.ac.il .

cycle	complexity		cycle	complexity	
C_3	$E^{1.41}$	, $E \cdot d(G)$	C_7	$E^{1.75}$	, $E^{3/2} \cdot d(G)$
C_4	$E^{1.5}$	, $E \cdot d(G)$	C_8	$E^{1.75}$	, $E^{3/2} \cdot d(G)$
C_5	$E^{1.67}$	, $E \cdot d(G)^2$	C_9	$E^{1.8}$	, $E^{3/2} \cdot d(G)^{3/2}$
C_6	$E^{1.67}$	, $E^{3/2} \cdot d(G)^{1/2}$	C_{10}	$E^{1.8}$	, $E^{5/3} \cdot d(G)^{2/3}$

Table 1. Finding small cycles in directed graphs – some of the new results

cycle	complexity	cycle	complexity
C_4	$E^{1.34}$	C_8	$E^{1.7}$
C_6	$E^{1.63}$	C_{10}	$E^{1.78}$

Table 2. Finding small cycles in undirected graphs – some of the new results

The *degeneracy* $d(G)$ of an undirected graph $G = (V, E)$ is the largest minimal degree among the minimal degrees of all the subgraphs G' of G (see Bollobás [Bol78], p. 222). The degeneracy $d(G)$ of a graph G is linearly related to the *arboricity* $a(G)$ of the graph, i.e., $a(G) = \Theta(d(G))$, where $a(G)$ is the minimal number of forests needed to cover all the edges of G. The degeneracy of a directed graph $G = (V, E)$ is defined to be the degeneracy of the undirected version of G. The degeneracy of a graph is an important parameter of the graph that appears in many combinatorial results. It is easy to see that for any graph $G = (V, E)$ we have $d(G) \leq 2E^{1/2}$. For graphs with relatively low degeneracy we can improve upon the previously stated results. A C_{4k} in a directed or undirected graph $G = (V, E)$ that contains one can be found in $O(E^{2-\frac{1}{k}} \cdot d(G))$ time. A C_{4k+1}, if one exists, can be found in $O(E^{2-\frac{1}{k}} \cdot d(G)^{1+\frac{1}{k}})$ time. Similar results are obtained for finding C_{4k-2}'s and C_{4k-1}'s. In particular, C_3's and C_4's can be found in $O(E \cdot d(G))$ time and C_5's in $O(E \cdot d(G)^2)$ time. Some of the results mentioned are summarized in Tables 1 and 2.

As any planar graph has a vertex whose degree is at most 5, the degeneracy of any planar graph is at most 5. As a consequence of the above bounds we get, in particular, that C_3's, C_4's and C_5's in planar graphs can be found in $O(V)$ time. This in fact holds not only for planar graphs but for any non-trivial *minor-closed* family of graphs.

Another contribution of this paper is an $O(V^\omega)$ algorithm for *counting* the number of C_k's, for $k \leq 7$, in a graph $G = (V, E)$.

2 Comparison with previous works

Monien [Mon85] obtained, for any fixed $k \geq 3$, an $O(VE)$ algorithm for finding C_k's in a directed or undirected graph $G = (V, E)$. In a previous work [AYZ94]

we showed, using the *color-coding* method, that a C_k, for any fixed $k \geq 3$, if one exists, can also be found in $O(V^\omega)$ expected time or in $O(V^\omega \log V)$ worst-case time, where $\omega < 2.376$ is the exponent of matrix multiplication.

Our new $O(E^{2-\frac{2}{k}})$ algorithm is better than both the $O(VE)$ and the $O(V^\omega)$ algorithms when the input graph $G = (V, E)$ is sufficiently sparse. It is interesting to note that for $k \leq 6$, Monien's $O(VE)$ bound is superseded by either the $O(V^\omega)$ algorithm, when the graph is dense, or by the $O(E^{2-1/\lceil \frac{k}{2} \rceil})$ algorithm, when the graph is sparse. For every $k \geq 7$, each one of the four bounds (including the bound that involves the degeneracy) beats the others on an appropriate family of graphs.

In a previous work [YZ94] we have also shown that cycles of an *even* length in *undirected* graphs can be found even faster. Namely, for any even $k \geq 4$, if an undirected graph $G = (V, E)$ contains a C_k then such a C_k can be found in $O(V^2)$ time. Our $O(E^{2-\frac{1}{2k}(1+\frac{1}{k})})$ bound for C_{4k-2} and $O(E^{2-(\frac{1}{k}-\frac{1}{2k+1})})$ bound for C_{4k} are again better when the graph is sparse enough.

Itai and Rodeh [IR78] showed that a *triangle* (a C_3) in a graph $G = (V, E)$ that contains one can be found in $O(V^\omega)$ or $O(E^{3/2})$ time. We improve their second result and show that the same can be done, in directed or undirected graphs, in $O(E^{\frac{2\omega}{\omega+1}}) = O(E^{1.41})$ time.

Chiba and Nishizeki [CN85] showed that triangles (C_3's) and quadrilaterals (C_4's) in graphs that contain them can be found in $O(E \cdot d(G))$ time. As $d(G) = O(E^{1/2})$ for any graph G, this extends the result of Itai and Rodeh. We extend the result of Chiba and Nishizeki and show that C_{4k-1}'s and C_{4k}'s can be found in $O(E^{2-\frac{1}{k}} \cdot d(G))$ time. We also show that C_{4k+1}'s can be found in $O(E^{2-\frac{1}{k}} \cdot d(G)^{1+\frac{1}{k}})$ time. This gives, in particular, an $O(E \cdot d(G)^2)$ algorithm for finding pentagons (C_5's). Our results apply to both directed and undirected graphs.

Itai and Rodeh [IR78] and also Papadimitriou and Yannakakis [PY81] showed that C_3's in planar graphs can be found in $O(V)$ time. Chiba and Nishizeki [CN85] showed that C_3's as well as C_4's in planar graphs can be found in $O(V)$ time. Richards [Ric86] showed that C_5's and C_6's in planar graphs can be found in $O(V \log V)$ time. We improve upon the result of Richards and show that C_5's in planar graphs can be found in $O(V)$ time. In a previous work [AYZ94] we showed, using color-coding, that for any $k \geq 3$, a C_k in a planar graph, if one exists, can be found in either $O(V)$ *expected* time or $O(V \log V)$ worst case time.

Finally, the fact that the number of triangles in a graph can be counted in $O(V^\omega)$ time is trivial. In [AYZ94] we showed, using color-coding, that for any $k \geq 3$, a C_k, if one exists, can be found in either $O(V^\omega)$ *expected* time or in $O(V^\omega \log V)$ worst case time. Here we show that for any $k \leq 7$ the number of C_k's in a graph can be counted in $O(V^\omega)$ time. The counting method used here yields, in particular, a way of finding C_k's for $k \leq 7$, in $O(V^\omega)$ worst case time.

3 Finding cycles in sparse graphs

Monien [Mon85] obtained his $O(VE)$ algorithm by the use of *representative collections*. Such collections are also used by our algorithms. In the sequel, a *p-set* is a set of size p.

Definition 1 ([Mon85]) *Let $\mathcal{F}$ be a collection of p-sets. A sub-collection $\hat{\mathcal{F}} \subseteq \mathcal{F}$ is q-representative for $\mathcal{F}$, if for every q-set B, there exists a set $A \in \mathcal{F}$ such that $A \cap B = \emptyset$ if and only if there exists a set $A \in \hat{\mathcal{F}}$ with this property.*

It follows from a combinatorial lemma of Bollobás [Bol65] that any collection $\mathcal{F}$ of p-sets, no matter how large, has a q-representative sub-collection of size at most $\binom{p+q}{p}$. Monien [Mon85] describes an $O(pq \cdot \sum_{i=0}^{q} p^i \cdot |\mathcal{F}|)$ time algorithm for finding a q-representative sub-collection of $\mathcal{F}$ whose size is at most $\sum_{i=0}^{q} p^i$. Relying on Monien's result we obtain the following lemma:

Lemma 2. *Let $\mathcal{F}$ be a collection of p-sets and let $\mathcal{G}$ be a collection of q-sets. Consider p and q to be fixed. In $O(|\mathcal{F}| + |\mathcal{G}|)$ time, we can either find two sets $A \in \mathcal{F}$ and $B \in \mathcal{G}$ such that $A \cap B = \emptyset$ or decide that no two such sets exist.*

Proof. We use Monien's algorithm to find a q-representative sub-collection $\hat{\mathcal{F}}$ of $\mathcal{F}$ whose size is at most $\sum_{i=0}^{q} p^i$ and a p-representative sub-collection $\hat{\mathcal{G}}$ of $\mathcal{G}$ whose size is at most $\sum_{i=0}^{p} q^i$. This takes only $O(|\mathcal{F}| + |\mathcal{G}|)$ time (as p and q are constants).

It is easy to see that if there exist $A \in \mathcal{F}$ and $B \in \mathcal{G}$ such that $A \cap B = \emptyset$, then there also exist $A' \in \hat{\mathcal{F}}$ and $B' \in \hat{\mathcal{G}}$ such that $A' \cap B' = \emptyset$. To see this note that if $A \cap B = \emptyset$ then by the definition of q-representatives, there must exist a set $A' \in \hat{\mathcal{F}}$ such that $A' \cap B = \emptyset$ and then, there must exist a set $B' \in \hat{\mathcal{G}}$ such that $A' \cap B' = \emptyset$ as required.

After finding the representative collections $\hat{\mathcal{F}}$ and $\hat{\mathcal{G}}$ it is therefore enough to check whether they contain two disjoint sets. This can be easily done in constant time (as p and q are constants). $\qquad\square$

We also need the following lemma obtained by Monien [Mon85].

Lemma 3 [Mon85]. *Let $G = (V, E)$ be a directed or undirected graph, let $v \in V$ and let $k \geq 3$. A C_k that passes through v, if one exists, can be found in $O(E)$ time.*

We are finally able to present our improved algorithm.

Theorem 4. *Deciding whether a directed or undirected graph $G = (V, E)$ contains simple cycles of length exactly $2k - 1$ and of length exactly $2k$, and finding such cycles if it does, can be done in $O(E^{2-\frac{1}{k}})$ time.*

Proof. We describe an $O(E^{2-\frac{1}{k}})$ time algorithm for finding a C_{2k} in a directed graph $G = (V, E)$. The details of all the other cases are similar. Let $\Delta = E^{\frac{1}{k}}$. A vertex in G whose degree is at least Δ is said to be of *high degree*. The graph $G = (V, E)$ contains at most $2E/\Delta = O(E^{1-\frac{1}{k}})$ high degree vertices. We check, using Monien's algorithm (Lemma 3), whether any of these high degree vertices lies on a simple cycle of length $2k$. For each vertex this costs $O(E)$ operations and the total cost is $O(E^2/\Delta) = O(E^{2-\frac{1}{k}})$. If one of these vertices does lie on a cycle of length $2k$ we are done. Otherwise, we remove all the high degree vertices and all the edges adjacent to them from G and obtain a subgraph G' that contains a C_{2k} if and only if G does. The maximum degree of G' is at most $\Delta = E^{\frac{1}{k}}$ and there are therefore at most $E \cdot \Delta^{k-1} = E^{2-\frac{1}{k}}$ simple directed paths of length k in G'. We can find all these simple paths in $O(E^{2-\frac{1}{k}})$ time. We divide these paths into groups according to their endpoints. This can be done using radix sort in $O(E^{2-\frac{1}{k}})$ time and space. We get a list of all the pairs of vertices connected by simple directed paths of length exactly k. For each such pair u, v, we get a collection $\mathcal{F}_{u,v}$ of $k-1$-sets. Each $k-1$-set in $\mathcal{F}_{u,v}$ corresponds to the $k-1$ intermediate vertices that appear on simple directed paths of length k from u to v. For each pair u, v that appears on the list, we check whether there exist two directed paths of length k, one from u to v and the other from v to u, that meet only at their endpoints. Such two paths exist if there exist $A \in \mathcal{F}_{u,v}$ and $B \in F_{v,u}$ such that $A \cap B = \emptyset$. This can be checked, as shown in Lemma 2, in $O(|\mathcal{F}_{u,v}| + |\mathcal{F}_{v,u}|)$ time. As the sum of the sizes of all these collections is $O(E^{2-\frac{1}{k}})$, the total complexity is again $O(E^{2-\frac{1}{k}})$. This completes the proof. □

In the case of triangles, we can get a better result by using fast matrix multiplication.

Theorem 5. *Deciding whether a directed or an undirected graph $G = (V, E)$ contains a triangle, and finding one if it does, can be done is $O(E^{\frac{2\omega}{\omega+1}}) = O(E^{1.41})$ time.*

Proof. Let $\Delta = E^{\frac{\omega-1}{\omega+1}}$. A vertex is said to be of *high degree* if its degree is more than Δ and of *low degree* otherwise. Consider all directed paths of length two in G whose intermediate vertex is of low degree. There are at most $E \cdot \Delta$ such paths and they can be found in $O(E \cdot \Delta)$ time. For each such path, check whether its endpoints are connected by an edge in the appropriate direction. If no triangle is found in this way, then any triangle in G must be composed of three high degree vertices. As there are at most $2E/\Delta$ high degree vertices, we can check whether there exists such a triangle using matrix multiplication in $O((E/\Delta)^\omega)$ time. The total complexity of the algorithm is therefore

$$O(E \cdot \Delta + (E/\Delta)^\omega) = O(E^{\frac{2\omega}{\omega+1}}).$$

This completes the proof. □

We have not been able to utilize matrix multiplication to improve upon the result of Theorem 4 for $k \geq 4$. This constitutes an interesting open problem.

4 Finding cycles in graphs with low degeneracy

An undirected graph $G = (V, E)$ is *d-degenerate* (see Bollobás [Bol78], p. 222) if there exists an acyclic orientation of it in which $d_{out}(v) \leq d$ for every $v \in V$. The smallest d for which G is d-degenerate is called the *degeneracy* or the *max-min degree* of G and is denoted by $d(G)$. It can be easily seen (see again [Bol78]) that $d(G)$ is the maximum of the minimum degrees taken over all the subgraphs of G. The degeneracy $d(G)$ of a graph G is linearly related to the *arboricity* $a(G)$ of the graph, i.e., $a(G) = \Theta(d(G))$, where $a(G)$ is the minimal number of forests needed to cover all the edges of G. The degeneracy of a directed graph $G = (V, E)$ is defined to be the degeneracy of the undirected version of G. It is easy to see that the degeneracy of any planar graph is at most 5. Clearly, if G is d-degenerate then $|E| \le d \cdot |V|$. The following simple lemma, whose proof is omitted, is part of the folklore (see, e.g., [MB83]).

Lemma 6. *Let $G = (V, E)$ be a connected undirected graph $G = (V, E)$. An acyclic orientation of G such that for every $v \in V$ we have $d_{out}(v) \le d(G)$ can be found in $O(E)$ time.*

The main result of this section is the following Theorem:

Theorem 7. *Let $G = (V, E)$ be a directed or an undirected graph.*
(i) Deciding whether G contains a simple cycle of length exactly $4k - 2$, and finding such a cycle if it does, can be done in $O(E^{2-\frac{1}{k}} \cdot d(G)^{1-\frac{1}{k}})$ time.
(ii) Deciding whether G contains simple cycles of length exactly $4k - 1$ and exactly $4k$, and finding such cycles if it does, can be done in $O(E^{2-\frac{1}{k}} \cdot d(G))$ time.
(iii) Deciding whether G contains a simple cycle of length exactly $4k + 1$, and finding such a cycle if it does, can be done in $O(E^{2-\frac{1}{k}} \cdot d(G)^{1+\frac{1}{k}})$ time.

Proof. We show how to find a C_{4k+1} in a directed graph $G = (V, E)$, if one exists, in $O(E^{2-\frac{1}{k}} \cdot d(G)^{1+\frac{1}{k}})$ time. The proofs of the other claims are easier. If $d(G) \ge E^{\frac{1}{2k+1}}$, we can use the algorithm of Theorem 4 whose complexity is $O(E^{2-\frac{1}{2k+1}}) \le O(E^{2-\frac{1}{k}} \cdot d(G)^{1+\frac{1}{k}})$. Assume therefore that $d(G) \le E^{\frac{1}{2k+1}}$.

Let $\Delta = E^{\frac{1}{k}} / d(G)^{1+\frac{1}{k}}$. As $d(G) \le E^{\frac{1}{2k+1}}$, we have that $d(G) \le \Delta$. A vertex is said to be of *high degree* if its degree is more than Δ and of *low degree* otherwise. As in the proof of Theorem 4, we can check in $O(E^2/\Delta)$ time whether any of the high degree vertices lies on a C_{4k+1}. If none of them lies on a C_{4k+1}, we can remove all the high degree vertices along with the edges adjacent to them from G and obtain a graph whose maximal degree is at most Δ. The degeneracy of a graph can only decrease when vertices and edges are deleted and $d(G)$ is therefore an upper bound on the degeneracy of the graph obtained.

Suppose therefore that G is a graph with maximal degree Δ and degeneracy $d(G)$. To find a C_{4k+1} in G, it is enough to find all directed simple paths of length $2k$ and $2k + 1$ in G and then check, using the algorithm described in the proof of Lemma 2, whether there exist a path of length $2k$ and a path of length $2k + 1$ that meet only at their endpoints.

In $O(E)$ time we can get an acyclically oriented version G' of G in which the out-degree of each vertex is at most $d(G)$. The orientations of the edges in G and G' may be completely different. The number of paths, not necessarily directed, of length $2k + 1$ in G, is at most

$$2 \cdot 2E \cdot \sum_{i=0}^{k} \binom{2k}{i} \Delta^i d(G)^{2k-i} = O(E\Delta^k d(G)^k).$$

To see this, consider the orientations, in G', of the edges on a $2k+1$-path in G. In at least one direction, at most k of the edges are counter directed. The number of paths of length $2k + 1$ in which exactly i among the last $2k$ edges are counter directed is at most $2E \cdot \binom{2k}{i} \Delta^i d(G)^{2k-i}$. The binomial coefficient $\binom{2k}{i}$ stands for the possible choices for the position of the counter directed edges in the path. Similarly, the number of paths of length $2k$ in G is $O(E\Delta^k d(G)^{k-1})$.

We can lower the number of paths of length $2k+1$ and $2k$ we have to consider by utilizing the fact that a C_{4k+1} can be broken into two paths of length $2k + 1$ and $2k$ in many different ways. In particular, let C be a directed C_{4k+1} in G and consider the orientations of its edges in G'. As $4k + 1$ is odd and as G' is acyclic, C must contain three consecutive edges e_{2k}, e_{2k+1} and e_{2k+2}, the first two of which have the same orientation while the third one has an opposite orientation. It is therefore enough to consider all $2k + 1$-paths that start with at least two backward oriented edges and all $2k$-paths that start with at least one backward oriented edge. The orientations referred to here are in G'.

The number of paths of length $2k+1$ in G whose first two edges are backward oriented in G' is $O(E\Delta^{k-1}d(G)^{k+1})$. To see this, note that any such path is composed of a directed path $\{e_{2k}, e_{2k+1}\}$ of length two, attached to an arbitrarily oriented path $\{e_1, \ldots, e_{2k-1}\}$ of length $2k - 1$. The number of paths of length $2k-1$ is, as shown earlier, at most $O(E\Delta^{k-1}d(G)^{k-1})$ and the number of directed path of length two with a specified starting point is at most $d(G)^2$. Similarly, there are at most $O(E\Delta^{k-1}d(G)^k)$ $2k$-paths that start with a backward oriented edge.

It should be clear from the above discussion that all the required paths of length $2k + 1$ and $2k$, whose total number is $O(E\Delta^{k-1}d(G)^{k+1})$, can be found in $O(E\Delta^{k-1}d(G)^{k+1})$ time. Paths which are not properly directed, in G, are thrown away. Properly directed paths are sorted, using radix sort, according to their endpoints. Using Lemma 2 we then check whether there exist a directed $2k + 1$-path and a directed $2k$-path that close a directed simple cycle. All these operations can again be performed in $O(E\Delta^{k-1}d(G)^{k+1})$ time.

Recalling that $\Delta = E^{\frac{1}{k}}/d(G)^{1+\frac{1}{k}}$, we get that the over all complexity of the algorithm is

$$O(\frac{E^2}{\Delta} + E \cdot \Delta^{k-1}d(G)^{k+1}) = O(E^{2-\frac{1}{k}}d(G)^{1+\frac{1}{k}}) \quad .$$

This completes the proof of the Theorem. □

As an immediate corollary we get:

Corollary 8. *If a directed or undirected planar graph $G = (V, E)$ contains a pentagon (a C_5), then such a pentagon can be found in $O(V)$ worst-case time.*

By combining the ideas of this section, the $O(E^{\frac{2\omega}{\omega+1}})$ algorithm of Theorem 5 and the color-coding method [AYZ94] we can also obtain the following result whose proof is omitted.

Theorem 9. *Let $G = (V, E)$ be a directed or undirected graph. A C_6 in G, if one exists, can be found in either $O((E \cdot d(G))^{\frac{2\omega}{\omega+1}}) = O((E \cdot d(G))^{1.41})$ expected time or $O((E \cdot d(G))^{\frac{2\omega}{\omega+1}} \cdot \log V) = O((E \cdot d(G))^{1.41} \cdot \log V)$ worst case time.*

5 Finding cycles in sparse undirected graphs

To obtain the results of this section we rely on the following combinatorial lemma of Bondy and Simonovits [BS74].

Lemma 10 [BS74]. *Let $G = (V, E)$ be an undirected graph. If $|E| \geq 100k \cdot |V|^{1+\frac{1}{k}}$ then G contains a $C_{2\ell}$ for every integer $\ell \in [k, n^{1/k}]$.*

By combining the algorithm described in the proof of Theorem 7 with an algorithm given in [YZ94] we obtain the following theorem.

Theorem 11. *Let $G = (V, E)$ be an undirected graph.*
(i) A C_{4k-2} in G, if one exists, can be found in $O(E^{2-\frac{1}{2k}(1+\frac{1}{k})})$ time.
(ii) A C_{4k} in G, if one exists, can be found in $O(E^{2-(\frac{1}{k}-\frac{1}{2k+1})})$ time.

Proof. We prove the second claim. The proof of the first claim is similar. Let $d = 200k \cdot E^{\frac{1}{2k+1}}$. If $d(G) \geq d$ then, by the definition of degeneracy, there is a subgraph $G' = (V', E')$ of G in which the minimal degree is at least d. Such a subgraph can be easily found in $O(E)$ time (see, e.g., [MB83]). Clearly $E' \geq dV'/2 \geq 100k \cdot V' \cdot E'^{\frac{1}{2k+1}}$ and therefore $E' \geq (100k \cdot V')^{1+\frac{1}{2k}} \geq 100k \cdot (V')^{1+\frac{1}{2k}}$. By Lemma 10 we get that G' contains a C_{4k} and such a C_{4k} can be found in $O(V'^2) = O(E^{2-\frac{2}{2k+1}})$ time using the algorithm given in [YZ94]. If, on the other hand, $d(G) \leq d$, then a C_{4k} in G, if one exists, can be found in $O(E^{2-\frac{1}{k}} \cdot d) = O(E^{2-(\frac{1}{k}-\frac{1}{2k+1})})$ time using the algorithm of Theorem 7. It is easy to check that $E^{2-\frac{2}{2k+1}} \leq E^{2-(\frac{1}{k}-\frac{1}{2k+1})}$ with equality holding only if $k = 1$. In both cases, the complexity is therefore $O(E^{2-(\frac{1}{k}-\frac{1}{2k+1})})$ as required. $\square$

Corollary 12. *Let $G = (V, E)$ be an undirected graph.*
(i) A quadrilateral (C_4) in G, if one exists, can be found in $O(E^{4/3})$ time.
(ii) A hexagon (C_6) in G, if one exists, can be found in $O(E^{13/8})$ time.

6 Counting small cycles

Let $A = A_G$ be the adjacency matrix of a simple undirected graph $G = (V, E)$. Assume for simplicity that $V = \{1, \ldots, n\}$. Denote by $a_{ij}^{(k)} = (A^k)_{ij}$ the elements of the k-th power of A. The trace $tr(A^k)$ of A^k, which is the sum of the entries along the diagonal of A^k, gives us the number of closed paths of length k in G. If we could also compute the number of *non-simple* closed paths of length k in G we would easily obtain the number of *simple* closed paths of length k in G. This number is just $2k$ times the number of C_k's in G.

Every closed path of length 3 is simple. The number of triangles in G is therefore

$$\#(\triangle) = \frac{1}{6} \cdot tr(A^3) \quad .$$

Furthermore, the number of triangles that pass through vertex i is exactly $\frac{1}{2}a_{ii}^{(3)}$ and the number of triangles that use the edge $(i, j) \in E$ is exactly $a_{ij}^{(2)}$.

Closed paths of length 4 are not necessarily simple. Every non-simple closed path of length 4 however is of one of the forms shown in Fig. 1. The number of quadrilaterals (C_4's) in G is therefore

$$\#(\square) = \frac{1}{8} \cdot [tr(A^4) - 4 \cdot \#(\bullet\!\!-\!\!\bullet\!\!-\!\!\bullet) - 2 \cdot \#(\bullet\!\!-\!\!\bullet)] \quad .$$

The number of simple paths of length 2 in G, i.e., $\#(\bullet\!\!-\!\!\bullet\!\!-\!\!\bullet)$, is clearly $\sum_{i=1}^{n} \binom{d_i}{2}$ where $d_i = a_{ii}^{(2)}$ is the degree of vertex i, and $\#(\bullet\!\!-\!\!\bullet)$ is just $|E|$, the number of edges in the graph.

Non-simple closed paths of length 5 are of one of the forms shown in Fig. 2. The number of pentagons (C_5's) in G is therefore

$$\#(\pentagon) = \frac{1}{10} \cdot [tr(A^5) - 10 \cdot \#(\bullet\!\!-\!\!\triangleleft) - 30 \cdot \#(\triangle)] \quad .$$

The number of $\bullet\!\!-\!\!\triangleleft$'s in G is easily seen to be $\frac{1}{2} \sum_{i=1}^{n} (d_i - 2)a_{ii}^{(3)}$. The number of triangles in G was already computed so the number of pentagons in G can easily be calculated using the above formula.

The same method can also be used to count the number of hexagons (C_6's) and septagons (C_7's). The formulae become however much more complicated. To obtain the number of hexagons, for example, we should count first the number of occurrences in G of each of the subgraphs shown in Fig. 3. We omit the exact formulae.

It can be checked that the number of occurrences in G of each of the subgraphs shown in Fig. 3 and those needed to find the number of septagons (C_7's) in G can be calculated using only $O(V^2)$ operations, once the powers $A, \ldots, A^7$ of the adjacency matrix A of the graph are available. We obtain therefore the following result.

Theorem 13. *The number of C_k's, for $k \leq 7$ in a graph $G = (V, E)$ can be found in $O(V^\omega)$ time.*

Similar formulae can be obtained of course for the number of octagons (C_8's) and even larger cycles. However, to compute the number of octagons, we have

363

Fig. 1. Non-simple closed paths of length 4.

Fig. 2. Non-simple closed paths of length 5.

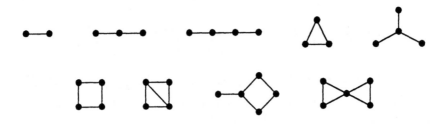

Fig. 3. Subgraphs that should be counted to obtain the number of hexagons

to find first the number of K_4's in the graph, where K_l denotes a clique of size l. We do not know how to do this in $O(V^\omega)$ time.

It is easy to count the number of K_4's in a graph in $O(V^{\omega+1})$ time: for each vertex, count the number of triangles among its neighbors, sum these numbers and divide by 4. Counting the number of K_4's in a graph, or in fact, deciding whether a graph contains a K_4, in $o(V^{\omega+1})$ time is an interesting open problem.

For counting the number of larger cycles using our method, we would have to count the number of larger cliques in the graph. Nešetřil and Poljak [NP85] give an $O(V^{\omega\lceil\frac{l}{3}\rceil})$ time algorithm for deciding whether a graph $G = (V, E)$ contains a K_l. It is easy to check that their method can also be used to count the number of such cliques contained in the graph. By combining the method of Nešetřil and Poljak [NP85] with the ideas used in Section 4, we get the following result.

Theorem 14. *The number of K_l's in an undirected graph $G = (V, E)$ can be counted in either $O(V \cdot (d(G))^{\omega \lceil \frac{l-1}{3} \rceil})$ or $O(E \cdot (d(G))^{\omega \lceil \frac{l-2}{3} \rceil})$ time.*

Using an idea similar to the one used in Theorem 5, we can also get an $O(E^{\frac{3\omega+3}{\omega+3}}) = O(E^{1.89})$ algorithm for deciding whether a graph $G = (V, E)$ contains a K_4.

References

[AYZ94] N. Alon, R. Yuster, and U. Zwick. Color-coding: a new method for finding simple paths, cycles and other small subgraphs within large graphs. In *Proceedings of the 26th Annual ACM Symposium on Theory of Computing, Montréal, Canada*, pages 326–335, 1994.

[Bol65] B. Bollobás. On generalized graphs. *Acta Math. Acad. Sci. Hungar.*, 16:447–452, 1965.

[Bol78] B. Bollobás. *Extremal graph theory*. Academic Press, 1978.

[BS74] J.A. Bondy and M. Simonovits. Cycles of even length in graphs. *Journal of Combinatorial Theory, Series B*, 16:97–105, 1974.

[CN85] N. Chiba and L. Nishizeki. Arboricity and subgraph listing algorithms. *SIAM Journal on Computing*, 14:210–223, 1985.

[IR78] A. Itai and M. Rodeh. Finding a minimum circuit in a graph. *SIAM Journal on Computing*, 7:413–423, 1978.

[MB83] D.W. Matula and L.L. Beck. Smallest-last ordering and clustering and graph coloring algorithms. *Journal of the ACM*, 30:417–427, 1983.

[Mon85] B. Monien. How to find long paths efficiently. *Annals of Discrete Mathematics*, 25:239–254, 1985.

[NP85] J. Nešetřil and S. Poljak. On the complexity of the subgraph problem. *Commentationes Mathematicae Universitatis Carolinae*, 26(2):415–419, 1985.

[PY81] C.H. Papadimitriou and M. Yannakakis. The clique problem for planar graphs. *Information Processing Letters*, 13:131–133, 1981.

[Ric86] D. Richards. Finding short cycles in a planar graph using separators. *Journal of Algorithms*, 7:382–394, 1986.

[YZ94] R. Yuster and U. Zwick. Finding even cycles even faster. In *Proceedings of the 21st International Colloquium on Automata, Languages and Programming, Jerusalem, Israel*, 1994. To appear.

Greedy Hot-Potato Routing on the Mesh[*]

Ishai Ben-Aroya and Assaf Schuster[1]

Computer Science Department, Technion, Haifa, Israel 32000. assaf@cs.technion.ac.il

Abstract. We propose hot-potato (or, deflection) packet routing algorithms on the two-dimensional mesh. The algorithms are strongly greedy in the sense that they attempt to send packets in good directions whenever possible. Furthermore, the routing operations are simple and independent of the time that has elapsed. These two features suggest that the algorithms are practical and may also be used for continuous routing. The first algorithm gives the best evacuation time known for delivering all the packets to their destinations. A batch of k packets with maximal source-to-destination distance d_{max} is delivered in $2(k-1) + d_{max}$. The second algorithm improves this bound to $k + d_{max}$ when all packets are destined to the same node. This also implies a new bound for the multi-target case, which is the first to take into account the number of in-edges of a node. The third algorithm is designed for routing permutations in which the maximal source-to-destination distance is small. In particular, when routing permutations for which $d_{max} = 3$, the algorithm terminates in at most seven steps. We also show a lower bound of five steps for this problem.

1 Introduction

In this work we propose greedy algorithms for packet routing in synchronous networks of processors in which at most one packet can traverse any directed link in each time step. We consider a routing mode known as *hot-potato* or *deflection* routing [AS91, GH92, Haj91, Max89, Szy90, ZA91]. The important characteristic of algorithms which assume this mode is that they use no buffer space for storing delayed packets. Each packet, unless it has already reached its destination, must leave the processor at the step following its arrival. Packets may reach a processor from all its neighbors and then have to be redirected, each one on a different outgoing link. This may cause some packets to be "deflected" away from their preferred direction. Such an unfortunate situation cannot happen in the "store-and-forward" routing mode, in which a packet can be stored at a processor until it can be transmitted to its preferred direction.

Variants of hot potato routing are used by parallel machines such as the HEP multiprocessor [Smi81], and by high-speed communication networks [Max89]. In particular, hot potato routing is very important in fine-grained massively-parallel computers, such as the Caltech Mosaic C [SBSS93]. For such machines, even the inclusion of a small sized storage buffer at each processor causes a substantial increase in the cost of the machine. Another domain in which deflection-type

---

[*] This work was supported in part by the French-Israeli grant for cooperation in Computer Science, and by a grant from the Israeli Ministry of Science.

routing is highly desirable is optical networks [AS91, Szy90, ZA91]. In such networks, storage must take the electronic form. Thus, packets that should be stored must be converted from (and back to) the optical form. In the current state of technology, this conversion is very slow, compared to optical transmission rates. It is more feasible (even if one pays with longer routes) to deflect the blocked messages.

Most of the recent work on hot potato routing focuses on *structured* routing. In structured routing "good behavior" is enforced on the packets in the network by sending them in pre-specified directions. Although this method was found to guarantee some asymptotically optimal results [NS92], it has a fundamental flaw: Consider a packet which originates very close to its destination, then obviously we expect it to reach its destination very fast. This, however, is not achieved in structured routing. A packet initially very close to its destination might find itself moved to a distant region of the network due to some obstinate policy which determines a fixed, pre-specified route. Moreover, long unnecessary routes are taken even when the actual number of packets is much smaller than the number of nodes, because the algorithms are not sensitive to the total load. In contrast, in *greedy* routing, a packet is bound to use an out-going link in the direction of its destination whenever such a link is available. In other words, whenever some packet is *deflected* away from its preferred direction, it is because some other packet is currently using this out-link. In this way, greediness guarantees that unless some global congestion forbids it, packets go in the shortest path to their destinations.

Another practical aspect of greedy routing is its simplicity. The amount of hardware and the cycle time of the routing mechanism (at every processor) depend to a large extent on the complexity of the routing algorithm. Structured algorithms are often designed in *phases*, where the algorithm changes at each phase. Moreover, the routing choices are usually complex, and might differ for different processors during the same phase. Thus, structured algorithms typically rely on complex routing mechanisms. Greedy algorithms, on the other hand, do not consist of phases. All nodes perform the same (simple) routing policy at each step of the algorithm, from the very first step to termination. This feature ensures the simplicity of the routing hardware, and moreover, suggests that the algorithm is suitable for *continuous routing* in which packets are injected into the network at different time steps.

Although greediness might cause congestion in certain regions of the network, deflection is hoped to "spread the load" so that the total routing time is decreased. Indeed, simulations of greedy hot potato routing algorithms show their superiority over the structured ones. Unfortunately, their analysis is considerably more difficult, since unexpected sequence of deflections may push the packet away from its destination. In fact, certain chains of deflections may eventually result back to the original configuration, thus raising the question whether the algorithm ever terminates. Such infinite loops are called *livelock*.

1.1 Related work and Results

The first greedy hot-potato algorithm was proposed by Baran [Bar64]. Borodin and Hopcroft suggested a greedy hot-potato algorithm for the hypercube [BH85]. Numerous experimental results on hot-potato routing have been published [AS91, GH92, Max89, Pra86]. Prager [Pra86] showed that the Borodin-Hopcroft algorithm terminates in n steps on the 2^n-nodes hypercube for a special class of permutations. Complete analysis, however, exists merely for *many-to-one* routing problems, in which many packets may be destined to reach the same target. Hajek [Haj91] presented a simple greedy algorithm for the hypercube that runs in $2k + n$ steps, where k is the number of packets in the system. The work of Hajek was simplified and generalized in a work by Brassil and Cruz [BC91]. They show a bound of $diam + P + 2(k-1)$ for any network, where $diam$ is the network diameter and P is the length of a walk connecting all destinations in a certain order. Ben-Dor, Halevi and Schuster gave a potential function analysis of greedy routing algorithms on d-dimensional arrays [BHS94]. Their results yield, for instance, a $8\sqrt{2}\sqrt{k}n$ bound for routing k packets on the two dimensional mesh. Recently, while preparing the camera-ready version of this work, we heard that Feige [Fei94], and Borodin, Rabani and Schieber [BRS94], match the bound that is given in this work for routing on the mesh.

Ben-Aroya, Newman and Schuster propose randomized single-target routing on n-dimensional arrays (including the n-dimensional hypercube) in $(2+o(n))\frac{k}{n} + poly(n)$ [BNS94, Ben94]. This single target result and the one given in this work suggest that the $d_{max} + 2(k-1)$ results are not the best possible. In fact they yield improved bounds for the multi-target case, especially when the number of targets is small relative to the number of packets.

Some recent results concern structured (non-greedy) hot potato *permutation routing*. Feige and Raghavan [FR92] presented an algorithm for the 2-dimensional torus that routes most of the routing problems in no more than $2n + O(\log n)$ steps. Newman and Schuster [NS92] presented an algorithm that is based on sorting for permutation routing in the 2-dimensional mesh. Their algorithm routes every permutation in $7n + o(n)$ steps, which was improved to $3.5n + o(n)$ by Kaufmann et. al. [KLS94]. Bar-Noy et al. [BRST93] present a relatively simple algorithm for the 2-dimensional mesh and torus that routes every routing problem in $O(n\sqrt{m})$ steps where m is the maximum number of packets destined to a single column. Kaklamanis, Krizanc and Rao [KKR93] presented an algorithm for permutation routing in the d-dimensional torus that routes most of the permutations within $\frac{1}{2}dn + O(\log^2 n)$ steps, and an algorithm for permutation routing in the 2-dimensional mesh that routes most of the permutations within $2n + O(\log^2 n)$ steps. Recently, Ben-Aroya and Schuster [BS94] proved a lower bound for deterministic algorithms that "stick" to the surrounding of the destination column or destination row once they get there. Their result yield an $\Omega(n^2)$ lower bound for permutation routing by a large class of algorithms.

Although the upper bounds stated above for permutation routing are optimal for the average case permutation, they are non adaptive to features that may speedup the routing. Such features include the maximal source-to-destination distance and the initial number of packets. We remark that the algorithms in [FR92, KKR93] have some "greedy tendency" included. However, we can show that they loose it when routing a worst case instance, and in that case their performance is very bad.

We consider the problem of greedy batch routing in the two-dimensional mesh. Suppose the system contains k packets and the maximal source-to-destination distance is d_{max}. We present an algorithm which completes routing of any request in $2(k-1) + d_{max}$ steps (Section 3). The algorithm is robust, in the sense that the number of packets originating at every node may be as high as its out-degree. We then consider single-target routing in which all packets are destined to the same node (Section 4). It is shown that the algorithm evacuates all the packets in at most $k + d_{max}$ steps which is also the lower bound. This result is the first to take advantage of the fact that the number of entries to every node is at least two. Using the single target result, we reconsider multi-target routing when the number of targets M is small, relative to the number of packets k, and the destinations are ordered so that there is a path of length P connecting them in order (Section 5). A weakly greedy algorithm is shown to route all packets to their destinations in at most $d_{max} + k + P + 2(M-1)$ steps. Finally, we consider short distance permutation routing in which each node is the origin and destination of up to one packet, and in which d_{max} is small (Section 6). When $d_{max} = 3$ the algorithm terminates in seven steps (where five is a lower bound).

2 Model and Definitions

In this work, we present greedy algorithms for batch packet routing on the two-dimensional mesh. There is a set of packets that are originated at time $t = 0$. Every packet has an origin node and a destination node in the network. The networks we consider are synchronous. That is, packets are sent in discrete time steps. The time it takes for a routing algorithm to solve a routing problem is the number of steps that elapse until the last packet reaches its destination.

In the *hot-potato* routing style, a packet cannot stay in a node (other than its destination node), so it must leave every intermediate node in the step that follows its arrival. Hence, every node in the network performs the following scheme in every step: (1) Get the packets that were sent to you in the previous step from your incoming arcs. (This sub-step is not performed at the first step of the algorithm). (2) Make a local computation that depends on the origins and destinations of the packets that arrived at the beginning of this step and on the arcs through which they have entered. (3) According to the results of the local computation, assign an outgoing arc for each of these packets. A *hot-potato routing algorithm* is a collection of such schemes (one for every node in the network) guiding the local computations in these nodes.

Definition 1. Let S be a node in the mesh, and let p be a packet in S. We say that an arc that goes out of S is a *good arc* for p if it enters a node that is closer to p's destination. The direction of this arc is a *good direction* for p. Similarly, we say that an arc that goes out of S is a *bad arc* for p which goes in a *bad direction* if it enters a node that is farther from p's destination. We say that a packet p

advances in step t if it gets closer to its destination in that step, i.e., it moves on a good arc at that step. Otherwise, we say that p is *being deflected*.

In this work we focus on *greedy* hot-potato routing algorithms, in which a packet always tries to advance. The definitions were originally given in [BHS94].

Definition 2. A hot-potato routing algorithm is said to be *greedy* if it satisfies the following condition: Whenever a packet p leaves its current node via a bad arc, all its good arcs must be used by other packets. Moreover, every arc that is good for p must also be good for the packet that uses it (that is - all these "other packets" must actually advance towards their destinations).

We say that an algorithm is *strongly greedy* when, in addition to being greedy, it also maximizes the number of advancing packets from every node. We say that the algorithm is *weakly greedy* when, from every non-empty node, at least one packet advances.

Definition 3. A packet in the 2-dimensional mesh is said to be *restricted* if it has only one good direction. This means that the packet is either in the row or in the column of its destination node.

3 A greedy hot-potato algorithm

In this section we present a greedy hot-potato algorithm for routing general routing problems on the two-dimensional mesh.

Algorithm General_Request. *The algorithm is described in terms of the behavior of packets.*

Priority to restricted packets: *Packets that are restricted (i.e., they have only one good direction) always get priority. (Among several restricted packets and a certain direction, arbitrarily one of them advance.)*

Priority to inertial packets: *Packets that are already moving horizontally in rows towards their target columns (i.e., they made such a move in the previous step), or are already moving vertically towards their rows, get priority over (non-restricted) packets which would like to join this direction.*

Greedy principle: *In all other cases packets are sent according to the greedy principle.*

Note that applying the above rules may involve conflicts which are not resolved by the specified rules. This does not disturb the analysis, i.e., the following claims hold for any algorithm which is consistent with the above rules.

Theorem 4. *Let k denote the initial total number of packets in the mesh, and let d_{max} denote the maximal distance from any source to any destination. Then, for any routing request, the GENERAL_REQUEST algorithm completes routing in at most $2(k-1) + d_{max}$ steps.*

We remark that the theorem is not sensitive to the initial number of packets in a node in the sense that the same result holds even if up to four packets originate at each processor. In this case, of course, there might be some additional conflicts to resolve in the first step. Again, the theorem holds regardless of the way these conflicts are resolved provided it is consistent with the algorithm (i.e., restricted packets get priority).

Another remark is that there are routing problems which require $k + d_{max}$ steps for any algorithm. For instance, if all k packets are destined to the upper left corner and all of them have the same distance to that corner. For such single-destination routing problems Algorithm GENERAL_REQUEST can be refined in order to utilize the additional structure so that this lower bound is matched, see Section 4. For general routing problems Theorem 4 is stronger than the related results in [Haj91], [BC91] and [BHS94], and its proof is simpler.

For the proof of Theorem 4 we use a "blaming" mechanism. In other words, we let the packets "blame" each other for being deflected. For each deflection of a packet we create a *token* which will be carried by advancing packets until the carrier packet arrives to its destination with a load of tokens. This packet will be "blamed for" the deflections for which it carries tokens. At this stage the tokens are removed. We make all this formal in the following rules governing the creation, delivery and removal of tokens.

The Token Rules.

- *If a packet p is deflected at step t of the algorithm then it produces a token $T(p, t)$.*
  - *If p was advancing at $t - 1$ or if p is restricted at the end of that step, then there is a restricted packet q which advances at t from the deflecting node in the direction p was advancing, and $T(p, t)$ is given to q.*
  - *Otherwise, $T(p, t)$ is given to the packet that proceeds in the direction opposite to the direction in which p is deflected. Clearly, there is such a packet that advances in a direction good for p, otherwise p would have not been deflected.*
- *A packet that carries tokens and is deflected will move these tokens to the advancing packet which gets the newly created token.*
- *When a packet arrives at its destination, all the tokens that are carried by it are removed.*

The most important fact in the proof of Theorem 4 is that out of every node from which a packet p is deflected there is another packet that advances and has either the same or a subset of the good directions of p. This implies that two tokens that are created by the same packet never meet, which let us "blame" the deflection which created the token on the packet which carries it when arriving to its destination. The details may be found in [BS93].

It can be easily verified that for a strongly greedy algorithm which gives priority to restricted packets this fact always holds and so we get the following corollary.

Corollary 5. *If in Algorithm* GENERAL_REQUEST *the priority to inertial packets rule is removed, and the greedy rule is replaced by a strongly greedy rule requirement, then the resulting algorithm completes the routing in* $d_{max} + 2(k-1)$ *steps.*

4   Single Target Routing

In this section we assume that all packets in the system are destined to reach the same target. Thus, each arc of every node is either always good for all the packets or bad for all the packets. A greedy routing algorithm can be seen as giving priority to arcs. When there are less packets than good arcs, the algorithm decides which good arcs to use. When there are more packets than good arcs, it decides which arcs will be used for deflections (since all the packets are going to the same destination all that matters is which arcs are used regardless of the identity of the packets).

For a node z, let $dx(z)$ be the horizontal distance from z to the target, and let $dy(z)$ be the vertical distance from z to the target. A packet is *restricted* if $dx(z) = 0$ or $dy(z) = 0$. A packet is *on the diagonal* if $dx(z) = dy(z)$. Consider the following algorithm.

Algorithm Single_Target. *The algorithm is described from the point of view of a node z and the routing operations it performs.*

– *If $dx(z) > dy(z)$, a packet is sent along the row in the good direction.*
– *If $dx(z) < dy(z)$, a packet is sent along the column in the good direction.*
– *If $dx(z) = dy(z)$ and an advancing packet entered via a row-arc, a packet will be sent along the good row-arc.*
– *If $dx(z) = dy(z)$ and an advancing packet entered via a column-arc, a packet will be sent along the good column-arc.*
– *When a restricted packet is deflected, arcs that will make it non-restricted are preferred.*
– *The algorithm is greedy – a bad arc will be used only if all the good arcs of the node are used.*

Our first observation concerning Algorithm SINGLE_TARGET is that when all packets are destined to a single target its rules do not conflict with those of Algorithm GENERAL_REQUEST. Thus, we can augment it so that it terminates fast for multi-destination routing problems.

Definition 6 Even distance. The even distance packets destined to a node z of a network are *m-dense around z* if there exists an integer $j > 0$ such that for all $1 \le i \le j$ there are at least m packets at distance $2i$ to z and there are no packets at a distance greater than $2(j + 1)$.

A similar definition applies for odd distance packets. When $m > 0$ is unimportant we simply say that the packets are *dense around z*. Note that if the packets are dense, then after two steps a packet will arrive at the destination, and the network remains at least 1-dense. In other words, when the packets are dense, at least one packet arrives on every second step.

Lemma 7. *For any network with bi-directional links, and any greedy algorithm, if all packets are destined to a single node z, then after d_{max} steps the packets will be 1-dense around z.*

Theorem 8. *For any network with bi-directional links, and any greedy algorithm, if all k packets are destined to a single node z, then after $2(k-1) + d_{max}$ steps, all packets will reach the destination.*

This theorem was given originally by Hajek [Haj91]. The notion of *greedy routing* makes it easier to prove in our framework. We proceed to show a stronger result for the two dimensional mesh. Let $d_{max\ restricted}$ be the maximum origin-to-destination of any initially restricted packet. If no packet is initially restricted then $d_{max\ restricted} = 0$.

Claim 9. *Let p_0 be a restricted packet at a distance of at least 2 from the destination after at least $d_{max\ restricted}$ steps of Algorithm SINGLE_TARGET. There exists another packet p_1 with the same distance to the destination as p_0, such that p_0 and p_1 are supposed to take minimal paths of the same length that are disjoint.*

Lemma 10. *After $t \geq d_{max}$ steps of Algorithm SINGLE_TARGET, the packets are 2-dense around the destination.*

Theorem 11. *Algorithm SINGLE_TARGET will route k packets to a single target in no more than $k + d_{max}$ steps.*

Proof: After d_{max} steps, at least one packet arrives at the destination. From the d_{max}-th step and at every even step by Lemma 10, there are two packets at distance 2. If both of these packets reside in the non-restricted node in distance 2, then in two steps both packets will arrive at the destination (since both packets will be sent to two different nodes of distance 1). On the other hand, if one of these packets is restricted, then by Claim 9, we know that there is another packet in distance 2, and that the two packets are supposed to take disjoint minimal paths (each with respect to itself). In any case, both packets at distance two arrive at the destination in two steps.

Since after d_{max} steps there are no more than $k - 1$ live packets, and at each second step, while there are at least two live packets, two packets enter the destination, the number of steps to total evacuation is bounded by

$$d_{max} + 2\lceil \frac{k-1}{2}\rceil = \begin{cases} d_{max} + k - 1 & \text{if } k \text{ is odd} \\ d_{max} + k & \text{if } k \text{ is even} \end{cases}$$

$\square$

Note that if the k packets origins are split to k_{even} of even distance and k_{odd} of odd distance with corresponding maximal distances $d_{max\ even}$ and $d_{max\ odd}$, then the theorem bounds the routing time by $\max\{k_{even} + d_{max\ even}, k_{odd} + d_{max\ odd}\}$.

5 Routing to a Small Set of Targets

Theorem 4 gives an upper bound of $2(k-1) + d_{max}$ on the termination time for multi-target situations. In this section we use Theorem 11 to tighten the bound when k is large relative to the number of targets M. Unfortunately, the resulting algorithm is only weakly greedy.

Let k packets be destined for each one of M targets. We begin by ordering the M destinations in an arbitrary fashion and assume that this ordering is known to each node. Let k_j, $j = 1, 2, \cdots, M$ be the number of packets initially admitted and destined for the jth destination, denoted z_j, so $k = \sum_{j=1}^{M} k_j$. We refer to packets destined to node z_j as *class j* packets. The following greedy algorithm was suggested by Brassil and Cruz [BC91].

Algorithm BC_Multi_Target.

> *Let each node order arriving packets by class from the lowest to the highest numbered class. Starting from the lowest numbered class, packets are greedily assigned to good exits, if available, in an arbitrary fashion. Then other packets are deflected in an arbitrary fashion.*

Theorem 12. *[BC91] Suppose a batch of k packets is to be delivered to a set of M targets. If the destinations are ordered so a walk connecting them in order has length P, then Algorithm* BC_MULTI_TARGET *terminates in $d_{max} + P + 2(k-1)$ steps on any network.*

Note that for the mesh, we already have a better bound by Theorem 4. Using Algorithm SINGLE_TARGET, however, we may get further acceleration by using the following algorithm.

Algorithm Mesh_Multi_Target.

> *Let each node of the mesh order arriving packets by class from the lowest to the highest numbered class. Starting from the lowest numbered class, packets are assigned to a preferred exit, if available, according to Algorithm* SINGLE_TARGET.

The algorithm may not follow the greedy principle when packets from a prioritized class must be deflected in directions which are good for a lower priority class. Notice, however, that it is weakly greedy. As the following theorem shows, it has a good termination time under certain conditions.

Theorem 13. *Suppose a batch of k packets is to be delivered to a set of M targets. If the destinations are ordered so a walk connecting them in order has length P, then Algorithm* MESH_MULTI_TARGET *terminates in $d_{max} + P + k + 2(M-1)$ steps.*

We remark that this bound is somewhat strange, as it *improves* when the number of destinations M becomes *smaller*. Moreover, it is somewhat unrealistic to assume that the ordering of the destinations minimizes P. In the more realistic case, the destinations are not known in advance. Nevertheless, when P and M are small relative to k, this bound becomes almost twice as good as that of Algorithm GENERAL_REQUEST which, in turn, is better than the bound of Theorem 12.

The proof of Theorem 13 can be found in [BS93]. We remark that one may deduce a proof for Theorem 12 with respect to any network by repeating the considerations in the proof of Theorem 13 with respect to the mesh.

6 Routing Short Distances

In this section we present a greedy algorithm which can be shown to terminate in seven steps when the routing request is a permutation (i.e., each node is the origin and destination of up to one packet), and when the maximal source-to-destination distance is three.

Algorithm Short_Distance. *The algorithm is described in terms of the behavior of packets.*

- *All packets attempt to get to the destination on a shortest path that first goes along the row to the correct column and then travels vertically.*
- *Packets that are destined to odd numbered rows will be high priority (We show that there will be no conflicts between high priority packets).*
- *Among lower priority packets maximize the number of packets that advance giving priority to packets that have longer minimum path to their destination (i.e., a low priority packet is deflected only if all its good arcs are used by higher priority packets).*
- *A packet that is deflected will choose (whenever possible) to be deflected on an arc that will not make it restricted. In other words, a packet that is deflected will try to have two good arcs for the next step.*

Theorem 14. Algorithm SHORT_DISTANCE *will route any permutation in which all the source-to-destination distances are at most 3 in at most 7 steps.*

Figure 1 shows paths of 3 packets as arrows from the source to the destination. Assume, without loss of generality, that packet a first moves horizontally. In the case that the algorithm chooses to route this packet via the vertical arc, a similar configuration of the other packets can be made such that the position after the first step will be the same position rotated 90 degrees. In the second step, the packets a and b are both restricted to use the same arc. Deflecting b will give the lower bound of 5 steps. Therefore, assume that a is deflected and b continues. The only direction in which a can be deflected without violating the greedy rule is to the right. After the second step, a, c and d are in the same node, with only two good exits. Any deflection now will cause the deflected packet to take 5 or 6 steps. This example can be extended to show a bound of $d + 2$ steps for packets in distance $\leq d$ (for $d \geq 3$).

7 Conclusions

One motivation for this work comes from considerations that everybody are doing when driving their cars. Driving on a highway, it is unlikely that one will

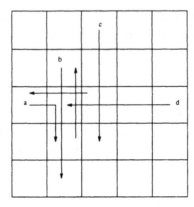

Fig. 1. If a moves to the right, a greedy hot-potato algorithm cannot terminate in less than 5 steps.

not take a free exit leading to one's destination. And yet, traffic flow in big cities is usually fast unless some global overload congests everything. Furthermore, when there are several possible roads to drive along the good direction, all of the same length, one would typically try to avoid committing to one of them since it may be blocked a few miles ahead. With these real-life examples, we have defined the notion of a *greedy algorithm* so that no packet is sent in the wrong direction unless all good directions are blocked.

Notice that the above discussion leads to (traffic) routing policies that are implemented in a distributed fashion in real life. This observation gave us the feeling that the greedy rule is very natural and relatively easy to implement, while the strongly greedy rule seemed more complex. This motivated the definitions as given in Section 2. Another motivation is our feeling that practically a non-greedy algorithm seems inefficient, while the benefits of strongly greedy algorithms over simple greedy ones are doubtful. Of course, this point is the focus of further research.

The notion of a *restricted packet* captures the feeling that it is not a good idea to commit oneself as one of two restricted packets will always be deflected. In fact, we believe that in adaptive routing, the best way to navigate is to avoid being restricted by staying on the "diagonal".

References

[AS91]   A.S. Acampora and S.I.A. Shah. Multihop lightwave networks: a comparison of store-and-forward and hot-potato routing. In *IEEE INFOCOM*, pages 10–19, 1991.

[Bar64]   P. Baran. On distributed communications networks. *IEEE Transactions on Communications*, 12:1–9, 1964.

[BC91]   J.T. Brassil and R.L. Cruz. Bounds on maximum delay in networks with deflection routing. In *29th annual Allerton conf. on Communication, Control and Computing*, pages 571–580, 1991.

[Ben94]   Ishai Ben-Aroya. Algorithms and bounds for deflection routing. Master's thesis (In Hebrew), April 1994.

[BH85]   A. Borodin and J.E. Hopcroft. Routing, merging, and sorting on parallel models of computation. *Journal of Computer and System Sciences*, 30:130–145, 1985.

[BHS94]   A. Ben-Dor, S. Halevi, and A. Schuster. Potential function analysis of greedy hot-potato routing. In *Proc. of 13th ACM Symp. on Principles of Distr. Comput.*, Los Angeles, August 1994. To appear. (Also: LPCR TR #9204, CS Dept., Technion, Jan. 1993).

[BNS94]   I. Ben-Aroya, I. Newman, and A. Schuster. Randomized single target hot potato routing. Manuscript, 1994.

[BRS94]   A. Borodin, Y. Rabani, and B. Schieber. Deterministic many-to-many hot potato routing. Manuscript, 1994.

[BRST93]   A. Bar-Noy, P. Raghavan, B. Schieber, and H. Tamaki. Fast deflection routing for packets and worms. In *Proc. 12th ACM symp. on Principles Of Distributed Computing*, 1993.

[BS93]   I. Ben-Aroya and A. Schuster. Greedy hot-potato routing on the two-dimensional mesh. Technical Report LPCR TR #9320, Technion, CS, November 1993.

[BS94]   I. Ben-Aroya and A. Schuster. A CLT-type lower bound for routing permutations on the mesh by hot-potato algorithms. Technical Report LPCR #9405, CS dept. Technion, May 1994.

[Fei94]   U. Feige. Observations on hot potato routing. Manuscript, 1994.

[FR92]   U. Feige and P. Raghavan. Exact analysis of hot-potato routing. In *proc. IEEE symp. on foundations of computer Science*, November 1992.

[GH92]   A.G. Greenberg and B. Hajek. Deflection routing in hypercube networks. *IEEE Transactions on Communications*, June 1992.

[Haj91]   B. Hajek. Bounds on evacuation time for deflection routing. *Distributed Computing*, 5:1–6, 1991.

[KKR93]   C. Kaklamanis, D. Krizanc, and S. Rao. Hot-potato routing on processor arrays. In *Symp. of Parallel Algorithms and Architectures*, 1993.

[KLS94]   M. Kaufmann, H. Lauer, and H. Schroder. Fast deterministic hot-potato routing on processor arrays. In *ISAAC*, 1994. To appear.

[Max89]   N.F. Maxemchuk. Comparison of deflection and store and forward techniques in the manhattan street and shuffle exchange networks. In *IEEE INFOCOM*, pages 800–809, 1989.

[NS92]   I. Newman and A. Schuster. Hot-potato algorithms for permutation routing. Technical Report LPCR #9201, CS dept. Technion, November 1992.

[Pra86]   R. Prager. An algorithm for routing in hypercube networks. Master's thesis, 1986.

[SBSS93]   C.L. Seitz, N. Boden, J. Seizovic, and W. Su. The design of the Caltech Mosaic C multicomputer. In *Proc. Symp. on Integrated Systems*, pages 1–22, 1993.

[Smi81]   B. Smith. Architecture and applications of the HEP multiprocessor computer system. In *Proc. (SPIE) Real Time Signal Processing IV*, pages 241–248, 1981.

[Szy90]   T. Szymanski. An analysis of hot potato routing in a fiber optic packet switched hypercube. In *Proc. IEEE INFOCOM*, pages 918–926, 1990.

[ZA91]   Z. Zhang and A.S. Acampora. performance analysis of multihop lightwave networks with hot potato routing and distance age priorities. In *Proc. IEEE INFOCOM*, pages 1012–1021, 1991.

Desnakification of Mesh Sorting Algorithms

Jop F. Sibeyn[*]

Max-Planck-Institut für Informatik
Im Stadtwald, 66123 Saarbrücken, Germany.
E-mail: jopsi@mpi-sb.mpg.de.

Abstract. In all recent near-optimal sorting algorithms for meshes, the packets are sorted with respect to some snake-like indexing. In this paper we present deterministic algorithms for sorting with respect to the more natural row-major indexing. For 1-1 sorting on an $n \times n$ mesh, we give an algorithm that runs in $2 \cdot n + o(n)$ steps, matching the distance bound, with maximal queue size five. It is considerably simpler than earlier algorithms. Another algorithm performs k-k sorting in $k \cdot n/2 + o(k \cdot n)$ steps, matching the bisection bound. Furthermore, we present *uni-axial* algorithms for row-major sorting. Uni-axial algorithms have clear practical and theoretical advantages over bi-axial algorithms. We show that 1-1 sorting can be performed in $2^{1}/_2 \cdot n + o(n)$ steps. Alternatively, this problem is solved with maximal queue size five in $4^{1}/_3 \cdot n$ steps, without any additional terms. For practically important values of n, this algorithm is much faster than any algorithm with good *asymptotical* performance. A hot-potato sorting algorithm runs in $5^{1}/_2 \cdot n$ steps.

1 Introduction

Various models for parallel machines have been considered. One of the best studied machines with a fixed interconnection network, is the *mesh*. In this model the processing units, *PUs*, form an array of size $n \times n$ and are connected by a two-dimensional grid of communication links.

Problems. The problems concerning the exchange of packets among the PUs are called *communication problems*. The packets must be sent to their destinations such that at most one packet passes through any wire during a single step. The quality of a communication algorithm is determined by (1) its *run time*, the maximum number of steps T a packet may need to reach its destination, and (2) its *queue size* Q, the maximum number of packets any PU may have to store. *Routing* is the basic communication problem. In this problem the packets have a known destination. We speak of *permutation*

---

[*] This research was partially supported by EC Cooperative Action IC-1000 (Project AL-TEC: Algorithms for Future Technologies).

routing if all PUs initially hold one packet, and if every PU is the destination of precisely one packet. The routing problem in which every PU is source and destination of k packets is called the *k-k routing* problem. *Sorting* is, next to routing, one of the most considered communication problems. Several variants of the problem have been studied. In the 1-1 sorting problem, each PU initially holds a single packet, where each packet contains a key drawn from a totally ordered set. The packets have to be rearranged such that the packet with the key of rank i is moved to the PU with index i, for all i. In the *k-k* sorting problem, each PU is the source and destination of k packets. *Scattering* is the problem of rearranging packets holding keys from a totally ordered set such that as little as possible packets with the same key stand in the same column. It is important in deterministic routing algorithms.

Models and Indexings. We assume the *MIMD* mesh model in which a PU can send in every step one packet to each of its neighbors and receive one packet from each of them. Algorithms that fully use these routing possibilities are called *bi-axial*. In a *uni-axial* algorithm all PUs communicate only along either the horizontal or the vertical connections. Uni-axial algorithms are important because two of these algorithms can be applied in parallel and because they can be used in more restricted routing models.

Several recent sorting algorithms [1, 5, 3] were designed for snake-like row-major indexings. In many cases it is desirable to have the packets in the more natural row- or column-major order. Furthermore, sorting in snake-like order is unsuited for scattering. In the 'one-packet' model considered by Schnorr and Shamir [7], the best known upper bound for row-major sorting is higher than for sorting in snake-like row-major order. In our model a PU may hold a constant number of packets and packets may be copied. From the results of this paper it follows that in this model, sorting in row-major order is not substantially harder than sorting in snake-like row-major order.

Results. This paper gives numerous improvements for row-major sorting:

	Uni-Axial				Bi-Axial	
	T	Q	T	Q	T	Q
1	$4^{1}/_{3} \cdot n$	5	$2^{1}/_{2} \cdot n + \mathcal{O}(n^{5/6})$	5	$2 \cdot n + \mathcal{O}(n^{5/6})$	5
2	$5^{1}/_{4} \cdot n$	4	$3 \cdot n + \mathcal{O}(n^{5/6})$	5	$2^{1}/_{2} \cdot n + \mathcal{O}(n^{5/6})$	9
k	$(7/4 \cdot k + 6) \cdot n$	$k + 2$	$k \cdot n + \mathcal{O}((k \cdot n)^{5/6})$	k	$k/2 \cdot n + \mathcal{O}((k \cdot n)^{5/6})$	k

Theoretically the result for large n (those with lower-order terms) are the most appealing. So far, the fastest bi-axial row-major sorting algorithm has $T = 2^{1}/_{4} \cdot n + o(n)$ and $Q = \mathcal{O}(1)$. It was recently designed by Krizanc and Narayanan [4]. However, it only works for the subproblem that all the keys are 0 or 1. The first near-optimal sorting algorithm, $T = 2 \cdot n + o(n)$, was presented by Kaklamanis and Krizanc [1]. The algorithm is randomized and sorts the packets in blocked snake-like row-major order. In [3] a deterministic version is presented. These algorithms are considerably more involved then the algorithm of this paper, and have queue sizes around 20. The best uni-axial row-major sorting algorithm so far appears to be a modification of the algorithm of Schnorr and Shamir [7]. It takes $4 \cdot n + o(n)$ steps. The first near-optimal algorithm for k-k sorting was discovered by Kaufmann and Sibeyn [2]. Then in [5] by Kunde and in [3], deterministic versions of this randomized algorithm were described. All use blocked snake-like row-major indexings. We present the first near-optimal algorithm for k-k sorting in row-major order.

Most current communication algorithms strive for $T = \alpha \cdot n + o(n)$, with α as small as possible. This completely neglects the fact that actual meshes tend to be of fairly moderate sizes, for which the $o(n)$ term dominates. Typically this term gives the number of steps for several sorting and rearrangement operations in submeshes of size $n^{2/3} \times n^{2/3}$ or $n^{3/4} \times n^{3/4}$. Even when it is just $10 \cdot n^{2/3}$, then still it exceeds n for all $n < 1000$. This clearly expresses the utmost importance of algorithms with a routing time not involving any hidden terms. Our uni-axial row-major sorting algorithm with $T = 4^{1}/_{3} \cdot n$ and $Q = 5$ is therefore of great *practical* importance. A sorting time which can be expressed as $T \leq \alpha \cdot n$, for all n, is even relevant in a *theoretical* setting: in recursive or divide-and-conquer algorithms, the submeshes on which they are applied are small. Minimizing Q at moderate expense, we obtain a 1-1 sorting algorithm with $Q = 1$ and $T = 5^{1}/_{2} \cdot n + \log n$. One of the algorithms is turned into the first practical 'hot-potato' sorting algorithm: $T = 5^{1}/_{2} \cdot n$.

In Section 3 we give the algorithms for uni-axial row-major sorting for all n. Then we introduce in Section 4 the 'desnakification' for k-k sorting and for near-optimal 1-1 and 2-2 sorting. For the many missing details and proofs, and for a complete list of references, we refer to the [8].

2   Preliminaries

Basics of Routing and Sorting. When several packets residing in a PU have to be routed over the same connection, then the packet that has to go farthest gets priority. For one-dimensional sorting we apply a suitable variant of odd-even transposition sort. For a given distribution of packets over the PUs, let $h_{\text{right}}(i, j) = \#\{packets\ passing\ from\ left\ to\ right\ through\ both\ P_i\ and\ P_j\}$, where P_i denotes the PU with index i. Define $h_{\text{left}}(j, i)$ analogously. For the analysis of the routing on one-dimensional meshes we use

Lemma 1 *Routing a distribution of packets on a linear array with n PUs, using the farthest-first strategy, takes $\max_{i<j}\{\max\{h_{\text{right}}(i, j), h_{\text{left}}(j, i)\} + j - i - 1\}$ steps. This bound is sharp. When the packets are evenly distributed, then the same bound can be achieved for sorting.*

Because of the distance a packet may have to go, $2 \cdot n - 2$ steps is a lower bound for any routing or sorting problem on the two-dimensional mesh: the *distance bound*. Because of the number of packets that may have to pass from one half of the mesh into the other half over only n connections, $k \cdot n/2$ steps is a lower bound for k-k routing or sorting: the *bisection bound*.

A *0-1 distribution*, is a distribution of packets that all have key zero or one. A row is called *dirty*, if it contains both zeros and ones. In our analyses we frequently use the '0-1 lemma', which states that under in our case satisfied conditions a sorting algorithm is correct iff it sorts any 0-1 distribution.

Indexings and Subdivisions. The PUs are indicated by giving their coordinates within the mesh, the PU at position (i, j), $0 \le i, j < n$ and $(0, 0)$ in the upper-left corner, is denoted $P_{i,j}$. In the common row-major indexing $P_{i,j}$ has index $i \cdot n + j$. In the column-major indexing $P_{i,j}$ has index $i + j \cdot n$, in the reversed row-major indexing $i \cdot n + (n - j)$. For a given indexing we denote the PU with index i, $0 \le i < n^2$, by P_i. For k-k sorting our default is a *non-layered* indexing, in which location r in P_i, $0 \le r < k$, $0 \le i < n^2$, has index $k \cdot i + r$. A non-layered row-major indexing is the indexing as if we have an $n \times k \cdot n$ mesh in row-major order. For row-major sorting we also use a *semi-layered* indexing, under which location $P_{i,j}$, has index $(i + r) \cdot n + j$, as if we have a $k \cdot n \times n$ mesh in row-major order, see Figure 1.

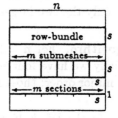

Fig. 1. Non-layered indexing (left) and semi-layered indexing (middle), for $k = 2$, $n = 4$. Subdivisions for the case $s = n/6$, $m = 6$ (right).

In our algorithms the mesh is divided in regular $s \times s$ submeshes. Let $m = n/s$. The submeshes are indexed as the PUs: starting with $(0, 0)$ in the upper-left corner. We refer to submesh (i, j) by $B_{i,j}$. Define *row-bundle i* to consist of the PUs in $\cup_{j=0}^{m-1} B_{i,j}$. Likewise, *column-bundle j* consists of $\cup_{i=0}^{m-1} B_{i,j}$. Additionally the mesh is subdivided in *sections*. A section is a subset of the PUs with consecutive indices. Section l is denoted S_l. If the sections have length s, then S_l consists of the PUs with indices $s \cdot l, \ldots, s \cdot (l+1) - 1$. Under a row-major indexing these sections regularly subdivide the rows and the submeshes. See Figure 1.

Definition 1 *An indexing is called* piecewise-continuous *with parameter s if for every i, $0 \le i < n^2$, there is an interval $\mathcal{I}_i \subset [0, n^2 - 1]$, with $i \in \mathcal{I}_i$ and*

$\#\mathcal{I}_i \geq s$, such that for all $j \in \mathcal{I}_i$ P_j is adjacent to P_{j-1} and P_{j+1}, whenever $j-1, j+1 \in \mathcal{I}_i$.

The row-major indexing is piecewise-continuous with parameter n. One of the achievements of this paper is to show that for efficient sorting it is sufficient to have *piecewise*-continuous indexings. A row i is said to be *sorted rightwards* if the packets stand in increasing order from $P_{i,0}$ to $P_{i,n-1}$. Analogously, rows can be sorted *leftwards* and columns *downwards* and *upwards*.

Definition 2 *An m-way merge is a procedure that turns a mesh that is divided in m^2 sorted $s \times s$ submeshes into a sorted $n \times n$ mesh.*

3 Uni-Axial Sorting for Small n

3.1 Powers of Two

Lemma 2 *Uni-axial sorting in arbitrary order can be performed on 2×2 meshes in 3 steps, with queue size two.*

Proof: Perform gossiping (all-to-all routing) along rows and then along columns. This takes three steps. A PU that finally should hold the packet with rank 0 or 1, needs to conserve only the two smallest packets, the other PUs only the two largest packets. □

Algorithm. For $n = 2^l$, $l > 1$, we use an optimized merge-sort algorithm. Initially we have four sorted $n/2 \times n/2$ submeshes: those in the left half in row-major order; those in the right half in reversed row-major order.

Algorithm MERGE

1. In the left half, shift the packets $n/4$ steps to the right. In the right half, shift the packets $n/4$ steps to the left.

2. In the central $n/2$ columns, sort the packets downwards.

3. Copy the smallest packet in every $P_{i,j}$, $0 < i \leq n-1$, $n/4 \leq j \leq 3/4 \cdot n - 1$, to $P_{i-1,j}$. Copy the largest packet in every $P_{i,j}$, $0 \leq i < n-1$, $n/4 \leq j \leq 3/4 \cdot n - 1$, to $P_{i+1,j}$.

4. In every row, sort the section of the row that lies in the central $n/2$ columns. If this submesh is going to be the right half of a larger mesh in the next merge, then the sorting is leftwards, otherwise rightwards.

5. Throw away the packets in $P_{i,j}$ with $j \in [n/4, 3/8 \cdot n - 1] \cup [5/8 \cdot n, 3/4 \cdot n - 1]$. For any $P_{i,j}$, with $3/8 \cdot n \leq j \leq 5/8 \cdot n - 1$, send the packet with rank r, $0 \leq r \leq 3$, to $P_{i,4 \cdot (j - 3/8 \cdot n) + r}$.

Step 1 takes $n/4$ steps, Step 2 can be performed in n steps, and Step 3 takes a single step but can easily be made to coincide with the last step of the sorting.

Lemma 3 *After Step 2 all packets that actually should be in a row can be found either in the row itself, or among the smallest packets of the row below, or among the largest packets of the row above.*

Proof: First we consider a modified problem. Suppose that initially four $n/2 \times n/2$ submeshes stand above each other in an $2 \cdot n \times n/2$ mesh. Two of these submeshes are sorted in row-major order, the other two in reversed row-major order. Consider a 0-1 distribution. It is easy to check that after sorting the columns of this mesh, there are at most two dirty rows. These dirty rows can be resolved as follows: copy every row to the row above and the row below; sort the rows; spread the packets from the central $n/3$ columns. In the real problem every two rows of the high and narrow mesh are compressed in one row in which every PU in the center holds two packets. $\square$

Lemma 4 *Step 4 can be performed in $3/4 \cdot n$ steps.*

Proof: The worst possible 0-1 distributions after Step 3 are like

According to Lemma 1, sorting this row takes $3/4 \cdot n$ steps. $\square$

Finally, Step 5 takes $3/8 \cdot n$ steps.

Overlapping. In MERGE Step 4 and 5 involve routing along the same axis. So, we might overlap these steps, without impairing the uni-axiality of the algorithm. The central observation is that the packets to throw away, are known well before the end of Step 4. After throwing them away, we can proceed with a combination of odd-even transposition sort and routing packets outwards: in the same step that we are sure which packets to throw away, we also know the largest surviving packet. One step later we know the second largest, and so on. Without further comparison these packets can be routed to their destinations, reducing the maximal distance the packets may have to travel after the end of the sorting.

Lemma 5 *In Step 4 all packets that will be thrown away have reached their destination region after $5/8 \cdot n$ steps.*

Now it is easy to see that we can use the following modified steps:

4′. For all i, $0 \leq i < n$, until step $5/8 \cdot n$, sort the packets in the central $n/2$ PUs of row i.

Throw away the packets that stand outside the central $n/4$ columns.

A PU that holds more than one packet continues to sort. A PU in the left (right) half that holds only one packet sends it leftwards (rightwards).

5'. Route the packets in row i to their destinations.

Step 4' takes as long as before: the sorting is influenced in no way by the action going on in the periphery. On the other hand, during Step 4' the packets that have to go farthest already covered $3/4 \cdot n - 5/8 \cdot n = n/8$ of the distance they had to cover in Step 5, and therefore Step 5' only takes $n/4$ steps.

Lemma 6 MERGE *takes at most* $2^{1/4} \cdot n$ *steps. The queue size is four.*

Sorting. Starting with sorted 2×2 meshes, MERGE can be used repeatedly for sorting on an $n \times n$ mesh. Call this algorithm SORT.

Theorem 1 *For all* $n = 2^l$, SORT *performs row-major sorting on an* $n \times n$ *mesh in* $4^{1/2} \cdot n$ *steps.* SORT *is uni-axial, and the queue size is four.*

Proof: $3 + 2^{1/4} \cdot (4 + 8 + \cdots + n) < 2^{1/4} \cdot n \cdot \sum_{i=0} 2^{-i}$. □

3.2 Powers of Two, Three, ...

We derived an efficient 1-1 sorting algorithm for $n = 2^l$. However, in practice, processor networks may not have such beautiful side lengths. Furthermore, some algorithms in which sorting is used as a subroutine, e.g., the algorithms of [9] specifically require that $n = m^l$, with $m \neq 2$. In principle we could use SORT by rounding n up to the nearest power of 2. But, this might give sorting times that are almost twice as large as necessary. In this section we present m-way merge algorithms, which perform well for $m \leq 5$. By combining them, we can efficiently sort $n \times n$ meshes for arbitrary n.

Lemma 7 *Uni-axial sorting in (reversed) row-major order on* $m \times m$ *meshes can be performed in* $m^2/2 + m$ *steps, for* m *even, and* $m \cdot (m+1)/2$ *steps, for* m *odd, with queue size* m.

Proof: Concentrate the packets in a central column, sort the column, and spread the packets along the rows. □

Algorithm. For performing an m-way merge for $m \geq 3$, an algorithm of the type of MERGE: leads to large queue size and rapidly growing time consumption in Step 4. It is better to first sort the row-bundles, then to merge the sorted row-bundles. In this way the number of dirty rows is limited to $\lceil m/2 \rceil$. Initially the $n/m \times n/m$ submeshes are appropriately sorted.

For even m, in every row-bundle $m/2$ submeshes are sorted in row-major order and $m/2$ in reversed row-major order. For odd m, $\lceil m/2 \rceil$ submeshes are sorted in row-major order in the highest $\lceil m/2 \rceil$ row-bundles, and $\lfloor m/2 \rfloor$ in the lowest $\lfloor m/2 \rfloor$ row-bundles. The other submeshes are sorted in reversed row-major order. The *central column-bundle*, consists of columns j, with $(m-1)/(2 \cdot m) \cdot n \leq j < (m+1)/(2 \cdot m) \cdot n$.

Algorithm MERGE'

1. Shift the packets in the column-bundles as blocks to central column-bundle.

2. In the central column-bundle, sort the sections of the columns that lie within the row-bundles downwards.

3. Let $q = \lceil m/2 \rceil - 1$. In the central column-bundle, for all i, $0 \leq i < n$, $i \neq 0, n/m, \ldots, (m-1)/m \cdot n$, copy the smallest q packets in $P_{i,j}$, to $P_{i-1,j}$; for all $i \neq n/m - 1, 2/m \cdot n - 1, \ldots, n-1$, copy the largest q packets in $P_{i,j}$, to $P_{i+1,j}$.

4. In the central column-bundle, for all i, $0 \leq i < n$, sort the section of row i. If $i < \lceil m/2 \rceil / m \cdot n$, then sort rightwards, else sort leftwards.

5. In every row, throw away the $q \cdot n/m$ packets with the smallest and with the largest keys. The remaining n packets stand in the central $\lceil n/(m+2 \cdot q) \rceil$ PUs. Spread these packets over the row. Overlap with Step 4.

6. In every column, sort the packets downwards.

7. In every column, for all $0 < i \leq n-1$, copy the smallest q packets in $P_{i,j}$, to $P_{i-1,j}$; for all $0 \leq i < n-1$, copy the largest q packets in $P_{i,j}$, to $P_{i+1,j}$.

8. Sort all rows rightwards.

9. In every row i, $0 \leq i < n$, throw away the $q \cdot n/2$ packets with the smallest and with the largest keys. The remaining n packets stand in the central $\lceil n/(1 + 2 \cdot q) \rceil$ PUs. Send the packet with rank j to $P_{i,j}$. Overlap with Step 8.

Step 3 and 7 are overlapped with the last steps of Step 2 and 6, respectively. There are two pairs of consecutive steps that involve routing along the same axis: Step 4 and 5, and Step 8 and 9. As in MERGE these steps are partially overlapped. This overlapping is essential for the performance of MERGE'. In [8] we give an algorithm in which the packets are spread only over the central $n/2$ columns in Step 5. It performs better for $m > 3$.

The correctness of MERGE-m is obvious: after Step 2 and Step 6, there are $q + 1$ dirty rows in a 0-1 distribution. These are resolved by the steps that follow. We give the time consumptions of the steps for $m = 2$ and, taking the overlapping into account, for $m = 3$ (omitting factors n):

step	1	2	4	5	6	8	9	total
$m = 2$	1/4	1/4	1/2	1/4	1/2	1	0	$2^3/4$
$m = 3$	1/3	2/9	1/2	1/6	2/3	1	0	$2^8/9$

For Step 2 it is important that the submeshes are already sorted, and in Step 6 that the row-bundles are sorted, this puts bounds on the distance packets may have to travel and the number of packets that may have to pass over a connection. Step 4 and 8 can be analized as in Lemma 4. By the end of Step 8 all packets have just reached their destinations.

Sorting. Let SORT' be the sorting algorithm based on MERGE'.

Theorem 2 *For all $n = 2^l$, SORT' performs row-major sorting on an $n \times n$ mesh in $5^1/2 \cdot n$ steps with queue size two. For all $n = 3^l$, SORT' requires $4^1/3 \cdot n$ steps with queue size five.*

3.3 Further Results

Mixed Powers. Suitably combining several merge algorithms we get

Theorem 3 *Uni-axial row-major sorting on $n \times n$ meshes can be performed in less than $4.75 \cdot n$ steps, for all n. The queue size is at most nine.*

k-k Sorting. The merging is almost the same as MERGE of Section 3.1. Initially the four $n/2 \times n/2$ submeshes are sorted in semi-layered row-major order on the left, and semi-layered reversed row-major order on the right. By the semi layered indexing, the merging corresponds to a 1-1 merge on a $k \cdot n \times n$ mesh. Step 1 is modified such that the packets from the submeshes are interleaved:

1. $P_{i,j}$, $0 \leq i, j < n$, sends its packet with rank r, $0 \leq r < k$, to $P_{i,(j+n/2)\bmod n}$ if $\mathrm{odd}(k \cdot i + r + j)$.

Theorem 4 *For all $k \geq 2$, uni-axial k-k sorting in row-major order on $n \times n$ meshes can be performed in $(7 \cdot k^2 + 26 \cdot k + 8 - \min\{k^2, 2 \cdot k + 8\})/(4 \cdot k + 8) \cdot n$ steps, with queue size $k + 2$.*

Minimizing the Queue Sizes. The queue sizes of our merge algorithms depend on the degree of concentration in the central columns and the number of dirty rows. By not cleaning away all dirt in a single operation, the queue sizes can be minimized. An additional advantage is that packets do not have to be copied anymore. Applying this technique in the k-k merging algorithm,

Theorem 5 *For all $k \geq 4$, uni-axial k-k sorting in row-major order on $n \times n$ meshes can be performed in $2 \cdot k \cdot n + 5/2 \cdot n - 3 \cdot n/(2 \cdot k) + 2 \cdot \log n$ steps, with queue size k. For $k = 2, 3$ it takes $k \cdot n + 6 \cdot n - 3 \cdot n/(2 \cdot k)$ steps. For $k = 1$, $5^1/2 \cdot n + \log n$ steps are required.*

Hot-Potato Sorting. In *hot-potato routing* the PUs have no queues at all: after the packets have been released, they continue to move until they have reach their destinations. Hot-potato routing is inspired by practice, and of particular importance in fine-grained parallel computers and in optical networks (see [6] for references). Though one has to be precise, it should in principle be possible to turn any uni-axial communication algorithm with queue size two or less into a corresponding bi-axial hot-potato algorithm: the at most two packets stored in a PU can 'vibrate' [6] over the connections orthogonal to the current routing direction. MERGE$'$ of Section 3.2 can be transformed into a hot-potato algorithm which takes $2\frac{1}{2} \cdot n + \lceil n/4 \rceil$ steps.

Theorem 6 *For $n = 2^l$ hot-potato sorting can be performed in $5\frac{1}{2} \cdot n$ steps.*

4 Sorting for Large n

4.1 k-k Sorting

Earlier k-k sorting algorithms [2, 5, 3] work according to the following scheme:

1. Route all packets to 'random' destinations.
2. Estimate the ranks of the packets by local comparisons.
3. Route all packets to their preliminary destinations.
4. Rearrange the packets locally to bring them to their final destinations.

These algorithms require a continuous indexing for sorting together pairs of submeshes with consecutive indices in Step 4. This is necessary and sufficient because the estimate of the rank in Step 2 is accurate up to one submesh. So, it may happen that after Step 3, a packet is not present in its destination submesh B_i, but resides in the preceding or succeeding submesh B_{i-1} or B_{i+1}. We introduce a novel technique, *desnakification*, to overcome this: send for all packets p, of which the destination submesh is not uniquely determined, a copy to both submeshes in which its destination may lie. Now it is sufficient to sort within the submeshes. If for B_i the numbers cl, of packets that actually belong in B_{i-1}, and ch, of packets that belong in B_{i+1}, are exactly known, then the smallest cl and largest ch packets in B_i are thrown away, and the remaining packets are redistributed within B_i. All this is very similar to the way dirty rows are resolved in the algorithms of Section 3.

The desnakification of the k-k sorting algorithm from [3] is particularly easy. In order to bound the number of copies, the $s \times s$ submeshes in which the local sorting is performed, should be taken large enough: $s = n^{5/6}/k^{1/6}$. The indexing must be piecewise-continuous with parameter at least s. In this case every submesh B_i receives in Step 3 from every section exactly the same number of packets that belong in B_{i-1} and B_{i+1}.

Theorem 7 *Let $s = n^{5/6}/k^{1/6}$. Bi-axial k-k sorting with respect to a piecewise-continuous indexing with parameter s can be performed in $\max\{4 \cdot n, k \cdot n/2\} + \mathcal{O}(k \cdot s)$ steps. The queue size is $k + 2$. Uni-axial k-k sorting takes $\max\{4 \cdot n, k \cdot n\} + \mathcal{O}(k \cdot s)$ steps, with queue size $k + 1$.*

4.2   1-1 Sorting

We start with a uni-axial algorithm for 1-1 sorting in row-major order. This algorithm is obtained by combining our new insight in merge sorting and the desnakification technique, with old knowledge about sorting with splitters. Without loss of generality, we assume that all packets have different keys.

Uni-Axial Sorting. The mesh is divided in $s \times s$ submeshes, $s = n^{5/6}$, and $m = n/s = n^{1/6}$. We distinguish packets and *splitters*. The splitters are copies of a small subset of the packets. They are broadcast and the packets estimate their ranks by comparison with the splitters. The splitters allow a rapid spread of the necessary information, while the packets are involved in more useful operations. They are selected as follows, and then broadcast:

In every submesh, sort the packets. Copy the packets with ranks $i \cdot m^2$, $0 \le i \le s^2/m^2 - 1$: the splitters.

Lemma 8 *After $2 \cdot n + \mathcal{O}(s)$ steps the splitters have reached all submeshes. No connection has to transfer more than $\mathcal{O}(s)$ packets.*

For the packets we perform a kind of m-way merge algorithm:

Algorithm 11SORT

1. In every submesh, sort the packets in row-major order.

2. In every submesh $B_{i,j}$, $0 \le i, j < m$, shift the packets in row l, $0 \le l < s$ to row l of $B_{i,(j+l)\mathrm{mod}(m/2)}$, and copies to row l of $B_{i,(j+l)\mathrm{mod}(m/2)+m/2}$.

3. In all columns, sort the packets downwards.

After Step 3, there are in a 0-1 distribution at most m^2 dirty rows. For a general distribution this means that a packet resides at most $m^2 - 1$ rows away from its destination row. These three steps take $2 \cdot n + \mathcal{O}(s)$ steps. So, after Step 3 the splitters are available in every submesh.

4. In every submesh, determine for every packet its 'rank', the number r, $0 \le r \le s^2$ of splitters that are smaller. The preliminary destination of a packet p with rank r, lies in section S_l, with $l = \lfloor r \cdot m^2/s \rfloor$. If $\lfloor (r \cdot m^2 - m^4)/s \rfloor = l - 1$, then create a copy p' of p with preliminary destination in S_{l-1}. If $\lfloor (r \cdot m^2 + m^4)/s \rfloor = l + 1$, then create a copy p' of p with preliminary destination in S_{l+1}. Discard the splitters, and the (copies of) packets that have preliminary destination in the other half of the mesh.

5. In every submesh, sort the packets in column-major order on their preliminary destination column-bundles.

6. In every row, route the packets to the first PUs in their preliminary destination column-bundles that hold less than two packets.

7. In each submesh, sort the packets in row-major order on their preliminary destination section.

8. In every column, route the packets to the sections of their preliminary destinations.

9. In every section, sort the packets.

10. In every section S_l, $0 < l \leq m \cdot n - 1$, throw away the m^4 packets with the smallest keys, and in each S_l, $0 \leq l < m \cdot n - 1$, throw away the m^4 packets with the largest keys. Redistribute the remaining $k \cdot s$ packets within S_l.

As the algorithm is given, it is not entirely correct. It is *not* true that, as in the algorithm for k-k sorting, exactly m^4 packets must be thrown away on both sides of every section: 11SORT orders the packets, but the sections do not necessarily hold exactly s packets. Fortunately, the numbers of packets that must be thrown away in a section on the low and high side, respectively, can be determined in an elegant way. We give a detailed description. Consider some section S and the sections from which it may receive packets after Step 3:

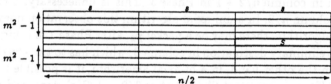

$(m^2 - 1) \cdot n/2$ packets are stored in these sections, among which the s packets with destination in S. After Step 8, these s packets all reside in S, but also some packets that do not belong in S. How can we figure out which packets to keep, and which packets to throw away? Suppose that S is the l-th section, $(m^2 - 1) \cdot n/s \leq l < m^2 \cdot n/s$, in the involved (whole) rows. Then finally S should hold the packets with ranks r, $l \cdot s \leq r < (l+1) \cdot s$ from among the $(2 \cdot m^2 - 1) \cdot n$ packets. Analogously to the merge algorithms of Section 3, we could copy *all* packets to S, sort them, and throw away the smallest $l \cdot s$ packets and the largest $(2 \cdot m^2 - 1) \cdot n - (l+1) \cdot s$ packets. This gives a correct but very inefficient algorithm. However, it is not necessary to copy all packets to S. It is sufficient if for each contributing section i the counters, the numbers $unders_{,i}$ and $overs_{,i}$ of packets that are *not* sent to S because they are definitely too small or definitely too large, respectively, are known in S. The counters can easily be determined in Step 4. They can be transferred to S during the subsequent steps, in parallel with the packets. As every section sends and receives only $\mathcal{O}(m^3)$ counters in total, they can be routed without causing substantial delay. $Unders = \sum_i unders_{,i}$ and $Overs = \sum_i overs_{,i}$ can be computed in Step 9. In Step 10, the smallest $l \cdot s - Unders$ packets and the largest $(2 \cdot m^2 - 1) \cdot n - (l+1) \cdot s - Overs$ packets in S are thrown away, leaving exactly the s packets belonging in S.

Theorem 8 *Uni-axial* 1-1 *sorting in row-major order can be performed in* $2^{1/2} \cdot n + \mathcal{O}(n^{5/6})$ *steps. The queue size is five.*

Proof: For the time consumption and correctness, it remains to show that Step 6 can be performed as specified in $n/2 + \mathcal{O}(s)$ steps. The central observation is that the estimate of the rank of a packet is accurate up to m^4. Hence, for some section S_l, only packets with destination in some PU P_k with $k \in [l \cdot s - 2 \cdot m^4, (l+1) \cdot s + 2 \cdot m^4]$, may get preliminary destination in S_l. For the queue size, we notice that a PU may hold up to four (copies of) packets during Step 4 and Step 5. Step 4 can be organized such that a PU holds at most one splitter or counter. □

Bi-Axial Sorting. Essentially 11SORT consists of three main routing phases: horizontal, vertical and horizontal (Step 2, Step 3 and Step 6). These phases take n, n and $n/2$ steps, respectively. The connections between the left and right half are not used anymore after step $n/2$. Thus it may happen that a packet p_1 that stands in column 0 after Phase 1 is routed to a preliminary destination in column $n/2 - 1$ in Phase 3. This is unnecessary: a copy of p_1 stands in column $n/2$. In a uni-axial algorithm this observation does not lead to a faster algorithm: there may be a packet p_2, after Phase 1 in column $n/2-1$ and with preliminary destination in column 0, which has to travel $n/2$ steps in Phase 3. On the other hand, in a bi-axial algorithm, it is possible to coalesce the phases. Then p_2 can start Phase 3 after $3/2 \cdot n + \mathcal{O}(s)$ steps, and will reach its preliminary destination after $2 \cdot n + \mathcal{O}(s)$ steps. Only Step 4 is changed:

> 4'. In all columns j, $0 \le j < n/2$, discard the (copies of) packets that have preliminary destination in some column j', with $j' \ge 2 \cdot j$. For $n/2 \le j < n$, discard the packets with $j' < 2 \cdot j - n$.

By this rule again exactly one of the copies of a packet reaches every possible destination section. The steps are coalesced: Step 3 begins in column j after $n/2 + |n/2 - j|$ steps, and Step 6 after $3/2 \cdot n + |n/2 - j|$ steps. Notice that the obtained algorithm is still *locally* uni-axial: every PU uses only horizontal or vertical connections.

Theorem 9 *Bi-axial* 1-1 *sorting in row-major order can be performed in* $2 \cdot n + \mathcal{O}(n^{5/6})$ *steps. The queue size is five.*

2-2 Sorting. For uni-axial 2-2 sorting, we modify the uni-axial version of 11SORT. For bi-axial 2-2 sorting, we apply two orthogonal versions of 11SORT:

Theorem 10 *Uni-axial 2-2 sorting in row-major order can be performed in* $3 \cdot n + O(n^{5/6})$ *steps. The queue size is five. Bi-axially this takes* $2^{1}/_{2} \cdot n + O(n^{5/6})$ *steps with queue size nine.*

Acknowledgement

The questions of Uli Meyer urged me to investigate row-major sorting for *all* n. Michael Kaufmann and Torsten Suel suggested several improvements.

5 Conclusion

We presented novel uni-axial and bi-axial row-major algorithms for sorting on two-dimensional meshes. They are considerably faster than existing algorithms. A tremendous improvement is our near-optimal algorithm for 1-1 sorting: it is much simpler than the earlier algorithm, it is suited for more useful indexings, it is locally uni-axial, and it has queue size five.

References

1. Kaklamanis, C., D. Krizanc, 'Optimal Sorting on Mesh-Connected Processor Arrays,' *Proc. 4th Symp. on Parallel Algorithms and Architectures*, pp. 50–59, ACM, 1992.

2. Kaufmann, M., J.F. Sibeyn, 'Randomized k-k Sorting on Meshes and Tori,' manuscript, 1992.

3. Kaufmann, M., J.F. Sibeyn, T. Suel, 'Derandomizing Algorithms for Routing and Sorting on Meshes,' *Proc. 5th Symp. on Discrete Algorithms*, pp 669–679 ACM-SIAM, 1994.

4. Krizanc, D., L. Narayanan, 'Zero-One Sorting on the Mesh,' *Proc. 5th Symp. on Parallel and Distributed Processing*, IEEE, pp. 641–647, 1993.

5. Kunde, M., 'Block Gossiping on Grids and Tori: Deterministic Sorting and Routing Match the Bisection Bound,' *Proc. European Symp. on Algorithms*, LNCS 726, pp. 272–283, Springer-Verlag, 1993.

6. Newman, I., A. Schuster, 'Hot-Potato Algorithms for Permutation Routing,' *Proc. ISTCS*, June 1993.

7. Schnorr, C.P., A. Shamir, 'An Optimal Sorting Algorithm for Mesh Connected Computers,' *Proc. 18th Symp. on Theory of Comp.*, pp. 255–263, ACM, 1986.

8. Sibeyn, J.F., 'Desnakification of Mesh Sorting Algorithms,' *Techn. Rep. MPI-I-94-102*, revised and extended version, Max-Planck Institut für Informatik, Saarbrücken, Germany, 1994.

9. Sibeyn, J.F., B.S. Chlebus, M. Kaufmann, 'Permutation Routing on Meshes with Small Queues,' *Proc. 19th Mathematical Foundations of Computer Science*, LNCS, Springer Verlag, 1994, to appear.

Tight Bounds on Deterministic PRAM Emulations with Constant Redundancy[*]

Andrea Pietracaprina and Geppino Pucci

Dipartimento di Elettronica e Informatica,
Università di Padova,
Via Gradenigo 6/A, I35131 Padova, Italy

Abstract. In this paper we present lower and upper bounds for the deterministic emulation of a Parallel Random Access Machine (PRAM) with n processors and m variables on a Distributed Memory Machine (DMM) with n processors. The bounds are expressed as a function of the redundancy r of the scheme (i.e., the number of copies used to represent each PRAM variable in the DMM), and become tight for any m polynomial in n and $r = \Theta(1)$.

1 Introduction

An (n, m)-PRAM consists of n processors that have direct access to m shared *variables*. The important feature of this model, which makes it very attractive for the design of parallel algorithms, is that in a PRAM *step*, executed in unit time, any set of n variables can be read or written in parallel by the processors. For large values of m and n, however, this assumption represents a serious obstacle to any realistic implementation. In practice, one has to emulate the PRAM on a more feasible machine, where the shared variables are distributed among memory modules local to the processors. In such a machine, the modules become a bottleneck, since only one item per module can be accessed in unit time.

In this paper we study the complexity of emulating an (n, m)-PRAM on a n-DMM, which consists of n processors, each provided with a local memory module, communicating through a complete interconnection. In a DMM *step*, each processor can issue an access request for an arbitrary module, but only one request per module is served. Although this model is still unfeasible, in that it idealizes interprocessor communication, it provides a good tool to assess the difficulty of memory distribution in a topology-independent setting.

This paper deals with deterministic PRAM emulations. In order to avoid trivial worst-case scenarios, in which a few modules are overloaded with requests, it is necessary to replicate each variable into a number of *copies* stored in distinct modules, so that, during the emulation of a PRAM step, contention at the modules can be reduced by carefully choosing the copies that have to be accessed. The number of copies used for each variable is called the *redundancy*

---

[*] This research was supported in part by the ESPRIT Basic Research Project 9072 *GEPPCOM: Foundations of GEneral Purpose Parallel COMputing.*

of the emulation. We say that an (n, m)-PRAM can be emulated on an n-DMM with *slowdown* s, if any $T \geq 1$ PRAM steps can be emulated in $O(Ts)$ DMM steps, in the worst case.

1.1 Previous Work

In the last decade, a large number of randomized and deterministic PRAM emulation schemes have been developed in the literature. All randomized schemes are based on the use of universal classes of hash functions to allocate the variables to the modules. The distribution properties of these functions yield very efficient emulations in the probabilistic sense. For instance, using few copies per variable, doubly logarithmic slowdown can be achieved with high probability [5, 2].

In contrast, the development of fast deterministic PRAM emulations appears to be much harder. The pioneering work of Mehlhorn and Vishkin introduced the idea of representing each variable by several copies, so that a read operation needs to access only one (the most convenient) copy. For $m = O(n^r)$, they present a scheme that uses r copies per variable and allows a set of n reads to be satisfied in time $O(rn^{1-1/r})$. However, the execution of write operations, where all the copies of the variables must be accessed, is penalized and requires $O(rn)$ time in the worst case.

Later, Upfal and Wigderson [12] proposed a more balanced protocol requiring that, in order to read or write a variable, only a majority of its copies be accessed. They also represent the allocation of the copies to the modules (*Memory Organization Scheme* or, for short, MOS) by means of a bipartite graph $G = (V, U; E)$, where V is the set of PRAM variables, U is the set of DMM modules, and r edges connect each variable to the modules storing its r copies. For m polynomial in n and $r = \Theta(\log n)$, the authors show that there exist suitably expanding graphs that guarantee a worst-case $O\left(\log n \left(\log \log n\right)^2\right)$ slowdown. Using a more complex access strategy, [1] improved the bound to $O(\log n)$. In all these schemes, the existence of the underlying MOS graphs is only proved by counting arguments, but no efficient constructions are known. Several authors devised similar, nonconstructive schemes for bounded-degree network machines [3, 4, 7].

Recently, Pietracaprina and Preparata gave the first constructive deterministic schemes for the DMM that exhibit sublinear slowdown for both read and write operations. In [9], two PRAM emulations with $m = O(n^2)$ and $m = O(n^3)$ variables are given, respectively with $O(n^{1/2})$ and $O(n^{2/3})$ slowdown and redundancy $r = 3$ and $r = 5$. In [10], an $O(n^{1/3})$ slowdown is achieved with $m = O(n^{3/2})$ and $r = 3$. A constructive scheme with nearly optimal slowdown has been recently developed for the mesh topology in [11].

A lower bound for PRAM emulations on the DMM appears in [12]. By a slight modification of their argument, it can be shown that any emulation scheme with redundancy r and m polynomial in n requires

$$\Omega\left(\frac{\log(m/n)}{\log\log(m/n)} + \left(\frac{m}{n}\right)^{\frac{1}{r}}\right)$$

slowdown. Lower bounds have been also developed for networks of bounded degree [1, 6, 3]. However, the techniques used to prove such bounds are of a slightly different nature since communication issues have to be taken into account.

1.2 New Results

The goal of this paper is to study the complexity of deterministic PRAM emulations on the DMM as a function of the redundancy. We improve the result of [12], tightening the lower bound and showing that there exist suitable memory organization schemes that achieve optimal slowdown for any value of the redundancy fixed independently of n and m. In Sect. 2 we prove the following

Result 1. *Let $m > n$, $r \geq 1$ and $\mu = \lfloor r/2 \rfloor + 1$. Any algorithm which emulates an (n, m)-PRAM on an n-DMM with redundancy r, has worst-case slowdown at least*

$$\Omega \left(\min \left\{ \frac{\log (m/n)}{\log \log (m/n)}, \sqrt{n} \right\} + \min \left\{ \frac{n}{r}, \left(\frac{m}{n} \right)^{\frac{1}{\mu}} \right\} \right) \ .$$

Note that this bound improves quadratically upon the one in [12] for any $r = o(\log (m/n) / \log \log (m/n))$. Moreover, its proof reveals that the majority protocol captures the best trade-off between the complexity of read and write operations, as suggested by intuition. As a by-product, the bound also shows that the explicit schemes of [9] are optimal.

The lower bound can be always matched in the case of constant redundancy. In Sect. 3 we prove the following

Result 2. *For any $m = n^{1+\epsilon}$, with constant $\epsilon > 0$, and any constant $r \geq 2\epsilon$ there is a scheme to emulate an (n, m)-PRAM on an n-DMM with redundancy r and optimal slowdown*

$$\Theta \left(n^{\frac{\epsilon}{\mu}} \right) \ ,$$

where $\mu = \lfloor r/2 \rfloor + 1$.

2 Lower Bound

Our lower bound is proved by refining the techniques used in [12, 6, 3], and relies upon the same basic assumptions, namely:

1. For each variable, the number and locations of the copies do not vary with time. Moreover, without loss of generality, we assume that copies of the same variables are stored in distinct modules.
2. Emulations are *on-line*, that is, the emulation of a step starts only after the emulation of the previous step is completed.

We first need two technical lemmas based on a well known combinatorial argument.

Lemma 1. *Given $m \geq n$ PRAM variables, each with at most $\rho \geq 1$ updated copies distributed among n DMM modules, there exists a set of n variables requiring*

$$\Omega \left(\min \left\{ \frac{n}{\rho}, \left(\frac{m}{n} \right)^{\frac{1}{\rho}} \right\} \right)$$

DMM steps to be read.

Proof. It is sufficient to show that there exist n variables whose updated copies are all concentrated into at most $\max \left\{ 2\rho, n \left(\frac{n}{m} \right)^{\frac{1}{\rho}} \right\}$ modules. Let t be the maximum value for which no set of t modules stores all the updated copies of n variables. If $t < 2\rho$ we are done. Let $t \geq 2\rho$. Consider a matrix with $\binom{n}{t}$ rows indexed by all subsets of t modules, and m columns indexed by the variables. Entry (i, j) of this matrix is 1 if the j-th variable has all its updated copies stored in modules of the i-th subset, 0 otherwise. Each row accounts for at most $n - 1$ 1's. Each column accounts for at least $\binom{n - \rho}{t - \rho}$ 1's. Thus,

$$\binom{n}{t}(n - 1) \geq m \binom{n - \rho}{t - \rho} ,$$

which implies that $t = O \left(n \left(\frac{n}{m} \right)^{\frac{1}{\rho}} \right)$. $\qquad \square$

Lemma 2. *Consider $m \geq n$ PRAM variables, each represented by $r \geq 1$ copies distributed among n DMM modules, and let $\mu = \lfloor r/2 \rfloor + 1$. There exists a set of n variables Φ, and a set of modules S such that each variable in Φ has at least μ copies stored in modules of S, and*

$$|S| = O \left(\max \left\{ r, n \left(\frac{n}{m} \right)^{\frac{1}{\mu}} \right\} \right) .$$

Proof. Let t be the maximum value for which no set of t modules stores μ copies of each of n variables. If $t < r$ we are done. Let $t \geq r$. Consider a matrix with $\binom{n}{t}$ rows indexed by all subsets of t modules, and m columns indexed by the variables. Entry (i, j) of this matrix is 1 if the j-th variable has at least μ copies stored in modules of the i-th subset, 0 otherwise. Each row accounts for at most $n - 1$ 1's. Each column accounts for exactly $\sum_{y=\mu}^{r} \binom{r}{y} \binom{n - r}{t - y}$ 1's. Therefore, it must be

$$\binom{n}{t}(n - 1) \geq m \sum_{y=\mu}^{r} \binom{r}{y} \binom{n - r}{t - y} ,$$

which implies that $t = O \left(n \left(\frac{n}{m} \right)^{\frac{1}{\mu}} \right)$. $\qquad \square$

Theorem 3. *Let $m > n$, $r \geq 1$ and $\mu = \lfloor r/2 \rfloor + 1$. For any algorithm which emulates an (n, m)-PRAM on an n-DMM with redundancy r, there exists a sequence of $T = \Omega(m/n)$ PRAM steps requiring emulation time*

$$\Omega\left(T\left(\min\left\{\frac{\log(m/n)}{\log\log(m/n)}, \sqrt{n}\right\} + \min\left\{\frac{n}{r}, \left(\frac{m}{n}\right)^{\frac{1}{\mu}}\right\}\right)\right) .$$

Proof. We first determine a sequence of $\Theta(m/n)$ *hard writes* that update a constant fraction of all the variables. Either these writes are expensive to emulate, since they update many copies of most of the variables, or we are able to determine $\Theta(m/n)$ *hard reads* that access variables with few updated copies, therefore establishing the stipulated bound.

The hard writes are determined as follows. Let V be the set of m PRAM variables. By repeatedly applying Lemma 2, we can find $k = \lceil \frac{m}{2n} \rceil$ subsets $\Phi_i \subseteq V$, $1 \leq i \leq k$, of variables, with the following properties:

1. $|\Phi_i| = n$, for $1 \leq i \leq k$;
2. $\Phi_i \cap \Phi_j = \emptyset$ for $1 \leq i \neq j \leq k$;
3. For each Φ_i, with $1 \leq i \leq k$, there exists a set of modules S_i with

$$|S_i| = O\left(\max\left\{r, n\left(\frac{n}{m}\right)^{\frac{1}{\mu}}\right\}\right) ,$$

such that each variable in Φ_i has at least μ copies stored in modules of S_i.

Observe that each Φ_i is chosen from the set $V - \cup_{j=1}^{i-1}\Phi_j$ containing at least $m/2$ variables.

Consider k PRAM write steps, where the i-th step writes the variables in Φ_i. After the emulation of these steps, partition the set $\cup_{i=1}^{k}\Phi_i$ into three sets V_1, V_2 and V_3 as follows. A variable $v \in \cup_{i=1}^{k}\Phi_i$ is in V_1 if it has less than $(\mu - 1)/\lambda$ updated copies (the actual value of λ will be determined later); v is in V_2 if it has at least $(\mu - 1)/\lambda$ and at most $\mu - 1$ updated copies; v is in V_3 otherwise. Clearly, one of these three sets has cardinality $\Theta(m)$. We distinguish among the following three cases.

Case 1: $|V_1| = \Theta(m)$. By repeatedly applying Lemma 1 we determine $\Theta(m/n)$ PRAM steps that read a constant fraction of the variables in V_1 each requiring $\Omega\left(\min\left\{n\lambda/\mu, (m/n)^{\lambda/(\mu-1)}\right\}\right)$ emulation time. Thus, the overall time needed to emulate the $T = \Theta(m/n)$ write and read steps is

$$\Omega\left(T\min\left\{\frac{n\lambda}{\mu}, \left(\frac{m}{n}\right)^{\frac{\lambda}{\mu-1}}\right\}\right) . \tag{1}$$

Case 2: $|V_2| = \Theta(m)$. Since at least $\Omega(m\mu/\lambda)$ copies have been updated during the emulation of the write steps, the overall time required by these steps is $\Omega((m/n)\mu/\lambda)$. Moreover, by repeatedly applying Lemma 1, we can determine $\Theta(m/n)$ PRAM steps that read a constant fraction of variables in V_2, each step requiring emulation time $\Omega\left(\min\left\{n/\mu, (m/n)^{1/(\mu-1)}\right\}\right)$. Thus, the overall time needed to emulate the $T = \Theta(m/n)$ write and read steps is

$$\Omega\left(T\left(\frac{\mu}{\lambda} + \min\left\{\frac{n}{\mu}, \left(\frac{m}{n}\right)^{\frac{1}{\mu-1}}\right\}\right)\right) . \tag{2}$$

Case 3: $|V_3| = \Theta(m)$. At least $\Omega(m\mu)$ copies are updated during the emulation of the write steps, hence, the overall time required by these steps is $\Omega((m/n)\mu)$. Moreover, there must be $\ell = \Theta(m/n)$ sets $\Phi_{i_1}, \Phi_{i_2}, \ldots, \Phi_{i_\ell}$, with $|\Phi_{i_j} \cap V_3| = \Theta(n)$, $1 \le j \le \ell$. Since each variable in $\Phi_{i_j} \cap V_3$ has μ updated copies, then one of these copies must belong to a module in S_{i_j}. Therefore, the i_j-th write step requires

$$\Omega\left(\frac{n}{|S_{i_j}|}\right) = \Omega\left(\min\left\{\frac{n}{\mu}, \left(\frac{m}{n}\right)^{\frac{1}{\mu}}\right\}\right)$$

emulation time, for $1 \le j \le \ell$. Therefore, the overall time needed to emulate the $T = \Theta(m/n)$ write steps is

$$\Omega\left(T\left(\mu + \min\left\{\frac{n}{\mu}, \left(\frac{m}{n}\right)^{\frac{1}{\mu}}\right\}\right)\right) . \tag{3}$$

The theorem follows by choosing

$$\lambda = \max\left\{1, (\mu-1)\,\alpha\,\frac{\log\log(m/n)}{\log(m/n)}, \frac{\mu-1}{\sqrt{n}}\right\} ,$$

for some fixed constant $\alpha > 1$, and computing the minimum among (1), (2) and (3). $\square$

3   Upper Bound

Suppose we want to emulate an (n, m)-PRAM on an n-DMM. Consider a PRAM step where the processors read/write n distinct variables, and assume that each variable is replicated into r copies. Our emulation schemes are based on the simple access protocol of [12]: namely, we prove that there exist suitable ways of distributing the copies of the variables among the modules of the DMM so that the protocol is efficient. In particular, its running time becomes optimal for constant redundancy.

Let us briefly recall how the access protocol in [12] works. Every copy is stored together with a time-stamp, which is incremented every time the copy is updated. In order to read or write a variable, at least a majority $\mu = \lfloor r/2 \rfloor + 1$ of its copies have to be accessed (for read operations, the value stored in the copy with the most recent time-stamp is returned). Note that this is sufficient to guarantee consistency. Accesses to the n variables are completed in two *stages*. Stage 1 satisfies all the read/write requests, but for at most n/r variables; Stage 2 satisfies the remaining requests. The processors of the DMM are partitioned into n/r *clusters* of r processors each. At any single step of the protocol, a cluster is in charge of a single variable v, with each processor in the cluster issuing a request for a distinct copy of v.

The first stage is divided into r *phases*. In each phase, the n/r clusters are in charge of a distinct set of n/r variables. A phase consists of a number of iterations, where the processors continuously try to access their assigned copy unless they previously succeeded or other μ copies of the same variable have already been accessed. A phase terminates when at most n/r^2 variables remain to be accessed. The second stage starts by distributing the $O(n/r)$ leftover variables among the n/r clusters of the DMM. A number of iterations is then executed by the processors until all the requests have been satisfied.

Let Φ_1 be the maximum number of iterations executed in any phase of Stage 1, and Φ_2 be the number of iterations needed in Stage 2. Each iteration requires $O(\log r)$ time (due to bookkeeping operations inside the clusters), and variable redistribution after Stage 1 can be accomplished in $O(\log n)$ time. Thus, the overall time complexity of the protocol is

$$T = O\left(\Phi_1 r \log r + \log n + \Phi_2 \log r\right) .$$

The values of Φ_1 and Φ_2 depend on the properties of the underlying Memory Organization Scheme. Recall that the MOS is modeled by a bipartite graph $G = (V, U; E)$ of degree r, where V is the set of PRAM variables, U the set of DMM processors, and the r edges connect a variable v to the processors storing its copies. We derive bounds on Φ_1, Φ_2 in terms of the expansion capabilities of G.

Definition 4. Let $G = (V, U; E)$ be an r-regular, bipartite graph, and let $\mu = \lfloor r/2 \rfloor + 1$. For $\alpha > 0$ and $0 < \delta < 1$, G is said to have (α, δ)-expansion if for any subset $S \subset V$, $|S| \leq |U|/r$, and any choice of μ outgoing edges for each node in S, the set $\Gamma^\mu(S) \subseteq U$ reached by the chosen edges has size

$$|\Gamma^\mu(S)| > \alpha r |S|^{1-\delta} .$$

We have:

Theorem 5. *If the MOS graph G has (α, δ)-expansion, then*

$$\Phi_1 = O\left(\frac{1}{\alpha}\left(\frac{n}{r}\right)^\delta \log r\right) ;$$

$$\Phi_2 = O\left(\frac{1}{\alpha}\left(\frac{n}{r}\right)^\delta \frac{1}{\delta}\right) .$$

Therefore the running time of the access protocol is altogether

$$T = O\left(\log n + \frac{1}{\alpha}\left(\frac{n}{r}\right)^\delta \left(r \log^2 r + \frac{\log r}{\delta}\right)\right) .$$

Proof. During the execution of the access protocol, we say that a copy is *alive* if it has not yet been accessed. Also, a variable is *alive* if at least a majority μ of its copies is alive. Consider now a single phase in Stage 1, and let R_k be the number of copies (of the n/r variables treated in this phase) still alive after k iterations. Since the number of live variables at the beginning of the k-th iteration is at least R_{k-1}/r, the expansion property of G yields the following recurrence:

$$R_k \leq R_{k-1} \left(1 - \frac{c}{R_{k-1}^\delta} \right) , \tag{4}$$

where $c = \alpha r^\delta$. Letting $R_0 = n$, (4) implies that

$$R_k \leq n \left(1 - \frac{c}{n^\delta} \right)^k .$$

For

$$k = \left\lceil \frac{n^\delta}{c} \ln(2r) \right\rceil = O\left(\frac{1}{\alpha} \left(\frac{n}{r} \right)^\delta \log r \right) ,$$

we get $R_k \leq n/(2r)$, which are the live copies of at most $n/(2r\mu) \leq n/r^2$ live variables.

The analysis of Stage 2 requires a more careful argument. Let $\Phi_2 = k_0 + k_1 + \ldots + k_{J-1}$, where k_i iterations are used to reduce the number of live copies from R_i to $R_{i+1} \leq R_i/e$, with $R_0 = n$. Here, R_i denotes the number of copies still alive after $\sum_{j=0}^{i-1} k_j$ iterations. As for (4), the expansion property of G yields

$$R_{i+1} \leq R_i \left(1 - \frac{c}{R_i^\delta} \right)^{k_i} ,$$

and therefore $k_i \leq R_i^\delta/c$. Since $J \leq \lceil \ln(n/r) \rceil$, we have

$$\Phi_2 = \sum_{i=0}^{J-1} k_i \leq \frac{1}{\alpha} \left(\frac{n}{r} \right)^\delta \sum_{i=0}^{J-1} e^{-i\delta} = O\left(\frac{1}{\alpha} \left(\frac{n}{r} \right)^\delta \frac{1}{\delta} \right) .$$

$\square$

We now show that, for m polynomial in n, using constant redundancy, it is always possible to emulate an (n,m)-PRAM on an n-DMM with optimal slowdown.

Theorem 6. *Let $m = n^{1+\epsilon}$, with constant $\epsilon > 0$. For any constant $r \geq 2\epsilon$, there exists an r-regular bipartite graph $G = (V, U; E)$ with $|V| = m$, $|U| = n$ which has (α, δ)-expansion with $\alpha = \Theta(1)$ and $\delta = \epsilon/\mu$, where $\mu = \lfloor r/2 \rfloor + 1$.*

Proof (sketch). A graph $G = (V, U; E)$ does not have (α, δ)-expansion if there is a set $S \subset V$ of at most n/r nodes such that some choice of μ outgoing edges for each node in S yields $|\Gamma^\mu(S)| \leq \alpha r|S|^{1-\delta}$. As in [12], we use the probabilistic method showing that there is a suitable constant $\alpha > 0$ such that the fraction of

r-regular bipartite graphs G that do not have $(\alpha, \epsilon/\mu)$-expansion is very small. This fraction is at most

$$\sum_{s=1}^{n/r} \binom{m}{s} \binom{n}{\alpha r s^{1-\delta}} \binom{r}{\mu}^s \left(\frac{\alpha r s^{1-\delta}}{n}\right)^{\mu s} .$$

Tedious but standard manipulations show that the above summation goes to 0 as n grows, for suitable values of the constant α. □

Corollary 7. *For any $m = n^{1+\epsilon}$, with constant $\epsilon > 0$, and any constant $r \geq 2\epsilon$ there is a scheme to emulate an (n, m)-PRAM on an n-DMM with redundancy r and optimal slowdown*

$$\Theta\left(n^{\frac{\epsilon}{\mu}}\right) ,$$

where $\mu = \lfloor r/2 \rfloor + 1$.

4 Conclusions

The efficient emulation of shared memory in distributed systems is a central problem for parallel computation. Our paper improves the understanding of the deterministic complexity of this problem, showing that optimal emulations are achievable using constant replication of the data.

Our lower bound reveals that for a shared memory of size $m = n^{1+\epsilon}$, a value of the redundancy greater than $\lceil 2(\epsilon - 1) \rceil$ is necessary to achieve sublinear slowdown. This immediately demonstrates that the deterministic scheme of [8] could not possibly achieve better than linear time for the write operations.

As for the upper bound, it must be remarked that the proof of Theorem 6 does not provide an explicit construction for the MOS graph, a limitation shared by most emulation schemes in the literature. For specific values of memory size and redundancy, explicit memory organizations have been proposed in [9, 10]. It remains a challenging open problem to extend these results to an entire range of values of the parameters.

References

1. H. Alt, T. Hagerup, K. Mehlhorn, and F.P. Preparata. Deterministic simulation of idealized parallel computers on more realistic ones. *SIAM Journal on Computing*, 16(5):808–835, 1987.
2. M. Dietzfelbinger and F. Meyer auf der Heide. Simple, efficient shared memory simulations. *Proceedings of the 5th ACM Symposium on Parallel Algorithms and Architectures*, pages 110–118, 1993.
3. K.T. Herley and G. Bilardi. Deterministic simulations of PRAMs on bounded degree networks. *SIAM Journal on Computing*, 1994. To appear. (See also *Proceedings of the 26th Annual Allerton Conference on Communication, Control and Computation*, pages 1084–1093, 1988.)

4. K.T. Herley. Efficient simulations of small shared memories on bounded degree networks. *Proceedings of the 30th IEEE Symposium on Foundations of Computer Science*, pages 390–395, 1989.
5. R. Karp, M. Luby, and F. Meyer auf der Heide. Efficient PRAM simulation on distributed machines. *Proceedings of the 24th ACM Symposium on Theory of Computing*, pages 318–326, 1992.
6. A.R. Karlin and E. Upfal. Parallel hashing: An efficient implementation of shared memory. *Journal of the ACM*, 35(4):876–892, 1988.
7. F. Luccio, A. Pietracaprina, and G. Pucci. A new scheme for the deterministic simulation of PRAMs in VLSI. *Algorithmica*, 5:529–544, 1990.
8. K. Mehlhorn and U. Vishkin. Randomized and deterministic simulations of PRAMs by parallel machines with restricted granularity of parallel memories. *Acta Informatica*, 9(1):29–59, 1984.
9. A. Pietracaprina and F.P. Preparata. An $O(\sqrt{n})$-worst-case-time solution to the granularity problem. *Proceedings of the 10th Symposium on Theoretical Aspects of Computer Science*, LNCS 665:110–119, 1993.
10. A. Pietracaprina and F.P. Preparata. A practical constructive scheme for deterministic shared-memory access. *Proceedings of the 5th ACM Symposium on Parallel Algorithms and Architectures*, pages 100–109, 1993.
11. A. Pietracaprina, G. Pucci, and J.F. Sibeyn. Constructive deterministic PRAM simulation on a mesh-connected computer. *Proceedings of the 6th ACM Symposium on Parallel Algorithms and Architectures*, 1994. To appear.
12. E. Upfal and A. Widgerson. How to share memory in a distributed system. *Journal of the ACM*, 34(1):116–127, 1987.

PRAM Computations
Resilient to Memory Faults*

B.S. Chlebus   A. Gambin   P. Indyk

Instytut Informatyki, Uniwersytet Warszawski, Banacha 2, 02-097 Warszawa, Poland.
E-mail: chlebus@mimuw.edu.pl, aniag@zaa.mimuw.edu.pl, indyk@mimuw.edu.pl

Abstract: PRAMs with faults in their shared memory are investigated. Efficient general simulations on such machines of algorithms designed for fully reliable PRAMs are developed.

The PRAM we work with is the Concurrent-Read Concurrent-Write (CRCW) variant. Two possible settings for error occurrence are considered: the errors may be either static (once a memory cell is checked to be operational it remains so during the computation) or dynamic (a potentially faulty cell may crash at any time, the total number of such cells being bounded). A simulation consists of two phases: memory formatting and the proper part done in a step-by-step way. For each error setting (static or dynamic), two simulations are presented: one with a $O(1)$-time per-step cost, the other with a $O(\log n)$-time per-step cost. The other parameters of these simulations (number of processors, memory size, formatting time) are shown in table 1 in section 6. The simulations are randomized and Monte Carlo: they always operate within the given time bounds, and are guaranteed to be correct with a large probability.

1   Introduction

Parallel Random Access Machine (PRAM) is a popular model to design parallel algorithms (see [10, 12]). It is a multiprocessor system in which every processor acts like a RAM, and all of them share the global memory. PRAM abstracts from real multiprocessor computers by disregarding the mechanism and cost of communication between the processors and the external memory. This facilitates the design and analysis of parallel algorithms.

The standard *ideal* (CRCW) PRAM has the following properties:

1. The processors are tightly synchronized, with no explicit cost of synchronization;

2. Every processor is always operational;

3. Every memory cell can be accessed by any processor;

4. Every memory cell can be accessed in one step;

5. Every memory cell can be read from or written to with no errors occurring.

Recently there has been a lot of research done concerning PRAM variants obtained by dropping or relaxing some of these properties. Asynchronous PRAMs were investigated in [6, 9, 17, 20]. PRAMs with faulty processors were studied by Kanellakis and Shvartsman [15] and Kedem *et al.* [16]. PRAMs with a differentiated cost of access to memory were considered by Aggarwal *et al.* [2] and Gibbons [9]. Valiant [24] considered the XPRAM model, where

---

* This research was partially supported by EC Cooperative Action IC-1000 (project ALTEC: *Algorithms for Future Technologies*).

processors have a direct access only to their local memory, and access other cells by passing messages to the respective processors. There is also a closely related model of *distributed memory machine*, see a survey paper by Meyer auf der Heide [8].

Issues of distributed computing with faulty shared memory have been investigated by Afek *et al* [1] and Jayanti *et al* [13]. The problem of exploring the use of randomization to tolerate memory failures in synchronous models was posed by Afek *et al.* [1], and in asynchronous models with large granularity by Aumann *et al.* [3].

In this paper we consider a *faulty-memory* PRAM. Except for possible memory faults, this model has all the remaining properties of the ideal PRAM: it is fully synchronized, each shared memory cell can be accessed in one step by any processor, and the processors are always operational. A read instruction places the read value in a designated register of the processor, and similarly, a write instruction places the value stored in a register into the accessed memory word, unless it is faulty. There is a mechanism to react to errors in read and write operations, as follows. If the accessed memory word is faulty, then some designated register shows a special "faulty" value, and the respective processor knows that the attempted instruction failed due to a memory error.

We present simulation algorithms, which emulate the ideal PRAM on the faulty-memory machine. They are Monte-Carlo algorithms, that is, they are randomized and correct with a high probability. More precisely, they always operate within the stated time bounds, but may produce incorrect results with a small probability.

The rest of the paper is organized as follows. In Section 2, we introduce notations and concepts, and discuss the models of faults occurrence. The operations of broadcasting and spreading are described in Section 3. Static faults are handled by the algorithms A and B presented in Section 4, and dynamic faults by the algorithms C and D of Section 5. Conclusions and further research are discussed in Section 6.

Proofs of the theorems will be described in the final version.

2 Preliminaries

The PRAM model considered in this paper is the Concurrent-Read Concurrent-Write (CRCW) one. There are two specific variants used. In Collision, if many (more than one) processors attempt to write to a memory cell then a special collision symbol gets written. In Collision$^+$, if many processors attempt to write to a memory cell, then there are two cases: if all the values of the processors are equal then this common value gets written, otherewise the collision symbol is written to the cell. The algorithms of section 3 and 4 are designed for Collision, and of section 4 for Collision$^+$. See [5] for more on the relative power of these variants of the CRCW PRAM.

We use the following notations. The simulated ideal PRAM is denoted by C_I, and the simulating faulty-memory PRAM by C_F. Two main parameters of a simulation algorithm are the size of memory and the number of processors. The machine C_I has n processors: $p_1, p_2, \ldots, p_n$, and n memory cells: $s_1, \ldots, s_n$. The machine C_F has N processors: $P_1, P_2, \ldots, P_N$, and M memory cells: $S_0, \ldots, S_{M-1}$. The number M is always assumed to be greater than n. A PRAM cell stores $O(\log n)$ bits. A processor of a PRAM has its own local memory, its cells are referred to as *registers*. The machine C_I has $O(1)$ registers per processor. All the registers of processors of C_F are assumed to be always operational and fully reliable. A processor of C_F has also $O(1)$ registers. Memory words of C_F are sometimes *marked*. This means setting some bits of them to specific values, the remaining bits to be used to simulate memory words of C_I.

There are two criteria by which (shared) memory errors are categorized.

Static versus Dynamic: if certain memory cells have become faulty before a computation starts, and no new faulty cells occur during the course of a computation, then the errors are *static*, otherwise the faults are called *dynamic*.

Deterministic versus Probabilistic: In the probabilistic case, each memory cell can be faulty with some fixed probability q, and any two cells are faulty independently of each other. In the deterministic version, at most $q \cdot M$ memory cells of C_F are faulty, for a constant parameter q, where $0 < q < 1$.

These two classifications are independent of each other, and this creates four settings for errors: deterministic static, deterministic dynamic, probabilistic static, and probabilistic dynamic. These combinations require further explanations. Conceptually, a setting can be imagined to be created in two steps. First, it is decided which cells are (potentially) faulty: in the deterministic case by selecting a subset of $q \cdot M$ elements from among the memory cells of C_F, and in the probabilistic setting, by deciding randomly and independently for each cell, whether it can be faulty. In the static case, all the selected cells become faulty, and in the dynamic case, it is assumed that there is an adversary who selects the step of simulation at which a given cell becomes faulty.

All the algorithms developed in this paper use randomization, and they can handle deterministic errors. They are automatically good against probabilistic memory failures, and hence probabilistic errors are not discussed as a special case. In what follows, it is always assumed that the faults are *deterministic*.

Similar models of error occurrence have been considered in the literature in the case of faulty processors. Probabilistic and dynamic faults of processors were studied by Kedem *et al.* [16] and Diks and Pelc [7]. Martel *et al.* [17] designed randomized simulations for deterministic processor errors.

For every presented simulation, the simulating machine C_F has to have the size of its memory within certain bounds for the simulation to run in the specified time with a large probability. To express this concisely, the notation $g(n) = \Theta^*(f(n))$ is used. It means that the inequalities $c_1 \cdot f(n) \leq g(n) \leq c_2 \cdot f(n)$ hold, for a *sufficiently large* constant c_1 and $c_2 > c_1$. The criterion for being "sufficiently large" depends on the context. For instance, if C_F is said to have the memory of size $M = \Theta^*(f(n))$, then "sufficiently large c_1" may be a function of both the constant q, determining the number of faulty memory cells of C_F, and the required bound $1 - n^{-\alpha}$ on the probability.

Simulations are divided into two phases: formatting and the simulation proper. During formatting the memory is explored and organized, typically to construct a mechanism to access (some of) the operational memory cells. Then follows the proper part of simulation, performed in a step-by-step fashion. A step of C_I consists of three parts: reading a memory cell, internal computation, writing to the same memory cell, and this is mimicked by C_F. The time of formatting is denoted by T_f, and the time of a one-step simulation by T_s, which is also referred to as the *step overhead*. The simulation algorithms are Monte Carlo and no synchronization between consecutive steps of C_I is needed. So algorithms operating in time t on C_I can be simulated on C_F to run in time $T_f + T_s \cdot t$, and be correct with a high probability.

A property $\Phi(n)$ is said to hold *with a high probability* (abbreviated to whp) if, for some $\alpha > 0$, $\Phi(n)$ holds with the probability at least $1 - n^{-\alpha}$, for a sufficiently large n. Throughout the paper, whenever this phrase is used, or the expression $1 - n^{-\alpha}$ is used explicitly as a bound on the probability, the value of α can be made arbitrarily large by manipulating other constants, for instance those involved either in the bounds on the number of memory cells, or the formatting time, or the step overhead.

3 Broadcasting and Spreading

In this section we consider two tasks that will be often performed during simulations. *Broadcasting* propagates a value known by one processor to the other processors. *Spreading* distributes $\log n$ values, known by $\log n$ processors, through the memory such that every value occupies $\Omega(M/\log n)$ cells.

ALGORITHM BROADCAST-1

Each processor repeats the following two steps $c_1 \cdot \log n$ times, for a constant c_1.

1. If the value v is not known then select randomly a memory cell S and read S. If S contains v then learn v.

2. If v is known then select randomly a memory cell S and write v to S.

Theorem 1. *If $N = \Theta(n)$ and $M = \Theta(n)$ then algorithm* BROADCAST-1 *operates in time $O(\log n)$ to propagate the value v among all the processors with the probability $1 - n^{-\alpha}$, for $\alpha > 0$.* □

The broadcasting algorithm may be also used when $M = O(n)$ does not hold. In this case each processor performs the body of algorithm BROADCAST-1, that is, its two steps, $O\left(\left(\sqrt{M/N} + 1\right) \cdot \log N\right)$ times. This algorithm is called BROADCAST-2.

Theorem 2. *Algorithm* BROADCAST-2 *propagates successfully the given value among all the processors, with the probability $1 - n^{-\alpha}$, for $\alpha > 0$.* □

Suppose that processor P_i knows value v_i, for $1 \leq i \leq \log n$, where $v_i \neq v_j$ for $i \neq j$. Then spreading is accomplished by the following algorithm:

ALGORITHM SPREAD

1. For each $1 \leq i \leq \log n$, processor P_i writes v_i into $c_2 \log n$ randomly chosen memory cells, for a constant c_2.

2. For each $1 \leq i \leq N$, processor P_i repeats $c_3 \log n$ times, for a constant c_3:

   2.1 If no value is known then select randomly a memory cell S and read it; if S contains v_j then learn v_j; otherwise

   2.2 If some value v_j is known then select randomly a memory cell S and write v_j to S.

Theorem 3. *If $N = \Theta(n)$ and $M = \Theta(n)$ then, after $O(\log n)$ steps of algorithm* SPREAD, *every v_i occurs in $\Omega(n/\log n)$ operational cells with the probability $1 - n^{-\alpha}$, for $\alpha > 0$.* □

4 Static Faults

In this section we consider static faults: once a memory cell is checked to be non-faulty, it is guaranteed to remain such through the whole computation. The first algorithm is used later as a subroutine. It builds a binary tree over n cells. Once this is done, the address of the root is made known to every processor, and accessing the ith cell is performed by traversing the tree to the ith leaf.

ALGORITHM T

1. Each processor P_i selects randomly an operational memory cell x_i: this is done by repeatedly reading random memory cells, until such an operational one is found that is not claimed by other processors. Then P_i marks x_i.

2. Processor P_1 broadcasts the address of x_1 to all the remaining processors.

3. All the processors build a binary tree: First, cell x_1 is used to store the addresses of x_2 and x_3. Then, iteratively, the cell x_i is used to store the addresses of x_{2i} and x_{2i+1}.

Lemma 4. ALGORITHM T *can be implemented to run successfully in time* $O(\log n)$ *with the probability* $1 - n^{-\alpha}$, *for* $\alpha > 0$. □

4.1 Algorithm A

Suppose that the simulating machine C_F has $N = n \log n$ processors and $M = \Theta^*(n)$ memory cells. The presented simulation has $O(1)$-time step overhead with a high probability.

Divide the processors of C_F into n groups: $\mathcal{P}_1, \mathcal{P}_2, \ldots, \mathcal{P}_n$, each consisting of $\log n$ processors. Group $\mathcal{P}_i$ is to simulate the ith processor p_i of C_I. Let U_l denote the memory cell of C_F simulating the lth memory cell s_l of C_I. To locate U_l, the processors in a group compute addresses $D_1(l), D_2(l), \ldots$, where the *address function* D_j is defined as $D_j(x) = x + d_j \pmod{M}$, for the numbers d_j being random elements from the interval $[0, M - 1]$. Only one of these addresses is designated as U_l, even if many are of operational memory cells. To visualize the underlying idea, consider the bipartite graph with edges (x, y), where x is the address of a memory cell in C_I and y is of the form $D_j(x)$, and the memory cell with address $D_j(x)$ in C_F is operational. Then a perfect matching in this graph gives a viable addressing scheme on C_F to simulate C_I. For every group $\mathcal{P}_i$ there is a special memory cell used for inter-processor communication, denoted by c_i. There are $d \log n$ address functions, where the number d is a parameter. The following algorithm initializes the memory such that among the memory words $D_j(l)$, for a fixed $1 \leq j \leq d \log n$, exactly one (operational) is marked as U_l.

ALGORITHM A (FORMATTING)

1. Processor P_1 generates $d \log n$ random numbers $d_1, d_2, \ldots, d_{d \log n}$ and places them in the shared memory organized as a list. This is performed by picking memory cells at random to place consecutive elements of the list.

2. The address of the header of the list is broadcast to all the processors by the algorithm BROADCAST-1.

3. The processors scan the list, and each processor, which is kth in his group, remembers the kth block of d values in the list.

4. A binary tree with n leaves is built by executing algorithm T. The ith leaf is the communication cell c_i for group $\mathcal{P}_i$.

5. For each $0 \leq k \leq m$, the kth cell U_k is selected from among the addresses $D_1(k), D_2(k), \ldots, D_{d \log n}(k)$, and marked as such. This is done as follows. First the addresses $D_1(1), D_1(2), \ldots, D_1(n)$ are examined, then $D_2(1), D_2(2), \ldots, D_2(n)$, and so on. For each l, the first $D_j(l)$, which is operational and not reserved already gets marked as U_l and then written to c_l to notify the remaining processors in $\mathcal{P}_l$.

In the step simulation, the task is to have the processors in some group $\mathcal{P}_i$ access a memory cell U_l.

ALGORITHM A (STEP SIMULATION)

1. Each processor in a group attempts to read the cell with address $D_j(l)$, for each j that it learned in Step 3 of formatting.

2. The processor that accessed the cell designated as U_l writes its address to c_i.

3. The remaining processors in $\mathcal{P}_i$ read c_i and then access U_l.

Theorem 5. *Algorithm A can be implemented in such a way that the time of formatting is $T_f = O(\log n)$ with the probability $1 - n^{-\alpha}$, and, once the formatting is successful, the step overhead is $T_s = O(1)$.* ☐

4.2 Algorithm B

Let the simulating PRAM C_F have $N = n/\log n$ processors and $M = \Theta^*(n)$ shared-memory cells. The algorithm B has $O(\log n)$-time step overhead, which is optimal for the available number of processors.

Suppose that $M = c_M \cdot n$, for a constant $c_M > 0$. Divide $c_M \cdot n$ memory cells into contiguous blocks of $\beta \cdot k$ cells, where $k = \log n + \sqrt{\log n} + 1$ and $\beta > 1$. A block containing at least k operational cells is said to be *good*. The following inequality estimates the number l of such blocks:

$$l \geq c_M \cdot n \frac{k(\beta(1-q)-1)+1}{\beta \cdot k(k(\beta-1)+1)} \geq c_M \cdot n \frac{\beta(1-q)-1}{\beta(k(\beta-1)+1)}$$

Choose c_M and β in such a way that the inequality $l \geq n/\log n$ holds. The first $n/\log n$ good blocks $B_0, B_1, \ldots$, are used to simulate the memory of C_I. A block B_i consists of the *root* $r(i)$, $\sqrt{\log n}$ *auxiliary cells*, and $\log n$ *normal cells*. Auxiliary cells are interspersed among normal cells and partition them into groups of $\sqrt{\log n}$ elements. The jth normal cell of B_i simulates the cell $i \cdot \log n + j$ of C_I. The bits of the root of a block are divided into $\sqrt{\log n}$ fields $F_1, F_2, \ldots$, a field F_j stores the offset of the jth auxiliary cell in the block. Similarly, an auxiliary cell has its bits divided into $\sqrt{\log n}$ fields to store offsets of the next $\sqrt{\log n}$ normal cells. The *offset* of an auxiliary cell x is the number of cells between the beginning of the block and x; and for a normal cell y, the *offset* of y is the number of cells between y and the auxiliary memory cell closest to the left. Notice that an offset is a number of size $O(\log n)$, hence it requires $\log \log n + O(1)$ bits to be stored. Since a memory word is assumed to be able to store $O(\log n)$ bits, all the offsets fit into the roots and auxiliary cells. Given the address of a root $r(i)$, the jth normal cell in B_i can be accessed in time $O(1)$ by first locating the respective auxiliary cell, and then the normal cell. To this end, the fields of root and the respective auxiliary cell are extracted by applying standard arithmetic and boolean bit operations. We show next how to make the roots accessible in time $O(1)$ with a high probability. The address functions $D_1, D_2, \ldots$ are defined similarly as in algorithm A, the number d is a parameter.

ALGORITHM B (FORMATTING)

1. The blocks are divided into $N = n/\log n$ groups. Processor P_i counts the number of good blocks in the ith group.

2. A binary tree of $M/\beta \log n$ leaves is built by executing algorithm T. The tree is used next to assign consecutive numbers to good blocks as in the parallel prefix algorithm.

3. Processor P_i sets the root and the auxiliary cells in block B_i to their proper values.

4. Processor P_1 generates a list of $d \log n$ random numbers $d_1, \ldots, d_{d \log n}$ and organizes them as a list, similarly as in algorithm A.

5. Processor P_i evaluates the address functions $D_1(i), \ldots, D_{d \log n}(i)$. For every l, if the cell $D_l(i)$ is operational and unmarked, then P_i marks it and sets to store a pointer to the root $r(i)$.

6. Each processor P_i creates a list of $\log n$ memory cells to store the contents of registers of the processors that it is to simulate.

ALGORITHM B (STEP SIMULATION)

Each processor P_l simulates $\log n$ processors of C_I. The contents of registers are stored in the list that P_l built in the last step of formatting. P_l scans the list and retrieves the addresses $x_1, x_2, \ldots$ of memory cells that need to be accessed. To locate x_1, processor P_l tries addresses $D_1(i_1), D_2(i_1), \ldots, D_{t_1}(i_1)$, where i_1 is the number of block containing x_1. The search terminates when a memory cell with a pointer to the root is found. Then P_l is able to access x_1 in time $O(1)$. Processor P_l continues with $D_{t_1+1}(i_2), D_{t_1+2}(i_2), \ldots, D_{t_2}(i_2)$, terminating when a pointer to the second root is found. The remaining addresses are processed similarly, each time starting with the first unused yet address function.

Theorem 6. *Algorithm B can be implemented in such a way that the time of formatting is equal to $T_f = O(\log^{3/2} n)$ with the probability $1 - n^{-\alpha}$, and the step overhead is $T_s = O(\log n)$, with the probability $1 - n^{-\alpha}$ for each step, for $\alpha > 0$.* □

5 Dynamic Faults

Dynamic faults are more challenging. In such a setting some form of duplication or dispersal of information stored in the memory of C_F is inevitable, since a memory cell storing useful information may turn out to be faulty at any time. To simplify the presentation and analysis, we assume that a memory word of C_F may store the contents of several words of C_I.

5.1 Algorithm C

Let the simulating PRAM have $N = n \log^2 n$ processors and $M = \Theta^*(n \log n)$ memory cells. We present a simulation that has a $O(1)$-time step overhead.

The number d is a parameter. To simplify the notation, assume rather that there are exactly $N = d^2 \cdot n \cdot \log^2 n$ processors of C_F. Divide these processors into n groups $\mathcal{P}_1, \ldots, \mathcal{P}_n$ of $d^2 \cdot \log^2 n$ processors. Divide each group $\mathcal{P}_i$ into subgroups $\mathcal{P}_{i,j}$ of $d \cdot \log n$ elements. The task of $\mathcal{P}_i$ is to simulate the processor p_i of C_I. It is assumed that, when the simulation starts, all the processors from group $\mathcal{P}_i$ know the initial state of processor p_i. In the course of the simulation, typically only a fraction of processors in every group $\mathcal{P}_i$ know the current state of p_i, these processors are called *informed*. Each informed processor knows that it is informed, otherwise it is aware that it is not. After the computation of C_I terminates, it may still take some time for all the processors of C_F to get informed. This time period is called the *termination delay*.

ALGORITHM C (FORMATTING)

1. The processors generate $d \cdot \log n$ random numbers d_i in $[0, M - 1]$.

2. The numbers d_i are distributed through the shared memory by executing the algorithm SPREAD on $O(n \log n)$ processors.

3. Each processor P_k makes $O(\log n)$ attempts to learn some of the numbers d_i by reading memory cells selected at random. It stores the first $r = O(1)$ of them, denote the respective address functions by $D_1^k, \ldots, D_r^k$. The number r is called a *fan*, and is a parameter of the algorithm. If the processor is in $\mathcal{P}_{i,j}$ and it happens to get to know d_j, then it stores d_j.

4. Steps 1 through 3 are repeated, for a new set of $d \cdot \log n$ numbers denoted g_j. Processor P_k learns the functions $G_1^k, \ldots, G_r^k$.

In Step 2, notice that whp there are only $O(1)$ pairs $i \neq j$ such that $d_i = d_j$, so Theorem 3 still holds. The generated random numbers define address functions $D_i(x) = x + d_i \pmod{M}$, and $G_i(x) = x + g_i \pmod{M}$. A memory cell $D_j(x)$ ($G_j(x)$, resp.) which is operational and such that its number is the value of only one of the functions D or G for exactly one argument is said to be *D-useful* (*G-useful*, resp.). D-useful cells simulate cells of C_I, G-useful cells are used by groups of processors for communication. We can mark each written value with the number of the simulated cell, in the case of D-usefulness, or of the processor group, in the case of G-usefulness. Then a cell can be verified as not being D-useful or G-useful if it is either not operational or contains the collision symbol or does not contain the correct number.

Suppose that p_i needs to access s_l in the current step.

ALGORITHM C (STEP SIMULATION)

1. (Read) If a processor P_k in $\mathcal{P}_i$ is informed, that is, P_k knows l, then it attempts to read the memory cells $D_1^k(l), \ldots, D_r^k(l)$.

2. If a processor P_k in $\mathcal{P}_{i,j}$ succeeded in Step 1 (that is, read at least one D-useful cell), and P_k knows g_j, then P_k writes the contents of its registers to $G_j(i)$.

3. If a processor P_k in $\mathcal{P}_{i,j}$ failed to read from an operational memory cell in Step 1 (during this iteration, or in one of the previous iterations and has not succeeded in Step 3 since then) then it attempts to read (at least one of) $G_1^k(i), \ldots, G_r^k(i)$.

4. If a processor P_k in $\mathcal{P}_i$ failed in Steps 1 and 3, then it selects a memory cell C at random and attempts to read it. If C stores a number g_a, then P_k attempts to read $G_a(i)$. If the state of computation stored there means termination, then P_k stops.

5. (Write) If a processor P_k in $\mathcal{P}_{i,j}$ succeeded in either Step 1 or Step 3, then it first performs the internal computation of C_I, and next writes to $D_j(l)$.

Denote $D(x) = \{D_i(x) : 1 \leq i \leq d \log n\}$, for $0 \leq x \leq n - 1$, and define $G(x)$ similarly.

Lemma 7. *There are at least $\frac{(1-q)}{2} \cdot d \log n + 1$ cells in $D(x)$ that are D-useful, with the probability $1 - n^{-\alpha}$, for $\alpha > 0$, for a sufficiently large $M = \Theta^*(n \log n)$. The same fact holds for $G(x)$.* $\square$

Let $G = (A, B, E)$ be a bipartite graph, where E is the set of edges connecting elements of A and B. Graph G is said to have the (γ, β) *weak-expansion property*, for $0 < \gamma, \beta < 1$, if, for every set $X \subseteq A$ such that $|X| \geq \gamma|A|$, the set $\Gamma(X) = \{y \in B : (x, y) \in E\}$ satisfies $|\Gamma(X)| \geq \beta|B|$. We will consider specific bipartite graphs defined as follows: $A = \mathcal{P}_{i,j}$, $B = D(x)$, there is an edge connecting a processor P in A with $D_a(x)$ iff D_a is among the address functions known by P after formatting. This graph is denoted $\mathcal{D}_{x,i,j}$. A similar graph $\mathcal{G}_{i,j}$ is defined as follows: $A = \mathcal{P}_{i,j}$, $B = G(i)$, there is an edge connecting a processor P in A with $G_a(i)$ iff G_a is among the address functions known by P after formatting.

Lemma 8. *Graphs $\mathcal{D}_{x,i,j}$ and $\mathcal{G}_{i,j}$ have the (γ, β) weak-expansion property, for any $0 < \gamma, \beta < 1$, with the probability at least $1 - n^{-\alpha}$, for $\alpha > 0$ and a sufficiently large fan r.* ☐

The performance of algorithm C is estimated in the following theorem:

Theorem 9. *Algorithm C has the formatting time $T_f = O(\log n)$, the step overhead $T_s = O(1)$, and the termination delay $O(\log n)$, all with the probability at least $1 - n^{-\alpha}$, for $\alpha > 0$.* ☐

5.2 Algorithm D

Suppose that C_F has $N = n$ processors and $M = \Theta^*(n)$ memory cells. To be specific, $M = c_M \cdot n$. It is assumed that all the processors of C_F know a primitive element of $GF(u)$, where $u > \log n / \log \log n$ is a power of 2. During a simulation, the contents of a memory cell of C_I are encoded, divided into pieces, and then distributed among $d \log n$ memory cells of C_F, for a constant parameter d. The numbers a and c are also parameters.

ALGORITHM D (FORMATTING)

1. Processors P_i generate $d \log n$ random numbers d_i in $[0, M - 1]$, and $a \log n$ random numbers a_j in $[0, M - 1]$.

2. The numbers a_j are distributed in the memory by executing SPREAD.

3. Each processor P_i, for $1 \leq i \leq d \log n$, repeats $c \log n$ times of the following two steps: select randomly memory cell and read it; if a_j was read then write d_i to the cell number $(a_j + i) \bmod M$.

We recall some facts from the theory of error correcting codes, consult [4, 19] for more information. Let C be sequence of codes $C = C_1 \ldots C_n \ldots$, where $C_n \subseteq \Sigma^n$, for some alphabet Σ of size s. Let $C_n(j)$ denote the jth codeword of code C_n. C is called *asymptotically good* if the lengths n, sizes $M_n = |C_n|$, dimensions $m_n = \lfloor \log_s M_n \rfloor$ and minimum Hamming distances d_n of codewords from C_n satisfy the following: the *rate* of the sequence $R = \liminf_{n \to \infty} \frac{m_n}{n}$, and the *relative minimum distance* $\delta = \liminf_{n \to \infty} \frac{d_n}{n}$ are both strictly greater then zero. By the Gilbert-Varshamov bound (see [4, 19]), for any $\delta \in [0, 1 - \frac{1}{s})$, there exists an asymptotically good code such that $R > 0$, that is, $R \geq 1 - H_s(\delta)$, where $H_s(x) = -x \log_s x - (1 - x) \log_s(1 - x) + x \log_s(s - 1)$. It is also known, see [19], that for $n = s - 1$ there exist codes, called Reed-Solomon (or simply RS) codes, with $\delta \geq 1 - R$. These codes can be quickly encoded and decoded, and are used in our algorithm. We apply a technique called "concatenation" of codes, in which the final code is obtained by first encoding by a code with a large alphabet size (outer code), and then encoding each symbol using another code (inner code). Notice that if a code C_n has the minimum distance d_n then it can correct n_ϵ erasures (eliminations of symbols) and n_e errors (changes of symbols) provided that $2n_e + n_\epsilon < d_n$. Suppose that processor P needs to simulate the instruction

algorithm	errors	#processors	memory size	formatting time	step overhead
A	static	$n \log n$	$\Theta^*(n)$	$O(\log n)$	$O(1)$
B	static	$n / \log n$	$\Theta^*(n)$	$O(\log^{3/2} n)$	$O(\log n)$
C	dynamic	$n \log^2 n$	$\Theta^*(n \log n)$	$O(\log n)$	$O(1)$
D	dynamic	n	$\Theta^*(n)$	$O(\log n)$	$O(\log n)$

Table 1. A comparison of the four simulations developed, in terms of their resources available and time performance. For the definition and explanation of notation Θ^* see section 2.

of writing some value w of $\log n$ bits into s_l. The following is a high level description of the algorithm. Coding and decoding need to be performed as bit operations on words. The details will be given in the final version.

ALGORITHM D (STEP SIMULATION, WRITE)

1. Divide the word w into blocks w_1, w_2, ... of $\log u$ consecutive bits and encode it as $v_1, v_2, \ldots$ by the RS code over $GF(u)$ with a relative minimum distance β, for a suitable β.

2. Encode each v_i by a suitable asymptotically good code C.

3. Let $k = \lfloor l / \log n \rfloor$. Attempt to store the consecutive symbols of the word $C(v_1), C(v_2) \ldots$ in $D_1(k), D_2(k), \ldots$, on the position $l \bmod \log n$. In every trial select $r = O(1)$ random cells. If a value a_j was found, try to read d_i from a cell number $(a_j + i) \bmod M$.

ALGORITHM D (STEP SIMULATION, READ)

1. As in Step 3 of the write part, read the cells $D_1(k), D_2(k) \ldots$ and form a sequence of codewords $z_1, z_2 \ldots$ (possibly with errors and erasures).

2. Decode every z_i.

3. Apply the RS decoding algorithm to the sequence obtained in Step 2.

The performance of algorithm D is estimated by the following theorem:

Theorem 10. *Algorithm D can be implemented in such a way that the formatting time T_f and the step overhead T_s are both $O(\log n)$, with the probability $1 - n^{-\alpha}$, for $\alpha > 0$.* $\qquad \Box$

6 Remarks

Four simulations of an ideal fully-reliable PRAM on a faulty-memory PRAM have been developed. Two settings of error occurrence are considered: static and dynamic. Given the kind of errors, there are two parameters of simulations: the number of processors, and the size of memory of the simulating PRAM. The performance of a simulation is measured by the formatting time and the step overhead. All this information is collected in Table 1. The simulation B is close to optimal, in the sense that the work done after formatting is $O(n \cdot t)$, where t is the time of the simulated algorithm on C_I. In general, the optimality of the presented algorithms, for the given resources, is an open problem. One could set some of the performance measures as targets, and try to minimize the other ones. Our choice was to design $O(1)$-time and $O(\log n)$-time step-overhead simulations while minimizing the formatting time, the number of processors and the size of the shared memory. It seems that the time cost of any simulation must be at least logarithmic, and proving such a lower bound would be interesting. The data compiled in Table 1 are consistent with this hypothesis, since the two rightmost columns contain at least one logarithm in every row.

There is an alternative method to algorithm D in which the information dispersal of Rabin [21] is used instead of the RS codes and asymptotically good codes. It requires $\log n$ registers per processor to have a time performance comparable to algorithm D. All the algorithms described in this paper can be adapted to a situation when the simulated PRAM has $m > n$ memory cells, where n is the number of processors. Then the time of a simulation of one step remains the same, and the formatting time is multiplied by at most $O(m/n)$. By checking the correctness of computations after each simulated step, the presented algorithms may be converted to be Las Vegas. Such checking may be performed by counting all the processors that performed the simulation correctly. Details will be presented in the full version of this paper.

This research is in the line of studying the PRAM model with weaker properties than the classical ideal version, for instance by allowing faults in hardware. We concentrated on the CRCW PRAM. It would be interesting to study weaker models, like CREW or EREW, and also the case when both the processors and memory cells may be faulty.

References

1. Y. Afek, D.S. Greenberg, M. Merrit, and G.Taubenfeld, Computing with Faulty Shared Memory, Proc. 11th Ann. Symposium on Principles of Distributed Computing (1992), 47-58.

2. A. Aggarwal, A.K. Chandra, and M. Snir, On Communication Latency in PRAM Computations, Proc. 1st Ann. ACM Symposium on Parallel Algorithms and Architectures (1989), 11-21.

3. Y. Aumann, Z.M. Kedem, K.V. Palem, and M.O. Rabin, Highly Efficient Asynchronous Execution of Large-Grained Parallel Programs, Proc. 34th Ann. Symposium on Foundations of Computer Science (1993), 271-280.

4. N. Alon, J. Bruck, J. Naor, M. Naor, and R.M. Roth, Construction of Asymptotically Good Low-Rate Error-Correcting Codes through Pseudo-Random Graphs, IEEE Trans. Inf. Theory, 38 (1992), 509-516.

5. B.S. Chlebus, K. Diks, T. Hagerup, and T. Radzik, New Simulations between CRCW PRAMs, Proc. 7th International Conference on Fundamentals of Computation Theory (1989), 95-104, Springer LNCS 380.

6. R. Cole, and O. Zajicek, The APRAM: Incorporating Asynchrony into the PRAM Model, Proc. 2nd Ann. ACM Symposium on Parallel Algorithms and Architectures (1990), 158-168.

7. K. Diks, and A. Pelc, Reliable Computations on Faulty EREW PRAM, manuscript, 1993.

8. F. Meyer auf der Heide, Hashing Strategies for Simulating Shared Memory on Distributed Memory Machines, Proc. of the 1st Heinz Nixdorf Symposium "Parallel Architectures and their Efficient Use,"(1992), 20-29, Springer LNCS 678.

9. P.B. Gibbons, A More Practical PRAM Model, Proc. 2nd Ann. ACM Symposium on Parallel Algorithms and Architectures (1990), 169-178.

10. A.M. Gibbons and W. Rytter, "Efficient Parallel Algorithms," Cambridge University Press, 1988.

11. T. Hagerup and Ch. Rüb, A Guided Tour of Chernoff Bounds, Inf. Proc. Letters 33 (1989/90), 305-308.

12. JáJá, "An Introduction to Parallel Algorithms," Addison-Wesley, 1992.

13. P. Jayanti, T.D. Chandra, and S. Toueg, Fault-tolerant Wait-free Shared Objects, Proc. 33rd Ann. Symposium on Foundations of Computer Science (1992), 157-166.

14. J. Justesen, On the Complexity of Decoding Reed-Solomon Codes, IEEE Trans. Inf. Theory, 22 (1976), 237-238.

15. P.C. Kanellakis and A.A. Shvartsman, Efficient Parallel Algorithms Can Be Made Robust, Distributed Computing, 5 (1992), 201-217.

16. Z.M. Kedem, K.V. Palem, and P.G. Spirakis, Efficient Robust Parallel Computations, Proc. 22nd ACM Symp. on Theory of Computing (1990), 138-148.

17. Ch. Martel, R. Subramonian, and A. Park, Asynchronous PRAMs Are (Almost) as Good as Synchronous PRAMs, Proc. 31st Ann. Symposium on Foundations of Computer Science (1990), 590-599.

18. Mc Diarmid, On the Method of Bounded Differences, in J. Siemon, ed., "Surveys in Combinatorics,", 148 - 188, Cambridge University Press, 1989, London Math. Soc. Lecture Note Series 141.

19. F.J. MacWilliams, and N.J.A Sloane, "The Theory of Error-Correcting Codes," North-Holland, 1977.

20. N. Nishimura, Asynchronous Shared Memory Parallel Computation, Proc. 1st Ann. ACM Symposium on Parallel Algorithms and Architectures (1989), 76-84.

21. M.O. Rabin, Efficient Dispersal of Information for Security, Load Balancing, and Fault Tolerance, Journal of ACM, 36 (1989), 335-348.

22. D.V. Sarwate, On the Complexity of Decoding Goppa Codes, IEEE Trans. Inf. Theory, 23 (1976), 515-516.

23. Y. Sugiyama, M. Kosahara, S. Hirasawa, and T. Namekawa, An Erasures and Error Decoding Algorithm for Goppa Codes, IEEE Trans. Inf. Theory, 22 (1976), 238-241.

24. L.G. Valiant, General Purpose Parallel Architectures, in J. van Leeuwen, ed., "Handbook of Theoretical Computer Science," vol. A, 943-971, Elsevier, 1990.

An Area Lower Bound for a Class of Fat-Trees[*]

(Extended Abstract)

Gianfranco Bilardi[1] and Paul Bay[2]

[1] Dipartimento di Elettronica e Informatica,
Università di Padova,
Via Gradenigo 6/A, I35131 Padova, Italy
[2] Thinking Machines Corporation,
245 First Street, Cambridge, MA 02142, USA

Abstract. A graph-theoretic definition is proposed to make precise the sense in which a "fat-tree" is a tree-like interconnection of subnetworks whose bandwidth is adequately described by the capacity of the channels between subnetworks. The definition is shown to encompass a number of known networks such as the concentrator fat-tree, the pruned butterfly, and the tree-of-meshes. In the established framework, a non-trivial $\Omega(N \log^2 N)$ lower bound is derived for the area of a class of fat-trees, with implications for their area-universality.

1 Introduction

A number of networks have been introduced in the literature and referred to as *fat-trees* with some further specifier such as concentrator fat-tree, pruned-butterfly fat-tree, and sorting fat-tree. Fat-trees have important *universality* properties in VLSI and form the basis for number of universal routers [11, 6, 12, 2, 7, 1] and universal circuits [3]. The CM-5 parallel supercomputer uses a fat-tree as its interconnection pattern [9].

Loosely speaking, a fat-tree is a tree whose leaves act as input/output terminals, whose internal nodes are subnetworks with switching capability, and whose edges are channels of appropriate capacity. Proposed fat-trees differ in node structure and channel capacities. In spite of its wide use, the term fat-tree has not been defined precisely. Here, we develop a graph-theoretic definition that captures an important class of networks of the fat-tree type, providing a framework for a general investigation of their properties, such as their layout area.

In Section 2 we introduce the notion of *tree-structured* network to model tree-like interconnections of subnetworks. For such a network, we call *reference tree* a weighted tree whose nodes represent the subnetworks and whose edges

---

[*] The research of the first author was supported in part by the ESPRIT Basic Research Project 9072 *GEPPCOM: Foundations of GEneralPurpose Parallel COMputing*, by the Italian National Research Council, and by the Italian Ministry of University and Research.

represent the channels between them. Edges are weighted by the capacities of these channels.

We then introduce γ-*channel-sufficient* (tree-structured) networks, where the number of edge-disjoint paths between two sets of terminals is determined by the maximum flow that can be pushed between those two sets in the reference tree. Using network-flow arguments and Menger's theorem on edge-disjoint paths, we characterize channel-sufficient networks as those for which the load factor of a message set can be estimated (in linear time) to within a multiplicative constant by considering only the capacities of the tree channels. Thus, the reference tree provides an adequate description of network bandwidth.

The *pruned butterfly* [2], the *concentrator fat-tree* [11], and the *tree-of-meshes* [5] are shown to be γ-channel-sufficient in Section 3. The proof is not straightforward, indicating that a non-obvious property is being exposed.

We are interested in the *layout area* of area-universal fat-trees, where, typically, channel capacity is constant at the N leaves and doubles every other level to become become $\Theta(\sqrt{N})$ at the root (*standard capacities*). These fat-trees admit layouts of area $O(N \log^2 N)$. Since bisection [14] or bifurcator [5] techniques yield only trivial $\Omega(N)$ lower bounds, it is natural to ask whether there exist smaller layouts.

In Section 4, γ-channel-sufficient fat-trees with standard capacities (satisfying a further technical assumption) are shown to require $\Omega(N \log^2 N)$ wire area [10]. This result provides us with an entire class of graphs which have the maximum area compatible with their bifurcator. The only previously known graphs exhibiting this behavior were the mesh-of-trees [10] and the expander-connected mesh-of-trees [5] (and graphs that can efficiently embed these ones).

Our lower bound applies, in particular, to the concentrator fat-tree (for which we do not know of previous results) and to the pruned-butterfly and the tree-of-meshes (for which the result was known from mesh-of-trees embeddings). All of these graphs admit optimal layouts of area $\Theta(N \log^2 N)$.

The fact that the bisection width of fat-trees is a factor $\Theta(\log N)$ smaller than the square root of the area implies a logarithmic slowdown in the simulation of some networks (e.g., a mesh) of the same area. The existence of area-universal networks capable to simulate any other network of the same area with only constant slowdown remains an open question. Our results indicate that such a network is unlikely to be a fat-tree.

2   A Formal Notion of Fat-tree

We begin by introducing the notion of *tree-structured* network to make precise the requirement that a fat-tree can be viewed as a tree-like connection of sub-networks.

Definition 1. Let N be a power of two. Let $R = (V, E)$ be an undirected graph with a distinguished subset of N vertices referred to as the *terminals* of R. A *tree-representation* of R is a partition of V into sets $\{V_{ij} : 0 \le i \le \log N, \ 0 \le j < 2^i\}$, such that:

- $V_{\log N, j}$ consists of exactly one terminal vertex, for $0 \leq j < N$;
- If $u \in V_{ij}$ and $(u, v) \in E$, then either $v \in V_{ij}$ or $v \in V_{i-1, \lfloor j/2 \rfloor}$.

We say that R is *tree-structured* if it has a tree-representation.

We denote by F_{ij} the set of edges of R with both endpoints in V_{ij} and by C_{ij} the set of edges with one endpoint in V_{ij} and the other in the parent node $V_{i-1, \lfloor j/2 \rfloor}$. Set C_{ij} is referred to as a *channel* and $cap(C_{ij}) = |C_{ij}|$ as its *capacity*.

Definition 2. Given a tree-representation of an N-terminal network R, we call *reference-tree*, denoted T_R, the complete ordered binary tree whose $(j + 1)$-st node at level i is set V_{ij}, and whose edge from node V_{ij} to its parent $V_{i-1, \lfloor j/2 \rfloor}$ is assigned capacity $cap(C_{ij})$.

With a slight abuse of terminology, we will refer to a set V_{ij} as to a node of R and will identify a terminal of R with the leaf of T_R containing that terminal.

Whereas the existence of a tree-representation is a prerequisite for a "fat-tree", it is not a characterizing property. Indeed, a set of $2N - 1$ isolated vertex does satisfy Definition 1. Large capacities do not guarantee a high degree of connectivity either. What needs to be formalized is the idea that bandwidth constrains for R come essentially from the channels rather than from bottlenecks within the nodes. It is natural to attempt a characterization in terms of some connectivity of the nodes themselves, such as a concentrating property. However, in some fat-trees as the pruned butterfly [2], there are no edges within nodes, and yet overall connectivity results from synergism among nodes. Therefore, a characterization must be based on global properties.

We will use some network flow concepts, briefly reviewed here for convenience [13]. Let $G = (V, E)$ be an undirected graph whose vertex pairs have an integer capacity $cap(v, w)$, positive when (v, w) is an edge and null otherwise. Let A be a distinguished set of *source* vertices and let B be a distinguished set of *sink* vertices. A *flow* is a real-valued function f such that $f(v, w) = -f(w, v)$, $f(v, w) \leq cap(v, w)$, and $\sum_{w \in V} f(v, w) = 0$ for every vertex v other than sources and sinks. The *value* $|f|$ of the flow is the net flow out of the sources, $\sum_{v \in A, w \in V} f(v, w)$. Let $s^*(R, A, B)$ denote the value of a maximum flow from A to B in R.

Definition 3. Let $0 < \gamma \leq 1$. A tree-structured network R with reference tree T_R is γ-*channel-sufficient* if for any disjoint sets of leaves A and B there exist at least $\gamma s^*(T_R, A, B)$ edge-disjoint paths from A to B in R.

The term "channel-sufficient" is motivated by Theorem 4 below, where R models a routing network whose terminals represent interface points with agents, say processors, that need to exchange messages. If edge e of R has a positive integer capacity $cap(e)$, the time for R to route a set M of messages is at least proportional to the *load factor* $\lambda(R, M)$, defined as follows. A *cut* between two subsets of terminals A and B is a set of edges intersecting every path from A to B. The capacity of a cut S is the sum of the capacities of its edges. The *load*

$\ell(R, M, S)$ placed on cut S by M is the number of messages that must cross it. The *load factor* on cut S, $\lambda(R, M, S)$, is $\ell(R, M, S)/\sum_{e \in S} \text{cap}(e)$. The load factor on the entire network, $\lambda(R, M)$, is the maximum load factor of any cut.

The next result shows that a network is γ-channel-sufficient precisely when its reference tree gives an accurate measure of load factor, to within a factor γ.

Theorem 4. *A tree-structured network R where each edge has unit capacity is γ-channel-sufficient if and only if, for all M, $\lambda(R, M) \leq \lambda(T_R, M)/\gamma$.*

Proof. ($\Rightarrow$) Suppose R is γ-channel-sufficient. Let M be any message set, let S be any cut, and let A and B be the terminal sets separated by S. It suffices to show that $\lambda(R, M, S) \leq \lambda(T_R, M)/\gamma$.

Since there are at least $\gamma s^*(T_R, A, B)$ edge-disjoint paths from A to B, clearly $\text{cap}(S) \geq \gamma s^*(T_R, A, B)$. By the max-flow, min-cut theorem, there is a set E of edges of T_R that separates A from B and has total capacity equal to the value of the maximum flow, i.e., a set E such that $\sum_{e \in E} \text{cap}(e) = s^*(T_R, A, B)$. Therefore $\text{cap}(S) \geq \gamma \sum_{e \in E} \text{cap}(e)$. So

$$\lambda(R, M, S) = \frac{\ell(R, M, S)}{\text{cap}(S)} \leq \frac{\ell(R, M, S)}{\gamma \sum_{e \in E} \text{cap}(e)}$$
$$= \frac{\ell(T_R, M, E)}{\gamma \sum_{e \in E} \text{cap}(e)} = \lambda(T_R, M, E)/\gamma \leq \lambda(T_R, M)/\gamma.$$

($\Leftarrow$) Suppose $\lambda(R, M) \leq \lambda(T_R, M)/\gamma$, i.e., for every message set M and cut S, $\lambda(R, M, S) \leq \lambda(T_R, M)/\gamma$. Let A and B be any two disjoint sets of terminals and let m be the minimum number of edges whose removal separates A from B. By Menger's theorem (specifically, the "sets of points" variant of Theorem 5.11 in [8]), m edge-disjoint paths join A and B in R. Let S be some specific set of m edges whose removal separates A from B.

Given any integral flow in a tree, it is easy to construct a corresponding message set with tree load factor at most 1 and cardinality equal to the flow value. In particular, let M be a message set corresponding to a maximum integral flow from A to B in T_R. Message set M has sources in A, destinations in B, $\lambda(T_R, M) \leq 1$, and $|M| = s^*(T_R, A, B)$. By hypothesis $\lambda(R, M, S) \leq \lambda(T_R, M)/\gamma$. So

$$\lambda(R, M, S) = \frac{\ell(R, M, S)}{\text{cap}(S)} = \frac{|M|}{m} = \frac{s^*(T_R, A, B)}{m} \leq \lambda(T_R, M)/\gamma \leq 1/\gamma,$$

which implies that $m \geq \gamma s^*(T_R, A, B)$. □

In summary, our notion of γ-channel-sufficient, tree-structured network makes precise the sense in which a "fat-tree" is a tree-like interconnection of subnetworks whose bandwidth is adequately described by the capacity of the channel between those subnetworks. In general, however, the reference tree does not adequately describe the routing capability of a network, since long paths (causing high latency) are not ruled out. Interstingly, in Section 4, we shall see that channel-sufficiency alone is responsible for the high layout area of certain tree-structured networks.

3   Examples of γ-Channel-Sufficient Fat-trees

In this section, we show that some known networks are channel-sufficient. We begin with the pruned butterfly [2] (see Figure 1).

Definition 5. The *pruned butterfly* is the graph $G(V, E)$ defined as follows, for N a power of 4. The vertex set is

$$V = \{\langle i, j, k \rangle : 0 \leq i \leq \log N,\ 0 \leq j < 2^i,\ 0 \leq k < \sqrt{N} 2^{-\lfloor i/2 \rfloor}\}.$$

Vertices $\langle i, j, k \rangle$ with a given value of i and j form node V_{ij} of a tree representation. There are no edges within a node $(F_{ij} = \emptyset)$. The channel from node V_{ij} to its parent, with capacity $cap(C_{ij}) = \sqrt{N} 2^{-\lfloor (i-1)/2 \rfloor}$, is defined as

$$C_{ij} = \{(\langle i, j, k \rangle, \langle i - 1, \lfloor j/2 \rfloor, k \rangle) :\ 0 \leq k < \sqrt{N} 2^{-\lfloor i/2 \rfloor}\}, \text{ for odd } i, \text{ and}$$

$$C_{ij} = \{(\langle i, j, k \rangle, \langle i - 1, \lfloor j/2 \rfloor, k \rangle),\ (\langle i, j, k \rangle, \langle i - 1, \lfloor j/2 \rfloor, k + \sqrt{N} 2^{\lfloor -i/2 \rfloor} \rangle)) :\ 0 \leq k < \sqrt{N} 2^{-\lfloor i/2 \rfloor}\}, \text{ for even } i.$$

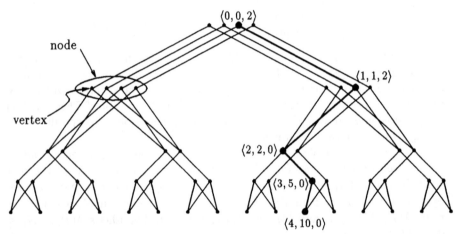

Fig. 1. The pruned butterfly when $N = 16$. The triples identify the vertices on the unique path from leaf 10 to root 2.

Theorem 6. *The pruned-butterfly fat-tree is 1-channel-sufficient.*

Proof. Let T_{PB} be the reference tree of the pruned butterfly (PB) and let A and B be two disjoint sets of terminals of PB. To show that there are $s^*(T_{PB}, A, B)$ edge-disjoint paths from A to B in PB, consider an integral maximum flow f^* from A to B in T_{PB}, i.e., the flow through every edge of T_{PB} is an integer. Such a maximum flow exists by Theorem 8.2 in [13]. We will construct the $|f^*| = s^*(T_{PB}, A, B)$ paths so that the number of paths through each channel is equal to the flow through the corresponding edge of T_{PB}.

The paths are constructed starting at the root and moving downward toward the leaves. The flows through the edges between the root and its two children are equal, and the paths are established consecutively on the smallest numbered edges. Let v be a generic pruned butterfly node, let z be v's parent, and let x and y be v's left and right children respectively. Assume inductively that $f^*(z,v)$ edge-disjoint paths enter node v from z on cyclically consecutive edges. As illustrated in Figure 2, there are essentially only two cases for the structure of the flow through node v. The flow direction is indicated by the arrows on the edges in the reference tree; the cases where the flow is in the opposite direction are obviously analogous.

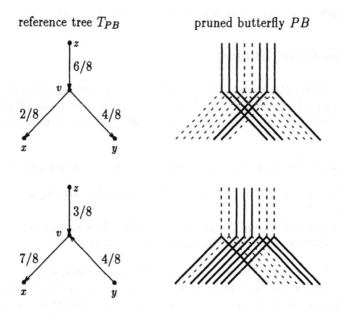

Fig. 2. Creating paths from a flow. For simplicity only an odd level of the pruned butterfly is illustrated; a similar diagram would apply for the even levels.

In the first case, the flow comes from the parent and goes to both children. In this case, the first $f^*(v,x)$ paths are made to continue downward on the (cyclically consecutive) edges to the left child; the remaining $f^*(z,v)-f^*(v,x) = f^*(v,y)$ paths continue downward on cyclically consecutive edges to the right child.

In the second case, the flow comes from the parent and from one child and goes toward the other child. Without loss of generality, assume the flow is directed out of the *right* subtree. In this case the $f^*(z,v)$ paths continue downward on cyclically consecutive edges to the left child. Then $f^*(y,v) =$

$f^*(v,x) - f^*(z,v)$ new paths are created going from right child to left on cyclically consecutive edges beginning with the edge immediately following the $f^*(z,v)$ paths already chosen.

Since the flow value of f^* on a given edge of T_{PB} never exceeds the edge's capacity, there are always enough edges in the pruned butterfly's channels to complete the path-creating step just described. □

Next, we state without proof the channel-sufficiency of two important networks.

Theorem 7. *A fat-tree whose nodes are α-partial concentrators [11, 3] is α-channel-sufficient.*

Theorem 8. *The tree-of-meshes [5] is 1-channel-sufficient.*

4 Area Lower Bound

We now turn our attention to the area of fat-trees. Let us say that a path between leaves u and v in a tree-structured network R is *tree-constrained* if it is entirely contained whithin the nodes of the simple path from u to v in the reference tree T_R.

Definition 9. An N-leaf tree-structured network F is a *natural* fat-tree if

- For $0 < i \le \log N$ and $0 \le j < 2^i$, channel C_{ij} has capacity $cap(C_{ij}) \ge \sqrt{N}2^{-\lfloor i/2 \rfloor}$.
- For any pair of disjoint sets of leaves A and B, there are at least $\gamma s^*(T_F, A, B)$ tree-constrained edge-disjoint paths from A to B.

From the proofs of Theorems 6, 7, and 8 it can be easily seen that the pruned butterfly, the concentrator fat-tree, and the tree-of-meshes are natural fat-trees.

In general, the tree-constrained requirement is a desirable one, as it allows routing on shortest paths of the underlying reference tree. Moreover, it guarantees that all sub-fat-trees of the network are natural, permitting partitions into non-interacting subnetworks (for example, to support multiple users). We conjecture that the tree-constrained assumption is not essential to the conclusion of next theorem, but it does play a crucial role in the present proof.

Theorem 10. *Any layout of an N-leaf natural fat-tree F takes $\Omega(N \log^2 N)$ wire area.*

Proof. The proof is by induction on N. Let $W(n)$ denote the minimum wire area of an n-leaf natural fat-tree. Assume inductively that $W(n) \ge \alpha n \log^2 n$ for all $n = 2^k$ where $n < N$, and where $\alpha = (\gamma/3 \cdot 2^{17.5})^2$.

Let F be an N-leaf natural fat-tree. The edges of F can be partitioned into levels $E_0, E_1, ..., E_{\log N-1}$, where E_i includes the edges in all nodes at depth i $(\cup_j F_{ij})$ and in the channels to their children $(\cup_j C_{i+1,j})$. Removal of the edges in $E_0, ..., E_{i-1}$ clearly yields 2^i natural fat-tree with the same parameter γ, each with $N/2^i$ leaves. Our analysis focuses on the total wire area L_i used in the layout by edges of E_i.

Lemma 11. *In a layout of F that achieves the minimum wire area $W(N)$, the following constraints are satisfied:*

$$W(N) = \sum_{i=0}^{\log N - 1} L_i, \tag{1}$$

$$\sum_{j=i+1}^{\log N - 1} L_j \geq \alpha N \log^2(N/2^{i+1}), \quad for\ i = 0, ..., \log N - 2, \tag{2}$$

$$\sum_{i=0}^{\log N - 1} 2^{-i/2} L_i \geq (\gamma \sqrt{\alpha}/2^{14.5}) N \log N. \tag{3}$$

Relation (1) follows from the definition of $W(N)$. Relation (2) follows by the inductive hypothesis applied to the 2^{i+1} natural fat-trees resulting from the removal of edges in $E_0, ..., E_i$, each of area at least $W(N/2^{i+1}) \geq \alpha(N/2^{i+1}) \log(N/2^{i+1})$. Relation (3) is considerably harder to establish. For clarity, the argument organized into Lemmas 12 (whose proof is omitted in this abstract) and 13 below. Finally, the constraints of Lemma 11 are shown to imply $W(N) \geq \alpha N \log N$.

Lemma 12. *In any layout of F, there are two sets of leaves C and D in the left and in the right subtree, respectively, with $|C|, |D| \geq N/16$, and such that any path in F from a leaf in C to a leaf in D has layout length at least $(\sqrt{\alpha N} \log N)/16$.*

Lemma 13. *There are subsets $A_1, ..., A_{\sqrt{N}/64}$ of C and $B_1, ..., B_{\sqrt{N}/64}$ of D such that:*

- *for each $q = 1, ..., \sqrt{N}/64$ there are at least $\gamma\sqrt{N}/16$ edge-disjoint paths from A_q to B_q;*
- *when these paths are all simultaneously embedded in F, the congestion of an edge in E_i is at most $\sqrt{N}/2^{(i-1)/2}$.*

Proof of Lemma 13: Let $r = \lceil (\log N)/2 \rceil - 1$ and $R = 2^r$. We define a partition $\mathcal{J} = \{J_0, ..., J_{R-1}\}$ of the leaves of the left subtree of F into subsets of size $N/2R$, as follows. For any r-bit string $x = x_{r-1}x_{r-2}\cdots x_0$, let J_x be the set of leaves whose id's have binary representation of the form $0 * x_{r-1} * x_{r-2} \cdots * x_0 *$ for even values of $\log N$, and of the form $0 * x_{r-1} * x_{r-2} \cdots * x_0$ for odd values of $\log N$. A $*$ denotes 0 or 1. It is tedious, but straightforward, to verify the following two useful properties of partition $\mathcal{J}$. For any $N/2^i$-leaf subtree T rooted at level i,

(a) at most $\sqrt{N}/2^{\lfloor i/2 \rfloor}$ members of a given J_x are leaves of T, and

(b) at most $\sqrt{N}/2^{(i-1)/2}$ different J_x's have some member in T.

We similarly partition the leaves of the right subtree of F by letting $K_x = \{j + N/2 : j \in J_x\}$. Analogous properties hold for this partition $\mathcal{K}$.

Given any $J_x \in \mathcal{J}$ and any $K_y \in \mathcal{K}$, by an arbitrary matching one can construct a set of $N/2R$ messages that all go through the root and have sources in J_x and destinations in K_y. By property (a) such a message set places a load factor of at most 1 on F's reference tree, so $s^*(T_F, J_x, K_y) = |J_x| = |K_y|$. Hence, if $A \subseteq J_x$ and $B \subseteq K_y$, then

$$s^*(T_F, A, B) = \min(|A|, |B|). \tag{4}$$

Let $\mathcal{A} = \{x : |J_x \cap C| \geq \sqrt{N}/16\}$. Since the J_x's form a partition and $|J_x| = N/2R$,

$$|C| = \sum_{x=0}^{R-1} |J_x \cap C| = \sum_{x \in \mathcal{A}} |J_x \cap C| + \sum_{x \notin \mathcal{A}} |J_x \cap C|$$

$$\leq |\mathcal{A}|N/2R + (R - |\mathcal{A}|)(\sqrt{N}/16).$$

Since $|C| \geq N/16$ and $\sqrt{N}/2 \leq R \leq \sqrt{N}/\sqrt{2}$, the above relation implies $|\mathcal{A}| \geq \sqrt{N}/64$. Let $A_1, ..., A_{\sqrt{N}/64}$ be distinct sets of the form $J_x \cap C$ where $x \in \mathcal{A}$. Similarly let $B_1, ..., B_{\sqrt{N}/64}$ be distinct sets of the form $K_y \cap D$ where $|K_y \cap D| \geq \sqrt{N}/16$. By (4), for each q,

$$s^*(T_F, A_q, B_q) = \min(|A_q|, |B_q|) \geq \sqrt{N}/16. \tag{5}$$

Since F is γ-channel-sufficient, the first condition of Lemma 13 is established.

To prove the second condition, consider an edge e in E_i in the left subtree of F. For any specific q, all the paths under consideration are edge-disjoint, so the congestion of e is no larger than the number of (A_q, B_q) pairs whose paths could potentially contain it. But since the paths are tree-constrained, a particular set of paths can contain edge e only if A_q includes a leaf in the subtree rooted at e. Recall that property (b) of partition $\mathcal{J}$ guarantees that at most $\sqrt{N}/2^{(i-1)/2}$ different sets J_x contain a leaf in the subtree rooted at level i containing e. Each A_q is a subset of exactly one J_x. Therefore at most $\sqrt{N}/2^{(i-1)/2}$ different pairs (A_q, B_q) have paths that contain e. A similar argument applies to edges in the right subtree of F. $\qquad \square$

Completion of the proof of Lemma 11: Consider the sum L of the path length over the $\sqrt{N}/64$ (A_q, B_q) pairs of Lemma 13 and over the at least $\gamma\sqrt{N}/16$ paths in each pair. Since, from Lemma 12, each path is at least $(\sqrt{\alpha N} \log N)/16$ long, we have

$$L \geq (\sqrt{N}/64)(\gamma\sqrt{N}/16)(\sqrt{\alpha N} \log N)/16 = (\gamma\sqrt{\alpha}/2^{14})N^{3/2} \log N. \tag{6}$$

On the other hand, by the second part of Lemma 13, since each edge in E_i is used at most $\sqrt{N}/2^{(i-1)/2}$ times by the above collection of paths, we have:

$$L \leq \sum_{i=0}^{\log N - 1} L_i \sqrt{N}/2^{(i-1)/2}. \tag{7}$$

Combining (6) and (7), we obtain (3). □

Next, we derive some consequences of Lemma 11. If, for $i = 0, ..., \log N - 2$, both sides of relation (2) are multiplied by the quantity $(2^{-i/2} - 2^{-(i+1)/2})$, and the resulting inequality is added side by side to (3), one obtains

$$\sum_{i=0}^{\log N-1} L_i \geq (\gamma\sqrt{\alpha}/2^{14.5})N \log N +$$
$$\alpha(1 - 1/\sqrt{2})N \sum_{i=0}^{\log N-2} (1/\sqrt{2})^i \log^2(N/2^{i+1}). \tag{8}$$

Observe now that $\log^2(N/2^{i+1}) \geq \log^2 N - 2\log N(i+1)$, hence

$$\sum_{i=0}^{\log N-2} (1/\sqrt{2})^i \log^2(N/2^{i+1})$$
$$\geq \log^2 N \sum_{i=0}^{\log N-2} (1/\sqrt{2})^i - 2\log N \sum_{i=0}^{\log N-2} (i+1)(1/\sqrt{2})^i$$
$$\geq (\log^2 N)/(1 - 1/\sqrt{2}) - 24\log N. \tag{9}$$

Using (9) in (8) and recalling that $\alpha = (\gamma/3 \cdot 2^{17.5})^2$, we have

$$\sum_{i=0}^{\log N-1} L_i \geq \alpha N \log^2 N + (\gamma\sqrt{\alpha}2^{-14.5} - 24\alpha)N \log N = \alpha N \log^2 N. \tag{10}$$

In view of (1), we conclude that $W(N) \geq \alpha N \log^2 N = \Omega(N \log^2 N)$. □

5 Conclusions

We have shown that, at least under some reasonable assumptions, the networks designed within the fat-tree paradigm exhibit $\Omega(N \log^2 N)$ area instead of the desirable $O(N)$ area. It remains to investigate whether there are routing networks with the same time performance of the known fat-trees but with smaller area (see [4] for a first step).

References

1. P. Bay. *Area-Universal Interconnection Networks for VLSI Parallel Computers*. Ph.D. dissertation, Cornell University, May 1992.
2. P. Bay and G. Bilardi. Deterministic on-line routing on area-universal networks. In *Proceedings of the 31st Annual Symposium on Foundations of Computer Science*, pages 297–306, October 1990.
3. P. Bay and G. Bilardi. An area-universal VLSI circuit. In *Proceedings of the 1993 Symposium on Integrated Systems*, pages 53–67, March 1993.

4. G. Bilardi, S. Chauduri, D. Dubashi, and K. Mehlhorn. A lower bound for area-universal graphs. Technical Report MPI-I-93-144, Max-Plank-Institut fur Informatik, October 1993.

5. S. N. Bhatt and F. T. Leighton. A framework for solving VLSI graph layout problems. *Journal of Computer and System Sciences*, 28(2):300–343, 1984.

6. R. I. Greenberg and C. E. Leiserson. Randomized routing on fat-trees. In S. Micali, editor, *Randomness and Computation*, pages 345–374. JAI Press, Inc., 1989.

7. R. I. Greenberg. The fat-pyramid: A robust network for parallel computation. In W. J. Dally, editor, *Advanced Research in VLSI: Proceedings of the Sixth MIT Conference*, pages 195–213, 1990.

8. F. Harary. *Graph Theory*. Addison-Wesley, Reading, MA, 1969.

9. C. E. Leiserson, Z. S. Abuhamdeh, D. C. Douglas, C. R. Feynman, M. N. Ganmukhi, J. V. Hill, W. D. Hillis, B. C. Kuszmaul, M. A. St. Pierre, D. S. Wells, M. C. Wong, S.-W. Yang, and R. Zak. The network architecture of the Connection Machine CM-5. In *Proceedings of the 4th Annual ACM Symposium on Parallel Algorithms and Architectures*, pages 272–285, July 1992.

10. F. T. Leighton. New lower bound techniques for VLSI. *Mathematical Systems Theory*, 17:47–70, 1984.

11. C. E. Leiserson. Fat-trees: Universal networks for hardware-efficient supercomputing. *IEEE Transactions on Computers*, C-34(10):892–900, October 1985.

12. T. Leighton, B. Maggs, and S. Rao. Universal packet routing algorithms. In *Proceedings of the 29th Annual Symposium on the Foundations of Computer Science*, White Plains, New York, October 1988.

13. R. E. Tarjan. *Data Structures and Network Algorithms*. Society for Industrial and Applied Mathematics, Philadelphia, PA, 1983.

14. C. D. Thompson. *A Complexity Theory for VLSI*. Ph.D. dissertation, Carnegie-Mellon University, August 1980.

A Unified Approach to Approximation Schemes for NP- and PSPACE-Hard Problems for Geometric Graphs[1,2,3]

H.B. Hunt III   M.V. Marathe   V. Radhakrishnan [4]
S.S. Ravi   D.J. Rosenkrantz   R.E. Stearns

Abstract

We present parallel approximation schemes for a number of graph problems when restricted to geometric graphs including unit disk graphs and graphs drawn in a civilized manner [CCJ90, MHR92, Te91]. Our NC-approximation schemes exhibit the same time versus performance trade-off as those of Baker [Ba83].

We also define the concept of λ-**precision** unit disk graphs and show that for such graphs our NC approximation schemes have better time versus performance trade-offs and several other graph problems have efficient approximation schemes.

Our parallel approximation schemes can also be extended to obtain efficient parallel approximation schemes for problems on unit disk graphs specified using a restricted version of the hierarchical specification language of Bentley, Ottmann and Widmayer [BOW83]. While we can show that many graph problems are PSPACE-hard even for this restricted form of specifications, nevertheless, these problems possess efficient approximation schemes.

1   Introduction

In this paper, we present efficient approximations and approximation schemes for NP-hard problems when restricted to geometric intersection graphs, particularly unit disk graphs. One important objective of this paper is to present a unified framework for obtaining efficient NC-approximation schemes for a large class of problems when restricted to intersection graphs of regular polygons. Our results also apply to graphs drawn in a civilized manner (See Definition 2.2).

---

[1] Dept. of Computer Science, University at Albany-SUNY, Albany, NY 12222.

[2] Email: {hunt,madhav,ravi,djr,res}@cs.albany.edu

[3] Supported by NSF Grants CCR 89-03319 and CCR 90-06396.

[4] Hewlett-Packard Company, 19447 Pruneridge Avenue, Cupertino, CA 95014. Email:rven@cup.hp.com.

Our approximation schemes can be extended so as to apply to several PSPACE-hard problems on unit disk graphs presented using the hierarchical specification language of Bentley, Ottmann and Widmayer [BOW83].

An undirected graph is a **unit disk graph** iff its vertices can be put in one to one correspondence with equal sized circles in a plane in such a way that two vertices are joined by an edge iff the corresponding circles intersect. (It is assumed that tangent circles intersect. We assume without loss of generality that the radius of each disk is 1.) These graphs have been used to model broadcast networks [CCJ90, Ha80, Ra93], image processing [FPT81, HM85], VLSI circuit design [MC80] and optimal facility location [MS84, WK88]. Consequently, unit disk graphs have been studied extensively in the literature [CCJ90, FPT81, MHR92, MS84, WK88].

As pointed out in [CCJ90], unit disk graphs need not be perfect (any odd cycle of length five or greater is a unit disk graph and is not perfect). Similarly, unit disk graphs need not be planar; in particular, any clique of size five or more is a unit disk graph but is not planar. Thus many of the known efficient algorithms for perfect graphs and planar graphs do not apply to unit disk graphs.

2 Summary of Results

It has been shown in [CCJ90, FPT81, MS84, WK88] that several standard graph theoretic problems are NP-hard even when restricted to unit disk graphs. Given this apparent intractability, we investigate whether these problems have efficient approximation algorithms and approximation schemes. Recall that an approximation algorithm for an optimization problem Π has a **performance guarantee** of ρ if for every instance I of Π, the solution value returned by the approximation algorithm is within a factor ρ of the optimal value for I. A **polynomial time approximation scheme** (PTAS) for problem Π is a polynomial time approximation algorithm which given any instance I of Π and an $\epsilon > 0$, returns a solution which is within a factor $(1 + \epsilon)$ of the optimal value for I. An approximation scheme which can be implemented in NC is referred to as **NC-approximation scheme**.

2.1 Parallel Approximation Schemes

2.1.1 Unit Disk Graphs and Extensions

In [MHR92], we have shown that several natural graph theoretic problems have approximation algorithms with constant performance guarantees when restricted to unit disk graphs. Extending those results we obtain NC-approximations for a number of these problems. Formally,

Theorem 2.1 *For unit disk graphs, there are NC-approximation schemes for the problems maximum independent set, minimum dominating set and minimum vertex cover.*

The NC-approximation schemes above are obtained by blending the techniques of Hochbaum and Maass [HM85] and Baker [Ba83].

Next, we introduce the notion of λ-precision unit disk graphs and show that for this restricted class of unit disk graphs, all the natural MAX SNP-complete graph problems in [PY91] and the problems in [Ba83] have efficient NC-approximation schemes.

Definition 2.1 *Let $\lambda > 0$ be a fixed rational number. Consider a finite set of disks each of radius 1 laid out in the plane where the centers of any two disks are at least λ apart. The λ-**precision unit disk graph** $G(V, E)$ corresponding to this layout of unit disks is defined as follows: The vertices of G are in one-to-one correspondence with the set of unit disks and two vertices are joined by an edge iff the corresponding disks intersect.*

As before, we assume that tangential disks intersect. Our definition of λ-precision unit disk graphs is motivated by the fact that in many instances of practical problems, when modeled as problems on unit disk graphs, seldom have unit disk centers placed in a continuous fashion. For example, in VLSI designs, λ is a parameter determined by the fabrication process.

For any $0 < \lambda \leq 2$, it can be seen that grid graphs[5] are λ-precision unit disk graphs. Since there are n node grid graphs with treewidth $\Theta(\sqrt{n})$, λ-precision unit disk graphs do not have constantly bounded treewidth. In fact, each unit disk graph is a λ-precision unit disk graph for some $0 < \lambda \leq 2$. It is also easy to see that λ- precision unit disk graphs are not necessarily planar. For any fixed $\lambda > 0$ given a λ-precision unit disk graph, we can show that many more problems have efficient NC-approximation schemes. Our results are summarized in Table 1.

2.1.2 Graphs drawn in a civilized manner

The results discussed in the previous subsection can also be extended so as to apply to graphs drawn in a civilized manner. We assume throughout this paper that the dimension (d) of the Euclidean space considered is at least 2. The following definition is from Teng [Te91].

Definition 2.2 *For each pair of positive reals $r > 0$ and $s > 0$, a graph G can be drawn in R^d in an (r, s)-civilized manner if its vertices can be embedded in R^d so that the length of each edge is $\leq r$. and the distance between any two points is $\geq s$.*

These graphs have been studied in the context of random walks by Doyle and Snell and finite element analysis by Vavasis (See [Te91]). It can be seen that a for any $\lambda > 0$, λ-precision unit disk graph can be drawn in $(2, \lambda)$-civilized manner. We show that for graphs drawn in a civilized manner, many basic graph

---

[5]A grid graph is a unit disk graph in which all the disks have radius $= 1/2$ and all the centers have integer coordinates.

problems have efficient NC approximation schemes. The approximation schemes assume that a civilized layout of the graph is given.

Summary of Results

Note: We assume that the performance guarantees for minimization as well as maximization problems are always at least 1. Problem defintions can be found in [Ba83]

Non-hierarchical				
Problem	λ-precision	civilized	unit disk	coin	isothetic squares
MIN VC	$(\frac{k+1}{k})$	$(\frac{k+1}{k})$	$(\frac{k+1}{k})^2$	1.66	$(\frac{k+1}{k})^2$
MAX IS	$(\frac{k+1}{k})$	$(\frac{k+1}{k})$	$(\frac{k+1}{k})^2$	5	$(\frac{k+1}{k})^2$
MIN DS	$(\frac{k+1}{k})$	$(\frac{k+1}{k})$	$(\frac{k+1}{k})^2$	?	$(\frac{k+1}{k})^2$
MAX Partition Into Triangles	$(\frac{k+1}{k})$	$(\frac{k+1}{k})$	3	3	3
MIN Edge DS	$(\frac{k+1}{k})$	$(\frac{k+1}{k})$	2	2	2
MAX H-matching	$(\frac{k+1}{k})$	$(\frac{k+1}{k})$	?	?	?

Table 1 : Summary of Results for Geometric Intersection Graph Problems.

k-near-consistent specifications				
Problem	λ-precision	civilized	unit disk	coin	isothetic squares
MIN VC	$(\frac{l+1}{l})^2$	$(\frac{l+1}{l})^2$	$(\frac{l+1}{l})^3$	1.66	$(\frac{l+1}{l})^2$
MAX IS	$(\frac{l+1}{l})^2$	$(\frac{l+1}{l})^2$	$(\frac{l+1}{l})^3$	5	$(\frac{l+1}{l})^3$
MIN DS	$(\frac{l+1}{l})^2$	$(\frac{l+1}{l})^2$	$(\frac{l+1}{l})^3$	?	$(\frac{l+1}{l})^3$
MAX Partition Into Triangles	$(\frac{l+1}{l})^2$	$(\frac{l+1}{l})^2$	3	3	3
MIN Edge DS	$(\frac{l+1}{l})^2$	$(\frac{l+1}{l})^2$	2	2	2
MAX H-matching	$(\frac{l+1}{l})^2$	$(\frac{l+1}{l})^2$	?	?	?

Table 2 : Summary of Results for Near-Consistent BOW specifications

2.2   Hierarchically Specified Geometric Intersection Graphs

Our results in the previous sections can be extended to geometric intersection graphs specified hierarchically. These extensions are summarized in Table 2. Our primary motivation for studying such hierarchically specified intersection graphs is that many VLSI design languages such as CIF [MC80] use geometric objects such as circles and/or rectangles as primitives. These languages are used to give succinct specifications of large designs. Hence it is natural to investigate the complexity of graph theoretic problems for intersection graphs specified hierarchically. The Hierarchical Input Language (HIL) of Bentley et al. [BOW83] is subset of CIF. Each HIL specification defines a set of geometric objects. A description in HIL can be interpreted naturally as specifying the intersection graph of the set of objects defined. It is in this sense that we view HIL as specifying geometric intersection graphs. In its full generality the specification language of [BOW83] is too powerful since even simple questions such as "Does there exist an intersecting pair of rectangles in the set?" become NP-hard. Bentley, Ottmann and Widmayer in [BOW83] state that

> "It will be important to identify families of restrictions that exclude only a few designs but admit to very rapid processing of remaining designs."

In an attempt to answer the above question, we define a family of syntactic restrictions on HIL descriptions and call the resulting specifications k-near-consistent specifications. While we can show that many graph problems are PSPACE-hard even for 1-near-consistent specifications, nevertheless, these problems possess efficient approximation schemes.

3   Related Work

The complexity of graph problems when restricted to unit disk graphs has been studied extensively in the past [CCJ90, FPT81, MS84]. In [MHR92], we showed that several natural graph problems such as maximum independent set, minimum vertex cover and minimum dominating set can be approximated to within a constant factor of the optimum for unit disk graphs. It was left open whether the performance guarantees for any of the problems could be improved.

It is easy to see that every λ-precision unit disk graph is the intersection graph of a k-ply neighborhood system [EMT93, Te91], where k depends on the precision factor λ. Similarly, the intersection graph of a k-ply neighborhood system with unit balls is a λ-precision unit disk graph, where λ is the minimum distance between the centers of any two balls. Using the geometric separator concept of Eppstein, Miller and Teng [EMT93], one can find an approximation scheme for problems restricted to λ-precision unit disk graphs in the same way as the planar separator theorem [LT79] can be used to find an approximation schemes for problems restricted to planar graphs. However, this approach has two main drawbacks. The first drawback is that as in the case of planar graphs, the approximation schemes apply only in the asymptotic sense and hence are not practical. (See Baker [Ba83]). The second drawback is that problems such as maximum independent set and minimum dominating set for which approximation schemes can be designed for arbitrary unit disk graphs by our method *cannot* be solved at all by the separator approach. This is because an arbitrary unit disk graph on n nodes can have a clique of size n. Hence, in general, it is not possible to find good separators for unit disk graphs.

In Baker [Ba83], polynomial time approximation schemes were provided for a large class of problems for planar graphs. Recently, [DST93, HM+93] showed how to parallelize the ideas in [Ba83] to obtain efficient NC approximation schemes for problems restricted to planar graphs. Hochbaum and Maass [HM85] used ideas similar to Baker [Ba83] to obtain polynomial time approximation schemes for covering and packing problems in the plane. Feder and Greene [FG88] use a similar technique to provide an approximation scheme for a geometric location problem related to clustering. Recently Jiang and Liang [JW94] have obtained a polynomial time approximation scheme for the Steiner tree problem for graphs drawn in civilzed manner using a technique similar to the one discussed in this paper.

In [MHR93, MR+93, MH+94] we have investigated the existence and/or non-existence of polynomial time approximations and approximation schemes for several PSPACE-hard problems for succinctly specified graphs. In [MH+94], we developed a general approach to prove hardness results for succinctly specified graphs. Recently, Condon, Feigenbaum, Lund and Shor [CF+93a, CF+93b] characterized PSPACE in terms of probabilistically checkable debate systems and used this characterization to investigate the existence and non-existence of polynomial time approximation algorithms for PSPACE-complete problems. Many of the problems shown to have efficient approximation schemes here are PSPACE-complete [MR+93]. Thus the results presented here along with our results in [MH+94] are the first ever approximation schemes for natural PSPACE-hard optimization problems.

4 Approximation Schemes for Unit Disk Graphs

Our technique of obtaining a polynomial time approximation scheme for a given problem for unit disk graphs relies on the following two crucial properties.

1. The ability to decompose the given set of unit disks into disjoint subsets of disks such that an optimal solution to each subset can be obtained in polynomial time.

2. The optimal solutions obtained for each subset can be merged into a near-optimal solution for the whole graph.

We illustrate the idea by giving an NC-approximation scheme for the maximum independent set problem. Given a set of unit disks in the plane, we divide the set first into horizontal strips of width two. (Recall that we assume without loss of generality that all the disks have unit radius). Given an $\epsilon > 0$, let

$k = \lceil 1/\epsilon \rceil - 1$. Next, for each i, $0 \le i \le k$, partition the set of disks into r disjoint sets $G_{i,1} \cdots G_{i,r}$ by removing all the disks in every horizontal strip congruent to $i \bmod (k+1)$. For each subgraph $G_{i,j}$, $1 \le j \le r$, we find an independent set of size at least $\frac{k}{k+1}$ times the optimal value of the independent set in $G_{i,j}$ as follows. For each i_1, $0 \le i_1 \le k$, partition the set of disks in the set $G_{i,j}$ into s_j disjoint sets $G_{i,j}^{i_1,1} \cdots G_{i,j}^{i_1,s_j}$ by removing every vertical strip congruent to $i_1 \bmod (k+1)$. Such a partition breaks the set of disks $G_{i,j}$ into small squares of side k, By a simple packing argument it can be shown that for any fixed $k > 0$, the maximum independent set of a unit disk graph, all of whose disks lie in a square of side k is $O(k^2)$. This immediately gives us a way of finding an optimal solution in NC for each small square. By an argument similar to the shifting lemma in [HM85], we can now prove the required performance guarantee.

We can get a more efficient algorithm in the sequential case by obtaining an optimal independent set for each $G_{i,j}$ using dynamic programming. As the subsequent analysis will indicate this gives an approximation algorithm with performance guarantee $\frac{k}{k+1}$.

The algorithm takes $O(n^k)$ work. It is easy to see that the algorithm can be implemented in NC. Other graph problems can also be solved similarly. In the case of minimization problems, instead of partitioning the set of unit disks, the subgraph G_l consists of the set of disks that lie between the $(l-1)^{st}$ horizontal strip congruent to $i \bmod (k+1)$ and the lth horizontal strip congruent to $i \bmod (k+1)$, including the disks in the horizontal strips. Details are omitted due to lack of space and can be found in [HM+94].

4.1 Performance Guarantee

Let $G_i = \bigcup_{1 \le j \le r} G_{i,j}$ and $G_{i,j}^{i_1} = \bigcup_{1 \le j_1 \le s_j} G_{i,j}^{i_1,j_1}$. Furthermore, let $IS(G_p)$ and $OPT(G_p)$ denote respectively the independent set obtained by the heuristic and an optimal independent set for the subgraph G_p. The performance guarantee of our approximation algorithm can be proven as a sequence of lemmas given below. The proof of these lemmas is similar to the proof for maximum independent set problem for planar graphs outlined in Baker [Ba83] and is omitted.

Lemma 4.1 $\max_{0 \le i \le k} |OPT(G_i)| \ge \frac{k}{(k+1)} |OPT(G)|$

Theorem 4.2 $|IS(G_{i,j})| \ge (\frac{k}{k+1}) \cdot |OPT(G_{i,j})|$.

By a repeated application of the above lemma we get

Theorem 4.3 $|IS(G)| \ge (\frac{k}{k+1})^2 \cdot |OPT(G)|$.

In case of graphs drawn in a civilized manner (and hence for λ-precision unit disk graphs), one can solve many more problems this way and also get better running times. This is because the treewidth of the vertex induced graph whose vertices (disks) lie in a strip of width k is $O(k)$. Given this, we can use efficient dynamic programming algorithms for treewidth-bounded graphs and get approximation algorithms which require $O(c^k n)$ work for some constant $c > 0$, and have a performance guarantee of $(\frac{k}{k+1})$.

5   Hierarchically Specified Geometric Intersection Graphs

Next, we discuss our ideas for obtaining approximation algorithms for the maximum independent set problem for a set of hierarchically specified unit disks. The hierarchical specification (referred as the BOW-specification) language used here to describe a set of unit disks is almost identical to that used by Bentley, Ottmann and Widmayer [BOW83] to describe a set of isothetic rectangles. The only difference is that instead of the BOX command we have the DISK command. The syntax of the DISK command is DISK (x, y, r) where (x, y) is the center of the disk and r is the radius. A **symbol** (also referred to as non-terminal) in this language represents a collection of unit disks and has a unique identifier (symbol number) which is a positive integer. The description for a symbol consists of zero or more DISK commands and DRAW commands. The syntax of the DRAW command is DRAW symbol# at (x, y). Here the symbol# is the identifier of a *previously defined* symbol and (x, y) specifies the translation factor to be applied to the centers of the disks defined in the specified symbol. A description in this language consists of a sequence of symbol definitions. The length of such a description is defined to be the total number of DISK and DRAW commands. The set of disks specified by such a description is the one corresponding to the symbol with the largest identifier.

With the set of unit disks defined as above, we associate an *intersection graph* which has one vertex per unit disk and two vertices are joined by an edge iff the corresponding disks intersect.

5.1   Near Consistent BOW-Specifications

In [BOW83], Bentley et al. show that the general HIL is too powerful in that, it makes most of the natural problems intractable. They then define the notion of *consistent HIL* which is easily realized by adding an attribute called the **MBR** (minimum bounding rectangle) to each symbol [BOW83]. This attribute denotes the smallest bounding rectangle enclosing the set of unit disks associated with the symbol. The formal syntax of a definition of a symbol is:
DEFINE symbol#
attribute: MBR([x-value],[y-value])([width],[height])
Sequence of DRAW and DISK commands

Here ([x-value],[y-value]) denote the leftmost corner of the minimum bounding rectangle(MBR) and ([width],[height]) denote the width and the height of MBR. Each disk specified using a DISK command in the definition of a symbol G will be referred to as an explicit disk.

Definition 5.1 *([BOW83]) A BOW specification is* **consistent** *if the MBR of symbols called within a symbol and the explicitly defined disks do not intersect.*

As observed in [BOW83], consistency is a very strong restriction as any set of rectangles containing an intersecting pair cannot be represented using a consistent BOW specification. Extending the definition of consistency, we define the concept of k-near-consistent specifications. We first give some additional notation. Associated with each BOW-specification $\Gamma = (G_1, ..., G_n)$ (where G_n denotes the largest symbol) is a tree structure depicting the sequence of calls made by the symbols defined in Γ. Following Lengauer and Wanke [LW87a] we call it the **hierarchy tree** $HT(\Gamma)$ associated with Γ. Intuitively, for a fixed $k \geq 0$, k-near-consistent BOW-specifications allow one to have intersections which occur between explicit symbols defined close to each other in the hierarchy tree of the specification. Given a hierarchy tree $HT(\Gamma)$, we can associate a level number with each node (a symbol) in the tree. The *level number* of a node in $HT(\Gamma)$ is the number of egdes in the path from the node to the root of the tree $HT(\Gamma)$.

We let $E(G_i)$ denote the set of unit disks obtained by expanding the hierarchy tree $HT(G_i)$. Thus $E(G_n)$ denotes the set of unit disks described by a given specification $\Gamma = (G_1, ..., G_n)$.

Definition 5.2 *A BOW specification is 1-near-consistent BOW-specification iff the following conditions hold:*

1. *For each symbol G_i, the MBR of G_i contains the MBR of all the symbols called in G_i.*

2. *The MBR of the symbols called in G_i do not intersect one another.*

3. *For each explicit disk u defined in G_i, u does not intersect the MBR of any symbol G_k such that G_k occurs in $HT(G_i)$ and level number of G_k in $HT(G_i)$ is ≥ 2.*

The above definition can be easily extended to k-near-consistent BOW-specifications, for any fixed $k \geq 1$. Note that this is a strict extension of consistent BOW-specifications[6]. Given the above syntactic restriction, a natural question to ask is the following: Given BOW specification, how hard is it to verify that the specification obeys the above restriction ? Our next theorem points out that the verification can be done in polynomial time.

Theorem 5.1 *For any fixed $k \geq 0$, there is a polynomial time algorithm to check if a given BOW-specification $\Gamma = (G_1, \cdots, G_n)$, is a k-near-consistent BOW-specification.*

5.2 Meaning of Approximation Algorithms for Succinct Specifications

Before discussing our algorithms, it is important to understand what we mean by a *polynomial time approximation algorithm* for a problem Π, when the instance is specified succinctly. Our definition of approximation algorithm is best understood by means of the following example from [MHR93].

---

[6] We remark that most of the problems solvable in polynomial time for consistent specifications [BOW83] are solvable in polynomial time for k-near-consistent specifications (for any fixed $k \geq 0$).

Example: Consider the minimum vertex cover problem, where the input is a succinct specification of a graph G and we wish to compute the size of a minimum vertex cover of G. Our polynomial time approximation algorithm for the vertex cover problem computes the size of an approximate vertex cover and runs in time *polynomial* in the *size* of the succinct description, rather than the *size* of G. Moreover, it also solves in polynomial time (in the size of the succinct specification) the following **query problem:** Given any vertex v of G and its position in the expanded specification, determine whether v belongs to the approximate vertex cover so computed. Moreover, for all the problems considered here, we can output in polynomial time a succinct specification of the approximate set computed. ■

5.3 The Basic Technique

We now give the basic technique behind all our approximation algorithms. for unit disk graphs specified using a near-consistent BOW-specification. For the sake of simplicity, we assume that the BOW-specification is 1-near-consistent. Given a maximization problem Π in Table 1, our approximation algorithm takes time $O(N \cdot T(N^l))$ to achieve a performance guarantee of $(\frac{l}{l+1}) \cdot FBEST$. Here, l is a fixed constant depending only on the required performance guarantee ϵ and $T(N^l)$ denotes the running time of a heuristic with performance guarantee $FBEST$, used to process flat specifications of size $O(N^l)$. During an iteration i we delete all the explicit vertices which belong to non-terminals defined at level j, $j = i \bmod(l+1)$ [7]. This breaks up the given hierarchy tree into a collection of disjoint trees. The heuristic finds a near optimal solution for the vertex induced subgraph[8] defined by each small tree and then outputs the union of all these solutions as the solution for the problem Π. It is important to observe that the hierarchy tree can have an exponential number of nodes and hence the deletion of non-terminals and finding near optimal solutions for each subtree has to be done in such a manner that the whole process takes only polynomial time. This is achieved by observing that the subtrees can be divided into n distinct equivalence classes and that it is easy to count the number of subtrees in each equivalence class.

Acknowledgements: We thank R. Ravi and Ravi Sundaram for constructive suggestions. We also thank Professor S.H. Teng and Dr. S. Ramanathan for making available copies of their theses and Professor Tao Jiang for making available a copy of their paper. The second author also thanks Thomas Lengauer and Egon Wanke for fruitful discussions on succinct specification languages.

References

[Ba83]   B.S. Baker, "Approximation Algorithms for NP-complete problems on

---

[7] For minimization problems instead of deleting the vertices in the level, we consider the vertices as a part of both the subtrees

[8] For a fixed l, the size of each subgraph is polynomial in the size of the specification.

Planar Graphs," *24th IEEE Symposium on Foundations of Computer Science (FOCS)*, 1983, pp 265-273.

[BOW83] J.L. Bentley, T. Ottmann and P. Widmayer, "The Complexity of Manipulating Hierarchically Defined set of Intervals," *Advances in Computing Research, ed. F.P. Preparata* Vol. 1, (1983), pp. 127-158.

[CF+93a] A. Condon, J. Feigenbaum, C. Lund and P. Shor, "Probabilistically Checkable Debate Systems and Approximation Algorithms for PSPACE-Hard Functions", in *Proc. 25th Annual ACM Symposium on Theory Of Computing, (STOC)*, 1993, pp. 305-313.

[CF+93b] A. Condon, J. Feigenbaum, C. Lund and P. Shor, "Random Debators and the Hardness of Approximating Stochastic functions for PSPACE-Hard Functions", to appear in *9th Annual IEEE Annual Conference on Structure in Complexity Theory*, June 1994.

[CCJ90] B.N. Clark, C.J. Colbourn, and D.S. Johnson, "Unit Disk Graphs," *Discrete Math.*, Vol. 86, 1990, pp. 165-177.

[DST93] J. Diaz, M.J. Serna and J. Toran, "Parallel Approximation Schemes for problems on planar graphs," *1st European Symposium on Algorithms (ESA '93)*, 1993, pp 145-156.

[EMT93] D. Eppstein, G.L. Miller and S.H. Teng, "A Deterministic Linear Time Algorithm for Geometric Separators and its Application," *9th ACM Symposium on Computational Geometry*, pp 99-108, 1993.

[FG88] T. Feder and D. Greene, "Optimal Algorithms for Approximate Clustering," *30th ACM Symposium on Theory Of Computing (STOC)*, pp 434-444, 1988.

[FPT81] R.J. Fowler, M.S. Paterson and S.L. Tanimoto, "Optimal Packing and Covering in the Plane are NP-Complete," *Inf. Proc. Letters*, Vol 12, No.3, June 1981, pp. 133-137.

[Ha80] W.K. Hale, "Frequency Assignment: Theory and Applications," *Proc. IEEE*, Vol. 68, 1980, pp 1497-1514.

[HM85] D.S. Hochbaum and W. Maass, "Approximation Schemes for Covering and Packing Problems in Image Processing and VLSI," *JACM*, Vol 32,No. 1, 1985, pp 130-136.

[JW94] T.Jiang and L. Wang, "An Approximation Scheme for Some Steiner Tree Problems in the Plane," to appear in *Fifth Annual International Symposium on Algorithms and Computation (ISAAC)*, 1994.

[HM+93] H.B. Hunt III, V. Radhakrishnan, S.S. Ravi, D.J. Rosenkrantz, R.E. Stearns, "Every problem in MAX SNP has a parallel approximation algorithm," Technical Report No 8, University at Albany, May 1993.

[HM+94] H.B. Hunt III, M.V. Marathe, V. Radhakrishnan, S.S. Ravi, D.J. Rosenkrantz and R.E. Stearns, "A Unified Approach to Approximation Schemes for NP- and PSPACE-Hard Problems for Geometric Graphs," Technical Report No 93-17, University at Albany, June 1994.

[LW87a] T. Lengauer and E. Wanke, "Efficient Solutions for Connectivity Problems for Hierarchically Defined Graphs ," *SIAM J. Computing*, Vol. 17, No. 6, 1988, pp. 1063-1080.

[LT79] R. Lipton and R. E. Tarjan, "A Separator Theorem for Planar Graphs," *SIAM J. Appl. Math.*, Vol. 32, No. 2, April 1979, pp 177-189.

[MHR92] M.V. Marathe H.B. Hunt III and S.S. Ravi, "Geometric Heuristics for Unit Disk Graphs", in the proceedings of *4th Canadian Conference on Computational Geometry*, 1992, pp. 244-249.

[MHR93] M.V. Marathe H.B. Hunt III and S.S. Ravi, "The Complexity of Approximating PSPACE-Complete Problems for Hierarchical Specifications", in the proceedings of *ICALP'93*, 1993, pp 76-87.

[MR+93] M.V. Marathe, V. Radhakrishnan, H.B. Hunt III and S.S. Ravi, "Hierarchically Specified Unit Disk Graphs", to appear in the proceedings of *WG'93*, 1993.

[MH+94] M.V. Marathe, H.B. Hunt III, R.E. Stearns and V. Radhakrishnan, "Approximation schemes for PSPACE-Complete problems for succinct graphs," to appear in *Proceedings of 26th Annual ACM Symposium on the Theory of Computing (STOC)*, May 1994.

[MC80] C. Mead and L. Conway, *Introduction to VLSI Systems*, Addison Wesley, 1980.

[MS84] N. Meggido and K Supowit, "On The Complexity Of Some Common Geometric Location Problems," *SIAM Journal Of Computing*, Vol 13, No.1, February 1984, pp. 182-196.

[PY91] C. Papadimitriou and M. Yannakakis, "Optimization, Approximation and Complexity Classes" *Journal of Computer and System Sciences* , No.43, 1991, pp. 425-440.

[Ra93] S. Ramanathan *Scheduling Algorithms for Multi-Hop Radio Networks*, Ph.D. thesis, Department of Computer Science, University of Delaware, Newark, 1993.

[Te91] S.H. Teng, *Points, Spheres, and Separators, A Unified Geometric Approach to Graph Separators*, Ph.D. thesis, School of Computer Science, Carnegie Mellon University, CMU-CS-91-184, Pittsburgh, August 1991.

[WK88] D.W. Wong and Y.S. Kuo, "A Study of Two Geometric Location Problems," *Inf. Proc. Letters*, Vol. 28, No. 6, Aug. 1988, pp 281-286.

The Parallel Complexity of Eden Growth, Solid-on-Solid Growth and Ballistic Deposition (Extended Abstract)

Raymond Greenlaw[1*] and Jonathan Machta[2**]

[1] Department of Computer Science, University of New Hampshire, Durham, New Hampshire 03824, e-mail address: greenlaw@cs.unh.edu
[2] Department of Physics and Astronomy, University of Massachusetts, Amherst, Massachusetts 01003, e-mail address: machta@phast.umass.edu

Abstract. In this interdisciplinary research we apply the tools of algorithmic complexity theory to three important non-equilibrium growth models that are used in statistical physics. Much of the insight into these models has been derived by numerical simulation; it is important to develop good parallel algorithms for them. *Eden growth, solid-on-solid growth* and *ballistic deposition* are all seemingly highly sequential processes. However, we are able to provide algorithms for the models that run in time $O(\log^2 N)$ using a polynomial number of processors on a randomized CREW P-RAM, where N is the system size. In addition to their potential practical value, our algorithms serve to classify these growth models as less complex than other growth models, such as *diffusion-limited aggregation*, for which fast parallel algorithms probably do not exist.

1 Introduction

Non-equilibrium growth phenomena are encountered in a diverse array of physical settings ranging from fluid flow in porous materials to the growth of cell colonies [11, 19, 2]. In many cases, non-equilibrium growth leads spontaneously to complex self-affine or self-similar patterns. A variety of models have been developed to gain an understanding of how complex patterns develop on large length scales due to the operation of simple microscopic rules. Although much is known about these models, most of our insight has been derived from numerical simulation. Developing fast parallel algorithms for conducting such simulations is an important line of research. To our knowledge, our research is the first to formalize, analyze and describe the parallel complexity of non-equilibrium growth models, beginning with the results contained in [12].

* This research was partially funded by the National Science Foundation Grant CCR-9209184.
** This research was partially funded by the National Science Foundation Grant DMR-9311580.

We examine several well-known growth models from the perspective of parallel computational complexity. Specifically we study *Eden growth, solid-on-solid growth* and *ballistic deposition*, and provide a fast parallel algorithm for each model. Here 'fast' means that the running time scales polylogarithmically ($O(\log^{k_1} N)$ for some constant k_1 and abbreviated *polylog*) in the size of the system and that the number of processors scales polynomially in the system size on a probabilistic CREW P-RAM. Note that we study *sampling problems* and use randomization. Thus it is only in an informal sense that our problems are in **NC**. It is interesting to compare our results with a previous one [12] that proves it is very unlikely there is a fast parallel algorithm for *diffusion-limited aggregation*.

There are several reasons for investigating growth models from the point of view of parallel computation. First, with the rapidly increasing availability of massively parallel computers it is important to develop approaches for simulating various physical systems in parallel. Although the algorithms presented here are for an idealized parallel model, they may serve as a starting point for the design of practical algorithms for large scale parallel machines. It should be possible to achieve significant speed up using the approaches described here; these techniques demonstrate the unexpected parallelism in the problems we consider.

On a more fundamental level, the existence, structure and complexity of the parallel algorithms provides a new perspective on the models themselves. One question that can be addressed is whether a growth process is intrinsically history dependent. All of the growth models discussed apparently are in the sense that the growth at a given time depends on the prior history of the system. For example, consider the growth rules for Eden clusters. At each time step a new particle is added randomly to the perimeter of the cluster. This particle modifies the cluster and its perimeter. The model is defined as a sequential procedure that requires K steps to create a cluster with K particles. It is not at all obvious that one can generate randomly chosen Eden clusters in polylog time even with many processors running in parallel. The existence of fast parallel algorithms for the growth models considered here shows that their apparent history dependence can be overcome.

A key step in constructing parallel algorithms for the non-equilibrium growth models is mapping them onto minimum weight path problems for which fast parallel algorithms already exist. Eden growth and the solid-on-solid model have previously been shown [9, 17, 18] to be equivalent to a *waiting time* growth model that can be mapped onto a minimum weight path problem. We give a new connection between the growth models and minimum weight paths that leads to a group of algorithms using a *random list* approach.

The remainder of this paper is organized as follows: in Section 2 we present some background material, in Section 3 we describe the growth models, in Section 4 fast parallel algorithms are developed for the growth models, and in Section 5 we discuss the results. Proof outlines and sketches of our main theorems are given in Section 4.

2 Parallel computation

The reader is referred to one of the following references for basic background on parallel computation [3, 6, 8]. The natural problems in computational statistical physics are *sampling* problems. The objective is to generate an unbiased random sample from a distribution of system states or histories. This is somewhat different from the *decision* problems usually considered in complexity theory. Decision problems have "yes" or "no" answers. The algorithms we describe here produce a list of integers as output.

In order to generate random objects one must have a supply of random or pseudo-random numbers. We assume a supply of perfect random numbers although in practice a good pseudo-random number generator would be used. The probabilistic model we adopt for studying sampling problems is a variant of the P-RAM. The *probabilistic P-RAM* is a P-RAM in which each processor is equipped with a register for generating random numbers. We use the concurrent read, exclusive write (CREW) P-RAM model in which only one processor is allowed to write to a given memory location at a time.

2.1 The minimum weight path parallel algorithm

At the heart of the fast parallel algorithms for growth models is a standard algorithm for finding minimum weight paths (MWPs) between each pair of vertices in an undirected graph [5]. Because we use variations of the standard algorithm for computing MWPs, we formally define the problem and provide the standard algorithm for solving it.

Input: An undirected graph, $G = (V, E)$, where V is a set of sites and E is a set of bonds connecting pairs of sites. Let $N = |V|$. The matrix of weights, $w(i, j)$, $1 \leq i, j \leq N$, with $w(i, i) = 0$, where $w(i, j)$ is the weight of bond $\{i, j\} \in E$.

Output: A matrix containing weights of the minimum weight *simple paths* between every pair of sites in V.

Parallel Minimum Weight Path Algorithm
    begin
        for i, j, $1 \leq i, j \leq N$, in parallel do
            if $(\{i, j\} \in E$ or $i = j)$ then $W(i, j) \leftarrow w(i, j)$ else $W(i, j) \leftarrow \infty$;
        for $l \leftarrow 1$ to $\lceil \log N \rceil$ do
            for all i, j, k, $1 \leq i, j, k \leq N$, in parallel do
                $W(i, j) \leftarrow \min_k [W(i, k) + W(k, j)]$;
    end.

At the lth step in the outer for loop, $W(i, j)$ contains the weight of the MWP from i to j having length 2^l or less. Each step allows the path length to double. Since finding the minimum over N numbers can be done in $\log N$ time and since the iteration is repeated $\log N$ times, the parallel MWP algorithm takes

$O(\log^2 N)$ time. It requires $N^3/\log N$ processors on the CREW P-RAM [5, page 26]. The resources required by the MWP algorithm dominate those needed for fast simulations of the growth models as we will see in Section 4.

The parallel MWP algorithm can be used to find path weights for which a quantity other than the *sum* of the weights is either minimized or maximized. Let $\oplus$ stand for an associative binary operation such as addition or taking the maximum. If the weight of a path is defined by the $\oplus$ of the weights along the path, the MWP problem can be solved by the parallel MWP algorithm by simply replacing $+$ by $\oplus$.

The parallel MWP algorithm can be used for directed graphs by changing the braces to parentheses in line three. It is also easily adapted to site weights. Here the weight of the destination is included in the path weight but the weight of the source is not included. The algorithm can also be used to find the *connected components* of a graph.

3 Growth models defined on general graphs

Growth models are conventionally defined on d-dimensional lattices, however, it is convenient for our purposes to define these models on more general structures, i.e. graphs. The graph theoretic viewpoint simplifies the presentation of the parallel algorithms and allows us to make connections with algorithms for other graph theory problems.

For the Eden model growth proceeds on an undirected graph, $G = (V, E)$. The output of the model is a connected set of sites called a *cluster* together with an ordering, $c : \mathbf{K} \to V$, of when sites are added to the cluster. Here $\mathbf{K} = \{0, 1, \ldots, K\}$, where K denotes the size of the cluster. In each case, K will denote the number of iterations of a growth rule. The function c will specify an ordering. In general, for a fixed graph, each growth model defines a different probability distribution on the cluster ordering. The distributions are defined by rules that specify how new sites are added to the cluster. The new site is chosen from the *perimeter* of the existing cluster. A perimeter site is a non-cluster site that is connected by some bond to an element of the cluster. Growth begins with an initial cluster S.

The solid-on-solid and ballistic deposition models describe directional growth. An undirected graph, $G = (V, E)$, defines a 'substrate' above which growth occurs. Growth occurs on a 'space-time' graph, $G_{st} = (V_{st}, E_{st})$, derived from G. The sites, $V_{st} = \{[i, h] \mid i \in V$ and $h = 0, 1, \ldots, H\}$, form columns and h is interpreted as the height above the substrate. The directed bonds for the space-time graph, E_{st}, are specified below in Section 3.2.

Each of the three models described above has been extensively studied and a wide range of variants have been introduced to model specific physical situations or to improve numerical results. Below we describe the basic versions of the growth rules for each model.

3.1 Eden growth

Eden growth was originally introduced as a model for tumor growth [1, 11]. The growth rule is

CLUSTER ← CLUSTER ∪ {j}, where j is chosen with equal
probability from the set of perimeter sites of the cluster;

Large Eden clusters grown on lattices are compact [16] and nearly spherical but exhibit statistically self-affine surface roughness [15, 4]. It is believed that the exponents characterizing this surface roughness are the same as those of the next two models we describe and also of the stochastic differential equation introduced in [7].

3.2 Restricted solid-on-solid growth

The solid-on-solid model [14, 10, 11] was originally devised to study crystal growth. The version presented here is most closely related to the model studied in [10] and referred to as the 'restricted solid-on-solid model' (RSOS). We begin with an undirected graph, $G = (V, E)$, and a maximum allowed height H. The RSOS clusters may be described by a set of column heights that can only be incremented if they are less than or equal to the height of all neighboring columns. The initial cluster consists of a set of sites together with their starting column heights (see below). The RSOS model can be viewed as a cluster growth model on a space-time graph with directed bonds,

$$E_{st} = \{([i, h-1], [j, h]) \mid \{i, j\} \in E \text{ and } h = 1, 2, \ldots, H\} \qquad (1)$$
$$\cup \ \{([j, h-1], [j, h]) \mid j \in V \text{ and } h = 1, 2, \ldots, H\}.$$

Technically, if the set of sites in the initial cluster S consists of i, $1 \leq i \leq n$, with corresponding initial heights $h(i, 0)$, then

$$S = \{[i, x] \mid 1 \leq i \leq n \text{ and } 1 \leq x \leq h(i, 0)\} \cup \{[i, 0] \mid \forall\, i \in V\}$$

forms the initial cluster in the space-time graph. Here h is the height function described previously. Typically, the first component in the union making up S will be empty. Allowed growth sites on the space-time graph have the property that *all* of their immediate predecessors are in the cluster. At each step, an allowed growth site is randomly chosen and added to the cluster. The interfaces created by RSOS growth are locally smoother than those created by Eden growth and ballistic deposition (see below).

3.3 Ballistic deposition

Ballistic deposition [20, 14, 11] was developed to model sedimentation, colloidal aggregation and vapor deposition of thin films. Ballistic deposition simulates growth above a substrate that is here taken as an undirected graph, $G = (V, E)$.

Each site of the graph defines a 'column.' A function, $h(i, n)$, is interpreted as the height of column i at step n. Initially, the height of each site is set to zero and at each step a site, i, is randomly chosen. The height of i is incremented according to the following rule:

$$h(i, n+1) \leftarrow \max[\{h(j, n) \mid \{i, j\} \in E\} \cup \{h(i, n) + 1\}];$$

If the substrate is a two-dimensional lattice, ballistic deposition simulates particles falling one at a time and sticking at the highest level at which they meet the growing cluster. In terms of the space-time graph, G_{st}, described in Section 3.2 (with H replaced by K), the initial cluster is the $h(i, 0) = 0$ 'plane.' At each step the cluster grows at the site where the height is incremented, i.e. $c(n) = [i, h(i, n)]$ if $h(i, n)$ is greater than $h(i, n-1)$. Ballistic deposition typically yields a delicate forest of closely packed trees.

4 Fast parallel algorithms for growth models

The probabilistic parallel algorithms presented below generate cluster orderings (and/or height functions) for each growth model. A single run of any of the algorithms requires polylog time and uses a polynomial number of processors on a probabilistic CREW P-RAM. The input to the algorithms is an undirected graph, $G = (V, E)$, and an initial cluster, S. For Eden growth S is a connected subset of V. For the RSOS model and ballistic deposition, S is a connected subset of the space-time graph that is an allowed cluster according to the rule of the given model. In each case, let N equal $|V|$.

The algorithms for the Eden and RSOS models presented below parallelize the following sequential strategy. At each step a site is chosen at random. If the site is an allowed growth site, it is added to the cluster; otherwise, it is discarded. To parallelize this method, the randomly chosen sites are first prepared as a list and then the MWP algorithm is used to determine whether each is an allowed growth site. A variant of this approach is used to sample the height function distribution for ballistic deposition.

4.1 Eden growth

The pseudo-code given below produces Eden clusters in parallel. The output of the algorithm is the cluster order.

Parallel Eden Growth Algorithm
> **begin**
>> 1. Generate a random list of sites, $(v_1, v_2, \ldots, v_T)$, chosen from $V - S$.
>> 2. Construct a directed graph, $G' = (\{0, 1, \ldots, T\}, E')$, where
>>> for each $1 \leq m, n \leq T$, if $(\{v_m, v_n\} \in E$ and $m < n)$ then
>>>> $(m, n) \in E'$, and
>>> for each $1 \leq n \leq T$, if v_n is on the perimeter of S then
>>>> $(0, n) \in E'$.

3. Let r be the number of sites in G' that can be reached from site 0 by a simple path.

4. Let $R[k] = j$, where j is the kth site in G' (in sorted order) reachable from site 0 ($1 \le k \le r$ and $j \in \{1, 2, \ldots, T\}$).

5. For $k \leftarrow 1$ to r do $L[k] \leftarrow v_{R[k]}$.

6. Compact L by deleting all but the first appearance of each site.

7. $c(n) \leftarrow L[n]$ is the nth site added to the cluster (n starts at 1).

end.

The following theorem proves that the algorithm is correct and runs in poly-log time.

Theorem 1. *Let $G = (V, E)$ be an undirected graph and S the initial cluster. The parallel Eden growth algorithm produces a cluster ordering for G according to the growth rules defining the Eden model. Suppose $T = O(N^{k_1})$ for some constant k_1. The algorithm runs in $O(\log^2 N)$ time using a polynomial number of processors on a probabilistic CREW P-RAM.*

Proof. (Idea) Correctness: The construction in Step 2 insures that G' reflects both the connectivity of the original graph G and the order of adding sites established by the list $(v_1, v_2, \ldots, v_T)$. If an element m is connected to site 0 in G' then its pre-image, v_m, is either a perimeter site at step $m - 1$, or it has already joined the cluster before step m. If m is connected to site 0, its pre-image is added to array L in Step 5. A site becomes part of the cluster when it makes its first appearance in L. In Step 6 additional appearances of every site are deleted from L resulting in the cluster order. Since the original list was random, each successive site in the cluster is chosen at random from the allowed growth sites as required by the definition of the Eden model.

Resource bounds: All of the steps in the algorithm can be parallelized using parallel sorting, parallel prefix computations [8] and parallel MWP computations. Each requires a polynomial number of processors. For a choice of T equal to $O(N^{k_1})$, the running time of the algorithm is dominated by the parallel MWP algorithm used in Step 3 (to test reachability) and is $O(\log^2 N)$. □

Remarks As presented the algorithm has a non-vanishing probability of failing to produce a cluster of a given size. The probability of obtaining a complete cluster of size K equal to N can be made close to one with the choice of T equal to $\Omega(N^2)$. Informally, this is the case since there is at least one perimeter site available at each time step and the probability of choosing that site is at least $1/N$. If failures are permitted, the Eden cluster distribution is not perfectly sampled since clusters with atypically large perimeters are less likely to fail and will be favored. However, one can generate the exact distribution by iterating the algorithm several times using the final cluster from one run as the initial cluster for the next run. The iterations are continued until a cluster of the desired size is obtained. This method produces an unbiased sample of Eden clusters.

For d-dimensional lattices Eden clusters are compact [16], so it is possible to estimate how large T must be and thus how many random numbers are required to make the failure probability small. The perimeter of a d-dimensional cluster of size N is expected to scale as $O(N^{\frac{d-1}{d}})$, so that it suffices to choose T equal to $\Omega(N^{2-\frac{d-1}{d}})$ to insure that the perimeter is hit N times. Therefore, the algorithm needs $O(N^{1+\frac{1}{d}} \log N)$ random bits.

4.2 Restricted solid-on-solid

The algorithm given below simulates the RSOS model in parallel. This algorithm builds and then operates on the space-time graph. The initial cluster S consists of all sites $[i, h]$ with $h \leq h(i, 0)$, where h with two arguments is the height function. We assume the initial heights are $O(N)$ and have a maximum achievable height of H. In most cases the initial heights are taken as zero so that $S = \{[i, 0] \mid i \in V\}$. The output of the algorithm is the space-time cluster order. The height function can easily be obtained from the cluster order.

Parallel Restricted Solid-on-solid Algorithm
> **begin**
> 1. Build a directed space-time graph $G_{st} = (V_{st}, E_{st})$, where
> $V_{st} \leftarrow \{[i, h] \mid i \in V \text{ and } h = 0, 1, \ldots, H\}$, and
> $E_{st} \leftarrow \{([i, h-1], [j, h]) \mid \{i, j\} \in E \text{ and } h = 1, 2, \ldots, H\}$
> $\qquad \cup \; \{([j, h-1], [j, h]) \mid j \in V \text{ and } h = 1, 2, \ldots, H\}$.
> 2. Generate a random list of size T of elements from V_{st},
> $L[m] \leftarrow [v_m, h_m], \; (1 \leq m \leq T)$.
> 3. Construct a directed graph, $G' = (\{0, 1, \ldots, T\}, E')$, where
>   for each $1 \leq m, n \leq T$, if $(([v_m, h_m], [v_n, h_n]) \in E_{st} \text{ and } m < n)$
>   then $(m, n) \in E'$, and
>   for each $1 \leq n \leq T$, if $[v_n, h_n]$ is on the perimeter of S then
> $(0, n) \in E'$.
> 4. For $k \leftarrow 1$ to T do $A'[k] \leftarrow \{n \mid \exists$ a directed path from n to k
>   in $G'\}$.
>   (Note, $A'[k]$ represents the set of ancestors of k in G'.)
> 5. For $k \leftarrow 1$ to T do $B[k] \leftarrow \{[i, h] \mid \exists$ a directed path from $[i, h]$ to
> $L[k]$ in $G_{st}\}$.
>   (Note, $B[k]$ represents the set of ancestors of $[v_k, h_k]$ in G_{st}.)
> 6. For $k \leftarrow 1$ to T do $A[k] \leftarrow \{[i, h] \mid n \in A'[k] \text{ and } [i, h] = L[n]\}$.
>   (Note, $L[0] = S$.)
> 7. For $k \leftarrow 1$ to T do if $B[k] = A[k]$ then $L[k] \leftarrow L[k]$ else $L[k] \leftarrow -1$.
> 8. Compact L by deleting -1's.
> 9. Compact L by deleting all but the first appearance of each site.
> 10. $c(n) \leftarrow L[n]$ is the nth site added to the cluster (n starts at 1).
> **end.**

Note that the initial cluster needs to be treated as a special case. The following theorem proves that the algorithm is correct and runs in polylog time.

Theorem 2. *Let $G = (V, E)$ be an undirected graph and S the initial cluster. The parallel restricted solid-on-solid algorithm produces the space-time cluster ordering for G according to the growth rules defining the restricted solid-on-solid model. Suppose $T = O(N^{k_1})$ for some constant k_1. The algorithm runs in $O(\log^2 N)$ time using a polynomial number of processors on a probabilistic CREW P-RAM.*

Proof. (Idea) Correctness: A random list of space-time sites, L, is generated in Step 2 and a directed graph, G', is constructed from this list. A directed bond exists in G' if it corresponds to a directed bond in G_{st} and if it is compatible with the ordering of the list L. The elements of L are potential growth sites and are accepted into the growing cluster if all of their ancestors in G_{st} are already in the cluster. Acceptance in the cluster is permitted if the set of ancestors of a site in G' includes all the ancestors of the pre-image of that site in G_{st}. This is as required by the defining rules for the RSOS model. All elements that fail this test are deleted from L. The ordering of L then determines the cluster order.

Resource bounds: All of the steps in the algorithm can be parallelized using parallel sorting, parallel prefix computations and parallel MWP computations. Each step requires a polynomial number of processors in N. For a choice of T equal to $O(N^{k_1})$, the running time of the algorithm is dominated by the parallel MWP subroutine used in Steps 4 and 5 (to determine ancestors) and is $O(\log^2 N)$. □

Remarks As is the case for the parallel Eden growth algorithm, the parallel RSOS algorithm may fail before the desired cluster size is obtained. If failures are allowed, smooth interfaces will be favored over rough interfaces because smooth interfaces have more allowed growth sites. Thus, the RSOS distribution will not be exactly sampled. If the exact distribution is required, it will occasionally be necessary to iterate the algorithm using the height function from one run as the initial cluster for the next run. Although this procedure is not guaranteed to halt in a fixed time, the probability of failure can be made small so that the expected running time of the iterated algorithm is polylog.

The algorithm may need $O(N_{st}^2 \log N_{st})$ random bits to form a complete cluster, where $N_{st} = N(H + 1)$ is the number of sites in the space-time graph. This is because each element in the list has a probability of at least $1/N_{st}$ to be a growth site and growth must occur N_{st} times.

4.3 Ballistic deposition

The algorithm described below simulates ballistic deposition in parallel. The output of the algorithm is a realization of the height function, $h(i, n)$, at site i at time step n, $n = 0, 1, \ldots, K$, as given by the defining algorithm for ballistic deposition.

Parallel Ballistic Deposition Algorithm
    begin

1. Generate a random list of sites $(v_1, v_2, \ldots, v_K)$.
2. In parallel create a directed space-time graph $G_{st} = (V_{st}, E_{st})$,
   where $V_{st} \leftarrow \{[i, n] \mid i \in V \text{ and } n = 0, 1, \ldots, K\}$ and
   $E_{st} \leftarrow \{([i, n-1], [j, n]) \mid \{i, j\} \in E \text{ and } n = 1, 2, \ldots, K\}$
   $\cup \{([j, n-1], [j, n]) \mid j \in V \text{ and } n = 1, 2, \ldots, K\}$.
3. Assign weights to the edges of the space-time graph as follows:
   for each $j \in V$ and $1 \leq n \leq K$ do
      if $j = v_n$ then $w(([j, n-1], [j, n])) \leftarrow 0$
      else $w(([j, n-1], [j, n])) \leftarrow 1$, and
   for each $\{i, j\} \in E$ and $1 \leq n \leq K$ do
      if $j = v_n$ then $w(([i, n-1], [j, n])) \leftarrow 1$
      else $w(([i, n-1], [j, n])) \leftarrow \infty$.
4. for each $j \in V$ and $1 \leq n \leq K$ do
   $h(j, n) \leftarrow n - $ the weight of the MWP from a site in
      $\{[i, 0] \mid i \in V\}$ to $[j, n]$.

end.

This algorithm builds and operates on a space-time graph, however, growth does not occur on this graph. A random list of sites is generated in the substrate. This list determines the order columns are chosen for growth. The MWP algorithm can be used to compute the height on each column. The following theorem proves that the above algorithm is correct and runs in polylog time.

Theorem 3. *Let $G = (V, E)$ be an undirected graph and S the initial cluster. The parallel ballistic deposition algorithm produces the height function for G according to the growth rules defining the ballistic deposition model. Suppose $K = O(N^{k_1})$ for some constant k_1. The algorithm runs in $O(\log^2 N)$ time using a polynomial number of processors on a probabilistic CREW P-RAM.*

Proof. (Sketch) Correctness: We argue the correctness of the algorithm below. The proof is by induction. After one step of growth, only site v_1 should have height one. All other sites should have height zero. Notice in Step 4 of the algorithm, the MWP from $[v_1, 0]$ to $[v_1, 1]$ will be zero because of the weighting in G_{st}. Therefore, $h(v_1, 1) = 1 - 0 = 1$ as required. For sites other than v_1, the MWP will have weight one and the computation in Step 4 correctly assigns them a height of zero.

Assume for the induction hypothesis that after l steps of growth, the heights of all sites are correctly computed by the algorithm. Consider step $l+1$. Suppose the MWP from $\{[i, 0] \mid i \in V\}$ to $[v_{l+1}, l+1]$ contains $[v_{l+1}, l]$. Then the MWP to any "neighbor" of $[v_{l+1}, l+1]$ from $\{[i, 0] \mid i \in V\}$ can be at most one less than the MWP to $[v_{l+1}, l+1]$. In G this means a neighbor of v_{l+1} can have height at most one greater than v_{l+1}. By the induction hypothesis the heights computed after l steps were correct. Step 4 has the effect of incrementing the height of v_{l+1} by one. This correctly simulates ballistic deposition.

Suppose $[v_{l+1}, l]$ is not on the MWP to $[v_{l+1}, l + 1]$. This implies there is a "neighbor," $[v, l]$, of $[v_{l+1}, l + 1]$ with shortest MWP among neighbors such that the MWP to $[v, l]$ plus one is the MWP to $[v_{l+1}, l + 1]$. By the induction hypothesis $h(v, l)$ was correctly computed. According to the rules for ballistic deposition, $h(v_{l+1}, l + 1)$ should be equal to $h(v, l)$. The arithmetic in Step 4 yields $(l + 1) - ($ MWP to $[v, l] + 1)$. This is the same as $l -$ MWP to $[v, l]$. Note, this value is the maximum height of a neighbor of v_{l+1} in G after l growth steps. Thus the algorithm is correct.

Resource bounds: The running time of the algorithm is dominated by the parallel MWP subroutine. If $K = O(N^{k_1})$, the running time is $O(\log^2 N)$. Only a polynomial number of instances of the MWP problem need to be solved in parallel in Step 4. These are solved on graphs larger than the original but still polynomial in size. Therefore, the number of processors required by the algorithm is polynomial.
□

Remarks In contrast to the growth rules for the Eden model and the RSOS model, in ballistic deposition every site is an allowed growth site. Thus a random list of length K yields a height function that is increased for a single site at each step. There are no failures and the maximum possible resulting height is K. Note, the randomness in this algorithm is used only in generating the initial random list.

5 Discussion

We have presented fast parallel algorithms for the following growth models: Eden growth, restricted solid-on-solid growth and ballistic deposition. All of the algorithms run in polylog time using a polynomial number of processors, although several of them may not produce complete clusters.

While fast algorithms exist for the growth models discussed here, it has been shown [12] that a natural sequential approach for diffusion-limited aggregation (DLA), a related growth model that produces a fractal cluster, defines a P-complete problem; there is almost certainly no fast parallel algorithm based on this approach. Although we have demonstrated here that Eden clusters can be generated in polylog time, there is a natural decision problem concerning Eden clusters on random inputs that is P-complete [13]. We conjecture that no fast parallel algorithm exists for DLA; this model seems fundamentally more complex than the models we consider.

The existence of parallel algorithms for simulating these models establishes a polylog upper bound on the parallel time required to simulate these models. It seems plausible that models such as these, which generate statistically self-similar or self-affine clusters by a local process, cannot be simulated in less than logarithmic time. Thus they are more complex than models that can be simulated in constant parallel time. However, it is a much more difficult question to establish lower bounds on computational resources. The general question of establishing lower bounds on the resources needed for sampling distributions is an interesting and difficult one.

References

1. M. Eden. A two-dimensional growth process. In F. Neyman, editor, *Proceeding of the Fourth Berkeley Symposium on Mathematical Statistics and Probability*, volume IV, page 223. University of California, 1961.

2. F. Family and T. Vicsek, editors. *Dynamics of Fractal Surfaces*. World Scientific, Singapore, 1991.

3. F. E. Fich. The complexity of computation on the parallel random access machine. In J. H. Reif, editor, *Synthesis of Parallel Algorithms*, chapter 20, pages 843–899. Morgan Kaufman, San Mateo, CA, 1993.

4. P. Freche, D. Stauffer, and H. E. Stanley. Surface structure and anisotropy of Eden clusters. *J. Phys. A: Math. Gen.*, 18:L1163, 1985.

5. A. Gibbons and W. Rytter. *Efficient Parallel Algorithms*. Cambridge University Press, 1988.

6. R. Greenlaw, H. J. Hoover, and W. L. Ruzzo. *Topics in Parallel Computation: A Guide to P-completeness Theory*. Computing Science Series, editor Z. Galil. Oxford University Press, to appear.

7. M. Kardar, G. Parisi, and Y.-C. Zheng. Dynamic scaling of growing interfaces. *Phys. Rev. Lett.*, 56:889, 1986.

8. R. M. Karp and V. Ramachandran. Parallel algorithms for shared-memory machines. In Jan van Leeuwan, editor, *Handbook of Theoretical Computer Science*, volume A: Algorithms and Complexity, chapter 17, pages 869–941. M.I.T. Press/Elsevier, 1990.

9. H. Kesten. *Percolation Theory for Mathematicians*. Birkhauser, Boston, 1982.

10. J. M. Kim and J. M. Kosterlitz. Growth in a restricted solid-on-solid model. *Phys. Rev. Lett.*, 62:2289, 1989.

11. J. Krug and H. Spohn. Kinetic roughening of growing surfaces. In C. Godreche, editor, *Solids Far From Equilibrium: Growth, Morphology and Defects*. Cambridge University Press, 1991.

12. J. Machta. The computational complexity of pattern formation. *J. Stat. Phys.*, 70:949, 1993.

13. J. Machta and R. Greenlaw. P-complete problems in statistical physics. Unpublished, 1994.

14. P. Meakin, P. Ramanlal, L. M. Sander, and R. C. Ball. Ballistic deposition on surfaces. *Phys. Rev. A*, 34:5091, 1986.

15. M. Plischke and Z. Racz. Active zone of growing clusters: Diffusion-limited aggregation and the Eden model. *Phys. Rev. Lett.*, 53:415, 1984.

16. D. Richardson. Random growth in a tessellation. *Proc. Camb. Phil. Soc.*, 74:515, 1973.

17. S. Roux, A. Hansen, and E. L. Hinrichsen. A direct mapping between Eden growth model and directed polymers in random media. *J. Phys. A: Math. Gen.*, 24:L295, 1991.

18. L.-H. Tang, J. Kertesz, and D. E. Wolf. Kinetic roughening with power-law waiting time distribution. *J. Phys. A: Math. Gen.*, 24:L1193, 1991.

19. T. Vicsek. *Fractal Growth Phenomena*. World Scientific, Singapore, 1992.

20. M. J. Vold. A numerical approach to the problem of sediment volume. *J. Colloid Sci.*, 14:168, 1959.

A New Approach to Resultant Computations and Other Algorithms with Exact Division

Arnold Schönhage and *Ekkehart Vetter**

Institut für Informatik II, Universität Bonn
Römerstraße 164, 53117 Bonn, Germany

Abstract. Computations like Collins' subresultant algorithm or Bareiss' method for the exact evaluation of determinants with integral entries spend a substantial amount of their running time in performing *exact divisions,* integer by an integer, or polynomial by a polynomial with remainder known to be zero. Here the main achievements are faster algorithms for exact division, their application to determinant evaluation and subresultant algorithms, thereby considerably reducing also the number of divisions and the other operation costs, and better asymptotical performance by fast computations (including divisions) modulo numbers of the form $2^N + 1$.

1 Introduction

This work is part of a long-term project concerned with the development, implementation, and use of fast integer arithmetic and other related multi-precision software, in particular including an efficient version of the asymptotically fast multiplication modulo integers of the form $2^N + 1$ with its numerous applications. Recently we have finished our book *Fast Algorithms* [SGV94] which summarizes the present state of these efforts and makes this software available for a wider community. With such powerful tools in our hands, we have of course been looking for rewarding applications. In this respect one of the challenging problems is to find faster algorithms for the computation of resultants $R = \text{res}_x(A(x), B(x))$ of polynomials $A(x), B(x) \in I[x]$ over integral domains like $I = \mathbb{Z}$, $I = F[y]$ with y as further indeterminate over a small finite field F, say, or for the important multivariate case with $I = \mathbb{Z}[y_1, \ldots, y_v]$.

Nowadays the method of choice for computing such resultants is Collins' subresultant algorithm (see [Co67], [Kn81, p. 410], and the overview [Lo82]) or one of the improved modular versions like [Co71] for the multivariate case, which shall serve as reference for comparing running times. To be specific, let us consider dense polynomials $A(x), B(x) \in \mathbb{Z}[y_1, \ldots, y_v][x]$ of degree $\leq n$ in x and of degree $\leq \tau$ in each of the "ballast" variables y_j, with integer coefficients bounded by 2^l, moreover also $n, \tau < 2^l$ shall hold. Then the bit length of the inputs is of order $n \cdot l \tau^v$, and the resultant R is a polynomial of degree $\leq 2n\tau$ in each of the y's with coefficient length $\mathcal{O} \cdot nl$, so the total length of R is $\mathcal{O} \cdot nl(n\tau)^v$ (here the symbol $\mathcal{O}$ stands for $O(1)$, cf. [SGV94, p. 4]).

---

* Supported by a grant from *Deutsche Forschungsgemeinschaft* (Az: Scho 415/2).

Under these assumptions the running time of the subresultant algorithm in its standard form (using standard integer and polynomial arithmetic) is bounded by $\mathcal{O} \cdot n^2 \cdot (nl)^2 (n\tau)^{2v}$, which becomes $\mathcal{O} \cdot n^{4+4v} \cdot l^2$ for $\tau = n$. The bounds reported for the modular version in [Co71] (adapted to our notation) are $\mathcal{O} \cdot n^{3+3v} l + \mathcal{O} \cdot n^{2+4v} l^2$ which gives a bound of order $n^6 l^2$ for the paradigmatic case $v = 1$.

In view of all what is known about "fast" algorithms for polynomials and integers, it is likely that there exists a resultant algorithm with an improved asymptotical time bound $\mathcal{O} \cdot n^{1+\varepsilon} \cdot (nl(n\tau)^v)^{1+\varepsilon}$, or $\mathcal{O} \cdot n^{2+2v+\varepsilon} \cdot l^{1+\varepsilon}$ for $\tau = n$, where $\varepsilon = o(1)$ absorbs any logarithmic factors, but for the time being, it is also rather likely that the break even points for such methods will lie beyond any practical bound. Looking for more practical solutions we keep to the outer frame of the subresultant algorithm, so there remains the bound of $\mathcal{O} \cdot n^2$ operations with long integers or multivariate polynomial operands of growing size (up to the length of the final R), but with our new methods we are able to achieve a substantial speed-up of these inner steps.

If it were for multiplications only, there would be no problem. We can simply invoke our fast integer multiplication, or a routine for the multiplication of integer polynomials by reduction to one long integer multiplication based on the method of binary segmentation as described in [Sc82]. The crucial problem is posed by the great number of (polynomial) divisions required by the subresultant algorithm. This brute force approach has been carried out in J. Klose's thesis [Kl93]. Our main innovation is a new method to perform these divisions as "divisions mod $(2^N + 1)$" with some N suitably chosen for each of the stages of the subresultant algorithm so that all operations of this stage can be carried out with the smaller operand size mod $(2^N + 1)$. Moreover, this approach admits to reduce the number of divisions by a factor of order n which leads to a further decrease of the implicit estimation constants. Altogether this new method yields the asymptotical time bound $\mathcal{O} \cdot n^2 \cdot (nl(n\tau)^v)^{1+\varepsilon}$, or $\mathcal{O} \cdot n^{3+2v+\varepsilon} \cdot l^{1+\varepsilon}$ for $\tau = n$. These bounds can also be taken for corresponding gcd computations. For that simpler task, however, interpolatory methods as in [Sc88] for the univariate case will give the better (deterministic) time bound $\mathcal{O} \cdot n^{2+2v+\varepsilon} \cdot l^{1+\varepsilon}$.

Surprisingly, the idea to reduce the number of divisions applies quite generally, without any asymptotical methods. Inspired by Jebelean's algorithm [Jeb93] for exact division by 2-adic methods, we have also found a very simple way to speed up the subresultant algorithm for small problems, usually by a factor greater than four. Furthermore, all these improvements apply to Bareiss' algorithm for exact Gaussian elimination as well. In order to avoid lengthy repetitions, we therefore proceed in the following way. After a brief exposition of the new method for exact division of integers (or univariate polynomials over a field) we demonstrate the small case techniques with a speed-up of Bareiss' algorithm by a factor of 4 or 6. Section 4 introduces our new asymptotical method for resultants over $\mathbb{Z}$, and in section 5 we finally show how to extend these ideas to the multivariate case by explicating the details for the case $I = \mathbb{Z}[y]$ with one further indeterminate y.

2 Exact Division of Integers and Polynomials

"Exact division" of integers (or in other integral domains) shall denote the computational task to find the quotient of two inputs when the remainder is known to be zero. Using standard methods, division of a $2m$-word integer by an m-word integer takes *typical time* $\simeq \gamma_o \cdot m^2$ (plus linear overhead), where γ_o measures the cost of an inner "multiply-and-add" loop. We want to show that exact division can be done about four times faster. The basic idea for such an improvement becomes especially clear if we consider the analogous problem of exact division in the polynomial ring $F[y]$ over some (small) finite field F, where constant time γ_o for operations of the form $d := a \cdot b \pm c$ is a realistic assumption. Let $f(y)/g(y) = q(y)$ be such an exact division with m coefficients in q to be computed from the coefficients of f and g. The computation of h high-end coefficients of q can be done with the upper h coefficients of f and the upper h coefficients of g in time $\simeq \gamma_o \cdot h(h+1)/2$. Similarly l low-end coefficients of q are found from the lower l coefficients of f and of g, now considered as division of initial segments of formal power series. A potential divisor $y^t | g(y)$ with (maximal) $t > 0$ must also divide $f(y)$ and can be removed first. Choosing $l = \lfloor m/2 \rfloor$ and $h = \lceil m/2 \rceil$ yields a time bound $\simeq \gamma_o \cdot h(h+1)/2 + \gamma_o \cdot l(l+1)/2 \leq (m+1)^2 \cdot \gamma_o/4$.

Exact division of integers $f/g = q$ can be done in a similar way with similar costs. If the denominator is given in k words, for example $g > 0$ with $\mathfrak{L}^{k-1} \leq g < \mathfrak{L}^k$ (where $\mathfrak{L}$ denotes the *wordbase*, e.g. $\mathfrak{L} = 2^{32}$), and the quotient q is known to fit into m words, then we choose $h \approx m/2$ and perform approximate division of the binary rationals $f/\mathfrak{L}^{k+h}$ by $g/\mathfrak{L}^k$ with precision bound $|f/\mathfrak{L}^{k+h} - z \cdot g/\mathfrak{L}^k| < \frac{1}{2}g/\mathfrak{L}^k$ (cf. routine *RDIV* in [SGV94, p. 256]), so this approximate quotient z satisfies $|q - z \cdot \mathfrak{L}^h| < \frac{1}{2} \cdot \mathfrak{L}^h$. Next we also compute the lower half quotient $w \equiv q \pmod{\mathfrak{L}^h}$, for instance by Jebelean's method [Jeb93]. Then, by inspection of the first fractional word of z, the precise result q can easily be found from z and w.

We do not know of any clue for a similar speed-up of exact division in the case of asymptotically fast integer division, but there are obvious savings with exact division of *polynomials*, when interpolation methods are used.—Section 9.2.2 of [SGV94] outlines "exact" polynomial division in the case of approximate computations over the complex numbers by means of discrete Fourier transforms, where it turns out that this is cheaper than general division roughly by a factor of two.

3 Gaussian Elimination with Integral Entries

Let us begin with a brief review of Bareiss' algorithm [Bar68] for size controlled Gaussian elimination over integral domains like $\mathbb{Z}$ or $F[y]$ by systematically dividing out predictable factors. Here it may suffice to consider the basic single-step version of the algorithm for the especially simple task of evaluating the determinant of an $n \times n$ matrix $A = A^{(1)}$ with integral entries $a_{i,j} = a_{i,j}^{(1)} \in \mathbb{Z}$, bounded by $|a_{i,j}| < 2^l$, or polynomials $a_{i,j} \in F[y]$ all of degree $\leq l$. Let us assume, for the moment, that no pivoting is needed. In its k-th stage, the algorithm processes the entries $a_{i,j} = a_{i,j}^{(k)}$, $k \leq i, j \leq n$, of the $(n-k+1) \times (n-k+1)$ matrix $A^{(k)}$ reached so far to compute the $(n-k) \times (n-k)$ matrix $A^{(k+1)}$ with the entries

$$b_{i,j} := (a_{k,k} \cdot a_{i,j} - a_{i,k} \cdot a_{k,j})/D \quad \text{for} \quad k+1 \leq i, j \leq n, \tag{1}$$

where the uniform denominator is $D = a_{k-1,k-1}^{(k-1)}$ (or $D = 1$ for $k = 1$). In the polynomial case these $a_{i,j}$ are of degree kl at most, and the $b_{i,j}$ have degree $\leq (k+1)l$. If (1) is done by straightforward application of elementary routines for polynomial multiplication and division (and assuming the dense case with these degree bounds usually attained), then the typical timing is $\simeq 2 \cdot \gamma_o \cdot (kl)^2$ for the two multiplications and $\simeq \gamma_o \cdot (k+1)(k-1)l^2$ for the division by D. By counting all these steps over all stages, we thus obtain a global time estimate of the form

$$T_1(l, n) \simeq \gamma_o \cdot l^2 \sum_{k=1}^{n-1} (n-k)^2 \cdot (3k^2 - 1) \sim l^2 n^5 \cdot \gamma_o / 10. \tag{2}$$

A similar estimate holds for the case of initial integer entries, $|a_{i,j}| < 2^l$, provided we use Hadamard's inequality to estimate the entries of $A^{(k)}$ as $k \times k$ minors by $(2^l \sqrt{n})^k$ and replace l in (2) by $l^+ = l + \frac{1}{2} \cdot \log n$.

Speed-up by a Factor of Four

Application of the improved exact division algorithm of section 2, in the k-th stage with at most $m = (k+1)l + 1$ coefficients per polynomial result $b_{i,j}$, reduces the time bound per element from $\simeq \gamma_o \cdot l^2 (3k^2 - 1)$ to $\gamma_o \cdot l^2 (2k^2 + (k+1)^2/4) \sim 2.25 \cdot \gamma_o \cdot l^2 k^2$, but we can do much better by performing all operations in (1) by the high-end/low-end technique of section 2 with respect to the predicted size bound $(k+1)l$ for the $b_{i,j}$. This leads to the improved time bound

$$T_2(l, n) \simeq \gamma_o \cdot l^2 \sum_{k=1}^{n-1} 3(n-k)^2 (k+1)^2/4 \sim l^2 n^5 \cdot \gamma_o / 40. \tag{3}$$

Compared with (2), we thus have gained a factor of four, at least in the limit, while moderate values of n give slightly weaker bounds; in case of $n = 10$, for instance, the sums $\sum_{k=1}^{n-1}$ in (2) and (3) take the values 9714, and 3951, respectively, which means a reduction to about 40 percent of the former time bound.

Pivoting for Fewer Divisions

Let us now consider the provisions required to guarantee $a_{k,k} \neq 0$, which implies $D \neq 0$ for the next stage. Aiming for another improvement, we want to make D a unit also for the low-end division. Let t_i denote the exponent of the least nonvanishing term in

$$a_{i,k}(y) = c_i \cdot y^{t_i} + \text{higher order terms}, \quad \text{with } c_i \neq 0$$

for all nonzero elements (else $t_i = \infty$) in the first column of $A^{(k)}$, and choose among these a pivot with minimal t-value. After appropriate exchange of rows we thus have $t_k \leq t_i$ for $i = k+1,\ldots,n$ (or $\det A^{(k)}$ is found to be zero). In case of $t_k > 0$ we can extract the common factor y^{t_k} from all entries of the first column (which becomes an extra factor for the determinant), so that the new $\tilde{a}_{k,k}(y) = a_{k,k}/y^{t_k}$ satisfies $\tilde{a}_{k,k}(0) = c_k \neq 0$. In case of integer entries, we can in the same way extract the factor 2^{t_k} leaving a new pivot $\tilde{a}_{k,k} \equiv 1 \pmod 2$.

By having used the same pivoting strategy in all previous stages, we are sure that also $D = \tilde{a}_{k-1,k-1}^{(k-1)}$ is a unit for polynomial low-end division (or 2-adic division, respectively). Therefore we can rearrange the computation of the b's (possibly reduced by the factor y^{t_k}) into

$$b_{i,j} := (\tilde{a}_{k,k}/D) \cdot a_{i,j} - (\tilde{a}_{i,k}/D) \cdot a_{k,j} \quad \text{for} \quad k+1 \leq i,j \leq n, \tag{4}$$

so that both the low-end and the high-end divisions by D are carried out first, but now with the $n - k + 1$ elements of the first column only! Accordingly the time bound for this version becomes

$$T_3(l,n) \simeq \gamma_o \cdot l^2 \sum_{k=1}^{n-1} \left(2(n-k)^2 + n - k + 1\right)(k+1)^2/4 \sim \frac{l^2 n^5 \cdot \gamma_o}{60}, \tag{5}$$

better than (2) by a factor of six in the limit. For the case $n = 10$, this inner sum has the value 3030, equal to about 31 percent of 9714 for (2).

4 Resultants

For the following we assume the reader to be familiar with the standard notations used in Chapter 4.6 of [Kn81]. Let I be an Euclidean domain and A, B polynomials over I, $A = A(x) = a_m x^m + \cdots + a_0$, $B = B(x) = b_n x^n + \cdots + b_0$, with $m \geq n$ and $a_m, b_n \neq 0$. We consider, extending Euclid's algorithm in $I[x]$, *polynomial remainder sequences* $F_1,\ldots,F_{K+1} \subset I[x]$ of A, B, defined by

$$F_1 = A, \quad F_2 = B, \tag{6}$$

$$H_k = \text{prem}(F_{k-2}, F_{k-1}), \quad F_k = H_k/\beta_k, \quad k = 3,\ldots,K+1, \tag{7}$$

where F_{K+1} is the first zero polynomial in the sequence. The parameters $\beta_k \in I \setminus \{0\}$ are arbitrary divisors of $\text{cont}(H_k)$. The above equations imply that $\gcd(A,B) = \gcd\left(\text{cont}(A), \text{cont}(B)\right) \cdot \text{pp}(F_K)$, so that for each method of choosing the β_k, equations (6) and (7) represent an algorithm for gcd computation over $I[x]$. Simply setting all $\beta_k = 1$ leads, e.g. for $I = \mathbf{Z}$, to exponential coefficient growth. However, this is an artificial effect only (cf. Habicht's theorem [Lo82]), caused by a large content of each polynomial F_k which can be removed by choosing an appropriate β_k. Computing these β_k via coefficient gcd's seems to be unacceptably costly. The *subresultant algorithm*, due to Collins [Co67] and independently to Brown [Br71], determines a large systematic factor of the content of H_k without gcd computations.

The Subresultant Algorithm. Let $n_k = \deg(F_k)$ for $k = 1, \dots, K$ and $\delta_k = n_k - n_{k+1}$ for $k = 1, \dots, K - 1$. Furthermore, let S_j $(0 \le j \le m)$ the j-th *subresultant* of A and B. We set

$$G_1 = A, \quad G_2 = B, \qquad G_k = S_{n_{k-1}-1}, \qquad k = 3, \dots, K, \tag{8}$$

and $g_k = \mathrm{lc}(G_k)$. Then $G_1, G_2, \dots, G_K, 0$ is a polynomial remainder sequence as above. The respective parameters β_k can easily be computed from $g_1, \dots, g_{k-2}$, (see [Kn81, p. 410] or [Br78]). In the resulting subresultant algorithm the coefficient length growth is, e.g. for $I = \mathbf{Z}$, nearly linear.

Besides the gcd of A and B, the algorithm computes the subresultants $S_{n-1}, \dots, S_{n_{K-1}-1}$ and especially the *resultant* $R = S_0$ of A and B (even in the case $n_{K-1} > 1$). Furthermore, it can be easily extended to compute all subresultants.

The determination of F_k in equation (7) can be achieved by $n_k + 1$ exact divisions. Thus the techniques of section 2 can be applied. But further improvements are possible. For that we use the subsequent modification of the subresultant algorithm.

A Modified Subresultant Algorithm. We consider the polynomial remainder sequence $F_1, \dots, F_{K+1}$ generated by the following recursive definition of the parameters $\beta_k = d_k/e_k$, where

$$d_3 = (-1)^{\delta_1+1}, \tag{9}$$

$$d_k = (-1)^{\delta_{k-2}+1} f_{k-2} h_{k-2}^{\delta_{k-2}} e_{k-1}^{\delta_{k-2}+1}, \qquad k = 4, \dots, K+1, \tag{10}$$

$$f_k = \mathrm{lc}(F_k), \qquad\qquad\qquad k = 1, \dots, K, \tag{11}$$

$$h_2 = f_2^{\delta_1}, \tag{12}$$

$$h_k = (f_k/e_k)^{\delta_{k-1}} h_{k-1}^{1-\delta_{k-1}}, \qquad k = 3, \dots, K, \tag{13}$$

and the *extra factors* $e_k \in I \setminus \{0\}$ are arbitrary divisors of d_k for $k = 3, \dots, K$ and $e_1 = e_2 = 1$. From above equations one obtains

$$F_k = e_k G_k, \qquad k = 1, \dots, K, \tag{14}$$

$$h_k = \mathrm{lc}(S_{n_k}), \qquad k = 2, \dots, K. \tag{15}$$

Equation (15) has two remarkable consequences. First, it follows $h_k \in I$. Second, it shows that, if A and B are relative prime, $R = \mathrm{res}(A, B) = h_K$. If all $e_k = 1$, equations (9–13) reduce to the common subresultant algorithm.

Pseudo Remainder. Knuth [Kn81, p. 407] computes the pseudo remainder $\text{prem}(F_{k-2}, F_{k-1})$ with the help of a sequence of polynomials $W_0, \ldots, W_{\delta_{k-2}+1} = \text{prem}(F_{k-2}, F_{k-1})$ recursively defined by $W_0 = F_{k-2}$ and

$$W_j = f_{k-1}W_{j-1} - \text{lc}(W_{j-1})F_{k-1}x^{\delta_{k-2}+1-j}, \qquad j = 1, \ldots, \delta_{k-2} + 1.$$

From that, one easily obtains

$$\text{prem}(F_{k-2}, F_{k-1}) = f_{k-1}^{\delta_{k-2}+1}F_{k-2} - \sum_{j=0}^{\delta_{k-2}} \phi_j F_{k-1}x^{\delta_{k-2}-j}, \tag{16}$$

where the $\phi_j \in I$ are computable from the highest $\delta_{k-2} + 1$ coefficients of F_{k-2} and F_{k-1}.

The Case $I = \mathbf{Z}$. Each coefficient of the polynomial $G_k = S_{n_{k-1}-1}$ ($k = 3, \ldots, K$) is a determinant of a submatrix of order $m_k = m + n - 2(n_{k-1} - 1)$ of the Sylvester matrix of A and B. With each coefficient of A and B bounded by 2^l, we obtain by Hadamard's inequality the bound 2^{l_k} for the coefficients of G_k, where $l_1 = l_2 = l$, and

$$l_k = m_k l + \tfrac{1}{2}m_k \log m_k, \qquad k = 3, \ldots, K. \tag{17}$$

Algorithm U. Computes the resultant and the greatest common divisor of two polynomials $A, B \in \mathbf{Z}[x]$.

1. [Initializations] Set $F_1 := A$, $F_2 := B$, and $k := 3$. Calculate h_2 using formula (12).

2. [k-loop] Calculate h_k and $D = d_k$ according to equations (13) and (10).

3. [Suitable number] Determine l_k according to (17). Call Algorithm S (see below) to determine a suitable $M = 2^N + 1$, so that

$$M > e_k \cdot 2^{l_k+1}, \tag{18}$$

where $e_k = \gcd(M, D)$. Calculate $u \in \mathbf{Z}$ with $e_k \equiv uD \bmod M$. The computations in steps 4 and 5 are done in the residue class ring $R_M = \mathbf{Z}/M\mathbf{Z}$ and in $R_M[x]$. With $^\circ$ we denote the canonical homomorphisms $\mathbf{Z} \overset{\circ}{\to} R_M$ and $\mathbf{Z}[x] \overset{\circ}{\to} R_M[x]$, respectively.

4. [Prepare elimination factors in R_M] Determine the elimination factors ϕ_j° for $j = 0, \ldots, \delta_{k-2}$ used in equation (16). Compute $u^\circ \cdot \left(f_{k-1}^{1+\delta_{k-2}}\right)^\circ$ and $u^\circ \cdot \phi_0^\circ, \ldots, u^\circ \cdot \phi_{\delta_{k-2}}^\circ$. This makes a final division step superfluous.

5. [Compute F_k] Use the above factors and equation (16) to compute

$$F_k^\circ = u^\circ \cdot \text{prem}(F_{k-2}, F_{k-1})^\circ.$$

Bound (18) ensures unique decoding of $F_k^\circ \in R_M[x]$ to $F_k \in \mathbf{Z}[x]$.

6. [Loop] While $F_k \neq 0$, increment k by 1 and continue with step 2.

7. [Finals] If $\deg(F_{k-1}) = 0$, return resultant $R = h_{k-1}$, else return $R = 0$. Additionally, return $\gcd(A, B) = \gcd\big(\text{cont}(A), \text{cont}(B)\big) \cdot \text{pp}(F_{k-1})$.

Fast multiplication with SML. To be able to use fast multiplication in R_M, its size M has to meet certain requirements. Here we want to use our routine SML from [SGV94], so that $M = 2^N + 1$ with a "suitable" number N. For more details, as the performance of SML, see [SGV94, p. 32 and 6.1.37].

Algorithm S. Inputs $D, L \in \mathbf{N}$ with $\log D < \mathcal{O} \cdot L$. Returns a suitable number $M = 2^N + 1$ with $N < L \cdot (1 + 1/\log^2 L) + \mathcal{O} \cdot L^{1/2}$ and $M > \gcd(M, D) \cdot 2^L$.

1. [Primes] Set $t := \max\left(\left\lceil \frac{1}{2}\log L \right\rceil, 11\right)$ and $h := \left\lceil L^{-1} \cdot \log^2 L \cdot \log D \right\rceil + 1$. Find the smallest prime numbers $p_1, \ldots, p_h$ greater than a bound L_1, defined by $L_1 := 2^{-t} \cdot (L + h^{-1} \cdot \log D) + 2$. Set $i := 0$.

2. [Suitable number] Increment i by 1 and set $N := 2^t \cdot p_i$ and $M := 2^N + 1$.

3. [Check] If $M > \gcd(M, D) \cdot 2^L$, return M. Otherwise goto step 2.

In practical applications this algorithm usually performs one or a few iterations only. For a theoretical worst case analysis, we note that $h = \mathcal{O} \cdot \log^2 L$ and $L_1 < \sqrt{L} \cdot (1 + 1/\log^2 L) + 2$. The h primes can be found by completely sieving the interval $[L_1, L_1 + H]$ (with $H = \mathcal{O} \cdot L^{0.3}$ for example). Let $M_i = 2^{2^t \cdot p_i} + 1$. Since $\gcd(M_i, M_j) = 2^{2^t \cdot \gcd(p_i, p_j)} + 1 = 2^{2^t} + 1$ for $i \neq j$, we can prove, by contradiction, the existence of an M_{i^*} $(1 \leq i^* \leq h)$ with $\gcd(M_{i^*}, D) \leq (2^{2^t} + 1) \cdot D^{1/h}$.

M_{i^*} has the desired property $\gcd(M_{i^*}, D) \cdot 2^L < 2^{2^t \cdot L_1} < M_{i^*}$, so that Algorithm S performs not more than i^* iterations. For all checked numbers $M = 2^N + 1$, we obtain $N < L \cdot (1 + 1/\log^2 L) + \mathcal{O} \cdot L^{1/2}$. Therefore, in the worst case at most

$$L^{1 + o(1)} \tag{19}$$

bit operations are needed for the computation of M.

Computing Time of Algorithm U. For simplicity, we give the time analysis for *normal* polynomial remainder sequences only, assuming $\delta_k = 1$ for $k = 3, \ldots, K$. Then $m_k \approx 2k$. Nevertheless, the bounds given below are valid in the abnormal case, too.

In the standard subresultant algorithm each polynomial G_k is calculated as pseudo remainder using about $3(n + 2 - k)$ integer multiplications with the result length bounded by $6kl$ and $n + 2 - k$ divisions with dividend length $6kl$ and divisor length $4kl$ for $k = 3, \ldots, n + 1$.

Our Algorithm U is faster because of two effects. First, all operations are performed with respect to the predicted size of $|F_k|_\infty$, thus avoiding larger intermediate results. Second, no division step is needed. Estimating the costs of Algorithm U, we can neglect all steps except step 5 which calculates at stage k the polynomials $F_k^\diamond$, or F_k, using $3(n + 2 - k)$ multiplications in the ring R_M, where $\log M \approx 2kl$. For small problems we proceed as in section 3 and obtain a time bound of the form

$$T_4(l, n) \simeq \gamma_0 \cdot l^2 \sum_{k=3}^{n+1} (n + 2 - k) \cdot 12k^2 \sim l^2 n^4 \cdot \gamma_0.$$

In the case of $n = 10$ the sum $\sum_{k=3}^{n+1}$ has the value 19980 which is about 35 percent of the respective value 57288 for the standard algorithm.

Using asymptotically fast integer arithmetic for larger values n, l, one obtains for the standard algorithm a time estimate of the form

$$T_5(l, n) \simeq \sum_{k=3}^{n+1} \left[3(n + 2 - k) \cdot 6kl \cdot L(6kl) + (n + 2 - k) \cdot 4kl \cdot \vartheta_1 L(4kl) \right]$$

$$\simeq n^3 \cdot l \cdot \left(3 + \tfrac{2}{3} \vartheta_1 \right) \cdot L(6nl), \tag{20}$$

where $L(n) = \gamma_1 \cdot \log(\lambda n) \cdot (1 + \log \log(\lambda n))$ (cf. [SGV94, 6.1.37]), and ϑ_1 measures the costs of divisions compared to multiplications. For Algorithm U the corresponding bound is

$$T_6(l, n) \simeq \sum_{k=3}^{n+1} 3(n + 2 - k) \cdot 2kl \cdot L(2kl) \simeq n^3 \cdot l \cdot L(2nl). \tag{21}$$

In comparison with the classical algorithm, Algorithm U is in the limit 11/3 times faster if we assume the same costs for multiplication and division by setting $\vartheta_1 = 1$. A more realistic estimation $\vartheta_1 = 3$ (cf. [SGV94, 6.1.53]) leads to the more realistic speed-up factor of five.

The description of Algorithm U given in this section is a sketch only. We mention just one of the various details for further improvement. Instead of the a priori bound l_k (17) determining the size of M at stage k, one can use the actual coefficients of F_{k-2} and F_{k-1}, to obtain an estimate of $|F_k|_\infty$ from equation (16). This "adapted" bound may be smaller than l_k thus leading to smaller rings R_M with cheaper arithmetic.

5 The Case $I = \mathbb{Z}[y]$

As announced, we discuss the multivariate case for one ballast variable y only. Let $\tau \in \mathbb{N}$ be a bound for the degree (in y) of the coefficients of A and B, and 2^l ($l \in \mathbb{Z}$) a bound for their size:

$$\deg a_\mu \leq \tau \quad \text{and} \quad |a_\mu|_\infty < 2^l, \qquad 0 \leq \mu \leq m,$$
$$\deg b_\nu \leq \tau \quad \text{and} \quad |b_\nu|_\infty < 2^l, \qquad 0 \leq \nu \leq n.$$

For technical reasons, let $l > \log m$ and $l > \log \tau$. Then, with arguments similar to those above, one obtains for each coefficient $g_k^{(\nu)}$ of G_k

$$\deg g_k^{(\nu)} \leq m_k \tau$$
$$\left\lceil \log \left| g_k^{(\nu)} \right|_\infty \right\rceil < l_k = \lfloor \vartheta_2 \cdot m_k \cdot l \rfloor, \qquad 0 \leq \nu \leq n_k, \ 3 \leq k \leq K, \tag{22}$$

with an effectively computable constant $\vartheta_2 \geq 1$.

Algorithm M. This algorithm computes the resultant and the gcd of two polynomials $A, B \in \mathbf{Z}[x, y]$.

1. [Initializations] Set $F_1 := A$, $F_2 := B$, and $k := 3$. Calculate h_2 using formula (12).

2. [k-loop] Determine l_k according to formula (22). As in the univariate case the a priori bound l_k might be replaced by an adaptive bound, obtained from F_{k-2} and F_{k-1} using equation (16).

3. [Find divisor] Choose a (small) integer $o_k \geq 0$ (e.g. $o_k := 2\lfloor \log n \rfloor$). Let $* : \mathbf{Z}[y] \to \mathbf{Z}$ be the evaluation homomorphism defined by $p^* = p\left(2^{l_k + o_k}\right)$ for $p \in \mathbf{Z}[y]$ or the induced homomorphism $\mathbf{Z}[x, y] \xrightarrow{*} \mathbf{Z}[x]$, respectively. Compute h_k^* and $D := d_k^*$ using equations (13) and (10).

4. [Suitable number] Let $L := \tau m_k(l_k + o_k) + 3l_k$. With equation (22) one obtains

$$\left|\left(g_k^{(\nu)}\right)^*\right| < 2^L, \qquad 0 \leq \nu \leq n_k. \tag{23}$$

Call Algorithm S to determine a suitable number $M = 2^N + 1$, so that

$$M > e_k \cdot 2^{L+1}, \tag{24}$$

where $e_k = \gcd(M, D)$. Compute $u \in \mathbf{Z}$ with $e_k \equiv uD \bmod M$. The computations in steps 5–7 are done in $R_M[x]$ with the homomorphism chain

$$\mathbf{Z}[x, y] \xrightarrow{*} \mathbf{Z}[x] \xrightarrow{\diamond} R_M[x].$$

5. [Prepare elimination factors] Use equation (16) to calculate the factors $u^{*\diamond} \cdot \left(f_{k-1}^{1+\delta_{k-2}}\right)^{*\diamond}$ and $u^{*\diamond} \cdot \phi_0^{*\diamond}, \ldots, u^{*\diamond} \cdot \phi_{\delta_{k-2}}^{*\diamond}$ for $0 \leq j \leq \delta_{k-2}$.

6. [Calculate F_k^*] Use the above factors to compute $F_k^{*\diamond}$. Because of (24), $F_k^{*\diamond}$ can be lifted to F_k^*.

7. [Calculate F_k] If $e_k \leq 2^{o_k}$, the bounds in (22) ensure a unique decoding of the coefficients of F_k^* to obtain F_k.

   Otherwise, if e_k is too large, calculate $G_k^* = F_k^*/e_k$ with n_k exact divisions and set $e_k = 1$. Note that these divisions can be carried out within multiplication time, since $e_k | M = 2^N + 1$. From G_k^* the polynomial G_k is decodable. Set $F_k := G_k$.

8. [Loop] While $F_k \neq 0$, increment k by 1 and continue with step 2.

9. [Finals] Proceed as in step 7 of Algorithm U to return resultant and gcd of A and B.

In the above algorithm additions and multiplications in $\mathbf{Z}[y]$ are performed in $\mathbf{Z}$ and in R_M via the evaluation homomorphism $*$. Parameter l_k covers the size of the subresultant coefficients $g_k^{(\nu)}$ while o_k is for swallowing the extra factors $e_k = \gcd(M, D)$ if $e_k \leq 2^{o_k}$.

For a worst case estimate we set $o_k = 0$ for $k = 3, \ldots, K$, since we do not know how to exclude the case $e_k > 2^{o_k}$ for moderate values o_k.

For an average case estimate, $o_k = 2 \log n$ might be used, justified by the following heuristic argument: Let the numbers M, D be independently chosen at random. The probability of $\gcd(M, D) \geq n^2$ is $6\pi^{-2} \cdot \sum_{i \geq n^2}^{\infty} i^{-2} = \mathcal{O}/n^2$. Therefore the probability, that none of n independent gcd's exceeds n^2, is $1 - \mathcal{O}/n$.

Computing Time of Algorithm M. Again we restrict the analysis of the algorithm to normal polynomial remainder sequences and omit the proof of our estimates for the abnormal case.

Let $\mathcal{S}_\mu(k)$ the costs for step μ of Algorithm M at stage k for $k = 3, \ldots, K$ and $\mu = 3, \ldots, 7$. Since $m_k \approx 2k$, one obtains the length bound

$$l_k = \mathcal{O} \cdot k \cdot l, \qquad k = 3, \ldots, K. \tag{25}$$

We assume $o_k = \mathcal{O} \cdot k \cdot l$. The costs of the ring operations are determined by

$$L = \mathcal{O} \cdot \tau \cdot k^2 \cdot l, \qquad k = 3, \ldots, K, \tag{26}$$

namely $\log M = L^{1+o(1)}$. The bound (19) for Algorithm S shows that $\mathcal{S}_4(k) = L^{1+o(1)}$. The same estimation applies to step 3 and to the preparation of the three elimination factors in step 5. Thus

$$\mathcal{S}_3(k) + \mathcal{S}_4(k) + \mathcal{S}_5(k) = L^{1+o(1)}, \qquad k = 3, \ldots, K. \tag{27}$$

In step 6 the polynomial remainder $F_k^\star$ is determined with $\mathcal{O} \cdot k$ ring multiplications and additions. So

$$\mathcal{S}_6(k) = k \cdot L^{1+o(1)}, \qquad k = 3, \ldots, K. \tag{28}$$

The division in step 7 by $e_k > 2^{o_k}$ (if required) can be carried out within the same time,

$$\mathcal{S}_7(k) = k \cdot L^{1+o(1)}, \qquad k = 3, \ldots, K. \tag{29}$$

Equations (27–29) give the bound $k \cdot L^{1+o(1)}$ for the costs of stage k. Summation over all stages leads, together with (25) and (26), to the following estimation for the overall costs of Algorithm M:

$$T_7(n, l, \tau) = \tau^{1+o(1)} \cdot n^{4+o(1)} \cdot l^{1+o(1)}. \tag{30}$$

From that we achieve with $\tau = n$, assuming a uniform bound n for the degrees of A and B in each of the variables x and y,

$$T_7(n, l, n) = n^{5+o(1)} \cdot l^{1+o(1)}. \tag{31}$$

As already pointed out in the Introduction, this result improves Collins' bound $\mathcal{O} \cdot n^6 \cdot l^2$ for the modular subresultant algorithm. It should be clear from the previous development how to generalize this method and the estimates to more than one ballast variable.

459

References

[Bar68]   E. H. Bareiss, *Sylvester's identity and multistep integer-preserving Gaussian elimination.* Math. Comp. **22** (1968), 565–578.

[Br71]   W. S. Brown, *On Euclid's algorithm and the computation of polynomial greatest common divisors.* J. ACM **18** (1971), 478–504.

[Br78]   W. S. Brown, *The Subresultant PRS Algorithm.* ACM TOMS **4** (1978), 237–249.

[Co67]   G. E. Collins, *Subresultants and reduced polynomial remainder sequences.* J. ACM **14** (1967), 128–142.

[Co71]   G. E. Collins, *The calculation of multivariate polynomial resultants.* J. ACM **18** (1971), 515–532.

[Jeb93]   T. Jebelean, *An algorithm for exact division.* J. Symbolic Computation **15** (1993), 169–180.

[Kl93]   J. Klose. *Schnelle Polynomarithmetik zur exakten Lösung des Fermat-Weber-Problems.* Dissertation Universiät Erlangen, 1993.

[Kn81]   D. E. Knuth. *The Art of Computer Programming, Vol. 2, Seminumerical Algorithms.* Addison-Wesley, Reading, MA, 2nd ed, 1981.

[Lo82]   R. Loos, *Generalized polynomial remainder sequences.* In *Computer Algebra, Symbolic and Algebraic Computation*, eds. B. Buchberger, G. E. Collins, R. Loos. Springer-Verlag, Wien, 2nd ed, 1982, 115–137.

[Sc82]   A. Schönhage. *Asymptotically fast algorithms for the numerical multiplication and division of polynomials with complex coefficients.* In *Computer Algebra EUROCAM '82 (Marseille 1982)*, ed. J. Calmet. Lect. Notes Comp. Sci. **144**, 3–15.

[Sc88]   A. Schönhage. *Probabilistic computation of integer polynomial GCDs.* J. of Algorithms **9** (1988), 365–371.

[SGV94]   A. Schönhage, A. F. W. Grotefeld, E. Vetter. *Fast Algorithms— A Multitape Turing Machine Implementation.* BI-Wissenschaftsverlag, Mannheim, 1994.

Testing equivalence of morphisms on context-free languages *

Wojciech Plandowski [1] **

Instytut Informatyki UW, 02-097 Banacha 2, Warszawa, Poland

Abstract. We present a polynomial time algorithm for testing if two morphisms are equal on every word of a context-free language. The input to the algorithm are a context-free grammar with constant size productions and two morphisms. The best previously known algorithm had exponential time complexity. Our algorithm can be also used to test in polynomial tiime whether or not n first elements of two sequences of words defined by recurrence formulae are the same. In particular, if the well known $2n$ conjecture for D0L sequences holds, the algorithm can test in polynomial time equivalence of two D0L sequences.

Additionally, we extend the result from [5] by proving the existence of polynomial size test sets for context-free languages not only in free monoids but in free groups as well. The main points of our proof are the same as in [6, 5]. The main change is a new short proof of the main lemma. The previous proof took 10 pages. It was complicated since it considered many cases and used advanced properties of periodicity of words. Our proof takes only 2 pages. The simplification is a consequence of embedding a free monoid into a free group.

1 Introduction

The problem of equivalence of morphisms on context-free languages (cfl, for short) consists in checking for given two morphisms f, g and a context-free grammar G (cfg, for short) if they coincide on $L(G)$ (i.e. if $f(w) = g(w)$ for $w \in L(G)$). The idea to solve the problem is connected to the notion of a test set. A set of words T is a test set for a language L iff T is a subset of L and for every two morphisms f, g if they coincide on T then they coincide on L. Hence, it is enough to compare morphisms on words from a test set to decide if they coincide on a whole language. The existence of finite test sets for cfls was proved in [2]. The number of words in the test set could be double exponential. This result was improved to single exponential [4] and later to polynomial, see [6]. The lengths of words in the test set however could be exponential. Moreover, the length of the shortest word in a cfl defined by a grammar with n productions can be exponential, see Example 1. Thus, for some cfls it is impossible to list in polynomial time all words from any test set. To avoid this problem we encode

* Supported by the grant EC Cooperation Action IC 1000 Algorithms for Future Technologies ALTEC.
** E-mail: wojtekpl@mimuw.edu.pl

every word from the test set by a small cfg. Then for each word w from the test set we find codes for $f(w)$, $g(w)$. Finally, we present a polynomial time algorithm to test if two codes represent the same word. The algorithm uses similar ideas as the Makanin algorithm (which is of exponential complexity) for testing whether or not a word equation has a solution, see [9] and later improvements [3, 7, 10, 11].

We generalize the definition of a test set to all monoids. A subset T of a language L is a test set for L in a monoid M iff for each two morphisms f, $g : \Sigma^* \to M$ if they coincide on T then they coincide on L. We prove the existence of polynomial size test sets for cfls in free groups. At one stroke we extend the result from [5] and obtain a shorter proof! As an example how it is possible we present two proofs of the fact which is used in the proof of Lemma 2. The proof in a free monoid considers two cases while the proof for groups uses inverse elements instead.

Lemma 1. $X = \{uv, \ uw, \ xv\}$ *is a test set for* $Y = \{uv, \ uw, \ xv, \ xw\}$ *in every group.*

Proof. (in a free monoid)
Let f, g be two morphisms. It is enough to prove that if $f(z) = g(z)$ for $z \in X$ then $f(xw) = g(xw)$. Denote $s' = f(s)$ and $s'' = g(s)$ for any word s. Since $u'v' = u''v''$ two cases are to be considered:
Case 1: u' is a prefix of u''.
There is a word σ such that $u'' = u'\sigma$. Using it in equations $u'v' = u''v''$ and $u'w' = u''w''$ and simplifying them we have $v' = \sigma v''$ and $w' = \sigma w''$, respectively. Since $x'v' = x''v''$ we obtain $x'' = x'\sigma$ and finally $x''w'' = x'\sigma w'' = x'w'$.
Case 2: u'' is a prefix of u'.
The proof in that case is symmetric to the one in the previous case.

Proof. (in a group)
Let f, g be two morphisms. It is enough to prove that if $f(z) = g(z)$ for $z \in X$ then $f(xw) = g(xw)$. Denote $s' = f(s)$ and $s'' = g(s)$ for any word s. We have

$$x'w' = x'v'(v')^{-1}(u')^{-1}u'w' = x'v'(u'v')^{-1}u'w' = x''v''(u''v'')^{-1}u''w'' = x''w''$$

and we are done.

The fact above is an extension of the main fact in [4] and allows to prove using the same construction as in [4] that in every group every regular language has a linear size test set.

2 Polynomial size test sets for context-free languages

Recall, that a cfg is in *Chomsky normal form* if each production of the grammar is either in form $A \to BC$ or in form $A \to w$ where A, B, C are nonterminal symbols and w is a terminal symbol. Since every cfg whose productions are of constant size can be transformed in polynomial time to a cfg in Chomsky normal form we restrict our considerations to cfgs which are in that form. A *linear context-free grammar* is a cfg whose productions are of the form $A \to uBv$ or $A \to u$ where u, v are terminal words and A, B nonterminal symbols. For every nonterminal A let w_A be any shortest word derivable from A. For a cfg G in Chomsky normal form we define a linear cfg $lin(G)$ by replacing every production of form $A \to BC$ by three productions $A \to w_B C$, $A \to B w_C$, $A \to w_B w_C$.

Lemma 2. *Let G be a cfg in Chomsky normal form. Then $L(lin(G))$ is a test set for $L(G)$ in every group.*

Proof. The proof is the same as the proof of Lemma 2 in [5], see also the proof of Theorem 1 in [6]

As we shall see in the next lemma to check if in a given group the size of a minimal test set can be bounded by a polynomial it is enough to consider test sets for finite languages L_k which are defined by the following linear cfg G_k:

$$A_i \to a_i A_{i-1} \bar{a}_i \mid b_i A_{i-1} \bar{b}_i \qquad \text{for } 1 < i \leq k$$
$$A_1 \to a_1 \mid b_1$$

where nonterminals are in capitals, terminal symbols in lower case letters and A_k is the starting symbol. The language L_k consists of 2^k words and it is defined over an alphabet consisting of $4k - 2$ symbols. Let μ_k be the only word from L_k consisting of letters b_i and $\bar{b}_i$. Define $T_k = L_k - \{\mu_k\}$.

A linear cfg G can be viewed as the digraph $graph(G)$ whose vertices correspond to nonterminals and there is an edge from A to B labeled by a pair of terminal words (u, v) if $A \to uBv$ is a production in G. Additionally we add a terminal node t. There is an edge from A to t labelled (u, v) if $v = \epsilon$ and there is a production $A \to u$ in G. In such a graph every path from the starting nonterminal S to t corresponds to a derivation in G of a word from $L(G)$. On the other hand each derivation of a word from $L(G)$ has its corresponding path in $graph(G)$ from S to t. A word w whose derivation corresponds to a path $\lambda = (S, v_1), \ldots, (v_k, t)$ can be obtained in the following way. If $(w_0, \bar{w}_0), \ldots, (w_k, \bar{w}_k)$ are labels corresponding to consecutive edges of λ then $w = w_0 \ldots w_k \bar{w}_k \ldots \bar{w}_0$.

For each node v of $graph(G)$ we build a tree $tree(v)$ rooted at v containing all vertices reachable from v in $graph(G)$. With each sequence of not necessarily adjacent edges $(u_1, v_1), \ldots, (u_k, v_k)$ we associate a path which starts at S goes in $tree(S)$ to u_1 then it traverses (u_1, v_1) and then goes in $tree(v_1)$ to a vertex u_2 runs through (u_2, v_2) etc. Finally, it goes via (u_k, v_k) and $tree(v_k)$ to t. Observe, that for some sequences of edges there are no paths which satisfy the conditions above and since there is exactly one path in a tree which starts at the root and ends with a given vertex of the tree the path if exists is unique. Let $F_k^{tree}(G)$ be a set of words corresponding to paths associated with a sequence of at most k edges. The index *tree* in $F_k^{tree}(G)$ means that the contents of this set depends on the function *tree*. The number of words in $F_k^{tree}(G)$ does not exceed the number of sequences of at most k edges. Since the number of edges of $graph(G)$ is equal to the number of productions in G, the size of $F_k^{tree}(G)$ does not exceed

$\sum_{i=0}^{k} n^i \leq (k+1)n^k$ where n is the number of productions in G. Suppose the sizes of productions from G are $O(1)$. Since the length of a path in a tree embedded in $graph(G)$ does not exceed n the length of a path associated with a sequence of k edges does not exceed kn and the words in $F_k(G)$ are not longer than $O(kn)$.

Lemma 3. *Let G be a linear cfg and Gr be a group. If T_k is a test set for L_k in Gr then $F_{2k-2}^{tree}(G)$ is a test set for $L(G)$ in Gr.*

Proof. The proof is a straightforward generalization of the proof of Lemma 3 from [5], see also the proof of Lemma 2 in [6].

Theorem 4. *Let Gr be a group. There is a polynomial upper bound in Gr for the size of minimal test sets for context-free languages which are defined by context-free grammars with constant size productions iff there is $k > 0$ such that T_k is a test set for L_k in Gr.*

Proof. Let G be a cfg in Chomsky normal form with n productions. If there exists $k > 0$ such that T_k is a test set for L_k then, by Lemma 3 and Lemma 2, $F_{2k-2}^{tree}(lin(G))$ is a test set for $L(G)$. Since $F_{2k-2}^{tree}(lin(G))$ contains at most $(2k-1)(3n)^{2k-2}$ words the polynomial upper bound exists. Suppose, there does not exist $k > 0$ such that T_k is a test set for L_k. Since the number of words in L_k is 2^k and it is defined by a grammar G_k which can be easily transformed to a grammar in Chomsky normal form with $O(k)$ productions the result follows.

By Theorem 4, to prove in a group the existence of polynomial size test set for cfl it is enough to find k such that T_k is a test set for L_k in the group. The next lemma proves that for $k = 4$ it is true in free groups.

Denote $w^a = a^{-1}wa$. Note, that $(w^a)^b = w^{ab}$

Lemma 5. *Consider morphisms f, g. Denote $s' = f(s)$ and $s'' = g(s)$ for each element s of a free group. Let τ, ρ, ω, σ, $\overline{\sigma}$, a, b, c, d be elements of the free group. Then*

1. *If $\omega\tau = \tau\omega$ and $\rho\tau = \tau\rho$ and $\tau \neq 1$ then $\omega\rho = \rho\omega$.*
2.

$$\begin{cases} \sigma^a\tau^c = \tau^c\sigma^a \\ \sigma^b\tau^c = \tau^c\sigma^b \\ \sigma^a\tau^d = \tau^d\sigma^a \end{cases} \implies \sigma^b\tau^d = \tau^d\sigma^b \ ,$$

3. *If $\sigma x' = x'\overline{\sigma}$ for $x \in T_3$ then $\sigma\mu'_3 = \mu'_3\overline{\sigma}$,*
4. *If $x' = x''$ for $x \in T_4$ then $\mu'_4 = \mu''_4$.*
5. *T_4 is a test set for L_4 in a free group.*

Proof.

    1. The proof of this point is a consequence of the fact that every subgroup of a free group is free. We omit details.

2. It is a consequence of point 1. We may assume that $\sigma \neq 1$ and $\tau \neq 1$. We apply point 1 to the first two equations to obtain $\sigma^a \sigma^b = \sigma^b \sigma^a$. Now, we apply point 1 again to obtained equation and the third equation.

3. We have $\bar{\sigma} = (x')^{-1} \sigma x' = \sigma^{x'}$ for $x \in T_3$. Hence $\sigma^{x'} = \sigma^{y'}$ for $x, y \in T_3$ and therefore $\sigma^{x'y'^{-1}} = \sigma$ for $x, y \in T_3$. For $x = a_3 a_2 a_1 \bar{a}_2 \bar{a}_3$, $y = a_3 a_2 b_1 \bar{a}_2 \bar{a}_3$ we have $\sigma^{a_3' a_2' a_1' b_1^{-1} a_2'^{-1} a_3'^{-1}} = \sigma$ (α). Similarly, we obtain $\sigma^{a_3' b_2' a_1' b_1^{-1} b_2'^{-1} a_3'^{-1}} = \sigma$ (β) and $\sigma^{b_3' a_2' a_1' b_1^{-1} a_2'^{-1} b_3'^{-1}} = \sigma$ (γ). Now, we transform the equation (α). Let $\rho = a_1' b_1'^{-1}$. We have $\sigma^{a_3' \rho^{a_2'^{-1}} a_3'^{-1}} = \sigma$. Thus $\sigma^{a_3' \rho^{a_2'^{-1}}} = \sigma^{a_3'}$ and therefore $(\sigma^{a_3'})^{\rho^{a_2'^{-1}}} = \sigma^{a_3'}$. Finally, we obtain $\sigma^{a_3' \rho^{a_2'^{-1}}} = \rho^{a_2'^{-1}} \sigma^{a_3'}$. Similarly, we transform equations (β), (γ) to have $\sigma^{a_3' \rho^{b_2'^{-1}}} = \rho^{b_2'^{-1}} \sigma^{a_3'}$, $\sigma^{b_3' \rho^{a_2'^{-1}}} = \rho^{a_2'^{-1}} \sigma^{b_3'}$, respectively. Now, we apply point 2 to obtain $\sigma^{b_3' \rho^{b_2'^{-1}}} = \rho^{b_2'^{-1}} \sigma^{b_3'}$. By transforming the last equation in the reverse order than previously we obtain $\sigma^{b_3' b_2' a_1' b_1^{-1} b_2'^{-1} b_3'} = \sigma$ which is equivalent to $\sigma^{\mu_3'} = \sigma^{\nu'}$ where $\nu = b_3 b_2 a_1 \bar{b}_2 \bar{b}_3$. Since $\nu \in T_3$ we have $\sigma^{\nu'} = \bar{\sigma}$ and finally $\sigma^{\mu_3'} = \bar{\sigma}$ which is equivalent to $\mu_3' \bar{\sigma} = \sigma \mu_3'$.

4. The system of equations $x' = x''$ for $x \in T_4$ can be written in form

$$a_4' x' \bar{a}_4' = a_4'' x'' \bar{a}_4'' \qquad \text{for } x \in L_3$$
$$b_4' x' \bar{b}_4' = b_4'' x'' \bar{b}_4'' \qquad \text{for } x \in T_3.$$

Hence for $x \in T_3$ we have
$a_4' x' \bar{a}_4' = a_4'' x'' \bar{a}_4'' = a_4'' b_4''^{-1} (b_4'' x'' \bar{b}_4'') \bar{b}_4''^{-1} \bar{a}_4'' = a_4'' b_4''^{-1} b_4' x' \bar{b}_4' \bar{b}_4''^{-1} \bar{a}_4''$
and therefore $(b_4'^{-1} b_4'' a_4''^{-1} a_4') x' = x' (\bar{b}_4' \bar{b}_4''^{-1} \bar{a}_4'' \bar{a}_4'^{-1})$. Now, we apply point 3 and obtain $(b_4'^{-1} b_4'' a_4''^{-1} a_4') \mu_3' = \mu_3' (\bar{b}_4' \bar{b}_4''^{-1} \bar{a}_4'' \bar{a}_4'^{-1})$.
Hence, we have $b_4'' a_4''^{-1} (a_4' \mu_3' \bar{a}_4') \bar{a}_4''^{-1} = b_4' \mu_3' \bar{b}_4' \bar{b}_4''^{-1}$. Since $a_4' \mu_3' \bar{a}_4' = a_4'' \mu_3' \bar{a}_4''$ we have $b_4' \mu_3' \bar{b}_4' = b_4'' \mu_3' \bar{b}_4''$ and finally $\mu_4' = \mu_4''$.

5. This point is a reformulation of point 4.

We say, that a grammar G *defines a word* w if it is a cfg in Chomsky normal form without useless nonterminals such that each nonterminal is on the left hand side of exactly one production and $L(G) = \{w\}$. In such a grammar the word w has exactly one derivation.

Example 1. Some grammars define a word of exponential size with respect to the number of their productions, e.g. the grammar with productions $A_i \rightarrow A_{i-1} A_{i-1}$ for $1 \leq i \leq k$, $A_0 \rightarrow a$ and the starting symbol A_k defines the word a^{2^k}. We treat a grammar defining w as a compressed representation of w.

Theorem 6. *In a free group every cfl generated by a grammar in Chomsky normal form with n productions has a test set of size at most $7(3n)^6$. Grammars defining every word from the test set can be found in polynomial time.*

Proof. Let G be a cfg in Chomsky normal form with n productions. By Lemma 2, Lemma 3 and Lemma 5 $F_6^{free}(lin(G))$ is a test set for $L(G)$. Since $lin(G)$ has at most $3n$ productions $F_6^{free}(lin(G))$ has at most $7(3n)^6$ words and the first part of the theorem is proved.

Since the shortest words derivable from nonterminals in grammar G can be of exponential size the productions of $lin(G)$ can be of exponential size. Fortunately, for each nonterminal A there is a shortest word derivable from A which is defined by a subgrammar of G. Take such a word as a w_A. Now, we treat all w_A as terminal symbols in the grammar $lin(G)$ remembering that those symbols represent words which are defined by grammars. On the basis of the graph for $lin(G)$ we find $F_6(lin(G))$. The words are of lengths $O(n)$ and they can be found in polynomial time. They consist of terminal symbols of $lin(G)$ which correspond to words defined by grammars. Now, for each word from the test set we can find a grammar defining this word.

3   A polynomial time algorithm for testing if two morphisms coincide on a context-free language

From now on we deal with a free monoid. Let f be a morphism and let G be a grammar defining w. A grammar defining the word $f(w)$ is easily derived from G by replacing every production of form $A \rightarrow u$ where u is a terminal symbol by $A \rightarrow f(u)$. By Theorem 6 to find a polynomial time algorithm for morphism equivalence problem for a cfl it is enough to find a polynomial time algorithm for testing if two grammars define the same word. We describe a polynomial time algorithm for a more general problem.

We say, that a grammar G *defines a set of words* S if the grammar is context-free in Chomsky normal form (there can be useless nonterminals), each nonterminal is on left hand side of exactly one production, every nonterminal generates exactly one terminal word and for each $w \in S$ there is a nonterminal A in G such that $A \rightarrow^* w$. In particular, if G defines a word then it defines a set of words derivable from nonterminals of G. Denote by w_A the terminal word derivable from A in G if A is a nonterminal and A if A is a terminal symbol.

Example 2. For fixed k consider a grammar G containing the following set of productions:

$$A_i \rightarrow A_{i-1}A_{i-1} \text{ for } 1 \le i \le k \qquad \begin{array}{l} A' \rightarrow A_0 A'_k \\ A'_i \rightarrow A'_{i-1}A_{i-1} \text{ for } 1 \le i \le k \\ A'_1 \rightarrow a \end{array}$$
$$A_0 \rightarrow a$$

Since $w_{A_i} = a^{2^i}$ for $0 \le i \le k$, $w_{A'_i} = a^{2^i - 1}$ for $1 \le i \le k$, $w_{A'} = a^{2^k}$ the grammar defines the set of words

$$\{a^{2^i} : 0 \le i \le k\} \cup \{a^{2^i - 1} : 1 \le i \le k\}.$$

We describe a polynomial time algorithm for the following problem:

The Problem

Let G be a grammar which defines a set of words and S be a set of pairs of nonteminals from G. Check if $w_A = w_B$ for every pair (A, B) from S.

Let G be a grammar with n productions which defines a set of words W. Since the shortest words derivable from a nonterminal of a grammar in Chomsky normal form with n productions are of size not exceeding 2^n the lengths of words from W are not longer than 2^n. Such numbers can be stored in n bits. Standard algorithms for basic operations (comparing, addition, substraction, division, multiplication) on such numbers work in polynomial time with respect to n. This allows to compute the length of every word from W in polynomial time.

The key point of our algorithm is to consider relations between words w_A. The relations are stored as sets of triples $(A, B.i)$ where A, B are nonterminal or terminal symbols and i is a nonnegative integer. We divide the triples into two groups which we call *suffix* and *subword* triples. Let $|w|$ be the length of the word w. A triple (A, B, i) is a suffix triple iff $i + |w_B| \geq |w_A|$ and it is a subword triple iff $i + |w_B| < |w_A|$ and $i \geq 1$. The basic relation in our considerations is $SUFSUB(G)$. Intuitively, a triple (A, B, i) is in $SUFSUB(G)$ iff starting at position $i + 1$ in word w_A consecutive symbols of w_A are the same as in w_B starting from the beginning. More precisely:

$(A, B, i) \in SUFSUB(G)$ iff
$\quad (A, B, i)$ is a suffix triple and $w_A[i + 1 \ldots |w_A|] = w_B[1 \ldots |w_A| - i]$ or
$\quad (A, B, i)$ is a subword triple and $w_A[i + 1 \ldots i + |w_B|] = w_B$

Now, The Problem can be refomulated as follows:
Given a grammar G defining a set of words and a set S of pairs of nonterminals from G. For each pair (A, B) from S check if $|w_A| = |w_B|$ and $(A, B, 0) \in SUFSUB(G)$.

Let rel be a set of triples. We define an operation $Split(A, rel)$ which eliminates occurrences of a nonterminal A from triples of rel and does not lose the information from rel. Assume, $A \to EF$ is a production of G. For one triple the operation $Split(A, (B, C, i))$ is defined as follows:

- Case $A \neq B$ and $A \neq C$

$$Split(A, (B, C, i)) = (B, C, i)$$

- Case $A = B$ and $A \neq C$

$$Split(A, (A, C, i)) = \begin{cases} (E, C, i) \cup (C, F, |w_E| - i) \\ \quad \text{if } |w_E| > i \text{ and } |w_C| + i > |w_E|, \\ (E, C, i) \\ \quad \text{if } |w_E| > i \text{ and } |w_C| + i \leq |w_E|, \\ (C, F, 0) \\ \quad \text{if } |w_E| = i \text{ and } |w_C| \leq |w_F|, \\ (F, C, 0) \\ \quad \text{if } |w_E| = i \text{ and } |w_C| > |w_F|, \\ (F, C, i - |w_E|) \\ \quad \text{if } |w_E| < i. \end{cases}$$

– Case $A \neq B$ and $A = C$

$$Split(A, (B, A, i)) = \begin{cases} (B, E, i) \\ \quad \text{if } |w_E| + i \geq |w_B|, \\ (B, E, i) \cup (B, F, |w_E| + i) \\ \quad \text{if } |w_E| + i < |w_B|. \end{cases}$$

– Case $A = B$ and $A = C$

$$Split(A, (A, A, i)) = \begin{cases} \emptyset \\ \quad \text{if } i = 0 \\ (E, E, i) \cup (E, F, |w_E| - i) \\ \quad \text{if } |w_E| > i \geq 1 \text{ and } i \geq |w_F|, \\ (E, E, i) \cup (F, F, i) \cup (E, F, |w_E| - i) \\ \quad \text{if } |w_E| > i \geq 1 \text{ and } i < |w_F|, \\ (E, F, 0) \\ \quad \text{if } |w_E| = i \text{ and } |w_E| \geq |w_F|, \\ (F, E, 0) \cup (F, F, i) \\ \quad \text{if } |w_E| = i \text{ and } |w_E| < |w_F|, \\ (F, E, i - |w_E|) \\ \quad \text{if } |w_E| < i \text{ and } i \geq |w_F|, \\ (F, E, i - |w_E|) \cup (F, F, i) \\ \quad \text{if } |w_E| < i \text{ and } i < |w_F|. \end{cases}$$

For a set of triples rel we define $Split(A, rel) = \bigcup_{(B,C,i) \in rel} Split(A, (B, C, i))$. The definition of $Split$ looks very complicated, but the idea is to eliminate nonterminal A from triples of rel without losing the information about the dependences between words in rel. Our next lemma states in formal way that the relation $Split(A, rel)$ is equivalent to rel.

Lemma 7. $rel \subseteq SUFSUB(G)$ iff $Split(A, rel) \subseteq SUFSUB(G)$ for any nonterminal A

Proof. The proof from left to right is a consequence of the fact that on the basis of the relation rel we define $Split$ and we do not add dependences which are not derivable from rel. The proof from right to left is a consequence of the fact that having $Split$ we can restore rel since all the information from rel were moved to $Split$. The formal proof of the above is simple but tedious. It consists of several cases corresponding to the cases in the definition of $Split$.

Let $\#suffix(rel)$ and $\#subword(rel)$ be the number of suffix and subword triples in rel, respectively. Let $\#sufsub(rel) = \#suffix(rel) + \#subword(rel)$. Clearly, $\#sufsub(rel)$ is the number of triples in rel. The following lemma will be useful later.

Lemma 8. *Let rel be a set of triples such that for every $(B, C, i) \in rel$ $|w_A| \geq \max\{|w_B|, |w_C|\}$. Then*

$$\#suffix(Split(A, rel)) \leq 3\#sufsub(rel)$$
$$\#subword(Split(A, rel)) \leq \#subword(rel) + 3\#suffix(rel)$$

Proof. The first equality is clear since one triple from *rel* corresponds to at most three triples in $Split(A, rel)$. The second one is a consequence of the fact that every triple from *rel* corresponds to at most one subword triple in *Split*. Look at the definition of *Split*. In most cases triples correspond to one triple. Since (H, H, i) for any H is always a suffix triple it remains to consider the case $A = B$ and $A \neq C$ for $|w_E| > i$ and $|w_C| + i > |w_E|$ and the case $A \neq B$ and $A = C$ for $i + |w_E| < |w_B|$. In the first case (E, C, i) is a suffix triple, in the second one since $|w_A| \geq |w_B|$ $(B, F, |w_E| + i)$ is a suffix triple.

The second operation we define for a set of triples *rel* is the operation $Compact(rel)$. It reduces the number of suffix triples in relation *rel*.

Lemma 9. (periodicity lemma, see [8])
If x, y are periods of a word w and $x + y \leq |w|$ then $gcd(x, y)$ is also a period of w where gcd stands for the greatest common divisor.

Computing the greatest common divisor of two n bit numbers takes polynomial time, see [1].
   Define the operation $SimpleCompact(rel)$ in the following way. If there are three suffix triples (A, B, i), (A, B, j), (A, B, k) in *rel* such that $(j - i) + (k - i) \leq |w_A| - i$ and $i < j < k$ then the operation replaces them by two triples (A, B, i), $(A, B, i + gcd(j - i, k - i))$. If there are no such three triples it fails. A $Compact(rel)$ becomes from *rel* by applying $SimpleCompact$ to *rel* until it fails.

Lemma 10. $rel \subseteq SUFSUB(G)$ *iff* $Compact(rel) \subseteq SUFSUB(G)$

Proof. It is enough to prove that $rel \subseteq SUFSUB(G)$ iff $SimpleCompact(rel) \subseteq SUFSUB(G)$. If $rel \subseteq SUFSUB(G)$ and $SimpleCompact$ does not fail then both $j - i$ and $k - i$ are periods of the word $w = w_B[1 \ldots |w_A| - i]$ and by periodicity lemma $gcd(j - i, k - i)$ is also a period of it. Hence, $(A, B, i + gcd(j - i, k - i))$ is in $SUFSUB(G)$. On the other hand if (A, B, i) and $(A, B, i + gcd(j - i, k - i))$ are in $SUFSUB(G)$ then $gcd(j - i, k - i)$ is a period of w. Then $j - i$, $k - i$ as multiplies of $gcd(j - i, k - i)$ are periods of w and therefore both (A, B, j), (A, B, k) are in $SUFSUB(G)$.

Lemma 11. *If rel is a set of triples then* $\#suffix(Compact(rel)) \leq (2n+1)n^2$.

Proof. Let (A, B, i_1), $(A, B, i_2), \ldots, (A, B, i_k)$ be a sequence of all suffix triples of form (A, B, i) from $Compact(rel)$ sorted on i. Since the operation $SimpleCompact$ fails we have $i_{r+2} - i_r + i_{r+1} - i_r > |w_A| - i_r$. Hence, $2i_{r+2} - i_r > |w_A|$ and therefore $\frac{1}{2}(|w_A| - i_r) > (|w_A| - i_{r+2})$. Since $|w_A| - i_1 \leq 2^n$, the sequence of numbers $|w_A| - i_1, \ldots, |w_A| - i_k$ has at most $2n + 1$ elements. Therefore the number of all suffix triples in $Compact(rel)$ does not exceed $(2n + 1) * n^2$.

Algorithm Test;
{ input:
   grammar G which defines a set of words
   set S of pairs of nonterminals from G

output:
    test if for each $(A, B) \in S$ $w_A = w_B$}
begin
    compute $|w_A|$ for each nonterminal A;
    if there is $(A, B) \in S$ such that $|w_A| \neq |w_B|$ **then return** false;
    $(A_1, \ldots, A_n)$:=sort nonterminals in descending order according to $|w_A|$;
    rel:=$\bigcup_{(A,B) \in S}(A, B, 0)$;
    for i:=1 **to** n **do**
    begin rel:=$Split(A_i$,rel); rel:=$Compact$(rel) **end**;
    {there are no nonterminals in triples of rel}
    if $\exists(a, b, 0) \in$ suffix and $a \neq b$ **then return** false **else return** true
end.

Theorem 12. *The worst-case performance of the algorithm Test is polynomial with respect to the size of input.*

Proof. Let rel_i be the value of the variable rel before the i-th iteration of the algorithm. The correctness of the algorithm is a consequence of Lemma 7 and Lemma 10. To prove the polynomial worst-case performance of the algorithm it is enough to prove that the $\#sufsub(rel_i)$ can be polynomially bounded. Initially, $\#subword(rel_0) = 0$ and $\#suffix(rel_0) = |S| \leq n^2$. By Lemma 8 we have $\#suffix(rel_i) \leq (2n+1)n^2$ for $i \geq 1$. Since the operation $Compress$ do not touch subword triples we have $\#subword(rel_i) \leq \#subword(rel_{i-1}) + 3\#suffix(rel_{i-1})$ for $i \geq 1$. Solving this recurrence we obtain $\#subword(rel_i) \leq 3i(2n+1)n^2 \leq 3n(2n+1)n^2$ and finally $\#sufsub(rel_i) \leq (3n+1)(2n+1)n^2$.

Theorem 13. *Testing of a set of equivalences between words from a set of words defined by a grammar G can be done in polynomial time.*

As a consequence of the above theorem we have

Theorem 14. *Testing mophism equivalence on a context-free language can be done in polynomial time*

Proof. To compare words $f(w)$, $g(w)$ defined by grammars G_f, G_g consider the grammar $G_f \cup G_g$ and the one element set of pairs of nonterminals $\{(A_f, A_g)\}$ where A_f, A_g are starting symbols of grammars G_f, G_g, respectively.

Because of shortage of space two other applications of algorithm Test are presented in examples.

Example 3. Using algorithm *Test* we can compare first n elements of two sequences of words which are defined by recurrence formulae of form:
$f_{n+s+1} = F(f_n, \ldots, f_{n+s})$ where F is a composition of concatenations and s is a constant. For example to compare first n elements of sequences $f_1 = a$, $f_2 = b$, $f_{k+2} = f_{k+1}f_k$ for $1 \leq k \leq n-2$ and $f_1' = a$, $f_2' = b$, $f_{k+2}' = f_k'f_{k+1}'$ for $1 \leq k \leq n-2$ we consider a grammar
$F_1 \rightarrow a$, $F_2 \rightarrow b$, $F_{k+2} \rightarrow F_{k+1}F_k$ for $1 \leq k \leq n-2$
$F_1' \rightarrow a$, $F_2' \rightarrow b$, $F_{k+2}' \rightarrow F_k'F_{k+1}'$ for $1 \leq k \leq n-2$
and a set of pairs $\{(F_k, F_k') : 1 \leq k \leq n\}$.

Example 4. A D0L sequence (w, f, Σ) is a sequence of words w, $f(w)$, $f^2(w)$, ...where w is a starting word, $f : \Sigma^* \to \Sigma^*$ is a morphism and Σ is an n letter alphabet. A well known $2n$ conjecture for D0L sequences states that if two D0L sequences coincide on first $2n$ elements then they are the same. Up to now it is not known whether or not the conjecture holds true. If it is true however then our algorithm can test in polynomial time whether or not two D0L sequences are identical. Consider for example two D0L sequences $(w, f, \{a, b\})$, $(w, g, \{a, b\})$ where $w = aba$, $f(a) = ab$, $f(b) = ba$ and $h(a) = aab$, $h(b) = baa$. In the construction of a grammar to our algorithm we use the identities $f^{k+1}(a) = f^k(a)f^k(b)$, $f^{k+1}(b) = f^k(b)f^k(a)$, $h^{k+1}(a) = h^k(a)h^k(a)h^k(b)$, $h^{n+1} = h^k(b)h^k(a)h^k(a)$. The grammar looks as follows

$F_a^{k+1} \to F_a^k F_b^k$, $F_b^{k+1} \to F_b^k F_a^k$, $F_{aba}^k \to F_a^k F_b^k F_a^k$ for $0 \le k \le 2n - 1 = 3$,

$H_a^{k+1} \to H_a^k H_a^k H_b^k$, $H_b^{k+1} \to H_b^k H_a^k H_a^k$, $H_{aba}^k \to H_a^k H_b^k H_a^k$ for $0 \le k \le 3$,

$F_a^0 \to a$, $F_b^0 \to b$, $H_a^0 \to a$, $H_b^0 \to b$,

and a set of pairs to test is $\{(F_{aba}^k, H_{aba}^k) : 0 \le k \le 2n - 1 = 3\}$.

Acknowledgements

The author would like to thank Prof. Wojciech Rytter from Instytut Informatyki UW for useful comments to the paper.

References

1. J. Aho, J. Hopcroft, J. Ullman, "The design and analysis of computer algorithms", Addison-Wesley, 1974.
2. J. Albert, K. Culik II, J. Karhumäki, Test sets for context-free languages and algebraic systems of equations, *Inform. Control* **52**(1982), 172–186.
3. J. Jaffar, Minimal and complete word unification, *JACM* **37**(1), 47–85.
4. S. Jarominek, J. Karhumäki, W. Rytter, Efficient construction of test sets for regular and context-free languages, *Theoret. Comp. Science* (to appear).
5. J. Karhumäki, W. Plandowski, W. Rytter, Polynomial size test sets for context-free languages, *JCSS* (to appear).
6. J. Karhumäki, W. Plandowski, W. Rytter, Polynomial size test sets for context-free languages, *in* Proceedings of ICALP'92, Lect. Notes in Comp. Science 623 (1992), 53–64.
7. A. Koscielski, L. Pacholski, Complexity of unification in free groups and free semigroups, *in* Proc. 3st Annual IEEE Symposium on Foundations of Computer Science, Los Alamitos 1990, 824–829.
8. M. Lothaire, "Combinatorics on words", Addison-Wesley Publishing Company, Massachussets, 1983.
9. G.S. Makanin, The problem of solvability of equations in a free semigroup, *Math. USSR Sbornik* **32**, 2(1977), 129–198.
10. J.P. Pecuchet, Equations avec constantes et algorithme de Makanin, These de doctorat, Laboratoire d'informatique, Rouen, 1981.
11. K.U. Schultz, Makanin's algorithm for word equations - two improvements and a generalization, CS Report 91-39, Centrum für Informations und Sprachverarbeitung, University of Munique, 1991.

Work-Time Optimal Parallel Prefix Matching (Extended Abstract)

Leszek Gąsieniec[1]* Kunsoo Park[2]**

[1] Instytut Informatyki, Uniwersytet Warszawski
Banacha 2, 02-097 Warszawa, Poland
[2] Department of Computer Engineering, Seoul National University
Seoul 151-742, Korea

Abstract. Consider the prefix matching problem: Given a pattern P of length m and a text T of length n, find for all positions i in T the longest prefix of P starting at i. We present a parallel algorithm for the prefix matching problem over general alphabets whose text search takes optimal $O(\alpha(m))$ time and preprocessing takes optimal $O(\log \log m)$ time, where $\alpha(m)$ is the inverse Ackermann function. An $\Omega(\log \log m)$ lower bound for the prefix matching problem is implied by the same lower bound for string matching. However, the lower bound is applied only to preprocessing of the pattern and the searching phase can be faster. We prove an $\Omega(\alpha(m))$ lower bound for any linear-work searching phase. Therefore our algorithm is work-time optimal in both preprocessing and text search. The idea of *leftmost* witnesses is introduced to obtain the algorithm.

1 Introduction

Consider the prefix matching problem: Given a pattern P of length m and a text T of length n, find for all positions i in T the longest prefix of P starting at i. This problem is a natural generalization of the string matching problem where occurrences of the entire pattern are sought. Note that any algorithm for the prefix matching problem solves the string matching problem, but not vice versa. The prefix matching problem was first considered by Main and Lorentz [15] in order to find all repetitions of a string, and a linear-time algorithm in [15] is a variation of the Knuth-Morris-Pratt algorithm [14] for string matching. Breslauer, Colussi and Toniolo [5] showed that prefix matching is harder than string matching in the sense of exact complexity.

In parallel computation all algorithms below work in the Common CRCW PRAM. A parallel algorithm is *semi-optimal* if it uses a linear number of processors. A parallel algorithm is *optimal* if its work is asymptotically equal to that

---

* Supported by the EC Cooperative Action IC 1000 Algorithms for Future Technologies "ALTEC". Email: lechu@mimuw.edu.pl.
** Supported by S.N.U. Posco Research Fund 94-15-1112.
Email: kpark@theory.snu.ac.kr

of the fastest sequential algorithm for the same problem. An optimal algorithm is *work-time* optimal if its time is the best possible.

Galil's string matching algorithm of $O(\log m)$ time and $O(n \log m)$ work [11] in fact solves the prefix matching problem in the same complexity. Breslauer [4] improved the time to $O(\log \log m)$ while the work is still $O(n \log m)$. Recently, Hariharan and Muthukrishnan [13] proposed the first optimal algorithms: an optimal $O(\log^2 m(\log \log m)^2)$-time algorithm over general alphabets and an optimal $O(\log m)$ time algorithm using more than linear space over alphabets whose size is polynomial in $n + m$.

We present a parallel algorithm for the prefix matching problem over general alphabets whose text search takes optimal $O(\alpha(m))$ time and preprocessing takes optimal $O(\log \log m)$ time, where $\alpha(m)$ is the inverse Ackermann function. The overall $O(\log \log m)$ time of our algorithm is the best possible for the prefix matching problem as implied by the $\Omega(\log \log m)$ lower bound for string matching due to Breslauer and Galil [7]. However, the lower bound is applied only to preprocessing of the pattern and the searching phase can be faster than $O(\log \log m)$ time as in string matching. Unlike string matching where an optimal constant-time searching phase was obtained [8], we prove an $\Omega(\alpha(m))$ lower bound for any linear-work searching phase. Therefore our algorithm is work-time optimal in both preprocessing and text search. As in the exact complexity of sequential computation, the $\Omega(\alpha(m))$ lower bound shows that prefix matching is harder than string matching in parallel computation.

The notion of *leftmost* witnesses is introduced to obtain the algorithm. In the original idea of witness [17] there may be many witnesses against a certain periodicity of the pattern, and anyone of them can be a witness. The leftmost witness is the first one, from the left, among the witnesses. The duels based on leftmost witnesses provide a new structure called the covering. It allows us to compare a collection of pattern prefixes with the text simultaneously. Finally, the KMP failure function can be computed optimally in $O(\log \log m)$ time from our algorithm, as it is shown in [4].

2 Preliminaries

All strings in the paper are built over a general alphabet Σ (without any restriction). Consider a *word* $w \in \Sigma^*$. By w_i we mean the symbol at the ith position of the word w. Positions in a word are enumerated from 0. A subword of length s of the word w is called an *s-block* of w. We say that the word w has *a period* q if and only if $w_i = w_{i+q}$ for all positions $0 \le i < |w| - q$. The shortest period of w is called *the period* of w. If the period $q \le |w|/2$, then the word w is called *periodic*; otherwise, w is *nonperiodic*.

Assume that the word w has the period p. Consider two copies of the word w, one over another, upper shifted by i, $i < p$, positions to the right. There is a position k in the lower copy where the two copies of w differ, because otherwise

i must be a period of w shorter than p, which contradicts the fact that p is the period of w. If there are more than one such positions, the position k is chosen arbitrarily. The position k is called a *witness* against periodicity i. We define the *leftmost* witness to be the first (from the left) witness against periodicity i.

Given pattern $|P| = m$ and text $|T| = n$. A position i, $0 \leq i \leq n - m$, is an *occurrence* of the pattern P in T if and only if for all $0 \leq j \leq m - 1$, $P_j = T_{i+j}$. If a position i is not an occurrence of P, there is a position $0 \leq k \leq m - 1$ such that $P_k \neq T_{i+k}$. The position $i + k$ is called a *witness to non-occurrence i*. The first (from the left) witness to non-occurrence i is called the *leftmost witness to non-occurrence i*. The string matching problem is to find all occurrences of the pattern P in the text T (and witnesses to all non-occurrences). The prefix matching problem is to find all occurrences of the pattern P in the text T and leftmost witnesses to all non-occurrences.

3 Consistency and Covering

Every position i in the text T against which the leftmost witness has not yet determined is called a *candidate*. Every candidate i in the text T has a bound $\tau(i)$ of the leftmost witness to its non-occurrence. At the beginning of the algorithm every candidate i in T is *bounded* by the position $i + m$, i.e., $\tau(i) = i + m$. The bound will remain $i + m$ if position i is an occurrence of the pattern. Consider two candidates $i < j$ such that k is a witness against $j - i$ (i.e., $P_k \neq P_{k-j+i}$).

- if $T_{i+k} \neq P_k$, $i + k$ is a witness to non-occurrence i,
- if $T_{i+k} \neq P_{k-j+i}$, $i + k$ is a witness to non-occurrence j.

The two tests above are called a *duel* between candidates i and j. After a duel, the lost candidate will change its bound to the position $i + k$ if $i + k$ is smaller than its previous bound.

Assume that we have precomputed leftmost witnesses for all positions $\leq \frac{m}{2}$ in the pattern P. All duels in the paper are based on leftmost witnesses. Consider two candidates which overlap. The two candidates are *consistent* over an overlapping segment if the overlapping segments of them match completely. We will construct a *covering* of a text segment which consists of pattern prefixes. A covering is *consistent* with a candidate i if the segment of the covering from i to $\tau(i) - 1$ match completely with the pattern prefix of length $\tau(i) - i$.

Lemma 1. *Candidates i and j ($i < j$), which have a duel at the position $i + k$ in the text T, are consistent over the text segment $T_j..T_{i+k-1}$.*

Consider two candidates i and j, $i < j$, in the text T. Assume that the duel between i and j was done at $i + k$ and candidate i won. Candidate j is going to be bounded by the position $i + k$, i.e, $\tau(j) = i + k$, and candidate i holds previous bound $\tau(i)$. The covering for both candidates is represented by the prefix of P of length $\tau(i) - i$ *associated* with candidate i (covering $T_i..T_{\tau(i)-1}$). Assume next that candidate j won. Then candidate i is going to be bounded by the position $i + k$, i.e., $\tau(i) = i + k$ and candidate j holds previous bound $\tau(j)$. This time the covering for both candidates is represented by the prefix of P of length $j - i$ associated with candidate i (covering $T_i..T_{j-1}$) and the prefix of P of length $\tau(j) - j$ associated with candidate j (covering $T_j..T_{\tau(j)-1}$). In case when both candidates i, j lose the duel at $i + k$, $\tau(i) = \tau(j) = i + k$ and the covering is represented by the prefix of P of length $\tau(i) - i$ *associated* with candidate i (covering $T_i..T_{\tau(i)-1}$). In all cases the coverings are consistent with candidates i and j. To find leftmost witnesses for candidates i and j, first compare the covering with the text segment, and then find the first mismatch for each candidate between the covering and the text. The leftmost witness against candidate i (j) is the minimum of the bound $\tau(i)$ ($\tau(j)$) and the first mismatch.

The above scheme we extend to more than two candidates covering the text by disjoint associated prefixes. Consider a set C of l candidates $c_1 < c_2 < \cdots < c_l$ with corresponding bounds $\tau(c_1), \tau(c_2), \ldots, \tau(c_l)$ after total dueling (duels in every pair of candidates). The rules for the covering construction, which will be called the *covering rules*, is as follows. For every candidate c_i, $1 \leq i \leq l$, there are three cases:

(C1) Candidate c_i is covered by another candidate c_j, i.e., $c_j < c_i$ and $\tau(c_i) \leq \tau(c_j)$. Candidate c_i does not contribute to the covering. In the following two cases candidate c_i has associated prefix in the covering.

(C2) Candidate c_i is not covered by other candidates, but there is a candidate c_j which covers $\tau(i)$, i.e., $c_i < c_j$ and $\tau(c_i) < \tau(c_j)$. Let c_k be the leftmost candidate satisfying the conditions. The length of the associated prefix with candidate c_i is equal to $c_k - c_i$.

(C3) There is no candidate which covers $\tau(c_i)$, i.e., for all $c_i < c_j$ we have $\tau(c_i) \leq c_j$ or $\tau(c_j) \leq \tau(c_i)$. The length of the associated prefix with candidate c_i is equal to $\tau(c_i) - c_i$.

Lemma 2. *The covering constructed by the covering rules is consistent with all candidates.*

4 Prefix Matching Algorithms

The idea of the algorithm is first to find a bound for every candidate in the text, holding consistency between associated prefixes. Then we construct a covering by associated prefixes according to the covering rules. When every text symbol is associated with a symbol in the covering, we compare the covering with the text. Now the length of the matched prefix at position i of the text is the minimum of the bound $\tau(i)$ and the first mismatch between the covering and the text.

First we introduce procedure Square which allows us to achieve a simple semi-optimal $O(\log \log m)$-time prefix matching algorithm. Then we show how to achieve an optimal $O(\log \log m)$-time algorithm for the entire problem. Finally we present an optimal $O(\alpha(m))$-time algorithm for the searching phase.

4.1 Semi-Optimal $O(\log\log m)$-Time Algorithm

Recall that a segment of the text T of length s is called an s-block. We say that an s-block is *sparse* if it contains at most one candidate in each of its $\sqrt{s}$-subblock. By the s_l-block we mean the s-block which starts at position $(l-1) \cdot s$ in the text T. Assume that leftmost witnesses for the pattern P are already computed. To simplify technical details, we also assume that all prefixes of the pattern P are nonperiodic. Procedure $\mathsf{Square}(s_l)$ performs duels between candidates in a given sparse s_l-block and generates a covering for the s_l-block (extended up to s_{l+1}-block). The candidate which won all duels at a given level and s_l-block does not have a bound determined by the lost duel. To such a candidate we give an *artificial* bound – the right end of s_{l+1}-block.

procedure $\mathsf{Square}(s_l)$;
1. Perform leftmost duels in every pair of candidates. For every candidate c_i determine a new bound $\tau(c_i)$ from the leftmost position of the lost duels, if such exists. By the nonperiodicity assumption, all but possibly one candidates in the s_l-block get new bounds in the s_l-block or s_{l+1}-block. The candidate which won all duels gets an artificial bound, the right end of s_{l+1}-block.
2. Construct a covering of the s_l-block (which can extend over s_{l+1}-block) according to the covering rules.

Lemma 3. *Procedure $\mathsf{Square}(s_l)$ performs duels and find the covering in constant time on s processors CRCW PRAM.*

The semi-optimal algorithm is divided into $O(\log\log m)$ consecutive stages. The invariant of the algorithm is that after the ith stage all leftmost witnesses shorter than 2^{2^i} have been determined.

algorithm $\mathsf{Search}(T)$;
1. $s \leftarrow 2$;
Determine all leftmost witnesses of length ≤ 2 naively. The positions of T without determined leftmost witnesses become candidates.
do $O(\log\log m)$ times for all s_l-blocks **in parallel**
    2. $s \leftarrow s^2$;
    Use the constant-time procedure $\mathsf{Square}(s_l)$ for every s_l-block independently to construct a covering for the s_l-block which is over s_l-block and s_{l+1}-block.
    3. For all s_l-blocks compare the covering with the text.
    The result is saved in the binary vector b_l.
    4. The left witness for every candidate in s_l-block is determined from the new bound or the first mismatch to its right in b_l.
    Surviving candidates go to the next round.
od

During every stage of algorithm Search we use n processors. Since every s_l-block at any level is sparse, the constant-time procedure Square finds the local bounds and coverings. The comparison between coverings and the text is made naively and for every candidate in the s_l-block the first mismatch to its right is found in constant time since the s_l-block is sparse. Thus algorithm Search works in $O(\log \log m)$-time on n processors CRCW PRAM.

In the preprocessing we find the leftmost witnesses for all positions of the pattern. We use as a black box the algorithm due to Breslauer [4]. The algorithm in [4] preprocesses the pattern in $O(\log \log m)$ time and $O(m \log m)$ total work.

algorithm Preprocessing(P);
1. Use Breslauer's algorithm to preprocess the $m/\log m$-size prefix of P;
2. Use algorithm Search to find all prefixes shorter than $O(m/\log m)$ in P; By the nonperiodicity assumption there are $O(\log m)$ positions (candidates) in P without leftmost witnesses.
3. Perform $O(\log \log m)$ phases of doubling the size of computed prefixes by naive checking;

The first two steps are implemented in $O(\log \log m)$ time and $O(m \log \log m)$ work. Every phase of the last step is easily implemented in constant time and $O(m)$ work. Hence the total work of the preprocessing is $O(m \log \log m)$.

Theorem 4. *There exists a semi-optimal $O(\log \log m)$-time prefix matching algorithm.*

4.2 Optimal $O(\log\log m)$-Time Algorithm

A *short* prefix is one of length $\leq \log \log m$. The general scheme of the optimal $O(\log \log m)$-time algorithm is as follows. First find all short prefixes in all $\log \log m$-blocks sequentially. Then construct the $O(\log \log m)$ different level coverings. The covering on the ith level corresponds to occurrences of prefixes not longer than $(\log \log m)^{2^{i+1}}$ but possibly longer than $(\log \log m)^{2^i}$. Candidates which are used to construct the covering on the ith level are determined by leftmost duels on the $(i-1)$th level. Duels between different levels are performed to get a single covering called the *common* covering. Then the common covering and the text T are compared and the result is stored in a binary vector B. The leftmost witness for every candidate is determined by its bound or the first mismatch to its right in the vector B.

Nonperiodic Case: We first design an optimal algorithm which works under the assumption that all prefixes of the pattern P are nonperiodic. The optimal algorithm is divided into $O(\log \log m)$ stages (i.e., levels). Every level has two kinds of additional memory: $n/\log \log m$-size memory M_1 corresponding to disjoint consecutive $\log \log m$-blocks and $\frac{n \log \log m}{(\log \log m)^{2^i}}$-size memory M_2 at level i. At each level M_1 is used to remember the position of the candidate whose associated prefix covers the given $\log \log m$-block and M_2 is used to remember the result of duels between different levels.

algorithm Search(T);
1. Find all short prefixes in every $\log \log m$-block sequentially.
2. Run procedure Square for $O(\log \log m)$ levels and construct a covering for

each level. When the text is divided into s-blocks in a level, we construct a level covering as follows. For each of two consecutive blocks s_l and s_{l+1} (except the last two blocks) we construct a covering using $\mathsf{Square}(s_l s_{l+1})$ and use only its first half for block s_l of the level covering. For the last two blocks, all of their covering is used for the level covering.

3. Perform duels between $O(\log \log m)$ different levels (every lower level performs duels with all upper ones) to find closer bounds and construct a common covering. Note that every candidate has a sequence of artificial bounds on consecutive levels followed by a real bound. For every candidate the bound from upper level dominates ones in lower levels.

3.a Perform duels between different levels. Lemma 5 below shows that every candidate from a lower level performs at most one duel at any upper level. The position of the candidate (to a duel) is read from additional memory M_1 at a given upper level. The results of duels are saved in additional memory M_2.

3.b Find new bounds for all candidates based on different level dueling. Candidates search sequentially additional memory M_2 and find a new bounds.

3.c According to the covering rules, construct the local level coverings based on the new bounds by different level dueling. The information about the covering is saved in the additional memory M_1.

3.d At each level, for every candidate c_i check sequentially if it is covered at any upper level, using additional memory M_1 at any upper level. If it is covered, it dies and will not belong to the common covering. If it is not covered, it belongs at least partially to the common covering. We look for the leftmost candidate c_j which extends over c_i's bound $\tau(c_i)$ by searching additional memory M_2. The length of the associated prefix with candidate c_i is equal to $c_j - c_i$. The common covering has been constructed.

4. Compare the common covering with the text symbols and construct a binary vector B where mismatches are marked by 0s.

5. For every position i in the text T the length of the longest prefix of P starting at i is determined by its bound or the first 0 to its right in vector B.

Lemma 5. *Every candidate c_i from a given level k have a duel with at most one candidate at any upper level $l > k$.*

Lemma 6. *The common covering constructed in algorithm Search is consistent with all candidates.*

Lemma 7. *Algorithm Search finds all prefixes of P in T optimally in $O(\log \log m)$ time.*

The scheme of the optimal $O(\log \log m)$-time preprocessing is given below.

algorithm $\mathsf{Preprocessing}(P)$;

1. Compute leftmost witnesses for the $m/\log m$-size prefix of the pattern P by Breslauer's algorithm [4] and find prefixes shorter then $m/\log m$ in the whole pattern P by algorithm Search. After this step there are at most $O(\log m)$ positions in P without leftmost witnesses (candidates).

2. Perform $\log\log m$ phases of doubling the size of computed prefix as follows. Starting from $k = m/\log m$, use computed prefix $P' = P_0P_1..P_{k-1}$ as the pattern and prefix $P'' = P_0P_1..P_{2k-1}$ as the text. Since there are $O(\log m)$ candidates in the text P'', we can perform duels and find a covering of P'' by using procedure **Square** only once. Compare the covering with text P'' and find leftmost witnesses of P''. In the next phase prefix P'' becomes a pattern and as the text a prefix of size $2|P''|$ is taken.

During the preprocessing we use $m/\log\log m$ processors. The first step is performed optimally in $O(\log\log m)$ time. All phases of step 2, when the size of computed prefix is $\leq m/\log\log m$, can be performed in constant time using $m/\log\log m$ processors. But the last $O(\log\log\log m)$ doubling phases, when the size of computed prefix is greater than $m/\log\log m$, are slowed down. The ith phase of the last $O(\log\log\log m)$ phases, in which the size of computed prefix is $O(2^i m/\log\log m)$, is slowed down to time 2^i by Brent's theorem [3]. The total slowdown is bounded by $O(\Sigma_{i\leq\log\log\log m} 2^i) = O(\log\log m)$.

Periodic Case: We assumed before that all prefixes of the pattern are nonperiodic. In fact, the assumption was used for some $O(\log\log m)$ prefixes, which correspond to block sizes on different levels of the covering construction. Let $\Pi = \{\pi_1, \pi_2, .., \pi_l\}$ be an ordered set of considered prefixes. Recall that $|\pi_1| = O(\log\log m)$, $|\pi_l| = O(m)$ and $|\pi_i|^2 = O(|\pi_{i+1}|)$ for all $1 \leq i < l$.

When some prefixes from the set Π are periodic, we transform Π to get a new sequence of nonperiodic prefixes as follows. Let $per(\pi_i)$ denote the size of the period of the prefix π_i.

Fact: $per(\pi_1) \leq per(\pi_2) \leq ... \leq per(\pi_l)$

We introduce two new notions $ext(\pi_i)$ and $nonper(\pi_i)$. Let $ext(\pi_i)$ be the extension of the prefix π_i to the shortest prefix with broken period $per(\pi_i)$, in the case prefix π_i is periodic. Otherwise $ext(\pi_i) = \pi_i$. Denote by $nonper(\pi_i)$ the prefix of size $2 * |per(\pi_i)| - 1$ in the case prefix π_i is periodic. Otherwise $nonper(\pi_i) = \pi_i$. Note that prefixes $nonper(\pi_i)$ and $ext(\pi_i)$ are nonperiodic for all $1 \leq i < l$. We can divide all prefixes from the set Π into $O(\log\log m)$ groups, such that every group consists of consecutive prefixes with the same period. Every group has its representative which is the shortest period in the group. Let $\bar{\Pi} = \{\pi_{i_1}, \pi_{i_2}, .., \pi_{i_r}\}$ be an ordered set of representatives.

Consider strings $nonper(\pi_{i_k})$ and $ext(\pi_{i_k})$ for all $\pi_{i_k} \in \bar{\Pi}$. The idea used in the periodic case is as follows. First recognize all prefixes of size from $ext(\pi_{i_k})$ to $nonper(\pi_{i_{k+1}})$ for all $1 \leq k < r$. The rest of prefixes (some periodic ones) can be obtained from (full and partial) occurrences of prefixes $nonper(\pi_{i_k})$.

algorithm Periodic Search: Part 1

1. First we choose candidates to blocks of size $|ext(\pi_{i_k})|/2$, for all $1 \leq k < r$ (one candidate in one block). This can be easily done by duels, because the prefix of size $|ext(\pi_{i_k})|$ is nonperiodic.

2. In every block of size $|nonper(\pi_{i_{k+1}})|/2$ perform duels between candidates from $|ext(\pi_{i_k})|/2$-blocks. Note that extension is at most squaring.

3. Local coverings, corresponding to $|nonper(\pi_{i_{k+1}})|/2$-blocks, are constructed according to the covering rules.

4. The common covering is constructed and its consistency with the text is checked and saved in vector B.

5. The lengths of prefixes are determined from the bounds and mismatches saved in B .

Remember that for some positions corresponding to periodic prefixes of size from $nonper(\pi_{i_k})$ to $ext(\pi_{i_k})$ (for all $1 \le k < r$), we computed shorter prefixes (determined by artificial bounds) than they really are. But all those prefixes can be reconstructed from the occurrences (full and partial) of the prefixes $nonper(\pi_{i_k})$ in the following way.

There are at most $O(\log \log m)$ (in fact r) levels of reconstructing periodic prefixes. At all levels the reconstruction is done independently. Consider the kth level on which we reconstruct the prefixes of size from $nonper(\pi_{i_k})$ to $ext(\pi_{i_k})$.

algorithm Periodic Search: **Part 2**

1. The kth level is divided into consecutive $|nonper(\pi_{i_k})|/2$-blocks.

2. At every block, in some additional memory, the offset of the occurrence of prefix $nonper(\pi_{i_k})$ is saved (if such exists)

Note that if there is an occurrence of longer prefix in a given block, we use it as a simple occurrence of $nonper(\pi_{i_k})$.

3. Connect into chains neighbouring blocks with the same offset. Every border between two blocks with different offsets is signed by 1, and then for every block we look for the first border marked with 1 to its right. This is the chaining problem which can be solved optimally in $O(\log \log m)$ time (or even faster $\alpha(m)$ time).

4. When every block knows its border marked with 1 in the chain, it is easy to compute (from the distance to the border and partial occurrence of $nonper(\pi_{i_k})$ at the border) the size of the extended $nonper(\pi_{i_k})$ prefix in constant time.

The computation of all leftmost witnesses is as follows:

algorithm Periodic Preprocessing

1. Preprocess prefix P' of size $m/\log m$ using Breslauer's algorithm [4]. If computed prefix P' is periodic, find $ext(P')$ by naive extending and looking for the first mismatch of period $per(P')$. Leftmost witnesses for all multiples of prefix $ext(P')$ are computed immediately, the same way as in Vishkin's string matching algorithm [17]; $P' \leftarrow ext(P')$

2. Find all prefixes shorter than $|P'|/2$ in the whole pattern by algorithm Periodic Search. After this step there are at most $O(\log m)$ positions in P without leftmost witnesses (candidates).

3. Perform $\log \log m$ phases of doubling the size of computed prefix, similar to the nonperiodic case. We do not need another approach because the number of candidates is small.

To get complete preprocessing we have to check periodicity of prefixes from the set Π. If some of prefixes are periodic we compute the set $\bar{\Pi}$ and prefixes $nonper(\pi_{i_k})$ and $ext(\pi_{i_k})$. All computations can be done optimally, because the total size of considered prefixes is linear in the size of pattern P.

Theorem 8. *The prefix matching problem can be solved in $O(\log\log m)$ time optimally on the* CRCW PRAM.

4.3 Optimal $O(\alpha(m))$-Time Search

Define a series of functions $f_i(r)$ recursively: $f_1(r) = r/2$; $f_i(r) = \min\{j|f_{i-1}^{(j)}(r) \leq 1\}$. Note that $f_2(r) = \log r$ and $f_3(r) = \log^* r$. Let $\alpha(r) = \min\{j|f_j(r) \leq j\}$.

We will use recursive star-trees introduced by Berkman and Vishkin [2]. For $2 \leq d \leq \alpha(r)$, a balanced tree $BT_d(r)$ with r leaves is defined. For $d = 2$, $BT_2(r)$ is a complete binary tree with r leaves. For $3 \leq d \leq \alpha(r)$, $BT_d(r)$ has $f_d(r) + 1$ levels. The root is at level 0, and r leaves are at level $f_d(r)$. A node v at level $0 \leq i < f_d(r)$ has $y = f_{d-1}^{(i)}(r)/f_{d-1}^{(i+1)}(r)$ children. In addition, node v recursively contains $BT_{d-1}(y)$.

Since the text can be divided into n/m overlapping $2m$-blocks, assume that the text size is $2m$ and we compute leftmost witnesses only for the first half. Thus in the following "text" means the first half. When the text is divided into blocks in a level, we construct a level covering as follows. For each of two consecutive blocks (except the last two blocks) we construct a covering and use only its first half for the level covering. For the last two blocks, all of their covering is used for the level covering.

Theorem 9. *For every integer $2 \leq d \leq \alpha(r)$, a covering for r candidates can be constructed in cd time for some constant c with $rf_d(r)$ processors.*

Proof. By induction on d. The computation is guided by the recursive star-trees as in [2] and [16]. Each node with x leaves corresponds to a block containing x candidates.

Induction base: $d = 2$ and $f_2(r) = \log r$.

1. Divide the text into $r/\log r$ blocks of size $\log r$. With $\log^2 r$ processors for each block, construct a covering in constant time.

2. Build $BT_2(r/\log r)$, a complete binary tree with $r/\log r$ leaves. For every level $0 \leq i < \log r$, find survivors (at most one for each node of $BT_2(r/\log r)$) by DS's in constant time with r processors [18, 8]. In every level build a level covering in constant time with r processors.

3. Construct a common covering from $\log r$ coverings in step 2. With $\log^2 r$ processors for each candidate, we can construct the common covering in constant time.

4. Combine the covering from step 1 and the covering from step 2 in constant time.

Therefore the theorem is true for $d = 2$. Assume that the theorem is true for $d = k \geq 2$. Let $b_0 = r$ and $b_{i+1} = f_k(b_i)$ (i.e., $b_i = f_k^{(i)}(r)$). We now prove the theorem for $d = k + 1$.

1. Divide the text into $r/f_{k+1}(r)$ blocks of size $f_{k+1}(r)$. With $(f_{k+1}(r))^2$ processors for each block, construct a covering in constant time.

2. Build $BT_{k+1}(r/f_{k+1}(r))$. There are $f_{k+1}(r) + 1$ levels: a block of size b_i in level i, $0 \leq i < f_{k+1}(r)$, is divided into b_i/b_{i+1} blocks of size b_{i+1} in level $i+1$. For every level $0 \leq i < f_{k+1}(r)$, find survivors (at most one for each node of $BT_{k+1}(r/f_{k+1}(r))$) by DS's in constant time with r processors.

3. For each node v in every level i, $0 \leq i < f_{k+1}(r)$, recursively construct a covering consisting of b_i/b_{i+1} candidates corresponding to the children of v. By induction hypothesis, each recursive construction takes ck time. Since we use two blocks at level i to construct a covering for v, the number of processors for each node at level i is $(2b_i/b_{i+1})f_k(2b_i/b_{i+1}) \leq (2b_i/b_{i+1})2f_k(b_i) = 4b_i$. So $4r/f_{k+1}(r)$ processors for each level and total $4r$ processors are required. We have enough processors, i.e., $rf_{k+1}(r)$.

4. Construct a common covering from $f_{k+1}(r)$ coverings in step 3. With $(f_{k+1}(r))^2$ processors for each candidate, we can construct the common covering in constant time.

5. Combine the covering from step 1 and the covering from step 2 in constant time.

Therefore the total time is $c(k + 1)$ for some c and the number of processors is $rf_{k+1}(r)$.

Theorem 10. *The searching phase of prefix matching can be done optimally in* $O(\alpha(m))$ *time.*

5 Concluding Remarks

We have presented a parallel algorithm for the prefix matching problem over general alphabets whose text search takes optimal $O(\alpha(m))$ time and preprocessing takes optimal $O(\log\log m)$ time. There is an $\Omega(\log\log m)$-time lower bound [7] for the string matching problem, which also applies to the prefix matching problem. But the lower bound for the string matching problem is valid only for preprocessing of the pattern. The searching phase of string matching can be done even in constant time, following $O(\log\log m)$-time preprocessing [8]. The following lemma shows that the $O(\alpha(m))$ time we obtained for the searching phase is the best possible.

Lemma 11. *The searching phase of any optimal prefix matching algorithm cannot be done faster than* $\Omega(\alpha(m))$ *time.*

Proof. Take as a pattern a word of the form $P = 0^m$ and as a text any binary vector $T \in \{0,1\}^*$. In that case the prefix matching problem is to find for every position in the text T the first 1 to the right, which solves the *chaining* problem: for every 1 in the text T find the first 1 to the right. There is an $\alpha(m)$-time lower bound for any optimal algorithm solving the chaining problem [9].

Given a string S of length m, the KMP failure function computes the period of every prefix of S. Once the leftmost witnesses of S are computed, the failure function can be computed from the leftmost witnesses optimally in $O(\log\log m)$ time as shown in [4]. Therefore, by the preprocessing in Section 4.3 the failure function can be computed optimally $O(\log\log m)$ time.

References

1. O. Berkman, B. Schieber, and U. Vishkin, Optimal doubly logarithmic parallel algorithms based on finding all nearest smaller values, *J. Algorithms* 14 (1993).
2. O. Berkman and U. Vishkin, Recursive star-tree parallel data structure, *SIAM J. Comput.* 22 (1993), 221–242.
3. R.P. Brent, The parallel evaluation of general arithmetic expressions. *J. Assoc. Comput. Mach.* 21 (1974), 201–206.
4. D. Breslauer, Fast parallel string prefix-matching, *Technical Reports, CWI*, 1991.
5. D. Breslauer, L. Colussi and L. Toniolo, Tight comparison bounds for the string prefix-matching problem, *Proc. 4th Symp. Combinatorial Pattern Matching* (1993), 11–19.
6. D. Breslauer and Z. Galil, An optimal $O(\log\log n)$ time parallel string matching algorithm, *SIAM J. Comput.* 19 (1990), 1051–1058.
7. D. Breslauer and Z. Galil, A lower bound for parallel string matching, *SIAM J. Comput.* 21 (1992), 856–862.
8. R. Cole, M. Crochemore, Z. Galil, L. Gąsieniec, R. Hariharan, S. Muthukrishnan, K. Park and W. Rytter, Optimally fast parallel algorithms for preprocessing and pattern matching in one and two dimensions, *Proc. 34th IEEE Symp. Found. Computer Science*, 1993.
9. S. Chauduri and J. Radhakrishnan, The complexity of parallel prefix problems on small domains, *Proc. 33rd IEEE Symp. Found. Computer Science* 1992, 638–647.
10. F.E. Fich, P. Ragde and A. Wigderson, Relations between concurrent-write models of parallel computation, *SIAM J. Comput.* 17 (1988), 606–627.
11. Z. Galil, Optimal parallel algorithms for string matching, *Inform. and Control* 67 (1985), 144–157.
12. Z. Galil, A constant-time optimal parallel string-matching algorithm, *Proc. 24th ACM Symp. Theory of Computing*, 1992.
13. R. Hariharan and S. Muthukrishnan, Optimal parallel algorithms for prefix matching, *Proc. 21st Int. Colloq. Automata Languages and Programming*, 1994.
14. D.E. Knuth, J.H. Morris, and V.B. Pratt, Fast pattern matching in strings, *SIAM J. Comput.* 6 (1977), 323–350.
15. M.G. Main and R.J. Lorentz, An $O(n\log n)$ algorithm for finding all repetitions in a string, *J. Algorithms* 5 (1984), 422–432.
16. P. Ragde, The parallel simplicity of compaction and chaining, *J. Algorithms* 14 (1993), 371–380.
17. U. Vishkin, Optimal parallel pattern matching in strings, *Inform. and Control* 67 (1985), 91–113.
18. U. Vishkin, Deterministic sampling–a new technique for fast pattern matching, *SIAM J. Comput.* 20 (1991), 22–40.

On the Exact Complexity of the String Prefix-Matching Problem (extended abstract)

Dany Breslauer[1]*, Livio Colussi[2]** and Laura Toniolo[3]***

[1] BRICS – Basic Research in Computer Science – a centre of the Danish National Research Foundation, Department of Computer Science, University of Aarhus, DK-8000 Aarhus C, Denmark.

[2] Dipartimento di Matematica Pura ed Applicata, Università di Padova, Via Belzoni 7, I-35131 Padova, Italy.

[3] Université de Marne-la-Vallée, 2 rue de la Butte Verte, F-93166 Noisy-le-Grand CEDEX, France.

Abstract. In this paper we study the exact comparison complexity of the string prefix-matching problem in the deterministic sequential comparison model with equality tests. We derive almost tight lower and upper bounds on the number of comparisons required in the worst case by on-line prefix-matching algorithms for any fixed pattern and variable text. Unlike previous results on the comparison complexity of string-matching and prefix-matching algorithms, our bounds are almost tight for any particular pattern.

We also consider the special case where the pattern and the text are the same string. This problem, which we call the *string self-prefix* problem, is similar to the pattern preprocessing step of the Knuth-Morris-Pratt string-matching algorithm that is used in several comparison efficient string-matching and prefix-matching algorithms, including in our new algorithm. We develop an algorithm for the self-prefix problem that requires at most $2m - \lceil 2\sqrt{m} \rceil$ comparisons, matching the lower bound that we gave in a previous work.

Our algorithms can be implemented in linear-time and space in the standard random-access-machine model.

* Partially supported by ESPRIT Basic Research Action Program of the EC under contract #7141 (ALCOM II). Part of the research reported in the paper was carried out while this author was visiting at the Istituto di Elaborazione dell'Informazione, Consiglio Nazionale delle Ricerche, Pisa, Italy, with the support of the European Research Consortium for Informatics and Mathematics postdoctoral fellowship.

** Partially supported by "Progetto Finalizzato Sistemi Informatici e Calcolo Parallelo" of the Italian National Research Councile under grant number 89.00026.69.

*** Partially supported by "Borsa di studi per attività di perfeziomento all'estero" from the University of Padua.

1 Introduction

In the *string prefix-matching* (SPM) problem one is interested in finding the longest prefix of a pattern string $\mathcal{P}[1..m]$ that starts at each position of a text string $\mathcal{T}[1..n]$. The output of the problem is an integer array $\Phi[1..n]$, $0 \leq \Phi[t] \leq \min(m, n-t+1)$, such that for each text position t, $\mathcal{T}[t..t+\Phi[t]-1] = \mathcal{P}[1..\Phi[t]]$ and if $\Phi[t] < m$ and $t + \Phi[t] \leq n$, then $\mathcal{T}[t + \Phi[t]] \neq \mathcal{P}[\Phi[t]+1]$.

The SPM problem is a natural generalization of the standard string matching (SM) problem where only complete occurrences of the pattern are sought. The classical linear-time SM algorithm of Knuth, Morris and Pratt [16] can be easily adapted to solve the SPM problem in the same time bounds without making additional symbol comparisons. This observation was first made by Main and Lorentz [18] who used a SPM algorithm to detect repetitions in strings. In the parallel setting, Galil's [9] SM algorithm also solves the SPM problem. Breslauer [2] and Hariharan and Muthukrishnan [15] gave more efficient parallel algorithms and recently Gąsieniec and Park [13] obtained an optimal parallel SPM algorithm. SPM algorithms have also been used in the sequential two-dimensional matching algorithm of Galil and Park [12] and in an early version of the parallel two-dimensional matching algorithm of Cole et al. [5].

In this paper we study the exact number of comparisons performed by deterministic sequential SPM algorithms that have access to the input strings by pairwise symbol comparisons that test for equality. This work was motivated by recent interest in the exact comparison complexity of the SM problem and is continuation of an earlier work by the same authors [3].

In a sequence of papers on the number of symbol comparisons required in SM algorithms Colussi [8], Galil and Giancarlo [11], Breslauer and Galil [4] and Cole and Hariharan [6] improved the upper bounds, while Galil and Giancarlo [10], Cole and Hariharan [6] and Cole et al. [7] tightened the lower bounds. Currently, the complexity of the SM problem is determined almost exactly with an upper bound of $n + (8/3m)(n - m)$ comparisons and a lower bound of $n + (9/4m)(n - m)$ comparisons. These bounds are on the number of comparisons made by on-line algorithms in the text processing step, after an unaccounted pattern preprocessing. There is a larger gap between the lower and upper bounds for off-line algorithms.

Boyer and Moore [1] showed that in the SM problem it is not always necessary to examine all n text symbols. On the other hand, Rivest [19] has proved that in the worst case any SM algorithm must examine at least $n - m + 1$ text symbols. Clearly, any algorithm that solves the SPM problem must examine all n text symbols since it must determine if each text symbol is equal to the first pattern symbol. Note that if the input alphabet is known to contain only two symbols, then the SPM problem requires at most n comparisons since inequality to one alphabet symbol implies equality to the other.

Breslauer, Colussi and Toniolo [3] defined on-line SPM algorithms and presented a family of algorithms that make at most $\lfloor (2 - 1/m)n \rfloor$ symbol comparisons. They also gave a tight lower bound for any SPM algorithm that has to match the pattern 'ab^{m-1}'. These results imply that the two similar SM and

SPM problems have intrinsically different asymptotic comparison complexities, approaching n and $2n$, respectively, as m grows. However, on-line SPM algorithms have many similarities to the on-line SM algorithms given by Colussi [8] and Breslauer and Galil [4]. The definition of on-line SPM algorithms also coincides with the finite automata approach to SM. In an analysis of Simon's automata based SM algorithm Hancart [14] independently obtained the same bounds that were given in [3] for on-line SPM algorithms. The new results about on-line SPM algorithms presented in this paper apply as well to automata based SM.

Our main effort in this paper is to determine $c_{on-line}^{\mathcal{P}}(n)$, the number of comparisons required in the worst case by on-line SPM algorithms to match a fixed pattern $\mathcal{P}[1..m]$ in a variable text of length n. We determine $c_{on-line}^{\mathcal{P}}(n)$ almost exactly by showing that for any pattern $\mathcal{P}[1..m]$, there exists a constant $C_{on-line}^{\mathcal{P}}$, $1 \leq C_{on-line}^{\mathcal{P}} \leq 2 - 1/m$, such that,

$$C_{on-line}^{\mathcal{P}} \times (n - m) + m \leq c_{on-line}^{\mathcal{P}}(n) \leq C_{on-line}^{\mathcal{P}} \times n.$$

The upper bound of at most $C_{on-line}^{\mathcal{P}} \times n$ comparisons is achieved by an algorithm that takes linear time and uses linear space.

One novel aspect of this work is that, unlike previous results on the exact comparison complexity of the SM and SPM problems, our bounds are almost tight for any particular pattern.

We then consider the special case where the pattern and the text are the same string. This problem, which we call the *string self-prefix* (SSPM) problem, is similar to the *failure function* that is computed in the preprocessing step of the Knuth-Morris-Pratt [16] SM algorithm. The Knuth-Morris-Pratt failure function is used in various SM algorithms and also in the pattern preprocessing step of our SPM algorithm. Using the techniques we develop for the on-line SPM algorithm, we derive an algorithm for the SSPM problem that makes at most $2m - \lceil 2\sqrt{m} \rceil$ comparisons. This upper bound matches a lower bound that was given by Breslauer, Colussi and Toniolo [3], and thus, it determines *exactly* the worst case comparison complexity of the SSPM problem and of computing the Knuth-Morris-Pratt failure function. The SSPM algorithm and the whole pattern preprocessing step of the SPM algorithm take linear time and use linear space. The details of the SSPM algorithm will be given in the full paper.

Finally, we consider general off-line SPM algorithms. Such algorithms are more difficult to analyze since they have more liberties about the way they might proceed comparing the input symbols. We were unable to obtain tight bounds for off-line algorithms. However, we show that there exist pattern strings for which off-line algorithms require significantly fewer comparisons than on-line algorithms.

2 On-line prefix-matching

The discussion below proceeds in the comparison model where only comparisons of input symbols are counted and all other computation is free. We assume that our algorithms can obtain complete information about the pattern $\mathcal{P}[1..m]$ in an unaccounted pattern preprocessing step that might compare even all $\binom{m}{2}$ pairs of pattern symbols. However, the algorithms, including the pattern preprocessing, can be implemented in the standard random-access-machine computational model in linear time and space. The implementation is similar to that of the SM algorithm given by Breslauer and Galil [4] and will be given in the full paper.

Recall the definition of on-line SPM algorithms as given by Breslauer, Colussi and Toniolo [3]:

Definition 1. A SPM algorithm is *on-line* if before comparing the text symbol $T[t]$ it has determined if the pattern prefixes that start at text positions l, for $1 \le l < t$, terminate before text position t.

Comparison efficient on-line SPM algorithms are restricted about the choice of comparisons they can make. It is not difficult to verify that on-line algorithms that compare pairs of text symbols are not more efficient than those that compare only pattern symbols to text symbols.

Let $\mathcal{K}^t = \{k_i^t \| t - m < k_0^t < \cdots < k_{q_t}^t = t\}$ be the set of all text positions for which $\Phi[k_i^t]$ can not be determined without examining $T[t]$. Namely, $T[k_i^t..t-1] = \mathcal{P}[1..t - k_i^t]$ and $T[t]$ must be compared to check whether $\Phi[k_i^t] = t - k_i^t$ or $\Phi[k_i^t] > t - k_i^t$. Then, all comparisons at text position t must be between $T[t]$ and the pattern symbols $\mathcal{P}[t - k_i^t + 1]$ or otherwise can be answered by an adversary as unequal without giving an algorithm any useful information, provided that the text alphabet is large enough.

Clearly, $T[t]$ has to be compared either until it is found to be equal to some symbol $\mathcal{P}[t - k_i^t + 1]$ or until it is known to be different from all these symbols. Thus, the only difference between the on-line comparison efficient SPM algorithms we consider next is the *order* according to which $T[t]$ is compared to the pattern symbols $\mathcal{P}[t - k_i^t + 1]$.

2.1 Periods in strings

Periods are regularities of strings that are exploited virtually in all efficient SM algorithms. In this section we give some basic properties of periods and define the notation that we use throughout the paper. For an extensive treatment of periods and their properties see Lothaire's book [17].

Definition 2. A string $\mathcal{S}[1..h]$ has a period of length π if $\mathcal{S}[g] = \mathcal{S}[g + \pi]$, for $g = 1, \ldots, h - \pi$. We define the set $\Pi^{\mathcal{S}[1..h]} = \{\pi_i^{\mathcal{S}} \| 0 = \pi_0^{\mathcal{S}} < \cdots < \pi_{p_s}^{\mathcal{S}} = h\}$ to be the set of all period lengths of $\mathcal{S}[1..h]$. $\pi_1^{\mathcal{S}}$, the smallest non-trivial period length of $\mathcal{S}[1..h]$ is called *the period* of $\mathcal{S}[1..h]$.

Periods are connected to on-line SPM algorithms by the following observation.

Lemma 3. *The members of the set $\mathcal{K}^t$ correspond to the periods of the pattern prefix $\mathcal{P}[1..t - k_0^t]$ by the relation $\mathcal{K}^t = \{k_0^t + \pi \| \pi \in \Pi^{\mathcal{P}[1..t-k_0^t]}\}$.*

We define $\Sigma^P_{t-k^t_0+1} = \{P[t - k^t_i + 1] \parallel k^t_i \in \mathcal{K}^t\}$ to be the *set* of all the symbols that $T[t]$ might be compared to. We can imagine to align the prefixes of P one under the other, starting at positions $k^t_i \in \mathcal{K}^t$; then the elements of $\Sigma^P_{t-k^t_0+1}$ are the pattern symbols appearing under $T[t]$. By Lemma 3, $\Sigma^P_l = \{P[l - \pi] \parallel \pi \in \Pi^{P[1..l-1]}\}$, for $l = t - k^t_0 + 1$. Note that Σ^P_l is a set, thus containing distinct symbols and identifying equal symbols. Given some symbol $\sigma \in \Sigma^P_l$, we define two functions that give the smallest and the largest period lengths of $P[1..l-1]$ that introduce σ into Σ^P_l:

$$\pi^l_{first}(\sigma) = \min\{\pi \parallel \pi \in \Pi^{P[1..l-1]} \text{ and } P[l - \pi] = \sigma\}, \text{ and}$$

$$\pi^l_{last}(\sigma) = \max\{\pi \parallel \pi \in \Pi^{P[1..l-1]} \text{ and } P[l - \pi] = \sigma\}.$$

The function $\pi^l_{first}(\sigma)$ plays an important role in on-line algorithms, since for $l = t - k^t_0 + 1$ it determines the position of k^{t+1}_0 as a function of k^t_0 and $T[t]$ as the following lemma shows.

Lemma 4. *If $T[t] \in \Sigma^P_{t-k^t_0+1}$, then $k^{t+1}_0 = k^t_0 + \pi^{t-k^t_0+1}_{first}(T[t])$. (Except if $t - k^t_0 + 1 = m$ and $T[t] = P[m]$, where $k^{t+1}_0 = k^t_0 + \pi^{P[1..m]}_1$.)*

The function $\pi^l_{last}(\sigma)$ will have important uses in the development of comparison efficient algorithms. Its main properties are summarized in the following lemma.

Lemma 5. *For any $l = 1, \ldots, m$ and $\pi \in \Pi^{P[1..l-1]}$,*

$$\begin{aligned}
\pi^l_{last}(\sigma) &= \pi + \pi^{l-\pi}_{last}(\sigma) && \text{for } \sigma \in \Sigma^P_{l-\pi} \\
\pi^l_{last}(\sigma) &< \pi^l_{last}(\tau); && \text{for } \sigma \in \Sigma^P_l \setminus \Sigma^P_{l-\pi} \text{ and } \tau \in \Sigma^P_{l-\pi}.
\end{aligned}$$

2.2 Static algorithms

We define a subclass of on-line algorithms that we call *static* algorithms. For each occurrence of a pattern prefix of length $l-1$, these algorithms do not change the order in which the symbols in Σ^P_l are compared to the text. Static algorithms are easier to analyze, but still general enough to draw conclusions about on-line algorithms from their performance.

Definition 6. An on-line SPM algorithm is said to be *static* if the order according to which the symbols in $\Sigma^P_{t-k^t_0+1}$ are compared to some text symbol $T[t]$ depends only on $t - k^t_0 + 1$.

Since in a static algorithm $\mathcal{A}$ the order of comparisons depends only on $l = t - k^t_0 + 1$, it will be defined by the functions

$$\Lambda_{\mathcal{A},l}(h) : \{1, \ldots, |\Sigma^P_l|\} \longmapsto \Sigma^P_l \quad \text{for } l = 1, \ldots, m,$$

where the algorithm $\mathcal{A}$ compares $T[t]$ first to $\Lambda_{\mathcal{A},l}(1)$, then to $\Lambda_{\mathcal{A},l}(2)$ and so on. The number of comparisons that $\mathcal{A}$ makes to discover that $T[t] = \sigma$, for $\sigma \in \Sigma^P_l$, is $\Lambda^{-1}_{\mathcal{A},l}(\sigma)$.

Example. The SM algorithm of Knuth-Morris-Pratt compares the symbols $\mathcal{P}[l - \pi] \in \Sigma_l^\mathcal{P}$, $\pi \in \Pi^{\mathcal{P}[1..l-1]}$, in *increasing* order of the periods π, sometimes repeating unnecessary comparisons. We define the static SPM algorithm KMP to proceed in the spirit of the Knuth-Morris-Pratt algorithm, comparing the symbols $\mathcal{P}[l - \pi]$ in *increasing* order of the periods π, skipping symbols that were already compared. The order of comparisons $\Lambda_{KMP,l}(h)$ is defined such that $\pi_{first}^l(\Lambda_{KMP,l}(h)) < \pi_{first}^l(\Lambda_{KMP,l}(g))$ for $l = 1, \ldots, m$, and $1 \leq h < g \leq |\Sigma_l^\mathcal{P}|$.

2.3 The optimization problem

Given a static SPM algorithm $\mathcal{A}$, we define the *cost* function $\Omega_\mathcal{A}(l)$ for matching the pattern prefix $\mathcal{P}[1..l]$ to reflect the number of comparisons the algorithm makes to match this prefix.

$$\Omega_\mathcal{A}(l) = \begin{cases} 0 & l = 0 \\ \Omega_\mathcal{A}(l-1) + \Lambda_{\mathcal{A},l}^{-1}(\mathcal{P}[l]) & l = 1, \ldots, m. \end{cases}$$

The goal is to bound the number of comparisons made by $\mathcal{A}$ by an expression of the form $C_\mathcal{A} \times n$, for some constant $C_\mathcal{A}$ that will be determined later. When the algorithm reaches text position t, just before comparing the text symbol $T[t]$, we maintain inductively that the number of comparisons made so far is at most $C_\mathcal{A} \times (k_0^t - 1) + \Omega_\mathcal{A}(t - k_0^t)$. This bound obviously holds initially at text position $t = 1$. However, when the algorithm advances to the next text position the term $\Omega(t + 1 - k_0^{t+1})$ might not account for all the comparisons. We shall bound the excess number of comparisons by $C_\mathcal{A} \times (k_0^{t+1} - k_0^t)$, therefore maintaining the inductive claim.

Let $l = t - k_0^t + 1$. If the algorithm discovers that $T[t] = \sigma$, for $\sigma \in \Sigma_l^\mathcal{P}$, then it has made $\Lambda_{\mathcal{A},l}^{-1}(\sigma)$ comparisons at this text position. If $\sigma = \mathcal{P}[l]$, then $k_0^{t+1} = k_0^t$, and the inductive hypothesis still holds as the cost function accounts for these comparisons. However, if $\sigma \neq \mathcal{P}[l]$, then by Lemma 4, $k_0^{t+1} = k_0^t + \pi_{first}^l(\sigma)$ and only $\Omega_\mathcal{A}(l - \pi_{first}^l(\sigma))$ comparisons will be accounted by the cost function. To maintain the inductive hypothesis, we require that the remaining $\Omega_\mathcal{A}(l-1) + \Lambda_{\mathcal{A},l}^{-1}(\sigma) - \Omega_\mathcal{A}(l - \pi_{first}^l(\sigma))$ comparisons are also accounted by imposing the following constraint on $C_\mathcal{A}$:

$$\frac{\Omega_\mathcal{A}(l-1) + \Lambda_{\mathcal{A},l}^{-1}(\sigma) - \Omega_\mathcal{A}(l - \pi_{first}^l(\sigma))}{\pi_{first}^l(\sigma)} \leq C_\mathcal{A}. \tag{1}$$

If the algorithm concludes that $T[t] \neq \sigma$, for all $\sigma \in \Sigma_l^\mathcal{P}$, then $k_0^{t+1} = k_0^t + l = t+1$ and there are $\Omega_\mathcal{A}(l-1) + |\Sigma_l^\mathcal{P}|$ comparisons that will not be accounted by the cost function. To maintain the inductive hypothesis in this case, we make certain that these comparisons are also accounted by requiring that:

$$\frac{\Omega_\mathcal{A}(l-1) + |\Sigma_l^\mathcal{P}|}{l} \leq C_\mathcal{A}. \tag{2}$$

In addition, if the algorithm discovers that $T[t] = \mathcal{P}[l]$ and the end of the pattern is reached, that is $l = m$, then $k_0^{t+1} = k_0^t + \pi_1^\mathcal{P}$ and only $\Omega_\mathcal{A}(m - \pi_1^\mathcal{P})$ comparisons will be accounted by the cost function. To maintain the inductive hypothesis we require also that:

$$\frac{\Omega_A(m) - \Omega_A(m - \pi_1^P)}{\pi_1^P} \leq C_A. \tag{3}$$

Finally, when the end of the text is reached, that is when $t = n+1$, the number of comparisons made is bounded by the inductive hypothesis by $C_A \times (k_0^t - 1) + \Omega_A(t - k_0^t)$. Since $\Omega_A(h) \leq \Omega_A(h-1) + |\Sigma_h^P| \leq C_A \times h$, for $h = n - k_0^{n+1} + 1$, by Inequality 2, we get that the number of comparisons made is at most $C_A \times n$, establishing the inductive claim.

For any static algorithm A let the characteristic constant C_A be the smallest constant satisfying the three inequalities above. Note that Inequality 1 has to be satisfied at pattern position $l = 1, \cdots, m$, and for all $\sigma \in \Sigma_l^P \setminus \{P[l]\}$; Inequality 2 at pattern positions $l = 1, \cdots, m$; Inequality 3 does not depend on the pattern position. Note that by the same line of reasoning, an adversary can force a static algorithm A to make at least $C_A \times (n-m) + m$ comparisons. Thus, since algorithms with smaller characteristic constants make fewer comparisons in the worst case as n grows, the characteristic constant can be used to compare the relative efficiency of static SPM algorithms. The constraints on C_A are summarized in the following lemma.

Lemma 7. *The number of comparisons an algorithm A makes while scanning the text $T[1..n]$ is at most $C_A \times n$, where the characteristic constant*

$$C_A = \max\{$$
$$\frac{\Omega_A(l-1) + \Lambda_{A,l}^{-1}(\sigma) - \Omega_A(l - \pi_{first}^l(\sigma))}{\pi_{first}^l(\sigma)} \,\|\, l = 1, \ldots, m, \text{ and } \sigma \in \Sigma_l^P \setminus \{P[l]\};$$
$$\frac{\Omega_A(l-1) + |\Sigma_l^P|}{l} \,\|\, l = 1, \ldots, m; \qquad \frac{\Omega_A(m) - \Omega_A(m - \pi_1^P)}{\pi_1^P}\}.$$

In the next sections we face the problem of efficiently finding some static SPM algorithm OPT that minimizes the characteristic constant C_{OPT}. Such an algorithm clearly exists as there is only a finite number of static SPM algorithms.

Example. Consider an instance of the SPM problem with the pattern string $P[1..m] = \text{`}a^\alpha b^\beta\text{'}$, for $\alpha \geq 2$, $\beta \geq 1$ and $m = \alpha + \beta$. If $\alpha = 0$ or $\beta = 0$, then the number of comparisons required is clearly n and if $\alpha = 1$ and $\beta \geq 1$, then the number of comparisons required is $\lfloor (2 - 1/m)n \rfloor$ as shown by Breslauer, Colussi and Toniolo [3] and Hancart [14].

Let us try to find an algorithm OPT that has the smallest possible constant C_{OPT}. We define two algorithms. The first, which we call algorithm AB (it compares first 'a' and then if necessary 'b'), is defined as:

$$\Lambda_{AB,l}(h) = \begin{cases} \text{`}a\text{'} & h = 1 \text{ and } 1 \leq l \leq m \\ \text{`}b\text{'} & h = 2 \text{ and } \alpha < l \leq m, \end{cases}$$

and the second algorithm, which we call algorithm X (identical to algorithm AB up to position $\alpha + 1$ and from there on it compares first 'b' and then 'a'), is defined as:

$$\Lambda_{\mathcal{X},l}(h) = \begin{cases} `a` & h = 1 \text{ and } 1 \leq l \leq \alpha + 1 \\ `b` & h = 1 \text{ and } \alpha + 1 < l \leq m \\ `a` & h = 2 \text{ and } \alpha + 1 < l \leq m \\ `b` & h = 2 \text{ and } l = \alpha + 1. \end{cases}$$

It is easy to verify that $C_{\mathcal{AB}} = 1 + (m - \alpha)/m$ and that for any algorithm $\mathcal{A}$ other than $\mathcal{AB}$ and $\mathcal{X}$, $C_{\mathcal{A}} > C_{\mathcal{X}}$. If $\beta = 1$, then the two algorithms are identical and if $\beta \geq 2$, then $C_{\mathcal{X}} = (\alpha + 3)/(\alpha + 1)$. Thus, the optimal static algorithm OPT can be chosen as the algorithm that has the smaller characteristic constant $C_{\mathcal{AB}}$ or $C_{\mathcal{X}}$, and $C_{OPT} = \min\{C_{\mathcal{AB}}, C_{\mathcal{X}}\}$. Note that there is a tie $C_{\mathcal{AB}} = C_{\mathcal{X}}$ only for the patterns '$a^{\alpha}b$', '$aabbbb$' and '$aaabbb$'.

In this example we have been able to reduce the number of candidates for an optimal static algorithm from the 2^{β} different algorithms to the two algorithms $\mathcal{AB}$ and $\mathcal{X}$. We show that in the general case it suffices to consider only few algorithms as candidates for an optimal algorithm. These algorithms are closely related to the generalized form of algorithm $\mathcal{AB}$ that we call algorithm $REVERSE$.

2.4 The algorithm $REVERSE$

In this section we define a static SPM algorithm that has some special properties. This algorithm, which we call $REVERSE$, or REV for short, will be the basis for the optimal static algorithm that is developed in Section 2.5.

The order of comparisons $\Lambda_{REV,l}(h)$ in algorithm REV is defined such that $\pi_{last}^{l}(\Lambda_{REV,l}(h)) > \pi_{last}^{l}(\Lambda_{REV,l}(g))$ for $l = 1, \ldots, m$ and $1 \leq h < g \leq |\Sigma_{l}^{\mathcal{P}}|$.

```
1 2 3 4 5 6 7 8 9 0 1 2 3 4 5 6 7 8 9 0 1 2 3 4 5 6 7 8 9 0 1 2
a c a b a c a a a c a b a c a c a c a b a c a a a c a b a c a D
                  a c a b a c a a a c a b a c a C
                                    a c a b a c a A
                                          a c a B
                                              a C
                                              A
```
$$\Sigma_{32}^{\mathcal{P}} = \{`a`, `b`, `c`, `d`\}$$

Fig. 1. Algorithm REV compares first 'a', then 'c', then 'b' and last 'd'. Algorithm KMP compares 'd', 'c', 'a' and 'b'. Note that these comparison orders are not opposite of each other.

More intuitively, if we recall that $\Sigma_{l}^{\mathcal{P}} = \{\mathcal{P}[l - \pi] \| \pi \in \Pi^{\mathcal{P}[1..l-1]}\}$, then algorithm REV compares the symbols $\mathcal{P}[l - \pi]$ in *decreasing* order of the periods π, skipping the symbols that were already compared. Notice that algorithm KMP compares the symbols $\mathcal{P}[l - \pi]$ in *increasing* order of the periods π, exactly the opposite of algorithm REV. See Figure 1.

The main property of algorithm REV that is used later in developing the optimal static algorithm is given in the following lemma.

Lemma 8. *The cost function of algorithm REV is additive. That is*

$$\Omega_{REV}(l-1) + \Lambda_{REV,l}^{-1}(\mathcal{P}[l-\pi]) = \Omega_{REV}(\pi) + \Omega_{REV}(l-\pi)$$

for $l = 1,\ldots,m$ and $\pi \in \Pi^{\mathcal{P}[1..l-1]}$.

Proof. By Lemma 5 and the definition of algorithm REV,

$$\Lambda_{REV,h}(g) = \Lambda_{REV,h-\pi}(g) \quad \text{for } h = \pi+1,\ldots,l, \text{ and } g = 1,\ldots,|\Sigma_{h-\pi}^{\mathcal{P}}|.$$

If $\pi = 0$, then by definition $\Omega_{REV}(l) = \Omega_{REV}(l-1) + \Lambda_{REV,l}^{-1}(\mathcal{P}[l])$. If $\pi > 0$, then we prove by induction on h that $\Omega_{REV}(h) = \Omega_{REV}(\pi) + \Omega_{REV}(h-\pi)$, for $h = \pi,\ldots,l-1$. The basis of the induction for $h = \pi$ clearly holds. Observing that $\mathcal{P}[h] = \mathcal{P}[h-\pi]$, for $h = \pi+1,\ldots,l-1$,

$$\Omega_{REV}(h) = \Omega_{REV}(h-1) + \Lambda_{REV,h}^{-1}(\mathcal{P}[h]) =$$
$$\Omega_{REV}(\pi) + \Omega_{REV}(h-\pi-1) + \Lambda_{REV,h-\pi}^{-1}(\mathcal{P}[h-\pi]) =$$
$$\Omega_{REV}(\pi) + \Omega_{REV}(h-\pi).$$

Finally, $\Lambda_{REV,l}^{-1}(\mathcal{P}[l-\pi])$ is defined since $\mathcal{P}[l-\pi] \in \Sigma_l^{\mathcal{P}}$, and thus,

$$\Omega_{REV}(l-1) + \Lambda_{REV,l}^{-1}(\mathcal{P}[l-\pi]) =$$
$$\Omega_{REV}(\pi) + \Omega_{REV}(l-\pi-1) + \Lambda_{REV,l-\pi}^{-1}(\mathcal{P}[l-\pi]) =$$
$$\Omega_{REV}(\pi) + \Omega_{REV}(l-\pi). \quad \square$$

The constraints on $\mathcal{C}_{REV}$ in Lemma 7 are redundant for algorithm REV.

Lemma 9. *The characteristic constant $\mathcal{C}_{REV}$ is given as,*

$$\mathcal{C}_{REV} = \max_{l=1,\ldots,m} \left\{ \frac{\Omega_{REV}(l)}{l} \right\}.$$

2.5 The optimal static algorithm

In order to define an optimal static algorithm, we consider variants of algorithm REV whose cost functions still satisfy properties similar to those of Ω_{REV}. The variants of algorithm REV that we call $\mathcal{R}:\theta$, for $2 \leq \theta \leq m$, such that $\mathcal{P}[\theta] \neq \mathcal{P}[1]$, are defined as follows:

$$\Lambda_{\mathcal{R}:\theta,l}(h) = \begin{cases} \mathcal{P}[l] & l = \theta \text{ and } h = \Lambda_{REV,l}^{-1}(\chi_\theta) \\ \chi_\theta & l = \theta \text{ and } h = \Lambda_{REV,l}^{-1}(\mathcal{P}[l]) \\ \mathcal{P}[l] & \theta < l \leq m \text{ and } h = 1 \\ \Lambda_{REV,l}(h-1) & \theta < l \leq m \text{ and } 2 \leq h \leq \Lambda_{REV,l}^{-1}(\mathcal{P}[l]) \\ \Lambda_{REV,l}(h) & \text{otherwise} \end{cases}$$

where the symbol χ_θ is defined for $\mathcal{R} : \theta$ as the symbol σ in the set

$$\{\Lambda_{REV,\theta}(h) \parallel h = 1, \ldots, \Lambda_{REV,\theta}^{-1}(\mathcal{P}[\theta]) - 1\}$$

that minimizes

$$\frac{\Omega_{REV}(\pi_{first}^\theta(\sigma)) + \Lambda_{REV,\theta}^{-1}(\mathcal{P}[\theta]) - \Lambda_{REV,\theta}^{-1}(\sigma)}{\pi_{first}^\theta(\sigma)}.$$

The algorithms $\mathcal{R} : \theta$ are constructed from three parts. The first part is used in positions $1, \ldots, \theta - 1$, and is exactly identical to algorithm REV. The second part is used in position θ and is identical to algorithm REV except that the order in which $\mathcal{P}[l]$ and χ_θ are compared is exchanged. In the third part that is used in positions $\theta + 1, \ldots, m$, the relative comparisons order of the symbols is the same as in algorithm REV, except that $\mathcal{P}[1]$ is compared first.

Due to the space limit, we are unable to include the proof of the general case and we only sketch the proof in the simpler case where the pattern string contain only two distinct symbols. Note that in this case $\chi_\theta = \mathcal{P}[1]$ and the algorithms $\mathcal{R} : \theta$ consist only of two parts as the second and third part coincide.

The algorithms $\mathcal{R} : \theta$ defined above are similar enough to algorithm REV to satisfy the following version of Lemma 8.

Lemma 10. *The cost function of algorithm $\mathcal{R} : \theta$ satisfies for $l = 1, \ldots, m$, and $\pi \in \Pi^{\mathcal{P}[1..l-1]}$:*

1. *If $\pi < \theta$, then,*

$$\Omega_{\mathcal{R}:\theta}(l - 1) + \Lambda_{\mathcal{R}:\theta,l}^{-1}(\mathcal{P}[l - \pi]) \leq \Omega_{\mathcal{R}:\theta}(\pi) + \Omega_{\mathcal{R}:\theta}(l - \pi),$$

except if $l = \theta$ and $\mathcal{P}[l - \pi] = \chi_\theta$, where,

$$\Omega_{\mathcal{R}:\theta}(\theta - 1) + \Lambda_{\mathcal{R}:\theta,\theta}^{-1}(\chi_\theta) = \Omega_{\mathcal{R}:\theta}(\pi) + \Omega_{\mathcal{R}:\theta}(\theta - \pi) + 1.$$

2. *If $\pi \geq \theta$, then,*

$$\Omega_{\mathcal{R}:\theta}(l - 1) + \Lambda_{\mathcal{R}:\theta,l}^{-1}(\mathcal{P}[l - \pi]) \leq \Omega_{REV}(\theta) + \pi - \theta + \Omega_{\mathcal{R}:\theta}(l - \pi).$$

The constraints in Lemma 7 are also redundant for algorithm $\mathcal{R} : \theta$. The proof is similar to the proof of Lemma 9, using Lemma 10 instead of Lemma 8.

Lemma 11. *For the algorithms $\mathcal{R} : \theta$ defined above,*

$$\mathcal{C}_{\mathcal{R}:\theta} = \max\left\{\frac{\Omega_{REV}(l)}{l} \parallel l = 1, \ldots, \theta - 1\right\} \cup \left\{\frac{\Omega_{REV}(\pi_{first}^\theta(\mathcal{P}[1])) + 1}{\pi_{first}^\theta(\mathcal{P}[1])}\right\}.$$

Given any static SPM algorithm $\mathcal{A}$, define

$$\rho(\mathcal{A}) = \min\{l \parallel 1 \leq l \leq m \text{ and } \Lambda_{\mathcal{A},l}^{-1}(\mathcal{P}[l]) < \Lambda_{REV,l}^{-1}(\mathcal{P}[l])\}.$$

If $\rho(\mathcal{A})$ is defined above, then $\mathcal{P}[\rho(\mathcal{A})] \neq \mathcal{P}[1]$, and algorithm $\mathcal{R} : \rho(\mathcal{A})$ is also defined.

Theorem 12. *Given any static SPM algorithm $\mathcal{A}$, then either*

$$C_{\mathcal{R}:\rho(\mathcal{A})} \leq C_{\mathcal{A}} \ \ or \ \ C_{REV} \leq C_{\mathcal{A}}.$$

Proof. Assume $\rho(\mathcal{A})$ is defined. Since $\Omega_{REV}(l) = \Omega_{\mathcal{A}}(l)$, for $l = 1, \dots, \rho(\mathcal{A}) - 1$, we get by Inequality 2 that $\Omega_{REV}(l)/l \leq C_{\mathcal{A}}$, for $l = 1, \dots, \rho(\mathcal{A}) - 1$. But $\Lambda_{\mathcal{A},\rho(\mathcal{A})}^{-1}(\mathcal{P}[1]) = 2$, and by Inequality 1, $(\Omega_{REV}(\pi_{first}^\theta(\mathcal{P}[1])) + 1)/\pi_{first}^\theta(\mathcal{P}[1]) \leq C_{\mathcal{A}}$. Similarly, if $\rho(\mathcal{A})$ is not defined then $C_{\mathcal{A}} \geq C_{REV}$.

If we take $\mathcal{A} = OPT$ in the last lemma for an optimal static algorithm OPT, then we can find an optimal algorithm among the m or fewer algorithms $\mathcal{R} : \theta, 2 \leq \theta \leq m$, and REV. Thus we define the static algorithm *REVERSE OPTIMAL*, or $\mathcal{RO}$ for short, to be the algorithm among $\mathcal{R} : \theta$ and REV that minimizes $C_{\mathcal{RO}}$. We define $C_{on-line}^{\mathcal{P}} = C_{\mathcal{RO}(\mathcal{P}[1..m])}$.

2.6 The lower bound

The class of static algorithms is general enough to draw conclusion from the performance of static algorithms to the performance of on-line algorithms, since the latter can be viewed as piecewise static algorithms. We can prove the following.

Theorem 13. *Any on-line SPM algorithm requires at least $C_{on-line}^{\mathcal{P}} \times (n-m) + m$ comparisons.*

3 The self-prefix problem

The techniques developed above for on-line algorithms can be used also for the SSPM problem. The bounds obtained hold as well for the pattern preprocessing step of the SPM algorithm.

Lemma 14. *There exists an algorithm for the SSPM problem that requires at most $2m - \lceil 2\sqrt{m} \rceil$ comparisons.*

4 Off-line prefix-matching

Off-line SPM algorithms have more liberties about the way they might proceed and thus are more difficult to analyze. We were unable to obtain tight bounds for off-line algorithms. However, we can prove the following.

Lemma 15. *There exists patterns for which off-line SPM algorithms require fewer comparisons than on-line algorithms.*

Proof. Given the pattern 'aaaabbb', on-line algorithms require about $\frac{7}{5}n$ comparisons and off-line algorithms only about $\frac{4}{3}n$ comparisons. The proof is omitted.

References

1. R.S. Boyer and J.S. Moore. A fast string searching algorithm. *Comm. of the ACM*, 20:762–772, 1977.
2. D. Breslauer. Fast Parallel String Prefix-Matching. Technical Report CUCS-041-92, Computer Science Dept., Columbia University, 1992.
3. D. Breslauer, L. Colussi, and L. Toniolo. Tight Comparison Bounds for the String Prefix-Matching Problem. *Inform. Process. Lett.*, 47(1):51–57, 1993.
4. D. Breslauer and Z. Galil. Efficient Comparison Based String Matching. *J. Complexity*, 9(3):339–365, 1993.
5. R. Cole, M. Crochemore, Z. Galil, L. Gąsieniec, R. Hariharan, S. Muthukrishnan, K. Park, and W. Rytter. Optimally fast parallel algorithms for preprocessing and pattern matching in one and two dimensions. In *Proc. 34th IEEE Symp. on Foundations of Computer Science*, pages 248–258, 1993.
6. R. Cole and R. Hariharan. Tighter Bounds on the Exact Complexity of String Matching. In *Proc. 33rd IEEE Symp. on Foundations of Computer Science*, pages 600–609, 1992.
7. R. Cole, R. Hariharan, M.S. Paterson, and U. Zwick. Which patterns are hard to find. In *Proc. 2nd Israeli Symp. on Theory of Computing and Systems*, pages 59–68, 1993.
8. L. Colussi. Correctness and efficiency of string matching algorithms. *Inform. and Control*, 95:225–251, 1991.
9. Z. Galil. Optimal parallel algorithms for string matching. *Inform. and Control*, 67:144–157, 1985.
10. Z. Galil and R. Giancarlo. On the exact complexity of string matching: lower bounds. *SIAM J. Comput.*, 20(6):1008–1020, 1991.
11. Z. Galil and R. Giancarlo. The exact complexity of string matching: upper bounds. *SIAM J. Comput.*, 21(3):407–437, 1992.
12. Z. Galil and K. Park. Truly Alphabet-Independent Two-Dimensional Pattern Matching. In *Proc. 33th IEEE Symp. on Foundations of Computer Science*, pages 247–256, 1992.
13. L. Gąsieniec and K. Park. Fully Optimal Parallel Prefix Matching. This conference proceedings.
14. C. Hancart. On Simon's string searching algorithm. *Inform. Process. Lett.*, 47(2):95–99, 1993.
15. R. Hariharan and S. Muthukrishnan. Optimal Parallel Algorithms for Prefix Matching. In *Proc. 21th International Colloquium on Automata, Languages, and Programming*. To appear.
16. D.E. Knuth, J.H. Morris, and V.R. Pratt. Fast pattern matching in strings. *SIAM J. Comput.*, 6:322–350, 1977.
17. M. Lothaire. *Combinatorics on Words*. Addison-Wesley, Reading, MA., U.S.A., 1983.
18. G.M. Main and R.J. Lorentz. An $O(n \log n)$ algorithm for finding all repetitions in a string. *J. Algorithms*, 5:422–432, 1984.
19. R.V. Rivest. On the Worst Case Behavior of String-Searching Algorithms. *SIAM J. Comput.*, 6:669–674, 1977.

Incremental Text Editing: a new data structure*

Paolo Ferragina

Dipartimento di Informatica, Università di Pisa,
Corso Italia, 40 - 56125 PISA, Italy
E-mail: ferragin@di.unipi.it

Abstract. We present new sequential and CRCW-PRAM parallel algorithms for the *incremental text editing problem*, in which a *text* string $T = \alpha\beta\gamma$ over an alphabet Σ is dynamically changed to a new text $T' = \alpha\delta\gamma$ by replacing the substring β with another string γ, where $\alpha, \beta, \gamma, \delta \in \Sigma^*$. The dynamically changes of the text are interleaved with on-line queries for finding the occurrences of a *pattern* string.

1 Introduction

Pattern matching on strings is the problem of finding all occurrences of a pattern string P as a substring of a text string T [5, 17]. A classical solution to pattern matching uses the *suffix tree* data structure [20], which is a digital search tree that stores compactly all the suffixes of T, so that all the possible substrings of T are represented by some unique path in the tree. The power of the suffix tree mainly lies in its ability to encode all the suffixes in linear space [7, 20].

Most of the applications considered so far in the known literature are *static*, in the sense that a *fixed* text is preprocessed, so that on-line queries about pattern occurrences can be quickly answered. In a more realistic situation, however, the text is read from left to right one symbol at time, so that an on-line construction of the suffix tree becomes relevant [16, 21]. Still, it may be required in common circumstances to treat changes successively performed on the text. For example in a text editor, a given text $T = \alpha\beta\gamma$ may be changed into the new text $T' = \alpha\delta\gamma$, where $\alpha, \beta, \gamma, \delta$ are (possibly empty) strings. Hence, we have to maintain a data structure that efficiently supports incremental changes of T, to answer quickly to the next queries on T'. McCreight [20] has considered this problem by providing a solution to perform incremental changes to the suffix tree in response to an incremental change in T. However, updating the tree with this method requires linear time in the size of the current text in the worst case, as for the reconstruction of the whole suffix tree from scratch.

Recently, Gu *et al.* [9] have introduced the field of on-line dynamic text indexing, providing an efficient algorithm that is based on a novel data structure, called *border tree*, which exploits string periodicity. The allowed edit operations are insertions and deletions of a *single* character, both performed in $O(\log |T^{(i)}|)$ time, where $T^{(i)}$ is the text after i changes. Each query operation can be performed on $T^{(i)}$ in $O(p + p_{occ} \log i + i \log p)$ time, where p_{occ} is the number of

---

* This paper has been supported in part by M.U.R.S.T. of Italy.

occurrences of P in $T^{(i)}$, and $p = |P|$. However, in many situations each edit operation consists of updating a substring rather a single character. Applying the single-character approach by Gu *et al.* produces a huge fragmentation of their data structure, and therefore increases the query time.

In this paper we present novel algorithmic techniques to combine the suffix tree with the naming technique on strings by Karp *et al.* [15, 4]. We achieve the following results. Let $T^{(upd)}$ be the text, $|T^{(upd)}| = n$, obtained after the application of *upd* edit operations (i.e., replacements of a substring of the text with another string). We give first a sequential algorithm to update $T^{(upd)}$ under a replacement of β with δ in $O(|\beta|\log l + |\delta|\log|\delta| + \log upd)$ time, where l is a parameter such that $l \leq n$. A search of a pattern P in $T^{(upd)}$ requires $O(p\log p + upd\log n + p_{occ})$ time, if $p \leq l$, where p_{occ} denotes the number of occurrences of P. Otherwise ($p > l$), it requires $O(p\log p + \min\{\frac{n}{l}, upd\} \cdot \log n + (upd + \frac{n}{l}) \cdot \log p + p_{occ})$ time.

Next, we provide the first efficient CRCW-PRAM algorithm for the incremental text editing problem. We introduce and solve the new "substring-substring" problem, and we provide the first work-optimal parallel construction of the border tree. Furthermore, we exploit a new logarithmic decomposition of a string into smaller substrings, obtaining the following bounds. The update of $T^{(upd)}$ takes $O(\log f + \log\log n + \log|\delta|)$ time and $O(|\beta|\log f + |\delta|\log^2|\delta| + \log n)$ work, where $f \leq upd \cdot \log \frac{n}{upd}$. The search of P requires $O(\log^2 p + p_{occ})$ time, using $O(\frac{p+l}{\log p})$ processors. The above bounds are obtained by using an $n \times n$ matrix BB for the naming, called Bulletin Board, which is not initialized. Such quadratic space can be avoided at a cost of an extra $O(\log n)$ slow-down factor in the naming.

Finally we show how to apply our algorithms to maintain a labeled tree L under deletions or insertions of subtrees, or renaming of nodes' labels. The pattern search consists in finding all the occurrences of another tree as a subtree of L.

2 Preliminaries

We briefly describe the border tree data structure, and its properties, which will be used later to solve the *prefix-suffix* matching problem (PSM) [9]. Due to the lack of space, we refer to [20, 4] for the labeling technique and the parallel and sequential algorithm to construct the suffix tree of a string. The PSM problem is formulated as follows: given a string $S = s_1 s_2 \ldots s_k$ and two indices i, j, such that $1 \leq j \leq i \leq n$, we have to compute all the occurrences of S into $S[1:i]S[j:n]$.

We say that $S[1:i] \prec S[1:j]$ if and only if $S[1:i]$ is a suffix of $S[1:j]$. We call $S[1:i]$ a *border* of $S[1:j]$, denoted as $S[1:i] \propto S[1:j]$, if $S[1:i] \prec S[1:j]$ and there no exists some k such that $S[1:i] \prec S[1:k] \prec S[1:j]$. We have the following two results:

Lemma 1. *[8] Given a string S, suppose $S[1:t] \prec S[1:t+d]$ are two prefixes of a string, with $t > 0$ and $d > 0$. Let $t = \alpha d + s$ with $0 \leq s \leq d-1$, then $S[1:t] = S[1:d]^\alpha S[1:s]$ and $S[1:t+d] = S[1:d]^{\alpha+1}S[1:s]$.*

Lemma 2. *[9] Let $S[1 : a_1] \propto \ldots \propto S[1 : a_l]$ be a chain of non-empty prefixes of a string S with $S[1 : a_j] = S[1 : a_{j-1}]D_{j-1}$. Then either $D_j = D_{j-1}$ or $|D_j| > |D_{j-1}|$ with $a_{j+1} > \frac{3}{2}a_j$.*

Lemma 2 describes the main structural property of borders, in fact it guarantees that while the number of prefixes in the prefix chain $S[1 : a_1] \propto \ldots \propto S[1 : a_l]$ can be as large as $O(p)$, the number of different D_j's in this chain can be at most $O(\log p)$. The idea is to use this result to compactly represent the tree formed by the *failure links* in the KMP automaton [17], which can be built in linear time even for unbounded alphabet. Let $\mathcal{L}(E(v))$ be the label of the arc in the failure tree connecting the father of v, denoted by $p(v)$, to v. The border tree T_S^b of a string S can be defined as a triple $(R, E, \mathcal{L})$ such that: (1) R is the set of nodes, which is a subset of prefixes of S, such that $v \in R$ if and only if either v has depth ≤ 1 or $\mathcal{L}(E(v)) \neq \mathcal{L}(E(p(v)))$ in the failure tree of S; (2) E is the set of arcs derived by setting $p(v) = u$ if and only if $u \prec v$ and there is no w such that $u \prec w \prec v$; (3) the edge label $\mathcal{L}(E(u))$ is defined as $(|u|, |D|, \alpha)$, where D is a non-empty string and α is the maximum integer such that PD^α is a prefix of S and $P \propto u = PD \propto PD^2 \propto \ldots \propto PD^\alpha$.

From Lemma 2 we immediately derive that T_S^b has depth $O(\log(|S|))$. Each pair of nodes (p', s') denotes a set of strings consisting of a chain of prefixes and suffixes, respectively. For example $\mathcal{L}(p') = (l_p, d_p, \alpha_p)$ with $l = l_p + \alpha d_p, 0 \leq \alpha \leq \alpha_p$ and $S[1 : l]$ is the chain denoted by p'; as well $\mathcal{L}(s') = (r_s, c_s, \beta_s)$ with $n - r + 1 = r_s - \beta c_s, 0 \leq \beta \leq \beta_s$ and $S[r : n]$ is the chain denoted by s'. We can find all prefix-suffix matches by consulting the border trees of strings S and S^R (reversal of S) and checking for length constraints walking up to shorten the prefixes and down to lengthen the suffixes. This at most requires constant time to solve one linear equation and therefore the PSM problem can be solved in $O(\log(|S|) + s_{occ})$, where s_{occ} is the total number of occurrences of S.

3   The data structure

We consider now our problem: Given a text T subject to substring deletions or insertions, answer efficiently to queries about the occurrences of arbitrary patterns. In the following, $D^{(upd)}$ denotes a data structure D after *upd* edit operations.

The string T is partitioned into segments S_i of length k_i, $T = S_1 S_2 \ldots S_r$, such that $l \leq k_i < 2l, \forall i = 1, 2, \ldots, r = \lceil \frac{|T|}{l} \rceil$, where the parameter l controls the trade-off between the edit and matching time, as it will be clear in the following. The last segment S_r is padded with endmarkers $\$ \notin \Sigma$, if $k_r < l$. The partition is lazy since we want to maintain it under the edit operations.

For each pair of adjacent segments $\tilde{S}_i = S_i S_{i+1}$ and $\tilde{S}_i^R = S_i^R S_{i-1}^R$, we compute the names of all substrings having a length $l' = 2^{q'} \leq l$, for each $q' \geq 0$ [4], requiring $O(n \log l)$ time in total. This labeling gives the names to all the substrings in T having a length that is a power of two smaller than l, because of the overlapping of $\tilde{S}_i$ and $\tilde{S}_i^R$.

We also build the *generalized suffix tree* $T_{GST}^{(upd)}$ (GST) for the set of strings $S = \{\tilde{S}_i : \forall 1 \leq i \leq r\}$ [11]. $T_{GST}^{(upd)}$ is the trie consisting of the suffixes of all the strings in S, and is built in linear time by "superimposing" the suffix trees T_i of all strings in S, identifying all edges with the same labels that do not connect leaves.

Since the text T can be updated under substring insertions and deletions, at each edit operation we would not recompute the above information for each affected segment from scratch. Thus, we maintain a data structure that contains a *description* of the operations performed on T, and which is reconstructed from scratch when segment $S_i^{(upd)}$ violates the constraint $l \leq |S_i^{(upd)}| < 2l$. For each segment S_i we build a bidirectional list L_i, possibly empty if S_i has not been affected by the previous edit operations. L_i consists of an ordered sequence of "intervals", denoted by $(\delta_k, a_k, b_k), 1 \leq k \leq s, s \in \mathbb{N}$, such that the segment $S_i^{(upd)}$ is equal to the string given by the concatenation of $\delta_1[a_1, b_1], \ldots, \delta_s[a_s, b_s]$, where $\delta_i[a_i, b_i]$ is a substring of some previously inserted string δ_i, and extends from position a_i to b_i of δ_i. If $\delta_i = T$ then $T[a_i, b_i]$ denotes a substring of the original text. Thus, $L = \bigcup_{j=1}^{r} L_j$ is the global list of intervals in which $T^{(upd)}$ is partitioned.

For each list L_i we maintain also bidirectional sublists of intervals corresponding to the same substring δ, $L_\delta = \{(\delta, a_{i_j}, b_{i_j}) \in L_i, 1 \leq i \leq r\}$. We build on each list L_δ a 2-3 tree to perform efficiently search, split and concatenate operations [2]. We assume that each character of $T^{(upd)}$ has associated a pair (δ, j) denoting that it corresponds to the j-th character of δ so that, we can compute the substring of L in which $\delta[j]$ is currently contained in $O(\log |L_\delta|)$ time, using the corresponding 2-3 tree.

Furthermore, for each inserted string δ we build a suffix tree T_δ. We augment each internal node of T_δ with two pointers to its leftmost and rightmost descending leaves. All the leaves are also numbered from left to right, and therefore, at each leaf u corresponds a pair (n_u, s_u), where n_u is the number given by the left to right numbering, and s_u is the suffix to which it refers to. Considering each pair of integers as a point in a plane, we can build for all the pairs a range tree RT_δ in $O(|\delta| \log |\delta|)$ time, requiring $O(\frac{|\delta| \log |\delta|}{\log \log \delta})$ space [22, 6]. In this way, we can give a novel geometric interpretation of the problem of finding occurrences of P internal into substrings of δ.

In the following we say that an occurrence of a string s into $T^{(upd)}$ *crosses* the left (resp., right) margin of an interval (δ, a, b), if the leftmost (resp., rightmost) position of that occurrence is to the left (resp., right) of $\delta[a, b]$. Note that, since we insert and remove substrings from the current text, it is simple to prove that $|L| \leq upd + 1$, where $|L|$ is the number of intervals in which $T^{(upd)}$ is partitioned.

4 Searching a pattern P

To detect the occurrences of a pattern P, with $p = |P|$, we compute the border trees and the $\log n$ refinement trees of P and P^R, by using the BB matrices built to label the strings $\tilde{S}_i, \tilde{S}_i^R$. The preprocessing phase requires $O(p \log p)$ time. On the suffix tree of P, we make a depth first visit to define for each node v, the pointer $lps(v)$ to its deepest ancestor v' having a \$-branch. This means that for each substring (node) of P we have a pointer to its longest prefix (ancestor) that is a suffix (there is a \$-branch) of the pattern P. The same computation performed on the suffix tree T_{P^R} of P^R allows to compute the longest suffix that is a prefix of P.

Searching "internal" occurrences of P. Given an interval $(\delta, a, b) \in L$, we need to search for occurrences of P entirely contained into $\delta[a, b]$. Using the set of refinement trees of δ, we search in $O(\log p)$ time the locus z of P in T_δ. All the leaves descending from z are occurrences of P starting into δ. We would not list all such leaves since, if P has period d then there could be $O(\frac{p}{d})$ occurrences overlapping the right margin b of (δ, a, b) or some occurrences could have been disappeared as the result of the previous edit operations.

Thus, we operate as follows. Each internal occurrence (if any) of P into $\delta[a, b]$ must start in $\delta[a, b - p + 1]$. Thus, each leaf is an occurrence of P into $\delta[a, b]$ if it has z as ancestor and it corresponds to a suffix starting in $\delta[a, b - p + 1]$. Let l^z and r^z be the ordering numbers associated to the leftmost and rightmost descending leaves of z, respectively. We check the two conditions above, finding in $O(\log |\delta| + K)$ time into RT_δ, all the K points (leaves) v with associated pair (n_v, s_v) such that, $r^z \leq n_v \leq l^z$ and $a \leq s_v \leq b - p + 1$ [22]. In this way we can list all the occurrences internal into the intervals of L_δ in $O(\log p + |L_\delta| \log |\delta| + p_{occ}^\delta)$ time, where p_{occ}^δ is the number of pattern occurrences into intervals of L_δ. To find the occurrences of P into the intervals of some segment S_i, we use the trie $T_{GST}^{(upd)}$, searching sequentially in $O(p)$ time the locus of P. As done for δ, we can list all the pattern occurrences starting from not deleted positions of the original text. Note that, here the occurrences of P may be listed twice because of the overlapping of $\tilde{S}_i$, but this can be easily managed and it does not affect the total time complexity. Summing up all the intervals we attain $O(p \log p + upd \log n + p_{occ}^{int})$ time, where p_{occ}^{int} is the total number of internal occurrences.

Searching "overlapping" occurrences of P. We recall that the PSM problem can be solved using the algorithm sketched in Section 2 [9]. We introduce now a subproblem whose solution will be useful to find all the overlapping occurrences.

Prefix-Substring problem

Given a string S, with $|S| = s$, and three indices $i, j, h \in [1, s]$, find the longest prefix of S that occurs as suffix of the string obtained concatenating $S[1 : i]$ to $S[j : k]$.

We provide a new solution to the Prefix-Substring problem for P, which uses the set of refinement trees of P^R and the names given to all the substrings of P^R of length a power of two. We search for $\bar{P} = P[j : k]^R P[1 : i]^R$ into the set of refinement trees of P^R. Note that we have not the names for all the substrings of $\bar{P}$, but using the searching algorithm shown in [4], we need to compute only other $O(\log p)$ names labeling the checked strings overlapping $P[j : k]^R$ and $P[1 : i]^R$.

Using the set of refinement trees, searching in T_{PR} requires $O(\log p)$ time and it returns the longest substring P' of P occurring as a suffix of $P[1:i]P[j:k]$. Then, using the pointer lps of the contracted locus of P' in T_{PR}, we compute the searched longest prefix.

For each interval $(\delta, a, b) \in L$, we want to find all the occurrences ending into this interval and crossing the left margin a. Using the border trees of P and P^R, we solve the PSM problem for the pair (P'', P_S), where P'' is the longest prefix of P occurring into $T^{(upd)}$ and ending at the left margin a, and P_S is the longest suffix of P that is a prefix of $\delta[a, b]$. Note that, as done before for P', P_S can be computed in $O(\log p)$ time, using the names and the lps pointers in T_P.

We give a sketch of the main idea in our searching algorithm. There are two cases. If $p \leq l$ then we detect the internal occurrences of P, which are found by using the algorithm shown before for all the intervals that compose $T^{(upd)}$. If $p > l$ then we search the overlapping occurrences of P as follows. We scan L and apply the before mentioned algorithm on the i-th interval of L under the following invariant. Let r_{i-1} be the rightmost position of the $(i-1)$-th interval of L in $T^{(upd)}$, and P_i' be the longest prefix of P that occurs as a suffix of $T^{(upd)}[1 : r_{i-1}]$. We solve the PSM problem for P_i' and the longest suffix of P occurring as prefix of the i-th interval of L. Before the execution of the $(i+1)$-th step, we maintain the invariant by searching, as done for P_S, in $O(\log p)$ time the longest substring $P[j_1, j_2]$ that occurs as suffix of $\delta[a, b]$. Then we solve the Prefix-Substring problem for the pair $(P_i', P[j_1, j_2])$, thus attaining P_{i+1}'.

It is simple to prove that setting $l = \sqrt{n}$, the total time of searching is $O(p \log p + upd \log n + p_{occ})$, if $p \leq l$, and $O(p \log p + \min\{p, upd\} \cdot \log n + upd \log p + p_{occ})$, otherwise.

The fragmentation of the text does not grow arbitrarily but it is controlled during the searching process, since we can reconstruct the data structure by compacting a sequence of k intervals of length $l' = O(k)$ in $O(l' \log l')$ time. Thus we amortize over the searching process.

5 Updating the text

Updating operations of text $T = \alpha\beta\gamma, \alpha, \beta, \gamma \in \Sigma^*$, can be defined as substituting β by $\delta \in \Sigma^*$, to obtain a new text $T' = \alpha\delta\gamma$. This will be attained by the deletion of β followed by the insertion of δ.

Deletion of β. We find the list L' of intervals that are overlapped by β, using the 2-3 trees available for each sublist L_δ. We search for the interval containing the starting position of β in $O(\log upd)$ time and then we scan L listing all the overlapped intervals $L' = I_1, I_2, \ldots, I_h$, $I_j = (\delta_j', a_j', b_j'), h \geq 1, 1 \leq j \leq h$. If $h = 1$ (i.e., β is entirely contained into an interval $I = I'\beta I'' \in L_i$) then we delete I and insert I' and I'' into $L_{\delta_i'}$ and L_i in $O(\log upd)$ time. Otherwise ($h > 1$), we maintain L_i and $L_{\delta_i'}$ by deleting the intervals $I_2, I_3, \ldots, I_{h-1}$, and updating the bidirectional links. To maintain the 2-3 trees associated to the lists affected by the deletion, we perform one split and one concatenate operation on each of them since the set of deleted intervals forms an adjacent sublist. Furthermore, we adjust b'_1 and a'_h according to the portion of β overlapping them, and we delete all the names corresponding to substrings of length a power of two starting into $\delta'[a'_i, b'_i]$ in $O(|\beta| \log l)$ time.

If β overlaps $k > 2$ segments, we can reconstruct all the affected segments requiring $O(1)$ amortized time. Some segment S_i may become inconsistent, because of the delete operation, and it has to be reconstructed to maintain the lazy partition, that is, $l \leq |S_j^{(upd)}| < 2l, j = 1, 2, \ldots, r$. This can be done by splitting the inconsistent segment or merging it to an adjacent *consistent* one. The reconstruction of the set of refinement trees and the labeling of all the substrings can be done in $O(l \log l)$ time, as well the construction of RT. Note that this process is performed on a segment after that the total length of the substrings deleted from it is $\Omega(l)$, hence the complexity of reconstruction can be amortized by the cost of the previous edit operations, attaining $O(1)$ amortized time.

Insertion of δ. Given a string δ to be inserted at position y of S_i, using the 2-3 tree for the list L'' we find in $O(\log upd)$ time the interval $I = I_1 I_2 \in L''$ containing y, such that $I_1 \delta I_2$ is formed. We delete the interval I from L_i and L''. We insert the intervals I_1, I_2 into L_i and L'', and δ into L_i and L_δ in $O(\log upd)$ time. We also build the set of refinement trees of δ and we label all the substrings of δ and δ^R having a length that is a power of two in $O(|\delta| \log |\delta|)$ time. We number the leaves of T_δ and compute the range tree RT_δ (see Section 3). If it is $|S_j^{(upd)}| > 2l$ during the insertion of δ into a segment S_j, then we have to maintain consistent the partition and split it, amortizing by the previous insertion operations. We label the substrings of δ using the same set of BB matrices employed to label T, and the other strings previously inserted. This naming process guarantees that two substrings are equal if and only if they have the same length and the same name.

When $upd > n$, $|T^{(upd)}| < \frac{|T^{(0)}|}{2}$ or $|T^{(upd)}| > 2 |T^{(0)}|$, we can reconstruct from scratch the whole data structure in $O(1)$ amortized time. Due to the strategy adopted in maintaining our data structure, the number of intervals forming the text does not depend on the size of the substrings deleted or inserted (in particular we have $|L| \leq upd + 1$). We extend the result in [9], which is heavily based on the number of single character insertions and deletions performed on T. Thus, for example, the deletion of a string β requires in our case $\Omega(1)$ time for each following search operation, while $\Omega(|\beta|)$ time is needed for each next search using the algorithm of [9]. Furthermore, the algorithm of [9] can amortize the cost of the previous edit operations only after $O(n)$ updates, reconstructing the suffix tree of the whole text. Instead, our partition strategy allows to reconstruct only a segment in $O(1)$ amortized time, thus taking advantage of more realistic situations in which the updates affect only some parts of the text. Note that, we can control the trade-off between the edit and matching time, varying the size l of the segments.

6 The parallel algorithm

The algorithm shown in [9] is intrinsically sequential since it needs P_i' to compute efficiently the longest prefix P_{i+1}' of P occurring as a suffix of $S_1 S_2 \ldots S_i$. Our CRCW-PRAM algorithm is obtained exploiting the following new ideas. Namely, a parallel solution for the new Substring-Substring problem, a *logarithmic decomposition* of a string, and a work-optimal parallel construction of the border tree.

The following results are useful for the design of our parallel algorithm. (For brevity, proofs are omitted)

Lemma 3. *Given a tree T, $\mid T \mid = n$, let us assume that some nodes in T are marked (the root is marked). For each node v of T, we can compute the pointer $lps(v)$ to its nearest marked ancestor in $O(\log n)$ time using $O(\frac{n}{\log n})$ processors on an EREW PRAM.*

Substring-Substring Problem. *Given a string S, with $\mid S \mid = s$, let $i, j, h, k \in [1, s]$ be indices such that $S[i, j]$ and $S[h, k]$ are (possibly empty) substrings of S. Find the longest substring of S occurring as suffix of the string obtained concatenating $S[i, j]$ to $S[h, k]$.*

Consider the Substring-Substring problem for $S = P$. We build the suffix tree of P^R, maintaining the set of all refinement trees, and we label all the substrings of P^R of length a power of two. We search for the string $P[h, k]^R P[i, j]^R$ into the set of refinement trees of P^R. We use the technique of [4] to perform this search. Therefore we need to compute other $O(\log p)$ names only, corresponding to the strings that overlap $P[h, k]^R$ and $P[i, j]^R$, and are of length a power of two. In this way we find the longest prefix of $P[h, k]^R P[i, j]^R$ occurring as a substring of P^R. The reversal of such prefix is the substring that we are searching.

Let us assume to have a list of intervals $L = \{I_1, I_2, \ldots, I_k\}$, each of them denoting a string x_i, such that $X = x_1 x_2 \cdots x_k$. Given a pattern P, we would list all the occurrences of P overlapping at least two intervals of X. For each x_i and x_i^R, we label all the substrings having length a power of two and we build the set of refinement trees and the border trees of P and P^R. Using the result of Lemma 3, we compute the pointer lps for each node of T_P and we execute the following steps:

1. Build a vector V of k positions, such that $V[i]$ is a pair (i_1, i_2) denoting the longest substring $P[i_1, i_2]$ occurring as suffix of x_i, computed using the names for x_i^R and the refinement trees of P^R.
2. on the vector V, perform a *parallel prefix*, computing at each step the following associative operation: $V[u] @ V[v] = P[u_1, u_2] @ P[v_1, v_2] = P[z_1, z_2]$, where $P[z_1, z_2]$ is the longest substring of P that occurs as suffix of the string obtained concatenating $P[u_1, u_2]$ to $P[v_1, v_2]$ (Substring-Substring problem).

Note that, at the i-th iteration of step 2, $V[j]$, for $1 \leq j \leq k$, contains the pair (j_1, j_2) such that $P[j_1, j_2]$ is the longest substring of P occurring as suffix of $x_{j-2^i+1} \cdots x_j$. It follows immediately that $O(\log p)$ iterations are sufficient to find the longest substring of P occurring as suffix of $x_1 \ldots x_j$. Hence, by using $O(\frac{k}{\log p})$ processors [14], we can perform both steps in $O(\log^2 p)$ time, since each @ computation requires $O(\log p)$ sequential time. In parallel, using the pointer lps of the contracted locus of each $P[j_1, j_2]$, we can find the longest prefix P'_j of P occurring as suffix of $x_1 \ldots x_j$, as shown for the sequential case.

Furthermore, each processor in $O(\log p)$ time, using the algorithm given for the sequential case, can compute the longest suffix P_j^S occurring as prefix of x_j. By simply using the border trees of P and P^R for the pair (P'_{j-1}, P_j^S), a processor assigned to the interval I_j can list all the p_{occ}^j occurrences overlapping I_{j-1} and ending in I_j in $O(\log p + p_{occ}^j)$ time. Finding all the overlapping occurrences of P requires $O(\log^2 p + p_{occ})$ time using $O(\frac{k}{\log p})$ processors, where p_{occ} is the total number of occurrences of P.

Logarithmic decomposition technique. Given a string S, with $|S| = t$, we consider the set S_S of strings s_k^q such that, each of them has length 2^q and occurs at position $k2^q$ of S, where $0 \le q \le \lfloor \log t \rfloor, 0 \le k \le \lfloor \frac{t}{2^q} \rfloor$, $k, q \in \mathbb{N}$.

Lemma 4. *Each substring $s = S[i, j]$ can be represented by using $O(\log |s|)$ strings of S_S.*

Proof. Let i be equal to $k_0 2^{i_0}$, where $i_0 = \max\{h : \frac{i}{2^h} \in \mathbb{N}\}$ and k_0 is obviously odd. We may represent the prefix of length 2^{i_0} of s by using $s_{k_0}^{i_0} \in S_S$. Note that the remaining suffix $S[i + 2^{i_0} : j]$ starts at a position which is divisible at least by 2^{i_0+1}, in particular we can compute $i_1 = \max\{h : \frac{i+2^{i_0}}{2^h} \in \mathbb{N}\} > i_0$, $k_1 = \frac{i+2^{i_0}}{2^{i_1}} \in \mathbb{N}$, and set $S[i + 2^{i_0} : j] = s_{k_1}^{i_1} \cdot S[i + 2^{i_0} + 2^{i_1} : j]$, and so on.

In this way, at each step we are able to represent the prefix of the remaining suffix of s by a string in S_S of larger (power of two) length. When $i + 2^{i_0} + 2^{i_1} + \ldots + 2^{i_k} > j$ and $i + 2^{i_0} + 2^{i_1} + \ldots + 2^{i_{k-1}} \le j$, then we use the same argument in a decreasing manner, to decompose the remaining suffix $S[i + 2^{i_0} + 2^{i_1} + \ldots + 2^{i_{k-1}} : j]$. Therefore, $O(\log |s|)$ strings of S_S are sufficient to represent s. □

In general, let us assume to have decomposed the string $S^{(upd)}$, updated by the previous upd delete operations, in $S_1'' S_2'' \ldots S_y''$, with $S_h'' \in S_S, 1 \le h \le y$. The deletion of a string s'' from $S^{(upd)}$ affects a subset of these strings, namely $S_{h_1}'', S_{h_1+1}'', \ldots, S_{h_2}''$. We cancel $S_{h_1+1}'', \ldots, S_{h_2-1}''$ and we decompose the prefix of S_{h_1}'' and the suffix of S_{h_2}'', remaining after the deletion of s'', using the strings of S_S (Lemma 4). It is simple to prove that we can maintain $S^{(upd)}$ as concatenation of at most $O(upd \log \frac{t}{upd})$ strings of S_S. Note that $O(t \log^2 t)$ is the total space required by the set of refinement trees computed for all the $O(t)$ strings of S_S. Furthermore, given a substring s of S, it is simple to compute its logarithmic decomposition work-optimally in $O(\log \log t)$ time.

Parallel construction of the border tree. We show now how to build the border tree of P in $O(\log p)$ time using $O(p)$ processors on a CRCW PRAM. The construction is work optimal if we use the algorithm of [12] to perform the Step 1. However, we apply the result of [4] to have a logarithmic time solution (for brevity, proofs are omitted).

1. Build the suffix tree T_{PR} for P^R. Let $len(u)$ be the length of the substring spelled out by the path in T_{PR} connecting the root to u.

2. Mark the root and the leaves of T_{PR}. Given a leaf z, the edge connecting z to its father $f(z)$ can have a label with \$. In this case, since the suffix tree is built on P^R\$, z and $f(z)$ denotes the same prefix of P. Thus, delete the mark of z and mark $f(z)$. Apply the algorithm shown in Lemma 3 to this labeled tree, computing for each node u the pointer $lps(u)$ to its deepest marked ancestor.

3. If u points to v (denoted by $u \to v$), s_v^R is the longest prefix of P occurring as suffix of s_u^R. Write $v \propto u$ to denote that $s_v^R \propto s_u^R$, where u and v are marked nodes, and s_u, s_v are the strings spelled out by the paths connecting the root of T_{PR} to u and v, respectively.

   Given two nodes p_i and p_j corresponding to prefixes of P, if $p_i \to p_j$ then we have $p_j \propto p_i$ and thus $s_{p_i}^R = s_{p_j}^R D_i$, where $|D_i| = len(p_i) - len(p_j)$. Construct a multiple list L as follows: given $p_i \to p_j$, define $point(p_i) = p_j$ if $D_i = D_j$, otherwise $point(p_i) = NIL$. Check $D_i = D_j$ in constant sequential time, by verifying $|D_i| = |D_j|$, since p_j is an ancestor of p_i in T_{PR} and thus, $s_{p_i}^R$ and $s_{p_j}^R$ share a suffix of length $len(p_j)$.

4. Do a multiple list ranking on L defining for each node u the pointer to the tail t_u of its chain;

5. If $point(p_i) = NIL$ and $p_i \to p_j$, set $point(p_i) = p_j$.

6. If $point(p_h) = p_k$ and $D_h = D_k$, set $point(p_h) = point(p_k)$. Compute for each node u the value $d_u = len(u) - len(point(u))$.

7. Define the border tree T_P^b for each marked node of T_{PR} (prefix of P) according to the pointers $point$ (computed in Step 5) and the values of len and d.

In Step 3, the chains formed by $point$ are disjoint, in fact (by contradiction) if we had $p_i \to p_j, p_h \to p_j, p_j \to p_k$, with $p_i = p_j D_i, p_h = p_j D_h, p_j = p_k D_j$, then if $point(p_i) = point(p_h) = p_j$, we would have $D_i = D_j = D_h$, that is $p_i = p_h$.

Steps 5 and 6 can be performed to attain the set of *maximal* prefix chains in the border tree of P (see Section 2). The other steps require to apply list ranking and tree contraction techniques that can be performed with optimal work [14].

The parallel searching algorithm

Preprocessing of T: Label all the substrings $x \in S_{T^{(0)}}$ and x^R, having length a power of two [15] and build in parallel the set of refinement trees for each substring of $S_{T^{(0)}}$, in $O(\log n)$ time using $O(n \log n)$ processors. We also build a list L, at the beginning formed by the interval $(T, 1, n)$. In general L is the list denoting at each step the set of intervals of $S_{T^{(upd)}}$ in which $T^{(upd)}$ is partitioned.

Preprocessing of P: Compute the border trees and the set of refinement trees of P and P^R by using the BB matrices adopted to label the substrings of T. Applying the result of Lemma 3, we compute for each node of the suffix tree, the pointer lps to its deepest ancestor having a \$-branch.

Searching: For each interval $I_i \in L$ do in parallel:

- Search for internal occurrences of P into I_i in $O(\log p)$ time, using the set of refinement trees available for the string corresponding to the interval I_i, since I_i denotes a string of $\mathcal{S}_{T(upd)}$;
- Compute the occurrences of P overlapping the string x_{i-1} corresponding to I_{i-1} and ending into x_i (corresponding to I_i), by using the parallel prefix algorithm applied to the list L. Note that we have the labeling of x_i^R and x_i, since $x_i \in \mathcal{S}_{T(upd)}$.

Therefore, searching for an arbitrary pattern P into a text T requires $O(\log^2 p + p_{occ})$ time using $O(\frac{p+f}{\log p})$ processors, where $f = |L| = O(upd \log \frac{n}{upd})$.

To maintain the data structure under substring insertions and deletions, we use the decomposition technique and we update, as done for the sequential case, the 2-3 trees of the lists affected by the updating process. Now, we can update all the affected lists in parallel requiring $O(\log f)$ time using $O(|\beta|)$ processors. Furthermore, to maintain consistent the representation of $T^{(upd)}$ as concatenation of strings of $\mathcal{S}_{T(upd)}$ we apply the logarithmic decomposition technique. Note that, after the deletion of β we have to decompose the remaining prefix of I_1 and suffix of I_k in $O(\log|I_1| + \log|I_k|)$ strings of $\mathcal{S}_{T(upd)}$ (see Section 5). The same holds for I_1 and I_2 in the case of insertion of δ.

Therefore, the updating process requires $O(\log f + \log|\delta| + \log\log n)$ time and $O(|\beta| \log f + |\delta| \log^2 |\delta| + \log n + \log f)$ work.

The "garbage" space occupied by the strings of $\mathcal{S}_{T(upd)}$ deleted from T is bounded by the total space of the parallel algorithm.

7 On-line dynamic subtree matching

In [19, 10] is shown that the following (maximal) subtree isomorphism problem is solvable in linear time on a RAM. Given two ordered (unlabeled) trees P and T, $|P| = m$, $|T| = n$, $m \le n$, decide whether P is isomorphic to any subtree of T (a subtree is defined as a node and all its descendants). The subtree isomorphism is reduced to exact string matching, by using a string coding for P and T, and applying a fast string matching algorithm. However, this solution is *static* since a fixed tree T is considered.

Let T be an ordered labeled tree over an alphabet Σ. A subtree of T is intended to be a node of T and *all* its descendants. In our problem, called *on-line dynamic subtree matching*, to distinguish it from the classical tree pattern matching problems [13, 18], T is not fixed at the beginning, but its structure may change under deletion or insertion of subtrees, or by renaming of some node. An on-line answer is required on the tree available at the moment, before the next subtree update is received. The operations allowed are:

Add(T',u,i): the subtree T' has to be added to T such that the root of T' becomes the i-th son of node u.

Delete(u,i): the descending subtree of the i-th son of node u in T is deleted.

Change(u,e): e becomes the new label of node u.

Query(B): list all the occurrences of B as subtree of T.

We use a simple list coding code(T) for an ordered labeled tree T of unbounded degree defined in [10]. A tree is coded in $O(|T|)$ time by visiting in preorder: symbol label(u) is appended to code(T) when a node u is encountered, and a symbol $\oplus$ is appended after that node u and its descendants have been visited. By an easy induction, it is possible to see that there is a one to one correspondence between ordered labeled trees T and strings code(T). Having this representation of trees, it is simple using the algorithms given in the previous sections to implement the operations listed above. Note that, the length of the string produced by the coding phase is twice the size of the coded tree, hence the complexity analysis made in the previous sections holds either for the sequential or for the parallel case. It seems very appropriate to apply our approach to this problem since the tree to be deleted or inserted into T is usually a string and not a single character.

Acknowledgments

I am grateful to Fabrizio Luccio for several comments on the early version of the paper. I warmly thank Roberto Grossi for hours of fruitful discussions.

References

1. A. V. Aho. *Algorithms for finding patterns in strings*, chapter A, 255–300. Handbook of Theoretical Computer Science. MIT press, Cambridge, 1990.
2. A. V. Aho, J. E. Hopcroft, and J. D. Ullman. *The design and analysis of computer algorithms*. Addison-Wesley, 1974.
3. A. Amir, M. Farach. Adaptive dictionary matching. *Proc. of IEEE Symposium on Foundations of Computer Science*, 760–766, 1991.
4. A. Apostolico, C. Iliopolus, G. M. Landau, B. Schieber, and U. Vishkin. Parallel construction of a suffix tree with applications. *Algorithmica*, 3:347–365, 1988.
5. R. S. Boyer and J. S. Moore. A fast string searching algorithm. *Communications of the ACM*, 20:762–772, 1977.
6. B. M. Chazelle. Filtering search: a new approach to query answering. *SIAM Journal of Computing*, 15:703–724, 1986.
7. M. T. Chen and J. Seiferas. Efficient and elegant subword tree construction. *Combinatorial algorithms in word*, 97–107. Springer-Verlag, 1985.
8. Z. Galil. Optimal parallel algorithms for string matching. *Information and Control*, 67:144–157, 1985.
9. M. Gu, M. Farach, and R. Beigel. An efficient algorithm for dynamic text editing. *ACM-SIAM Symposium on Discrete Algorithms*, 1994.
10. R. Grossi, F. Luccio, and L. Pagli. *Coding trees as strings for approximate tree matching*, 245–259. Sequences II: Methods in Communication, Security and Computer Science, 1992.
11. D. Gusfield, G. M. Landau, and B. Schieber. An efficient algorithm for all pairs suffix-prefix problem. *Information Processing Letters*, 41:181–185, 1992.
12. R. Hariharan. Optimal parallel suffix tree construction. *ACM Symposium on Theory of Computing*, 1994.

13. C. M. Hoffmann and M. J. O'Donnell. Pattern matching in trees. *Journal of the ACM*, 29:68–95, 1982.
14. R. M. Karp and V. Ramachandran. *A survey of parallel algorithms for shared memory machines*, chapter 17, 869–941. Handbook of Theoretical Computer Science. Elsevier Science Publisher B.V., j. van Leeuwen edition, 1990.
15. R. Karp, R. Miller, and A. Rosenberg. Rapid identification of repeated patterns in strings, arrays and trees. In *Proc. 4th Symposium on Theory of Computing*, pages 125–136. ACM, 1972.
16. M. Kempf, R. Bayer, and U. Guntzer. Time optimal left to right construction of position trees. *Acta Informatica*, 24:475–489, 1987.
17. D. E. Knuth, J. H. Morris, and V. R. Pratt. Fast pattern matching in strings. *SIAM Journal of Computing*, 6(2):63–78, 1977.
18. S.R. Kosaraju. Efficient tree pattern matching. *Proc. IEEE Symposium on Foundations of Computer Science*, 178–183, 1989.
19. E. Mäkinen. On the subtree isomorphism problem for ordered trees. *Information Processing Letters*, 32:271–273, 1988.
20. E. M. McCreight. A space-economical suffix tree construction algorithm. *Journal of the ACM*, 23(2):262–272, 1976.
21. E. Ukkonen. On-line construction of suffix tree. Technical Report A-1/93, Dept. of Computer Science, University of Helsinki, Finland, 1993.
22. D. E. Willard. New data structures for orthogonal range queries. *SIAM Journal of Computing*, 14:232–253, 1985.

Erratum to the ESA '93 Proceedings

We would like to announce that our paper **Computing Treewidth and Minimum Fill-In: All You Need are the Minimal Separators** published in the Proceedings of the First Annual European Symposium on Algorithms (Lecture Notes in Computer Science 726, Springer-Verlag, Berlin, 1993, pp. 260–271) is mistaken.

More precisely, the Lemma 18 is false which implies that the given algorithm is not correct. Currently we cannot give a proof of our Main Theorem (Theorem 21), although we do not have a disproof.

We like to mention that Section 4 remains correct.

Ton Kloks
Hans Bodlaender
Haiko Müller
Dieter Kratsch

Authors Index

N. Alon 354
A. Andersson 82
E.M. Arkin 36
S. Arya 48
T. Asano 215
P. Bay 413
I. Ben-Aroya 365
P. Berman 60
T. Biedl 24
G. Bilardi 413
J.-D. Boissonnat 254
D. Breslauer 483
A. Brodnik 72
C. Burnikel 227
S. Carlsson 106
B. Chen 300
J. Chen 106
V. Chepoi 159
Y.-J. Chiang 266
B.S. Chlebus 401
J.D. Cho 148
L. Colussi 483
J. Czyzowicz 254
S.K. Das 331
O. Devillers 254
F. Dragan 159
P. Ferragina 331, 495
U. Fössmeier 60
P.G. Franciosa 343
A. Gambin 401
G. Gambosi 343
A. Garg 12
L. Gąsieniec 471
A.V. Goldberg 1
G.H. Gonnet 10
R. Greenlaw 436
P. Gupta 278
M. Held 36
H.B. Hunt III 424
P. Indyk 401
A. Israeli 171
R. Janardan 278
G. Kant 24
M. Karpinski 60

N. Katoh 215
M. Kaufmann 60
C. Lund 202
J. Machta 436
M.V. Marathe 424
K. Mehlhorn 227
J.S.B. Mitchell 36
J.I. Munro 72, 94
U. Nanni 343
S. Nilsson 82
T. Nishizeki 118
M.H. Overmars 240
K. Park 471
G. Parlati 183
A. Pietracaprina 391
W. Plandowski 460
P.V. Poblete 94
G. Pucci 391
V. Radhakrishnan 424
S. Raje 148
S.S. Ravi 424
N. Reingold 202
J.-M. Robert 254
D.J. Rosenkrantz 424
M. Sarrafzadeh 148
I. Schiermeyer 290
S. Schirra 227
A. Schönhage 448
A. Schuster 365
A. Shirazi 171
J.F. Sibeyn 377
S.S. Skiena 36
M. Smid 48, 278
A. Srivastav 307
P. Stangier 307
R.E. Stearns 424
M. Stoer 141
R. Tamassia 12, 266
T. Tokuyama 215
L. Toniolo 483
A.F. van der Stappen 240
A. van Vliet 300
E. Vetter 448
A. Viola 94

F. Wagner 141
D. Wedelin 319
K. Weihe 130
J. Westbrook 202
G.J. Woeginger 300
D. Yan 202

M. Yung 183
R. Yuster 354
M. Yvinec 254
A. Zelikovsky 60
X. Zhou 118
U. Zwick 354

Springer-Verlag
and the Environment

We at Springer-Verlag firmly believe that an international science publisher has a special obligation to the environment, and our corporate policies consistently reflect this conviction.

We also expect our business partners – paper mills, printers, packaging manufacturers, etc. – to commit themselves to using environmentally friendly materials and production processes.

The paper in this book is made from low- or no-chlorine pulp and is acid free, in conformance with international standards for paper permanency.

Lecture Notes in Computer Science

For information about Vols. 1–774
please contact your bookseller or Springer-Verlag

Vol. 775: P. Enjalbert, E. W. Mayr, K. W. Wagner (Eds.), STACS 94. Proceedings, 1994. XIV, 782 pages. 1994.

Vol. 776: H. J. Schneider, H. Ehrig (Eds.), Graph Transformations in Computer Science. Proceedings, 1993. VIII, 395 pages. 1994.

Vol. 777: K. von Luck, H. Marburger (Eds.), Management and Processing of Complex Data Structures. Proceedings, 1994. VII, 220 pages. 1994.

Vol. 778: M. Bonuccelli, P. Crescenzi, R. Petreschi (Eds.), Algorithms and Complexity. Proceedings, 1994. VIII, 222 pages. 1994.

Vol. 779: M. Jarke, J. Bubenko, K. Jeffery (Eds.), Advances in Database Technology — EDBT '94. Proceedings, 1994. XII, 406 pages. 1994.

Vol. 780: J. J. Joyce, C.-J. H. Seger (Eds.), Higher Order Logic Theorem Proving and Its Applications. Proceedings, 1993. X, 518 pages. 1994.

Vol. 781: G. Cohen, S. Litsyn, A. Lobstein, G. Zémor (Eds.), Algebraic Coding. Proceedings, 1993. XII, 326 pages. 1994.

Vol. 782: J. Gutknecht (Ed.), Programming Languages and System Architectures. Proceedings, 1994. X, 344 pages. 1994.

Vol. 783: C. G. Günther (Ed.), Mobile Communications. Proceedings, 1994. XVI, 564 pages. 1994.

Vol. 784: F. Bergadano, L. De Raedt (Eds.), Machine Learning: ECML-94. Proceedings, 1994. XI, 439 pages. 1994. (Subseries LNAI).

Vol. 785: H. Ehrig, F. Orejas (Eds.), Recent Trends in Data Type Specification. Proceedings, 1992. VIII, 350 pages. 1994.

Vol. 786: P. A. Fritzson (Ed.), Compiler Construction. Proceedings, 1994. XI, 451 pages. 1994.

Vol. 787: S. Tison (Ed.), Trees in Algebra and Programming – CAAP '94. Proceedings, 1994. X, 351 pages. 1994.

Vol. 788: D. Sannella (Ed.), Programming Languages and Systems – ESOP '94. Proceedings, 1994. VIII, 516 pages. 1994.

Vol. 789: M. Hagiya, J. C. Mitchell (Eds.), Theoretical Aspects of Computer Software. Proceedings, 1994. XI, 887 pages. 1994.

Vol. 790: J. van Leeuwen (Ed.), Graph-Theoretic Concepts in Computer Science. Proceedings, 1993. IX, 431 pages. 1994.

Vol. 791: R. Guerraoui, O. Nierstrasz, M. Riveill (Eds.), Object-Based Distributed Programming. Proceedings, 1993. VII, 262 pages. 1994.

Vol. 792: N. D. Jones, M. Hagiya, M. Sato (Eds.), Logic, Language and Computation. XII, 269 pages. 1994.

Vol. 793: T. A. Gulliver, N. P. Secord (Eds.), Information Theory and Applications. Proceedings, 1993. XI, 394 pages. 1994.

Vol. 794: G. Haring, G. Kotsis (Eds.), Computer Performance Evaluation. Proceedings, 1994. X, 464 pages. 1994.

Vol. 795: W. A. Hunt, Jr., FM8501: A Verified Microprocessor. XIII, 333 pages. 1994.

Vol. 796: W. Gentzsch, U. Harms (Eds.), High-Performance Computing and Networking. Proceedings, 1994, Vol. I. XXI, 453 pages. 1994.

Vol. 797: W. Gentzsch, U. Harms (Eds.), High-Performance Computing and Networking. Proceedings, 1994, Vol. II. XXII, 519 pages. 1994.

Vol. 798: R. Dyckhoff (Ed.), Extensions of Logic Programming. Proceedings, 1993. VIII, 362 pages. 1994.

Vol. 799: M. P. Singh, Multiagent Systems. XXIII, 168 pages. 1994. (Subseries LNAI).

Vol. 800: J.-O. Eklundh (Ed.), Computer Vision – ECCV '94. Proceedings 1994, Vol. I. XVIII, 603 pages. 1994.

Vol. 801: J.-O. Eklundh (Ed.), Computer Vision – ECCV '94. Proceedings 1994, Vol. II. XV, 485 pages. 1994.

Vol. 802: S. Brookes, M. Main, A. Melton, M. Mislove, D. Schmidt (Eds.), Mathematical Foundations of Programming Semantics. Proceedings, 1993. IX, 647 pages. 1994.

Vol. 803: J. W. de Bakker, W.-P. de Roever, G. Rozenberg (Eds.), A Decade of Concurrency. Proceedings, 1993. VII, 683 pages. 1994.

Vol. 804: D. Hernández, Qualitative Representation of Spatial Knowledge. IX, 202 pages. 1994. (Subseries LNAI).

Vol. 805: M. Cosnard, A. Ferreira, J. Peters (Eds.), Parallel and Distributed Computing. Proceedings, 1994. X, 280 pages. 1994.

Vol. 806: H. Barendregt, T. Nipkow (Eds.), Types for Proofs and Programs. VIII, 383 pages. 1994.

Vol. 807: M. Crochemore, D. Gusfield (Eds.), Combinatorial Pattern Matching. Proceedings, 1994. VIII, 326 pages. 1994.

Vol. 808: M. Masuch, L. Pólos (Eds.), Knowledge Representation and Reasoning Under Uncertainty. VII, 237 pages. 1994. (Subseries LNAI).

Vol. 809: R. Anderson (Ed.), Fast Software Encryption. Proceedings, 1993. IX, 223 pages. 1994.

Vol. 810: G. Lakemeyer, B. Nebel (Eds.), Foundations of Knowledge Representation and Reasoning. VIII, 355 pages. 1994. (Subseries LNAI).

Vol. 811: G. Wijers, S. Brinkkemper, T. Wasserman (Eds.), Advanced Information Systems Engineering. Proceedings, 1994. XI, 420 pages. 1994.

Vol. 812: J. Karhumäki, H. Maurer, G. Rozenberg (Eds.), Results and Trends in Theoretical Computer Science. Proceedings, 1994. X, 445 pages. 1994.

Vol. 813: A. Nerode, Yu. N. Matiyasevich (Eds.), Logical Foundations of Computer Science. Proceedings, 1994. IX, 392 pages. 1994.

Vol. 814: A. Bundy (Ed.), Automated Deduction—CADE-12. Proceedings, 1994. XVI, 848 pages. 1994. (Subseries LNAI).

Vol. 815: R. Valette (Ed.), Application and Theory of Petri Nets 1994. Proceedings. IX, 587 pages. 1994.

Vol. 816: J. Heering, K. Meinke, B. Möller, T. Nipkow (Eds.), Higher-Order Algebra, Logic, and Term Rewriting. Proceedings, 1993. VII, 344 pages. 1994.

Vol. 817: C. Halatsis, D. Maritsas, G. Philokyprou, S. Theodoridis (Eds.), PARLE '94. Parallel Architectures and Languages Europe. Proceedings, 1994. XV, 837 pages. 1994.

Vol. 818: D. L. Dill (Ed.), Computer Aided Verification. Proceedings, 1994. IX, 480 pages. 1994.

Vol. 819: W. Litwin, T. Risch (Eds.), Applications of Databases. Proceedings, 1994. XII, 471 pages. 1994.

Vol. 820: S. Abiteboul, E. Shamir (Eds.), Automata, Languages and Programming. Proceedings, 1994. XIII, 644 pages. 1994.

Vol. 821: M. Tokoro, R. Pareschi (Eds.), Object-Oriented Programming. Proceedings, 1994. XI, 535 pages. 1994.

Vol. 822: F. Pfenning (Ed.), Logic Programming and Automated Reasoning. Proceedings, 1994. X, 345 pages. 1994. (Subseries LNAI).

Vol. 823: R. A. Elmasri, V. Kouramajian, B. Thalheim (Eds.), Entity-Relationship Approach — ER '93. Proceedings, 1993. X, 531 pages. 1994.

Vol. 824: E. M. Schmidt, S. Skyum (Eds.), Algorithm Theory – SWAT '94. Proceedings. IX, 383 pages. 1994.

Vol. 825: J. L. Mundy, A. Zisserman, D. Forsyth (Eds.), Applications of Invariance in Computer Vision. Proceedings, 1993. IX, 510 pages. 1994.

Vol. 826: D. S. Bowers (Ed.), Directions in Databases. Proceedings, 1994. X, 234 pages. 1994.

Vol. 827: D. M. Gabbay, H. J. Ohlbach (Eds.), Temporal Logic. Proceedings, 1994. XI, 546 pages. 1994. (Subseries LNAI).

Vol. 828: L. C. Paulson, Isabelle. XVII, 321 pages. 1994.

Vol. 829: A. Chmora, S. B. Wicker (Eds.), Error Control, Cryptology, and Speech Compression. Proceedings, 1993. VIII, 121 pages. 1994.

Vol. 830: C. Castelfranchi, E. Werner (Eds.), Artificial Social Systems. Proceedings, 1992. XVIII, 337 pages. 1994. (Subseries LNAI).

Vol. 831: V. Bouchitté, M. Morvan (Eds.), Orders, Algorithms, and Applications. Proceedings, 1994. IX, 204 pages. 1994.

Vol. 832: E. Börger, Y. Gurevich, K. Meinke (Eds.), Computer Science Logic. Proceedings, 1993. VIII, 336 pages. 1994.

Vol. 833: D. Driankov, P. W. Eklund, A. Ralescu (Eds.), Fuzzy Logic and Fuzzy Control. Proceedings, 1991. XII, 157 pages. 1994. (Subseries LNAI).

Vol. 834: D.-Z. Du, X.-S. Zhang (Eds.), Algorithms and Computation. Proceedings, 1994. XIII, 687 pages. 1994.

Vol. 835: W. M. Tepfenhart, J. P. Dick, J. F. Sowa (Eds.), Conceptual Structures: Current Practices. Proceedings, 1994. VIII, 331 pages. 1994. (Subseries LNAI).

Vol. 836: B. Jonsson, J. Parrow (Eds.), CONCUR '94: Concurrency Theory. Proceedings, 1994. IX, 529 pages. 1994.

Vol. 837: S. Wess, K.-D. Althoff, M. M. Richter (Eds.), Topics in Case-Based Reasoning. Proceedings, 1993. IX, 471 pages. 1994. (Subseries LNAI).

Vol. 838: C. MacNish, D. Pearce, L. Moniz Pereira (Eds.), Logics in Artificial Intelligence. Proceedings, 1994. IX, 413 pages. 1994. (Subseries LNAI).

Vol. 839: Y. G. Desmedt (Ed.), Advances in Cryptology - CRYPTO '94. Proceedings, 1994. XII, 439 pages. 1994.

Vol. 840: G. Reinelt, The Traveling Salesman. VIII, 223 pages. 1994.

Vol. 841: I. Prívara, B. Rovan, P. Ružička (Eds.), Mathematical Foundations of Computer Science 1994. Proceedings, 1994. X, 628 pages. 1994.

Vol. 842: T. Kloks, Treewidth. IX, 209 pages. 1994.

Vol. 843: A. Szepietowski, Turing Machines with Sublogarithmic Space. VIII, 115 pages. 1994.

Vol. 844: M. Hermenegildo, J. Penjam (Eds.), Programming Language Implementation and Logic Programming. Proceedings, 1994. XII, 469 pages. 1994.

Vol. 845: J.-P. Jouannaud (Ed.), Constraints in Computational Logics. Proceedings, 1994. VIII, 367 pages. 1994.

Vol. 846: D. Shepherd, G. Blair, G. Coulson, N. Davies, F. Garcia (Eds.), Network and Operating System Support for Digital Audio and Video. Proceedings, 1993. VIII, 269 pages. 1994.

Vol. 847: A. L. Ralescu (Ed.) Fuzzy Logic in Artificial Intelligence. Proceedings, 1993. VII, 128 pages. 1994. (Subseries LNAI).

Vol. 848: A. R. Krommer, C. W. Ueberhuber, Numerical Integration on Advanced Computer Systems. XIII, 341 pages. 1994.

Vol. 849: R. W. Hartenstein, M. Z. Servít (Eds.), Field-Programmable Logic. Proceedings, 1994. XI, 434 pages. 1994.

Vol. 850: G. Levi, M. Rodríguez-Artalejo (Eds.), Algebraic and Logic Programming. Proceedings, 1994. VIII, 304 pages. 1994.

Vol. 851: H.-J. Kugler, A. Mullery, N. Niebert (Eds.), Towards a Pan-European Telecommunication Service Infrastructure. Proceedings, 1994. XIII, 582 pages. 1994.

Vol. 853: K. Bolding, L. Snyder (Eds.), Parallel Computer Routing and Communication. Proceedings, 1994. IX, 317 pages. 1994.

Vol. 855: J. van Leeuwen (Ed.), Algorithms – ESA '94. Proceedings, 1994. X, 510 pages. 1994.

Vol. 856: D. Karagiannis (Ed.), Database and Expert Systems Applications. Proceedings, 1994. XVII, 807 pages. 1994.